W9-BZI-111

Human Development and Cognitive Processes

Human Development and Cognitive Processes

Edited by
John Eliot
University of Maryland

HOLT, RINEHART AND WINSTON, INC.
New York Chicago San Francisco Atlanta Dallas
Montreal Toronto London Sydney

To my father

"A wise man will hear, and will increase learning; and a man of understanding shall attain unto wise counsels. . . ."

Proverbs 1 : 5

Library of Congress Catalog Card Number: 73–129482
SBN: 03–077375–X
Printed in the United States of America
1 2 3 4 22 9 8 7 6 5 4 3 2 1

Foreword

The shifts of interest in a field of inquiry are difficult to predict in advance, but they become fairly explicable when we review the past. So it is with the recent upsurge of interest in cognitive psychology and, in particular, cognitive development.

While the dominant role of Jean Piaget is evident, his insights do not account for the widespread interest in cognitive growth, for much of his message was being asserted in the 1920s and 1930s, at the time listened to and rejected, only to be celebrated in the 1960s. The early neglect permitted William Brownell, writing in 1942, to deplore the lack of attention to Piaget then. Of Piaget he wrote: "Piaget's studies seem to provide the most illuminating single description of the way in which children attain power in problem-solving." But few heeded his call to notice Piaget.

The steps that led to a revived interest in cognition include the following:

1. The launching of the Russian satellite, Sputnik I, in 1957. The peculiar shock effect of the Russian intellectual and technological success catalyzed an interest in furthering a tough-minded education in the United States, one that would emphasize acquiring and using knowledge, hence cognition.

2. A renewed interest in the psychology of language and language development. Miller's book on *Language and Communication* appeared in 1951, Chomsky's *Syntactic Structures* in 1957, Skinner's *Verbal Behavior* in 1957, and Brown's *Words and Things* in 1958.

3. Advances in computer technology permitting new experimentation in problem solving and thinking. The field of thinking was already show-

ing some new stirrings, as in the Bruner, Goodnow, and Austin book in 1956. The important computer studies of Newell, Shaw, and Simon appeared in 1958.

4. Interest in the later effects of early experience. Although psychoanalysis had placed great stress on the later significance of early mothering, a growing lack of interest in psychoanalysis had weakened its impact. A reassertion of interest in the early years came about through laboratory studies of imprinting, as by Ramsay and Hess in 1954; the infant monkey studies of Harlow in 1958; and the effects of early stress in animals, as by Levine in 1958.

5. The broadened front of contemporary psychology. A growing dissatisfaction with a somewhat narrow and strident behaviorism had begun to be felt, with a growing interest in subjective processes, so that Miller, Galanter, and Pribram in 1960 could describe their position as a "subjective behaviorism." New interests included attention, imagery, planning, hypnosis, states of awareness in general. This was helped along by the new psychophysiology of sleeping and dreaming—and by the widespread use of drugs, leading to a scientific interest in their effects.

I find it impressive that so many of the foregoing events moved to some sort of intersecting climax around 1957 and 1958. When Bruner, Flavell, Hunt and others gave renewed prominence to Piaget's work in the early 1960s, a favorable atmosphere had been prepared.

Other events have contributed also. For example, when the public became sensitized to the disadvantaged child, a new practical outlet for cognitive development arose, as in connection with Head Start programs.

In any case, a book such as the present one is timely, and coherent with the concerns of an increasing body of contemporary psychologists and educators. I believe that Dr. Eliot has been wise in his selection of essays, including those already published and those prepared for this volume. As one of the official readers of his doctoral dissertation, I know from working with him how deep his involvement has been in these problems for many years, and this book is a fitting endeavor to invite a wider audience to share his sense of the importance and excitement of cognitive development.

ERNEST R. HILGARD

Stanford University
August 1970

Preface

Recent books about cognitive processes or cognitive development demand a fairly sophisticated background on the part of the reader. Unfortunately for many of us, there are no easy introductions to speculation about human knowledge or the problem of intellectual development. Similarly, there are no "instant" backgrounds to the technicalities of research design. There are available, however, essays which can help to provide the interested reader with the beginnings of a frame of reference. For example, there are some essays that present a survey of theoretical positions, some that attempt to show the interrelationships between knowledge and development, and some that summarize the current status of methodology and research interest.

The purpose of this book is to present a selection of such essays to serve as an introduction to a complex and abstract field of inquiry. The essays are not, in themselves, either exhaustive or comprehensive, nor, taken together, do they attempt to integrate definitively all that is known about cognitive processes within a developmental framework. They are of varying complexity. In some cases, comprehensiveness was desirable; in others, it was felt that readers might better get the flavor of actual research through a narrower and deeper approach. Hopefully, readers will be stimulated to seek out other writings in the field and eventually to create a useful, integrated viewpoint of their own.

The essays assembled in this book were not selected to prejudice the reader in favor of a particular theorist or theoretical position, although current research interests and the range of essays available may have introduced some bias. This is not a book about Piaget's theory or Skinner's views, although there are references to both men and their ideas throughout the book. The purpose of this book, to repeat, is to help *open* a field of inquiry, not to close it.

Inspiration for this book came from students who repeatedly asked for references which would help them to sort out theories, people, and research findings. The reprinted essays assembled in this book are, for the most part, those found most valuable by students in my courses.

The book contains both published and original essays. I am grateful to many authors for their permission to reprint essays, and I am especially grateful to the authors of original essays for their effort and their patience. Royalties from the sale of this volume are consigned, after expenses, to the American Cancer Society.

Many people were helpful in suggesting authors for original essays. I am indebted especially to Dr. Ethel Albert, Dr. Jerome Bruner, Dr. Eleanor Gibson, Dr. Ralph Haber, Dr. Irving Sigel, Dr. Boyd McCandless, and Dr. Joachim Wohlwill for their assistance.

I regret that because of circumstances beyond my control I was unable to include an original essay by Dr. Jeannete Gallagher in this volume. I am confident that readers will find her essay published elsewhere in the near future.

Acknowledgement is due to my colleagues, Dr. Gail Inlow, Dr. Gilbert Krulee, and Dr. Joe Park, for their encouragement and confidence. I am also grateful to Catherine Bell for her editorial help with this book.

Finally, special gratitude is due to Sylvia, my wife, who spent many hours and much effort in shaping this book. The making of this book has been an education for us both.

JOHN ELIOT

University of Maryland
August 1970

Contents

Introduction

Most of us have wondered, at one time or another, how we happen to possess whatever knowledge we have about the world. We may have wondered how we can be watching one series of events while simultaneously thinking about another. Again, we may have wondered how true it is that "you don't really know something until you can put it into words." Perhaps most intriguing of all, we may have wondered how we might control our thinking so as to use our knowledge, or help our children use their knowledge, more effectively.

In academic language, our questions about perception, language, and thinking are questions about knowledge or "cognitive processes." Similarly, when we talk about knowledge changing as we grow older, we are talking about the course of cognitive growth or "cognitive development."

Since 1950, philosophers and behavioral scientists have studied cognitive processes and cognitive development as never before. As a consequence, there has been a bewildering outpouring and accumulation of hypotheses and data in books and journals. Indeed, at the present time there are many indications that we may be on the verge of a significant advance in our understanding of the ancient problem of human knowledge—an advance that carries with it all the excitement and uncertainty which characterized physics in the 1920s when the study of quantum mechanics was just emerging.

The outpouring of hypotheses and data has also brought with it demands for pragmatic and authoritative statements by those faced with the practical problems of helping children to develop and to function more effectively. Unfortunately, most scholars have reasons to be reticent about making authoritative statements about cognitive development. In the first place, the task of coordinating a multidisciplinary effort that is producing masses of dissimilar data is very difficult. In the second place, we lack appropriate models to account for cognitive processes, and we lack adequate language to describe their cumulative interdependence. At the most, therefore, this book can draw attention to, and attempt to coordinate, hypotheses and data on major issues in a field of inquiry which remains fascinating in its controversy and its disarray.

Vertical Organization of the Book

The book is divided into four parts. In the first part, cognitive development is considered in the context of human development in general. Cognitive processes must be grounded in a developmental framework if we are to avoid sacrificing an understanding of the human being to the often narrow categories of logical analysis. Then too, a great many of the assumptions and presuppositions we have about development hold for the positions we take with respect to cognitive processes.

The remainder of the book is divided into three parts according to the traditional psychological categories of cognition: perception, language, and thinking. It seems reasonable to maintain this division and the order of this division since theoretical discussion and research in psychology continue to observe them. The second part of the book, then, centers upon the category of perception. Perception is a complex process which at least involves the ability to deploy and maintain selective attention, to organize perceptual elements logically, and to transport information to and from memory storage efficiently and accurately.

The third part of the book is devoted to the category of language. The emphasis is upon speculation about the origins of language, the relationship between child and adult language, and the complex interrelationship between language and thought. Finally, in the fourth part of the book, the emphasis is upon the various attempts by philosophers and psychologists to define thought and its components and, eventually, to characterize the development of thought.

Horizontal Organization of the Book

Each of the four parts of this book contains at least six sequentially arranged essays. The organization of these essays is strongly influenced by a contemporary philosopher, Israel Scheffler. In his *Conditions of Knowledge* (1965), Scheffler offers one of the clearest and most succinct statements available about the problem of knowledge. His statement contains a combination of five questions about knowledge and the criteria which must be met if these questions are to be adequately answered. Briefly, these questions are:

1. "What is knowledge?" How can we describe or define it? How can we provide it with some logical status?
2. "What knowledge is most reliable or important?" By what reasonable standard can we classify and rank various types of knowledge?
3. "How does knowledge arise?" How do men and animals compare with respect to knowing and the acquisition of knowledge? What processes and mechanisms do we ascribe to man to account for intellectual growth over his life span?
4. "How should the search for knowledge be conducted?" To what extent is our knowledge of cognitive development restricted by our logic or constrained by our experimental procedures?
5. "How is knowledge best taught?" How can parents and teachers foster the growth of knowledge and help children use it more effectively?

In many respects, Scheffler's five questions about knowledge neatly summarize the principal concerns of those who study cognitive development. Moreover, they provide a useful basis for organizing the diverse material which this book explores. Although it is recognized that psychological theories have been traditionally preoccupied with Scheffler's third question about how knowledge arises, it is hoped that the essays in this book will prompt future authors to respond more directly to all of Scheffler's questions.

The first essay in each of the four parts is a survey of issues and theoretical positions. These surveys are an attempt to define development and cognition and, indirectly, to respond to Scheffler's first question about the nature of knowledge. One must be clear about the nature of knowledge before talking about how knowledge arises. Ausubel's essay, for example, classifies different theoretical positions about the nature of development in terms of the heredity-environment issue. By contrast, O'Neil's essay begins Part II with a discussion of perceptual issues and a classification of theorists with respect to those issues. Although these essays do not provide a philosophical analysis of knowledge per se, they do present different theories of mind from a psychological viewpoint. Further consideration of these theories will lead one ultimately back to Scheffler's first question.

The second essay is intended to convey the complexity of studying human development or specific cognitive processes. In light of this complexity, it is difficult to ascertain the quality of knowledge which is gained through empirical testing. In this sense, we do not know whether the results of our empirical tests are giving us knowledge or, if they are, how worthwhile that knowledge is. It should be clear from these essays how far we are from achieving definitive answers to Scheffler's question about what knowledge is most valuable.

The third essay in each part attempts to place the processes involved in cognitive development within a phylogenetic context. If we are to say how knowledge arises, we must keep in mind the strategies which animals and humans use when adapting themselves to changing environments. As Bruner has observed, no theory of cognitive development can ignore man's primate ancestry or the manner in which evolution has imposed limits upon his growth (Bruner, 1966a and b).

The fourth essay in each part focuses upon the complex factors which influence our study and understanding of developmental or cognitive processes. Our thoughts, for example, do influence our perceptions; our language systems do reflect our culture. These essays underscore the problems encountered when we study one process (such as perception) when that process is itself embedded in a network of competing processes.

The fifth essay is concerned with infant cognition. In the past decade especially, infant studies have come to the forefront of research activity and have spawned a host of critical questions about the nature of inference and the constraints of research methodology. The fifth essays, consequently, touch on Scheffler's question about the conduct of the search for knowledge and, indirectly, refer to his question about the origins of knowledge.

The final essays are intended to introduce the reader to some of the current research in later cognitive development. The essays restate many of the themes from preceding essays and indicate something of the diversity of developmental research.

Scheffler's fifth question, "How is knowledge best taught?" is not considered explicitly in this book, although it is the question of ultimate practical interest. Logically, it can be argued that as long as we remain unsure about the nature of knowledge and how knowledge arises, we will be unsure about teaching it. So far, our definitions do not take into account the overwhelming complexity and diversity of home or school situations. On the other hand, incomplete attempts to provide limited answers to Scheffler's fifth question do exist in the recent works of Bruner (1966a and b), Gibson (1969), and in the efforts of those involved in such projects as Head Start.

The essays in this book were organized with readers from different disciplines in mind. Indeed, the book may be read either part by part, as it is printed, or topic by topic, reading corresponding essays in the

Table 1

ORGANIZATION OF BOOK

Ordering of Essays	Part One Development	Part Two Perception	Part Three Language	Part Four Thinking
1. Survey of theoretical positions and issues	Ausubel, "Historical Overview of Theoretical Trends"	O'Neil, "Basic Issues in Perceptual Theory"	Houston, "The Study of Language"	Hosper, "Knowledge: Concepts"
2. Complexity of process	Anderson, "Dynamics of Development: Systems in Process"	Dick, "Perception as Information Processing"	Furth, "Piaget's Theory of Knowledge"	Bourne, "Human Conceptual Behavior" Kendler, "The Concept of the Concept"
3. Comparative status	Schneirla, "The Concept of Development in Comparative Psychology"	Schiff, "Comparative Study of Sensory and Perceptual Processes"	Lenneberg, "The Natural History of Language"	Glucksberg, "Thinking: A Phylogenetic Perspective"
4. Interrelationships	Kessen, "Questions for a Theory of Cognitive Development"	Soltis, "Language of Perception" Wohlwill, "From Perception to Inference"	Bruner, "On Cognitive Growth II"	Mouton, "The Principal Elements of Thought"
5. Infant research	Horowitz, "Infant Learning and Development"	Aronson and Tronick, "Implications of Infant Research for Developmental Theory"	Kaplan and Kaplan, "The Prelinguistic Child"	Charlesworth, "Cognition in Infancy"
6. Later research	Gagné, "Contributions of Learning to Human Development"	Epstein, "Developmental Studies of Perception"	Sachs, "The Status of Developmental Studies of Language"	Elkind, "Cognition in Infancy and Early Childhood"

several parts. Those interested in perception, for example, might find Part II of interest. Similarly, those interested in infant research might concentrate upon the fifth essay in each of the four parts of the book. Hopefully, this organization of topics will help to provide readers with the beginnings of a useful frame of reference for interpreting a large and inordinately complex literature.

part one

HUMAN DEVELOPMENT

The term *human development* is commonly used to describe changes in behavior which occur with age. We all have observed, for example, that there is a time when a child is unable to talk and a later time when it is difficult to stop him from talking. Between these two stages the child acquires and uses language—an achievement of complex changes in behavior which we call development. Like the terms *stimulus* and *response,* however, the term *development* can be endowed with many connotations or it can be given limited meaning within a highly restrictive context. The precision of definition often depends upon whether the writer is more interested in describing the achievement of broad stages or plateaus of behavior or the mechanisms which apparently govern the transitions between stages.

How we define *development* subsequently limits what we then observe. David Ausubel's article from his *Theory and Problems of Child Development* begins this book by describing six concepts of development: two of which emphasize the contributions of endogenous or innate factors, and four of which stress the shaping influences of the environment. Although Ausubel does not discuss the strengths and weaknesses of a seventh concept which emphasizes the interaction of heredity and environment, readers will find references to interactionism (for example, Piaget's theory) throughout the book.

The second essay, John Anderson's "Dynamics of Development: System in Process," should convey something of the complexity of human development. Anderson unravels some characteristics of development and then reweaves these characteristics in a "living-systems" model of development. His analysis suggests the difficulties involved when we try to interrelate the various *biological* systems (some of which are hereditarily constrained in their function, others of which are extremely sensitive to environmental stress) to the myriad of *social* or *cultural* systems which also shape, pattern, and otherwise provide a sense of direction to our lives.

T. C. Schneirla's essay, "The Concept of Development in Comparative Psychology," indicates that there are a number of parallels between phyla in the mechanisms underlying adaptive behavior. However, the essay is principally concerned with the marked discontinuities between animal and human development. Schneirla's observations tie in with Ausubel's summary of biological trends in predeterminist thought and with Anderson's biologically oriented model of development.

William Kessen's essay, "Questions for a Theory of Cognitive Development," signals a shift in focus from general aspects of human development to specific problems with cognitive development. Kessen is especially concerned with the limits imposed upon psychological analysis by the preconceptions we bring to our observations. In particular, he argues that we must be more explicit about the nature of the cognitive processes we ascribe to man before we attempt to describe how these processes evolve or change with increasing age. The similarity between his questions about cognitive development and Scheffler's questions about the problem of knowledge are worth noting.

The fifth essay in Part One is Frances Horowitz's "Infant Learning and Development: Retrospect and Prospect." She begins her essay by reviewing infant learning research from 1956 to 1966. She then considers this research in terms of various strategies.

The final essay is Robert Gagné's "Contributions of Learning to Human Development." Gagné critically considers three models of human intellectual development—predeterminist, interactionist, and environmentalist. He then elaborates upon the environmentalist or cumulative learning model and, indirectly, responds to the questions which Kessen poses about the nature of cognition and the conditions governing its change or development.

The essays of Part One, when taken together, are addressed both to the problem of achieving broad stages of development and to the mechanisms which govern the transitions between stages. Gagné's three models obviously have their roots in Ausubel's continuum, for example. In any case, all six essays indicate why it is so difficult to find incidences where different theorists have concerned themselves with the same problem.

Suggested Topical Readings

SURVEY OF THEORETICAL POSITIONS

Baldwin, Alfred, "The Study of Child Behavior and Development," in Mussen, *Handbook of Research Methods in Child Development.* New York: Wiley, 1960.

Koch, S., "Psychology and Emerging Conceptions of Knowledge as Unitary," in T. W. Wann, *Behaviorism and Phenomenology.* Chicago: University of Chicago Press, 1964.

COMPLEXITY OF PROCESSES

Campbell, Donald T., "Evolutionary Epistemology," in Paul A. Schilpp, ed., *The Philosophy of Karl Popper.* La Salle, Ill.: Open Court Publishing, in press.

COMPARATIVE STATUS

Hodos, William, and C. B. G. Campbell, "Scala Naturae," *Psychol. Rev.*, 76 (1969), pp. 337–350.

LATER RESEARCH

Langer, Jonas, *Theories of Development.* New York: Holt, Rinehart and Winston, Inc., 1968.

INTERRELATIONSHIPS

Wright, John, "Cognitive Development," in Frank Falkner, ed., *Human Development.* Philadelphia: Saunders, 1966.

INFANT RESEARCH

Bruner, Jerome, "Eye, Hand, and Mind," in D. Elkind and J. Flavell, eds., *Studies in Cognitive Development.* New York: Oxford, 1969.

Additional Readings in Developmental Theory and Research

Bayley, N., "Research in Child Development: A Longitudinal Perspective," *Merrill-Palmer Quarterly* (1965), pp. 183–208.

Baldwin, A., *Theories of Child Development.* New York: Wiley, 1967.

Berlyne, D., "Delimitation of Cognitive Development," *Child Development Monograph* (1966), Serial 107.

Bijou, S., and D. Baer, *Child Development,* Vol. I, No. 2. New York: Appleton, 1965.

Bijou, S., and D. Baer, *Child Development: Readings in Experimental Analysis.* New York: Appleton, 1967.

Bloom, B., *Stability and Change in Human Characteristics.* New York: Wiley, 1964.

Brofenbrenner, U., "Developmental Theory in Transition," *NSSE Yearbook,* 1963, *Child Psychology.*

Campbell, D., and J. Stanley, *Experimental and Quasi-Experimental Designs for Research.* Chicago: Rand McNally, 1963.

Campbell, D., and W. R. Thompson, "Developmental Psychology," *Annual Review of Psychology,* Palo Alto (1968).

Cassari, E., "Genetic Endowment and Environment in the Determination of Human Behavior," *American Educational Research Journal,* 5, no. 1 (1968).

Clark, le Gros, *The Antecedents of Man.* New York: Harper & Row, 1963.

Elkind, D., and A. Sameroff, "Developmental Psychology," *Annual Review of Psychology,* Palo Alto (1970).

Endler, N., L. Boulter, and H. Osser, *Contemporary Issues in Developmental Psychology.* New York: Holt, Rinehart and Winston, Inc., 1968.

Flavell, J., and J. P. Hill, "Developmental Psychology," *Annual Review of Psychology,* Palo Alto (1969).

Fowler, H., "Cognitive Learning in Infancy and Early Childhood," *Psychological Bulletin,* 59 (1962).

Gale, R., *Developmental Behavior: A Phenomenological Approach.* New York: Macmillan, 1969.

Gollin, E., "Developmental Approach to Learning and Cognition," in L. Lipsitt and C. Spiker, *Advances in Child Development and Behavior.* New York: Academic Press, 1965.

Grinder, R., *History of Genetic Psychology.* New York: Wiley, 1968.

Harris, D., "Child Psychology and the Concept of Development," in D. Palermo and L. Lipsitt, eds., *Research Readings in Child Psychology.* New York: Holt, Rinehart and Winston, Inc., 1963.

Hill, J., *Minnesota Symposium on Child Psychology.* Minneapolis: University of Minnesota Press, 1967, 1968, 1969, 1970.

Jensen, A. R. *et al.*, "Environment, Heredity, and Intelligence," *Harvard Educational Review* (Spring 1969).

Kagan, J., and H. A. Moss, *Birth to Maturity: A study in Psychological Development.* New York: Wiley, 1962.

Kessen, W., "Research Design in the Study of Development Problems," in P. H. Mussen, ed., *Handbook of Research Methods in Child Development.* New York: Wiley, 1960.

Kessen, W., "'Stage' and 'Structure' in the Study of Children," *Child Development Monograph* (1962).

Kessen, W., *The Child.* New York: Wiley, 1965.

Kohlberg, L., "Early Education: A Cognitive Developmental Point of View," in *Child Development,* 39 (1968), p. 1013.
Lorenz, K., *King Solomon's Ring.* New York: Crowell, 1952.
Maier, H., *Three Theories of Child Development.* New York: Harper & Row, 1969.
Marx, M., *Theories in Contemporary Psychology.* New York: Macmillan, 1963.
Marx, M., *Learning Theories.* New York: Macmillan, 1970.
McCandless, B., *Children: Behavior, and Development.* New York: Holt, Rinehart and Winston, Inc., 1967.
McCandless, B., and C. Spiker, "Experimental Research in Child Psychology," *Child Development Monograph,* 27 (1956), pp. 75–80.
McClearn, G., "Genetics and Behavior Development," in Martin L. and Lois W. Hoffman, eds., *Review of Child Development Research.* New York: Russell Sage, 1964.
McNeil, E., *The Concept of Human Development.* Belmont, Calif.: Wadsworth, 1966.
Mussen, P., "European Research in Cognitive Development," *Child Development Monograph,* Serial 100 (1965).
Mussen, P., U. Langer, and M. Covington, *Trends and Issues in Developmental Psychology.* New York: Holt, Rinehart and Winston, Inc., 1969.
Mussen, P., *Carmichael's Manual of Child Psychology.* New York: Wiley, 1970.
Neisser, U., *Cognitive Psychology.* New York: Appleton, 1967.
Piaget, J., *Psychology of the Child.* New York: Basic Books, 1969.
Pikunas, J., *Human Development: Science of Growth.* New York: McGraw-Hill, 1969.
Robinson, D., *Heredity and Achievement.* New York: Oxford, 1970.
Scott, J., *Early Experience and the Organization of Behavior.* Belmont, Calif.: Brooks/Cole, 1968.
Spiker, C., "Concept of Development, Relevant and Irrelevant Issues," *Child Development Monograph,* Series 107 (1966).
Stevenson, H., "Developmental Psychology," *Annual Review of Psychology,* Palo Alto (1967).
Van de Geer, V. P., "Cognitive Functions," *Annual Review of Psychology,* Palo Alto (1966).
Vandenberg, S., *Methods and Goals in Human Behavior Genetics.* New York: Academic Press, 1965.
Wallach, M., and N. Kogan, *Modes of Thinking in Young Children.* New York: Holt, Rinehart and Winston, Inc., 1965.
Wohlwill, J., "The Age Variable," *Psychol. Rev.,* 77 (1970), pp. 49–64.

HISTORICAL OVERVIEW OF THEORETICAL TRENDS*

David Ausubel

The regulation of human development is still very much of a live and highly controversial issue. The nature-nurture controversy as such has abated somewhat in the sense that the two factors are now seldom regarded as mutually exclusive or as operative on an either-or basis; interaction between them in determining the direction of growth is widely accepted in many quarters. Nevertheless much disagreement still persists, regarding their relative influence with respect to particular aspects of development, and very little is known about the mechanisms through which such interaction is mediated. Furthermore, the same basic issue of the regulation of development turns up in relation to many other theoretical problems in which it is often only dimly perceived as being relevant, e.g., the doctrines of maturation and recapitulation, psychoanalytic theories of personality, hypotheses regarding the nature of drives, various conceptions of cultural relativism, etc. For these reasons, therefore, more explicit consideration of the historical roots of and relationships between different ideological trends bearing on this issue may prove rewarding.

Preformationist Approaches

Historical analysis shows that an interactional point of view regarding the role of internal and external determinants of development is of relatively recent origin. Over the past few centuries, and even in our own time, the theories of development that have wielded most influence

* David Ausubel, "Historical Overview of Theoretical Trends," in his *Theory and Problems of Child Development*, New York: Grune & Stratton, 1957, pp. 27–49.

have either stressed (a) an environmentally oriented *tabula rasa* approach or (b) a preformationist or predeterministic approach emphasizing the contributions of endogenous and innate factors.

The fundamental thesis of preformationism is a denial of the essential occurrence and importance of development in human ontogeny. The basic properties and behavioral capacities of man—his personality, values and motives, his perceptual, cognitive, emotional and social reaction tendencies—are not conceived as undergoing qualitative differentiation and transformation over the life span, but are presumed to exist preformed at birth. Nothing need develop as a result of the interaction between a largely undifferentiated organism with certain stipulated predispositions and his particular environment. Instead everything is already prestructured, and either undergoes limited quantitative modification with increasing age or merely unfolds sequentially on a prearranged schedule.

The origins of preformationist thinking are not difficult to trace. On the one hand, it is obviously related to theological conceptions of man's instantaneous creation and to the widespread belief in the innateness of the individual's personality and sense of unique identity as a person. A quaint prescientific embryological counterpart of this point of view is the formerly popular homuncular theory of human reproduction and prenatal gestation. It was seriously believed that a miniature but fully-formed little man (i.e., an homunculus) was embodied in the sperm, and when implanted in the uterus simply grew in bulk, without any differentiation of tissues or organs, until full-term fetal size was attained at the end of nine months.

On the other hand, the disposition to perceive the infant or child as a miniature adult is largely an outgrowth of the ubiquitous tendency toward extrapolation or anthropomorphism in interpreting phenomena remote from own experience or familiar explanatory models. What is easier than to explain the behavior of others in terms of one's own response potentialities? In order to extend this orientation to the interpretation of the child's behavior, it was necessary, of course, to endow him with the basic attributes of adult motivation, perceptual maturity and reaction capacities. Modern and extreme expressions of this tendency include such widely accepted psychoanalytic views that the prototype of all later anxiety lies in the psychological trauma of birth (Freud, 1936; Rank, 1929), that infantile and adult sexuality are qualitatively equivalent (Freud, 1935), and that infants are presumably sensitive to the subtlest shadings of parental attitudes.

The theological variety of preformationism, allied as it was to a conception of man as innately sinful, inspired a rigid, authoritarian and pessimistic approach to education. Since ultimate form was assumed to be prestructed and complete in all of its essential aspects, one could at best only improve slightly on what the individual already was or was fated to become. Hence, it was unnecessary to consider the child's developmental needs and status, the conditions propitious for development at a given stage of maturity, or readiness for particular experience. Because he was not perceived as qualitatively different from the adult or as making any significant contribution to his own development, the arbitrary imposition of adult standards was regarded as self-evidently defensible.

Innate Ideas

Philosophically, in the realm of cognition, preformationism was represented by the doctrine of innate ideas, i.e., ideas existing independently of individual experience.[1] Vigorously combatted, by John Locke (1632–1704) and other empiricists, this notion consistently waned in influence and all but disappeared from view until revived and made popular by psychoanalytic theorists. Jung (1928), for example, postulated the existence in the "racial unconscious" of such inborn ideas as eternality, omnipotence, reincarnation, male and female, mother and father. And Freud's analogous "phylogenetic unconscious" (1935) included—as the basis for resolving the Oedipus complex— an inherited identification with the like-sexed parent prior to any opportunity for actual interpersonal experience.

Human Instincts

On a behavioral level, preformationist doctrines have flourished in various theories of instincts and innate drives. Influenced by studies of inframammalian behavior and by the early nativistic implications of Mendelian genetics, psychologists, represented by such notable figures as McDougall (1914) and Thorndike (1919), devised elaborate lists of human instincts, e.g., mating, maternal, acquisitive, pugnacious, gregarious. These were conceived as unlearned, complexly patterned, sequentially organized responses, perfectly executed on initial performance, that either unfolded in due course or were triggered off by appropriate environmental cues. But undermined by the rising tide of behaviorism in the 1920's, by demonstrations of numerous forms of conditioned responses, and by research findings in infrahuman primate development, ethnology, and sociology which pointed to the acquired, experiential basis of such behavior, this variety of instinct theory as applied to human behavior has long since passed into oblivion.

[1] The content and validity of this and other historically important concepts will be considered more fully elsewhere in . . . [Ausubel, "Historical Overview of Theoretical Trends"]. Here we are only concerned with gaining historical perspective.

Primary and Libidinal Drives

Doctrines deeply rooted in cultural tradition do not die easily. Rejected in one guise, they subsequently gain reacceptance in other more palatable forms. Thus, instinctual theories reappeared respectively in stimulus or viscerogenic constructs of "primary drives" and in psychoanalytic conceptions of libidinal drives. The former notion, which was more compatible with the prevailing behavioristic and biologically oriented climate of psychological opinion, assumed the existence of a certain irreducible number of states of physiological disequilibrium which supposedly constituted in themselves the innate, energizing basis of motivated behavior. These states (i.e., primary drives) were, by definition, conceived as innate and inevitable since their operation was simply a function of the presence of persistent visceral or humoral stimuli within the organism or of intense external stimuli (e.g., pain) to which the organism invariably responded in predetermined ways.

Libidinal drives, in contrast, were conceptualized as innate, substantive sources of energy virtually independent of internal or external stimulation. Because the uninhibited expression of such drives seemed to engender a conflict of interest between biologically related needs of the individual and the mores of his culture, and because their sequential appearance was couched in terms of psychosexual "development," this point of view was more congruent than other instinct theories with the theoretical leanings of the more dynamically oriented psychiatrists, clinical psychologists and social anthropologists. Actually, however, no more development was envisaged than in any other orthodox, preformationist concept of instinct, for the energizing aspect of the libidinal drives, their locus, mode, and object of expression, and the sequential order of their appearance were all prestructured in advance; and although the emergence of later-appearing drives necessarily had to be latent at first, their eventual unfolding was assured without any intervening process of transformation or interaction with individual experience.

Thus, despite their differences with respect to the source and nature of the fundamental drives, both "primary" and libidinal drive theorists agreed that the energizing basis of human behavior was innate and inevitable. Furthermore, both groups tended to regard the preformed drives as original and all other drives as derived from them through the various mechanisms of conditioning, symbolic equivalence, and sublimation. In other words, the environment was not conceived as capable of independently generating drives but only as repressing, modifying, differentiating and re-channeling innate drives.

Freud and his followers also derived the other two layers of personality, the ego and superego (i.e., conscience), from the libidinal instincts as they came into contact with a repressive reality and had to adapt to its demands. Similarly, character traits were conceptualized as symbolical derivatives of fixated libido at one of the stages of infantile psychosexual development following inordinate experiences of frustration or gratification. In all of these latter instances, it is true, preformationist concepts as such were not employed. Environmental vicissitudes were granted a share in the developmental process, but only in the sense of modifying or accentuating a preformed product rather than of participating crucially in the directional regulation of new patterns.

Predeterministic Approaches

In contradistinction to preformationism, predeterministic doctrines satisfy the minimal criteria for a developmental approach. Successive stages of the organism are not merely regarded as reflective of a sequential unfolding of preformed structures or functions forever fixed at conception or birth, but as the outcome of a process of qualitative differentiation or evolution of form. Nevertheless, because the regulation of development is conceived as so prepotently determined by internal factors, the net effect is much the same as if preformationism were assumed. Interaction with the environment and the latter's influence on the course of development is not completely ruled out; but its directional role is so sharply curtailed that it never crucially affects eventual outcome, accounting at the very most for certain minor limiting or patterning effects.

Hence, insofar as the essential and distinguishing features of development are concerned, theoretical violence is not perpetrated by referring to this approach as basically predeterministic. As will be pointed out later, the truly objectionable aspect of this point of view is its blanket or wholesale application to all aspects of development. In some cases it undoubtedly comes close to approximating the actual state of affairs (e.g., infantile locomotor and prehensile development), but in most other areas of human development it is recommended by neither evidence nor self-evident validity.

Rousseau and the Educational Philosophers

The first definitive predeterministic theory of child development was elaborated by the famous French philosopher, J. J. Rousseau (1712–1778). Rousseau postulated that all development consists of a series of internally regulated sequential stages which are transformed, one into the other, in conformity with a prearranged order and design (1895). According to this conception of development, the only proper role of the environment is avoidance of serious interference with the processes of self-regulation and spontaneous maturation. It facilitates development best not by imposing restrictions or setting coercive goals and standards, but by providing a maximally permissive field in which, unhampered by the limiting and distortive influences of external constraints, the predetermined outcomes of growth are optimally realized. Consistent with this orientation was Rousseau's belief that the child is innately good, that society constitutes the source of all evil, and that a return to a less inhibited and less socially restrictive method of child rearing would necessarily result in the unfolding of the individual's inherently wholesome and virtuous developmental proclivities.

The educational implications of these doctrines, which in essence were shared by such distinguished followers of Rousseau as Pestalozzi (1895), (1746–1827) and Froebel (1896) (1782–1852), were in marked contrast to those of the preformationists. Prominent recognition was given to the child's contributions to his own development, to his developmental needs and status, to his expressed interests and spontaneously undertaken activities, and to the importance of an unstructured, noncoercive instructional climate. This point of view has, of course, exercised tremendous influence on all subsequent educational theory and practice, and is essentially identical and, in a sense, historically continuous with present-day movements advocating a nondirective and child-centered approach to the training, education and guidance of children.

The Doctrine of Recapitulation: G. Stanley Hall

An especially fanciful but historically significant facet of Rousseau's (and later of Froebel's) conception of development was the theory that the child, in progressing through the various stages of his growth toward maturity, recapitulates the phylogenetic and cultural history of the human race. The analogy was only crudely drawn, but it served the purpose of providing a seemingly plausible explanation (a) for the hypothesized internal regulation of development, and (b) for the predetermined inevitability of its outcome in a direction which presumably paralleled the ascending spiral of cultural evolution. More than a century later, G. Stanley Hall (1846–1924) elaborated and refined this theory in great detail, postulating many ingenious and specific parallelisms between various hypothetical epochs in the history of civilization (e.g., arboreal, cave-dwelling, pastoral and agricultural) and supposedly analagous stages in the development of the behavior and play interests of the child (1904).

These speculations, which were advanced with great skill, comprehensiveness and internal consistency, acquired considerable vogue and many enthusiastic adherents. Their initial favorable reception was attributable, perhaps, to the fact that they were in accord with the prevailing evolutionary approach to cultural anthropology, and superficially, at least, seemed congruous with certain broad generalizations linking embryology and biological evolution. They were also bolstered by the prevalent beliefs that the thought processes of the civilized child are comparable to those of a stereotyped "primitive" adult (the "primitive mind" fallacy) and that the cultures of contemporary primitive peoples are analagous to the early stages of more advanced civilizations. Later, following more searching examination in the light of emerging data in comparative child development, and of changing concepts regarding the complex interrelationships between cultural environment, genic endowment and individual development, this theoretical orientation was no longer accepted as a parsimonious and potentially fruitful approach to problems in developmental psychology (see p. 25).

Arnold Gesell: Theory of Maturation (Ausubel, 1954)

With the collapse of Hall's elaborate theory of recapitulation, predeterministic theories of development received a serious setback, but they by no means disappeared from the scene. They simply assumed other forms more compatible with the prevailing theoretical climate. Perhaps the most influential and widely accepted of all present-day predeterministic approaches is Arnold Gesell's theory of maturation which reiterates Rousseau's emphasis upon the internal control of development, but discards the specific parallelisms between cultural history and individual development which made Hall's position so vulnerable to attack.

Gesell's theory also capitalized on its general resemblance to the empirically demonstrable concept of maturation which had gained considerable acceptance among behavior scientists, educators and the lay public. Actually, the latter concept dealt with the non-learning (as distinguished from the learning) contributions to enhancement in capacity, rather than with the more general issue of the relative importance of internal and external regulatory factors in development irrespective of the role of learning. Operationally it merely referred to increments in functional capacity attributable to structural growth, physiological change or the cumulative impact of incidental experience, in contradistinction to increments attributable to specific practice experience (i.e., learning). Gesell, however, used the term *maturation* in a very special and more global sense to represent the endogenous regulatory mechanisms responsible for determining the essential direction of *all* development, including that conditioned in part by learning and enculturation.

In essence, Gesell proposed an embryological model for all aspects of human growth—structural, physiological, behavioral and psychological—which "are obedient to identical laws of developmental morphology" (1954). In *all* of these areas alike, a growth matrix consisting of endogenous factors supposedly determines the basic direction of differentiation and patterning, whereas "environmental factors [merely] support, inflect and modify, but . . . do not generate the progressions of development" (1954). These intrinsic regulatory factors correspond to "ancestral genes" which in general reflect the evolutionary adaptive achievements of the race, but neither refer to specific epochs in cultural history nor condition the development of analagous ontogenetic phases (1933).

Because phylogenetic genes by definition have a species-wide distribution and are unusually potent in their effects, Gesell theorized that developmental sequences are relatively invariable in all areas of growth, evolve more or less spontaneously and inevitably, and show basic uniformities even in strikingly different cultural settings. Like Hall before him, he taught that certain undesirable stages in behavioral development were inevitable by virtue of the child's phylogenetic inheritance, and could be handled best by allowing them spontaneously to run their natural course without interference. Since comparable endogenous factors assured the eventual unfolding of more acceptable behavior, permissiveness and patience on the part of the parents and self-regulation and self-discipline on the part of the child could be confidently relied upon to correct the situation. The application of parental expectations, demands, limitations and controls was not only regarded as unnecessary but also as calculated to increase negativism and to impair the parent-child relationship (1943).

This embryological model is basically tenable when applied to the development of structures, functions and behaviors which are phylogenetic in nature, i.e., which characterize all individuals of a given species. It would apply quite well to the total development of members of lower phyla, to the prenatal behavioral development of human beings and to much of the sensori-motor growth that occurs during human infancy. But as far as the greater part of postnatal psychological development in the human species is concerned, unique factors of individual experience and cultural environment make important contributions to the direction, patterning and sequential order of all developmental changes. Not only is there significantly greater variability in the content and sequence of development, but also the uniformities that do occur (both intra- and interculturally) largely reflect the existence of common problems of physical and social adaptation and common culturally derived solutions to these problems.

Supporting Biological Trends

Related biological trends influenced and bolstered predeterministic concepts of child development in at least two important ways. First, they helped create a general climate of scientific opinion which affected the acceptability of the latter theories. Second, various biological concepts suggested, modified or reinforced the specific content of predeterministic theories. That these supportive effects and conceptual resemblances were often based on popular misconceptions, outdated formulations, and even on basically irreconcilable contradictions with allegedly analagous biological models does not in the least minimize the historical fact or importance of their occurrence. The three fields of biology that exercised most influence on predeterministic theories of child development were evolutionary theory, embryology, and genetics—concerned respectively with the origin of species, prenatal development and the mechanisms of inheritance. In more recent years, advances in these fields have belatedly influenced a trend toward a more interactional approach to human development.

Biological Evolution. In 1859, Charles Darwin proposed the revolutionary theory that biological evolution was a consequence of gradual and cumulative developmental changes in species resulting from the selective survival and transmission of small inherited variations that furnished adaptive advantages in relation to prevailing environmental conditions. The environment, he believed, could not directly induce structural, functional or behavioral changes in the organism that were transmissible to its offspring;[2] it merely played a role in the determination of which of a number of naturally occurring variations was most adaptive and, hence, selectively favored for representation in future generations by virtue of a differential rate of survival and ultimate self-perpetuation. The impetus for and the regulatory mechanisms of organic evolution, resided, therefore, in existing and spontaneously occurring variability ascribable to hereditary endogenous factors rather than to environmental factors.

Applied to the development of human behavior, this latter principle was frequently misinterpreted in ways that supported the predeterministic position. It was not fully appreciated that although the environment could not directly induce changes that were transmissible to offspring and, hence, could not initiate phylogenetic differentiation (i.e., the development of new species), it could still influence ontogeny (i.e., developmental sequences in the life cycle of individual members of a species). Thus, the Darwinian position was frequently misrepresented to mean what was never intended, namely, that environmental factors

[2] According to J. B. Lamarck (1744–1829) and his followers, "acquired characters" *were* transmissible to offspring. This doctrine however is in conflict with modern genetic theory which holds that ordinary, environmentally-wrought changes in the phenotype are not accompanied by corresponding changes in the genotype. Although there is "no proof that Lamarckian inheritance is impossible . . . no incontrovertible evidence in favor of it has been brought forward." Weismann's famous experiments (1889), which have been widely accepted as conclusively disproving Lamarck's hypothesis, actually did not adequately test it. The acquired traits, which were hypothesized by Lamarck as inheritable, were conceived as products of prolonged exposure or adaptive exercise over many generations, and could hardly be equated with such artificial, instantaneous insults as Weismann's snipping off of several generations of animals' tails.

were also incapable of exercising a direct effect on ontogenesis. Unfortunately this misinterpretation was reinforced and rendered more credible by the fact that it was not too far afield in relation to lower organisms with a more or less stereotyped pattern of adaptive behavior in response to environmental vicissitudes. Here individual experience *is* actually not much more crucial for the ontogeny than for the phylogeny of behavior. However in higher species, particularly in man, adaptation is characterteristically a function of a learned and flexible organization of behavior modified by individual and cultural experience. Thus, the predeterministic tendency to discount the impact of experience on human ontogeny was seriously tainted with error and distortion. The preformationists, starting from the same position, denied behavioral development altogether, equating man's socially learned behavior in Western civilization with a catalogue of king-sized inframammalian instincts.

Embryology. Early knowledge of embryology also reinforced predeterministic doctrines by pointing to more or less invariable sequences of development regulated predominantly by endogenous factors. Even when slightly qualified by later research showing that gestational environment was not entirely inconsequential for the outcome of development, the embryological model as such, when projected into postnatal life, still proved highly inapplicable to most problems of human development. In the first place it dealt almost exclusively with developmental acquisitions that characterized the species as a whole. Second, it was concerned with development in a relatively constant physiological environment largely insulated from external stimulation. It was, therefore, an analogy heavily loaded in favor of predeterministic conceptions, thereby confirming the bias of theorists who, like Gesell, minimized the contribution of individual experience to ontogeny.

Actually these latter considerations interpreted in the light of numerous findings in experimental embryology should have led to precisely the opposite conclusions. All of the research of the past forty years indicates (a) that marked and even less extreme variations in the intrauterine environment such as rubella or advanced age of the mother are associated with developmental abnormalities in the fetus (David and Snyder, 1951); and (b) that structural growth and functional development of embryonic neural tissue are affected by many factors in the internal fetal and intrauterine environments. Much experimental work (Carmichael, 1954; McGraw, 1946) points to the conclusion that structural differentiation and the sequential development of function in different portions of the nervous system are influenced in part by differential concentrations of biocatalysts, by quantitative levels of various metabolites and hormones in the fetal blood stream, by mechanical and other external stimulation communicated to the child *in utero*, and by the presence and functioning of adjacent tissues. This last-mentioned effect has been variously explained in terms of regional differences in metabolic activity (gradients), the organizing potential of certain embryonic cells in promoting tissue differentiation (organizers), and the operation of electrodynamic fields induced by biological activity.

We may conclude, therefore, that the internal (intrafetal and gestational) environment plays an important contributory role in embryological development, and that the preservation of its constancy is important for uniformity of developmental outcome—even with respect to species characteristics. If this is so, it would be reasonable to expect that the directional influence of the environment on the development of intraspecies *differences* would be infinitely greater once the individual were exposed to the tremendously wide spectrum of extrauterine variability in stimulation.

Genetics. The rediscovery of Mendel's laws in 1900 and the early subsequent work of geneticists had tremendous repercussions on theories of human development. The demonstration of a physical basis for heredity in the form of relatively stable, self-reproducing, discrete genes, which were resistant to ordinary environmental influences and which seemingly exercised an inevitable, unconditional, and one-to-one effect on the determination of specific traits, naturally favored prevailing predeterministic conceptions regarding the development of human behavior. And, of course, quite apart from this latter influence, the new science of genetics provided an urgently needed model for explaining the mechanisms underlying (a) *phylogenetic* inheritance as manifested in both biological evolution and in the embryological development of the individual, and (b) the *familial inheritance* operative in numerous studies of animal breeding, of the recurrent incidence of various "hereditary" diseases in certain human families, and of trait relationships between individuals differing in degree of consanguinity.

Later research in genetics showed that the model of single major genes with gross effects on variability, completely and invariably influencing the development of specific traits, was greatly over-simplified. It was convincingly demonstrated that "the phenotype of an organism is not a mere mosaic of independently expressed single-gene effects . . . [but] depends on developmental interactions involving the entire aggregate of genic material" (David and Snyder, 1951). Thus it is now known that the effects of many single genes are modified by other genes, and that most normal (and less extreme pathological) genic variability in human beings is produced by constellations of polygenes. The latter exert "individually minute but cumulatively appreciable [and] . . . quantitatively equivalent effects" resulting in continuous rather than in conspicuously discontinuous distributions of phenotypic variability" (David and Snyder, 1951).

More important perhaps was the undermining of the older genetic view supporting the established belief (mistakenly derived from Darwinian theory and from exaggerated instinct approaches to animal behavior) that the environment does not appreciably influence ontogeny. Modern genetics fully supports the proposition that "the phenotype is always the resultant of the interaction between a certain genotype and a certain environment" (Boyd, 1953). This, of course, does not mean that environmental factors ordinarily alter genes but that they alter the expression of genes (Boyd, 1953). The effect of genes on the development and patterning of morphological traits is frequently contingent upon the presence of a restricted range of such environmental conditions as moisture, temperature and diet; in other cases environmental influences are similarly operative only within a restricted range of genotypes; and sometimes the effects of heredity and environment on development are more nearly independent, additive, or complementary. Finally, "the effects of certain genes appear to be expressed with great uniformity within any range of environmental conditions" (David and Snyder, 1951); and, conversely, the effects of certain environmental conditions are manifested in practically all genotypes.

According to modern conceptions of genetics, therefore, the influence of genes on development is never complete or absolute, but always reflects to a variable extent the influence of the intracellular, intercellular, gestational or external environments. The phenotypic consequences of genic action are presently conceived of in such terms as probabilities of determination, degrees of regularity and completeness of expression, and limiting and threshold values of response and attainment. As will be pointed out later, this shift in theoretical orientation played an important role in resolving dichotomous views of the nature-nurture controversy and in generating an interactional approach to problems of human development. Nevertheless, exaggerated notions of the simplicity, specificity, prepotency and inevitability of genic effects continued to flourish and influence predeterministic formulations. For example, over-enthusiastic supporters of the latter point of view uncritically accepted much fragmentary and unreliable evidence from pedigree studies which purported to show that an amazing variety of feeble-mindedness, mental disease, social delinquency, moral waywardness and personality inadequacy was exclusively or predominantly the effect of genes inherited from a single and remote defective ancestor (Goddard, 1912).[3]

The ultimate in emphasis upon the pre-eminent importance of genic factors in human development is embodied in the eugenics movement.

[3] In reaching these conclusions little attention was paid to such obvious considerations as the representativeness of the sample, comparisons with a control population, the accuracy and equivalence of diagnosis made over an extended time span, the reliability of hearsay evidence, and the influence of substandard family and social conditions which invariably accompany and compound such conditions.

Its program is predicated on the belief that the soundest and surest method for improving the lot of mankind lies in upgrading the genic endowment of large populations through the rigorous application of principles of selective and restrictive mating. However, even if men and women could be induced to choose their mates on the basis of eugenic considerations, it would still require vastly greater knowledge about the mechanisms of human genetics and evolution than is presently available before such a program could be successfully inaugurated. Furthermore, from the study of cultural history it is clearly evident that profound changes in the behavior of human beings and in the quality of their civilization can be effected by social, economic, technological, scientific and educational advances within several generations. On the other hand, comparable examination of human evolution indicates that significant changes in the genic basis of human behavior and capacity could only be expected over time periods measuring tens or hundreds of thousands of years. Negative eugenics, i.e., the reduction and elimination of physical and psychological abnormalities by sterilizing the grossly unfit, is unfortunately not much more realistically grounded. Most of the more common and less extreme human defects that have an appreciable hereditary component are polygenically determined; and the relatively few and uncommon defects that are attributable to the effects of single genes are extremely rare "recessives" the incidence of which would not be significantly altered by sterilization.

Relationships between Biological Evolution and Embryology. The existence of many obvious parallels between biological evolution and embryological development inevitably led to much speculation about how these two phenomena are related. As a result many biological and biocultural concepts of recapitulation were elaborated, varying greatly in degree of empirical substantiation and theoretical credibility.

The biological theory of recapitulation embodied in Ernst Haeckel's (1834–1919) famous proposition that ontogeny recapitulates phylogeny was predicated upon certain gross sequential parallelisms in morphogenesis between the biological evolution of a species in geologic time and the embryological development of its members. This proposition is compatible with the fact that biological evolution is characterized by both continuity and modification; that is, in addition to well-marked lines of divergence, there is also much structural and functional continuity between a given species and its evolutionary forbears. Genically speaking, therefore, it could be anticipated that each species would inherit and transmit genes reflecting such commonalities and divergences, and, hence, that its members would tend to recapitulate in their early ontogeny the course of its descent from earlier forms of animal life.

That such parallelisms are not exact and do not embody *all* previous stages is not at all surprising. In the first place, the line of descent typically zig-zags instead of following a directly vertical course. Second, considering the difference in the relative time scales involved in each process and the undoubted influence of the more recent genic material on older morphogenic sequences, considerable telescoping and modification of ontogenetic phases could reasonably be expected to occur.

However, *biogenetic* theories of recapitulation (e.g., Rousseau's, G. Stanley Hall's), although superficially resembling Haeckel's proposition, were actually an entirely different breed of cat. The analogy was extended to include the *cultural* history of the race and the *postembryological behavioral* development of the individual. We have already noted that the latter kind of development (in contrast to embryological morphogenesis) is both less insulated from environmental influence and is characterized more by intraspecies differences in ontogenesis resulting from unique individual experience. In addition, these theories are not substantiated by any convincing evidence, and are based on the untenable assumptions (a) that cultures universally undergo a parallel sequence of evolutionary changes, and (b) that such cultural acquisitions are genically transmissible, and hence universally recapitulated.

On both empirical and theoretical grounds, the once fashionable notion of universal stages of cultural evolution is now thoroughly discredited (Steward, 1953). Some gross developmental sequences may conceivably be parallel in different cultures because of "recurrent causal relationships in independent cultural traditions" (Steward, 1953). For example, the evolution of certain levels of social organization may almost universally be dependent upon the prior attainment of supporting levels of technology (Steward, 1953). However, apart from such limited parallels, and in the absence of significant cultural diffusion, the cumulative impact of differences in geography, climate, history, values, institutions, etc., typically leads to progressively greater divergence in the development of cultural forms. We must conclude, therefore, that all human beings, regardless of cultural membership, hold their biological descent in common and undergo the same embryological development but by no means share a cultural history which reflects the operation of substantially identical processes of social evolution.

But even if cultures everywhere did undergo the same evolutionary process, what effect would this have on the genic constitution of man? It will be remembered from the preceding discussion (see p. 20) that ordinarily only spontaneous, genically induced structural or behavioral variability is inheritable, and that the main contribution of the environment to phylogeny is its role in natural selection. Although environmental factors influence profoundly the development of the human individual during his lifetime,[4] the changes they induce do not affect his genes and, hence, are only culturally rather than genically transmissible to his offspring. It is clearly evident, therefore, that the genetic assumptions of biocultural recapitulation are incompatible with modern conceptions of biology.

Thus, despite the vast changes that have occurred in man's behavior and cultural level since the emergence of Homo sapiens a quarter to half a million years ago,

> it is hardly probable either on theoretic grounds or on the basis of inferences from human history and archaeology that the biologic basis of human abilities or behavioral potentialities has appreciably changed during this period (David and Snyder, 1951).

And it is even more certain that all contemporaneous groupings of human beings—irrespective of past cultural history—share the same genic potentialities for psychological and cultural development.

In a very limited and quite different sense of the term, concepts of psychocultural recapitulation might hold promise of manifesting somewhat greater face validity. If, for example, we conceive of a trend toward increased use of symbols and abstractions as characteristic of cultural development, it would appear that at later stages in the history of most cultures, the intellectual development of the individual would tend to be extended beyond the point which generally prevailed when the ideational level of the culture was lower. In a sense, therefore, the historically later-born individual might be said to be "recapitulating" the intellectual development of his culture as he gradually grows in intellectual capacity. However, parallel development would occur in this instance not because certain cultural sequences were written into his genes and merely needed to unfold, but (a) because a trend from concrete to symbolic ideation happens to characterize the course of intellectual development in both individual and culture, and (b) because the limits of individual growth are dependent in part upon the level of cultural achievement.

Thus, the greater attained intellectual capacity of individuals in more highly advanced civilizations would not be indicative of cultural alteration of genotypic endowment, but of the greater phenotypic achievement that is possible with a constant genotype under conditions of enriched cultural stimulation. Its occurrence would in no sense be inevitable but would be dependent upon actual ontogenetic exposure to the necessary

[4] It may be noted that predeterministic theorists tended to reject the quite modest proposition that the environment significantly influences human ontogeny. Yet, paradoxically enough, they accepted the primacy of certain internal regulatory factors, the very existence of which presupposed the validity of the much more extreme environmentalistic position that cultural experience directly influences phylogeny by affecting the genotype.

experience. Hence, if twentieth century American children were artificially insulated from all ideational stimulation, their prospects for advanced intellectual development would hardly be brighter than those of prehistoric men.

"Tabula Rasa" Approaches

In marked contrast to the preformationist and predeterministic doctrines we have been discussing are such movements as humanism, behaviorism, "situational determinism" and certain varieties of cultural relativism. If we consider the former approaches as constituting one extreme of a continuum embracing the various theories concerned with the regulation of human development, the latter ideological movements would have to be placed at the other extreme of the same continuum. They are referred to as *tabula rasa* (literally, "blank slate") approaches because they minimize the contributions of genic endowment and of directional factors coming from within the individual, and concomitantly emphasize the pre-eminent role of the environment in determining the outcome of development.[5] The analogy which likens the neonate to a *tabula rasa* is aptly representative of their general thesis that no fundamental predispositions are inherent in the raw material from which behavior and personality develop, and that human beings are infinitely malleable. All of the patterning, differentiation, integration, and elaboration of specific and general behavioral content that emerges during the course of development is accounted for in terms of the unique stimulus conditions to which the individual is or has been subjected.

It should be noted, however, that the term *tabula rasa* is being used here in a very general sense, and only to denote such extreme environmentalist positions as described above. In the more specific sense of the term, as employed by John Locke, the "blank slate" only referred to the *ideational* state of affairs at birth and not to the *complete* absence of developmental predispositions; as a matter of fact in his discourses on education he placed much emphasis on the need for restraining the natural impulses of children. Furthermore, in the light of modern conceptions of cognitive and behavioral development, neither Locke's *tabula*

[5] We have already referred to Lamarckianism as an example of the extreme environmentalist position in biology since it is based on the assumption that the genotype as well as the phenotype can be altered by prolonged exposure to certain environmental conditions. However, the most basic feature of the *tabula rasa* approach is its emphasis on the plasticity of human beings (i.e., the absence of significant or enduring predispositions) rather than on the importance of environmental determinants of development. Certain more recent *tabula rasa* orientations, e.g., client-centered therapy, stress the notion of plasticity, but assign the main directional control of significant personality change to self-directed cognitive and motivational processes.

rasa proposition nor more recent dissatisfaction with the notion of human instincts could be regarded as indicative of an extreme position with respect to the nature-nurture controversy. Hence, although theories of innate ideas, instincts and instinctual drives must be categorized as essentially preformationist in orientation, disavowal of these constructs does not necessarily constitute a *tabula rasa* approach to human development.

Humanism and Related Approaches

The humanistic movement in philosophy and education has consistently championed the environmentalist position that given proper conditions of nurturance, man's developmental potentialities are virtually unlimited in scope or direction. Implicit in this optimistic appraisal is (a) the belief that "human nature" is essentially amorphous and can be molded to whatever specifications man chooses to adopt as most compatible with his self-chosen destiny, and (b) unbounded confidence in the possibility of attaining this objective through appropriate educational procedures.

Of course, the humanistic conviction that man can deliberately select and take steps to insure the realization of whatever goals he chooses and, hence, is the master of his own fate, would be a perfectly defensible proposition *if* it were related to and qualified by the actual psychological capacities of human beings. More often than not, however, it is merely stated as an unqualified philosophical desideratum. This is especially detrimental to its acceptability since it is becoming increasingly more evident that the extent of developmental plasticity is no longer a question which can be settled by speculative fiat. Furthermore, it is extremely unlikely that one blanket generalization could ever suffice to cover all aspects of development. In the modern era this issue is more properly regarded as a matter for empirical determination. And regardless of the ultimate outcome of particularized research inquiry, any realistic statement of human objectives and potentialities should presently be formulated within the framework of the limitations imposed by man's genic endowment as currently conceived in the light of all relevant data.

Although predicated upon quite different theoretical premises, the humanistic program of education was strikingly similar in spirit and content to the preformationist (theological) approach already described above (see p. 14). Despite the fact that one school viewed the infant as a formless entity wholly at the mercy of his environment and the other conceived of him as essentially prestructured in advance, both were in agreement (a) that the individual himself contributes little to his own development, (b) that the child in essence is a miniature adult, and (c) that improvement of man's nature could be best effected through the imposition of a stern and rigorous regimen of training and education. Preformationists reached this conclusion by both denying that any significant developmental changes occur in the first place, and by conceding that some quantitative improvement of prestructured attributes could result if superimposed from without by proper authority. Humanists, on the other hand, arrived at the same position more directly by attributing all developmental changes in an originally amorphous creature to the all-important influence of environmental factors, and by conceiving of such changes as occurring in quantitative steps rather than by qualitative stages.

Typically, therefore, the humanistic approach to education was rigidly academic, traditional and authoritarian. Severe and arbitrary standards were imposed and strictly enforced by the application of physical punishment and other extrinsic motivational devices. If necessary, rationality and classical erudition were literally pounded into the resistive or reluctant individual. Age level differences in capacity and in developmental needs and status were largely ignored and little or no attention was paid to individual differences in ability or temperament. Humanistic educators did not seriously attempt to enlist the child's voluntary participation, encourage his spontaneity or appeal to endogenous motivation. The contribution of ongoing personality to behavior and cognitive development was regarded as inconsequential, and the learner was granted no directive role or responsibility in the educative process.

Behaviorism shared many of the environmentalist biases of humanism but conceptualized them in more psychological terms. Consistent with its *tabula rasa* emphasis upon behavioral plasticity was its denial of subjective experience (except as a form of subliminal behavior), its rejection of all developmental predispositions (except for reflexes and certain emotional responses) and its conception of the human organism as a non-cognitive response mechanism subservient to the control of conditioned stimuli (Watson, 1919). Similarly in the area of child care and education, its advocacy of impersonal handling, strictness, regularity, and the importance of habit training was strikingly reminiscent of humanistic practices (Watson, 1928).

However, it need not be thought that a *tabula rasa* conception of human nature is necessarily or inevitably associated with a concomitant emphasis upon the pre-eminence of environmental determinants of development. The currently flourishing school of client-centered therapy, for example, combines a clinical estimate of extreme human plasticity with an emphasis upon endogenously derived needs, goals, insights, responsibility, initiative for change, etc., in a maximally permissive and nonauthoritarian therapeutic environment. As long as this relationship between endogenous and exogenous influences prevails, the possibilities for reorganizing personality on a more wholesome and constructive basis are held to be virtually unlimited, irrespective of existing personality structure or previous developmental history.

This point of view is obviously very close to the educational position of those predeterminist theorists (e.g., Rousseau, Hall, Gesell) who stressed the importance of permissiveness and self-direction in child rearing. Its principal point of departure from the older approach is that it conceives of these latter conditions as essential for the active *self-creation* of a personality with almost limitless possibilities for self-realization (or for the therapeutic reconstruction of an environmentally distorted personality) rather than for the optimal unfolding of a developmentally prestructured personality. However, we must reiterate again that the plasticity of the human personality and its responsiveness to reorganization are not issues that can be resolved by doctrinal assertion, but are matters for explicit empirical determination. And although self-direction is undoubtedly important for many aspects of both therapeutically facilitated and more normative sequences of development, there is little reason to believe that directional influences originating in the environment are unnecessary, unimportant or typically detrimental.

Cultural and Situational Determinism

The growth of empirical cultural anthropology during the first four decades of this century led to the formulation of a more explicit environmentalist position in conjunction with the conception of infinite human plasticity. Studies of modal behavior, socialization and enculturation in different primitive cultures impressed ethnologists with the remarkable homogeneity of these phenomena within cultures, with their tremendous diversity from one culture to another, and with the apparent absence of intercultural uniformities. The almost inevitable outcome of such conclusions was the emergence of a concept of cultural determinism, i.e., the notion that the human being is "well-nigh an empty vase into which culture and social prescriptions are poured" (Sherif, 1951), and that his behavior and personality development, therefore, are simply a function of the particular sociocultural stimuli which impinge on him. The personality-culture and the individual-society dichotomies were thus "resolved" by the simple expedient of virtually abolishing the categories of individual and personality.[6]

As long as intracultural differences in behavior were ignored, there

[6] We have already referred to social instinct theories which resolved the same dichotomies in opposite fashion by deriving culture and society from the preformed patterned behavior of individuals. Spiro (1951) is representative of a new trend in social science theory which seeks to reduce *both* personality and culture to a unique configuration of cumulatively learned individual behavior in an interpersonal setting; hence, according to this view "there are as many cultures as there are personalities." This reductionism is based on the propositions that the *locus* of culture resides in the behavior of its individual members, that the acquisition of culture can only be conceived as a learning (internalization) process occurring in *particular* individuals, and that individuals typically modify their cultural heritage. However,

was no pressing need to acknowledge the contributions of enduring response tendencies, selective perceptual sensitivities, and differential thresholds of reactivity established by the interaction between the individual's unique genotype and experiential history; and likewise, as long as cross-cultural similarities in development were disregarded, it seemed quite unnecessary to search for those panhuman regularities (of genic, physiological, psychological or interpersonal origin) that serve to limit and channel the impact of cultural influences on the human growth matrix along ontogenetic lines that are roughly parallel in terms of process from one culture to another.

Fortunately, expressions of this extreme *tabula rasa* version of cultural determinism are less commonly heard today. Most anthropologists, although still not greatly impressed by intercultural uniformities, have become much more cognizant of the importance of intracultural differences. However, the battle line they abandoned is still vigorously manned by various sociologists and sociologically oriented social psychologists who explain *all* such differences on the basis of subcultural membership or situational variables, and steadfastly deny the existence of stable, enduring antecedent predispositions to behavior or development. The situational *determinism* they advocate shifts the locus of personality structure from an organized system of underlying behavioral predispositions ("under the skin") to a series of behavioral *acts* manifested under specified socio-situational conditions. Whatever needs or motives are required to initiate or sustain behavior are derived intracurrently from the situation itself. Personality, as the more extreme of these theorists conceive it, is not a continuing, self-consistent structure exhibiting generality over situations, but a transitory configuration of individual behavior that is purely a function of the particular social stimulus conditions which evoke it.

This view of personality is rationalized on the grounds that since an individual's behavior does in fact vary *every time* the situational context is altered, it must therefore be determined by the latter variable *alone*. It is hardly necessary to point out however, that the demonstration of

---

one can accept all three propositions as valid without necessarily reaching the conclusion that personality and culture are one and the same thing. Although "culture" as such is an abstraction derived from a non-homogenous totality of individual behaviors, their interactions and products (and can obviously enjoy no existence independently of the persons who comprise, internalize, influence and are influenced by it), it still is a conceptually (if not functionally) independent phenomenon external to personality. The consensus, commonalities and uniformities to which it refers are *real* (e.g., actual shared values, beliefs, social customs), distinguishable from those of other cultures, and are sufficiently stable to be studied as if existing in their own right (Herskovits, 1948). They affect and are acquired by the individual as a result of influencing and being internalized by the particular cultural representatives (e.g., parents, teachers, peers) with whom he interacts in the course of his enculturation.

behavioral change associated with variability in one factor does not necessarily preclude the possibility that other variables are simultaneously operative. In fact, by simply reversing the picture, i.e., keeping the situation constant and varying the individuals exposed to it, one could just as easily emerge with the equally one-sided conclusion that only personality factors determine behavioral change. However, when a number of persons are studied in a diversity of situations, it becomes quite evident that both factors contribute to the obtained variability in behavior. This is shown by the fact that intercorrelations among behavioral measures in different situations are neither zero nor unity but somewhere in between, and that they tend to become higher when either the situations themselves or the subjects' degree of ego-involvement in them are made comparable.

The situational approach to personality not only strips it of any explanatory implications, but also renders futile any search for the genotypic bases of behavior. If personality has neither stability nor generality, there is certainly little point in considering the possible effects it might have on behavior, and even less point in attempting to trace the course of its development. And similarly, if overt behavior cannot possibly be related to underlying predispositions in personality structure, behavioral taxonomy must be based entirely on phenotypic similarities and differences irrespective of their genotypic reference. . . .

It might also be noted at this point that in practice (if not theoretically) nondirective schools of therapy tend to support the situational concept of personality. Although they do not explicitly deny the influence of antecedent response tendencies on ongoing behavior, they minimize their importance (a) by regarding them as almost invariably reversible, (b) by considering the discovery of their developmental origins as irrelevant for therapy, and (c) by placing major emphasis on the current adjustive situation.

Cultural Relativism

Viewed in historical perspective, cultural relativism must undoubtedly be reckoned the outstanding component and moving force behind the concept of cultural determinism discussed above. However, for purposes of conceptual clarity, it would be desirable for several reasons to consider the former movement separately. First, cultural determinists need only assume that the behavior of human beings is both plastic and crucially influenced by cultural factors; they need *not* accept the relativistic position that it is *completely unique* in every culture. In fact, if significant parallels of custom and tradition are demonstrable from one culture to another, to be perfectly consistent with the logic of cultural determinism one would have to postulate a corresponding parallelism in the area of behavior and personality development. Second, cultural relativism has been associated historically with an empirical, field study approach to ethnology and with a non-evolutionary, non-individualistic interpretation of cultural change[7] that are by no means indigenous to the position of cultural determinism. Third, because of these historical associations cultural relativism has come to represent an extreme point of view with respect to such issues as the plasticity, cultural uniqueness, intracultural homogeneity and intercultural heterogeneity of behavior, that is not necessarily inherent even in a relativistic position. Thus, many theorists who would readily agree that behavior and development are relative to and determined by the cultural environment in many important respects still hold views on these latter issues that are much less extreme than is implied in a more doctrinaire statement of cultural relativism.

Cultural relativism, of course, provided a much needed corrective against the ethnocentric social instinct and biogenetic doctrines that flourished during the same and preceding decades. It rejected the notion that complex social behavior is ever innately patterned by virtue of universal instincts, or that intra- and intercultural uniformities ever reflect the operation of an identical species-wide genotype with prepotent and invariable directional influence on the content and sequence of development. In accounting for behavioral regularities within a culture, it pointed to the obvious importance of considering commonalities in social conditioning; but in explaining cross-cultural similarities, it advanced the less convincing hypothesis of cultural diffusion. Most important, however, by demonstrating that the cultural patterning of innumerable aspects of behavior and development is characterized by an extremely wide range of variability, it completely demolished the ethnocentric preformationist view that distinguishing features of personality structure in Western civilization are manifestations of an immutable "human nature" and hence must be universally distributed. Instead it advanced the thesis that the unique values, traditions, institutions and historical development of each culture gives rise to a distinctive personality type. In so doing it established the beginnings of the now flourishing research area concerned with empirical investigation of the personality-culture problem.

Of course, even cultural relativists could not assume complete be-

[7] Since we are only concerned here with individual development, this issue naturally lies outside the scope of our inquiry. However, it is important to point out that the methodological approach of the relativists, which emphasized the empirical study of *behavior* in particular cultures (as opposed to the logical analysis of cultural institutions and products in relation to a universal concept of cultural evolution) predisposed them toward a conception of uniqueness in considering the impact of culture on behavior and personality (Spiro, 1951). Their non-evolutionary view of cultural change similarly predisposed them in this direction; but since similarity in cultural development is only one of many factors affecting cross-cultural uniformities in personality, the two positions (the anti-evolutionary and the relativistic) are not necessarily co-extensive.

havioral plasticity. Certain limitations imposed by man's species membership and by his biological and psychological needs, capacities, and mechanisms were recognized as constraining the impact of culture on behavior. But within the framework of these highly general limits, all patterning, differentiation and selectivity in behavioral development was considered a function of cultural variables. Thus, for example culture was conceived as determining the kinds of stimuli evoking a particular emotion and the manner in which it is expressed, and as selecting through differential rewards and punishments which potential capacities and personality traits of man are either emphasized or neglected in a particular cultural setting.

Relationship to Psychoanalytic Theory. To a very great extent the full impact of cultural relativism on conceptions of personality development was blunted by the considerable influence exerted by psychoanalytically oriented ethnologists and by psychoanalytic theorists concerned with the problem of the individual in society. The partial fusion of these two currents of thought (i.e., psychoanalysis and cultural relativism) probably reflected the prevailing absence of a satisfactory body of competing psychological theory in the area of personality as well as some dissatisfaction with the extremeness of the relativistic view. In any event it occurred despite the presence of the following serious conceptual incompatibilities between the two positions:[8] First, psychoanalysis reintroduced the anthropologically suspect doctrine of instinct in the somewhat more palatable form of patterned psychosexual drives, and established the latter as the new basis of intercultural uniformities. Nevertheless, this conception of drives as innately prestructured and biogenetically transmitted entities was in direct conflict with the relativistic principle that all significant and detailed psychological patterning is determined by unique factors of cultural conditioning; and projected universally as it was from an unrepresentative sample of neurotic individuals in our own society, it naturally ran afoul of the rigorous relativistic strictures directed against ethnocentrism. Second, the psychoanalytic view of society as basically frustrating was incompatible with the proposition implicit in any form of cultural determinism that the social order not only provides the means of gratifying the individual's biologically instigated drives but is also capable of independently generating in its own right highly sig-

[8] Basic incompatibilities in viewpoint with respect to the development of cultural institutions are not pertinent to the present discussion. In general, however, Freudian interpretations of cultural forms and practices as institutionalized mechanisms of repressing or symbolically expressing psychosexual drives have not won as much acceptance from ethnologists as have psychoanalytic formulations regarding the influence of culture on personality development.

nificant drives of interpersonal origin. Lastly, the psychoanalytic school explained intercultural differences in personality structure almost exclusively on the basis of differential parental practices influencing the course of psychosexual development. Cultural relativists, on the other hand, have necessarily taken a much broader view of the potential range of interpersonal and sociocultural factors in a given society that are significant for personality development, and have recognized that aspects of personality structure other than erogenous impulses are subject to societal influence.

Implications of Historical Trends for Modern Conceptions of Development

We can summarize most helpfully the implications of the foregoing historical trends for modern conceptions of developmental regulation by indicating briefly in what general ways both predeterministic and *tabula rasa* approaches to human development are theoretically untenable. These considerations will point up the desirability of adopting the emerging interactional approach. . . . Since the main issue here is the extent of behavioral plasticity, we may properly subsume preformationist views under the predeterministic category.

Summary Critique of Predeterministic Approaches

1. Except for simple responses of a reflex nature, there is little substantial basis in either logic or empirical data for the belief that *any* psychological aspect of human functioning is preformed at birth, *completely* independent of subsequent environmental experience. Even the initial, unpatterned psychological repercussions of intense visceral and hormonal stimuli (i.e., drive states) are influenced by the effects of prior experience and by concurrent internal and external stimulation; and under extremely unfavorable social auspices, it sometimes happens that certain "primary" drives (e.g., sex) may *never* be generated, regardless of the adequacy of gonadal output. . . . Where complex patterning is involved, the possibility of prestructured psychological entities is, of course, still less credible. But although the existence of human instincts is no longer taken seriously by most behavior scientists, the quite comparable notion that patterned affectional-sex drives exist preformed in a biogenetically inherited id has won much acceptance in many quarters. The most anachronistic of all present-day preformationist thinking is exemplified in the psychoanalytic theory of innate ideas (e.g., cosmic

identification, reincarnation, omnipotence) lodged in a phylogenetic unconscious. Supporters of this doctrine point to the widespread occurrence of these themes in the mythologies of historically unrelated cultures and in the ideational outpourings of deeply regressed psychotics. The first phenomenon, however, is more parsimoniously explained by the independent cultural generation of common ideological solutions to such universal problems as death and supernatural control of the environment, and the second by regression to an earlier ontogenetic stage of ego development. . . .

2. Equally unsubstantiated is the embryological model of psychological development which is not predicated upon preformationism, but asserts nevertheless that developmental sequences and outcomes are basically predetermined and inevitable because of the prepotent influence of internal (genic) directional factors. Actually this conception only holds true for the relatively few and simple behavioral acquisitions which in terms of specificity of content and sequential appearance characterize every member of the human species (e.g., locomotion). For all other behavioral traits, the contribution of unique environmental conditions to developmental regulation is considerably greater; and, hence, both the kinds of growth changes that take place and the sequence in which they occur are more variable by far. It is quite erroneous, therefore, (a) to underestimate the impact of culture and individual experience on almost any psychologically significant aspect of human development; (b) to minimize the extent and significance of culturally conditioned diversity in individual development; and (c) to overlook cultural commonalities operative in the life histories of individuals, and attribute all observed developmental uniformities both within and between cultures solely to the influence of similar genic factors.

Biogenetic theories of recapitulation hypothesizing specific parallelisms between successive stages in the psychological development of the individual and various inferred stages in the cultural evolution of mankind are insupportable on both empirical and theoretical grounds. They rest on the discredited assumptions that cultures everywhere evolve in parallel sequence and that the cultural acquisitions of a people are genically transmissible to their offspring.

Summary Critique of "Tabula Rasa" Approaches

1. Not content with having successfully cast doubt on the validity of preformationist and predeterministic doctrines, *tabula rasa* theorists unfortunately veered to the opposite extreme and asserted that human behavior is infinitely plastic and malleable to environmental influences. Although they were probably correct in assuming that *some* aspects

of behavior (e.g., social roles and attitudes) are almost entirely determined by cultural variables, they stood on palpably less solid ground in refusing to recognize that other facets of psychological development are patterned in many significant ways by various selective predispositions, limitations, capacities and potentialities arising from within the individual. Because these internal factors (which either directly or indirectly have a genic basis) do not characteristically exercise solitary, highly specific and invariable effects on the content and sequence of development, *tabula rasa* theorists fallaciously concluded that they do not even operate as partial or general determinants.

2. Hence, for example, extreme cultural relativists and situational determinists failed to appreciate (a) that many intracultural differences in behavioral development are conditional by genotypic diversity as well as by subcultural, familial and individual differences in background experience; and (b) that numerous intercultural uniformities in psychological development are undoubtedly determined in part by various aspects of man's genic endowment which both relate him to and differentiate him from other species.[9] Thus, the unique ontogeny of human beings is more than a reflection of their uniqueness in being the only species in nature whose development happens to be systematically molded by a culture. It is also a reflection of the fact that they are the only species *genically capable* of responding to cultural stimuli in ways that characterize the development of a cultural organism. No amount of cultural stimulation could possibly make chimpanzees develop like human beings.

3. In addition to overlooking the genic basis of intercultural uniformities in behavioral development, cultural relativists also failed to appreciate that many of these uniformities (e.g., general stages in personality development) are induced by numerous "common denominators" in culture itself. The latter in turn are derived from universal features in man's physical and interpersonal environment and from his adaptations thereto, as well as from panhuman biological and psychological characteristics (Kluckhohn, 1953; Murdock, 1945).

4. As will be pointed out in later chapters, many cultural relativists (under the influence of psychoanalytic and stimulus theories of drive), paradoxically relapsed into some of the most serious errors they berated in their adversaries. For example, in assuming that sex drives are either preformed or inevitably generated by gonadal hormones, they surprisingly underestimated the characteristic plasticity of human beings in responding to factors that induce and pattern drive states. . . . And in defining a basic human capacity such as guilt in terms of the particular conditions

[9] See . . . [Ausubel, "Historical Overview of Trends," p. 65] for a listing of both genic and environmental factors contributing to intercultural uniformities in development.

under which it arises and the specific forms it adopted in our own culture, they reached the surprising ethnocentric conclusion that individuals in most other cultures exhibit shame rather than guilt. . . .

DYNAMICS OF DEVELOPMENT: SYSTEM IN PROCESS*

John E. Anderson

All of us are interested in development, in what happens to living beings and to social systems as they move forward in time. But we are only incidentally occupied with the transactions of the moment; we are much more concerned with long-time trends; with structuring, organization, system formation, whether we see it in species, in individuals, or in groups. Perhaps we can find a common language in the principles which underlie development.

For my part, as a student of human development, I have tried to break down the concept of development into dimensions which frame mutually exclusive principles, and then to explore each briefly. My former students will recognize a number of familiar principles, some of which appear here in the sections on Openness, Growth, and Selection. Let us then consider the age-bound behavior of the person as it takes on form or shape.

What manner of thing is the person whom we study from birth on to maturity? Even the most superficial observation reveals a very complex system moving forward in time by growing in complexity and size. This system receives much stimulation from the outside world and reacts to that stimulation in many and varied ways. As it grows, it contacts a wider and wider range of objects and persons; its ability to solve problems increases; it builds up many habits and skills; it gains in knowledge and self-control. It is a complex manifold of many characteristics and potentials which together make up the total shape or form we call a human being.

How shall we characterize this manifold and break it up into parts? Some idea of the complexity of our task can be gained from the recent

---

* John E. Anderson, "Dynamics of Development: System in Process," in Dale B. Harris (ed), *The Concept of Development, An Issue in the Study of Human Behavior*, University of Minnesota Press, Minneapolis. © Copyright 1957 by University of Minnesota.

work of Barker and Wright (1955) who have recorded the experiences of children from early morning until late at night. If you have any illusions that the developing system is simple, prepare to lose them. Barker and Wright in a midwestern town of 700 people find 2,030 different behavior settings to which children are called to react. Within these behavior settings there are some 1,200,000 behavior objects to which children build specific reactions. In this quiet community the flow of life is so great that the three children who were followed for 14.5, 13.6, and 13.0 hours at the ages of 8.7, 7.4, and 7.4 years respectively, reacted to a new behavior object every 1.5, 1.2, and 1.0 minutes. And in the course of a single day they reacted to 571,671, and 749 behavior objects, making 1,882, 2,282, and 2,490 different reactions, respectively. Long before these studies, research on language had shown the four-year-old child uttering a total of 10,000 words a day. Studies of social interactions gave similarly large figures per day, while estimates of the number of distinct visual images formed on the retina ranged from a low of 250,000 to a high of 500,000 for the waking portion of the day. We have then an amazing system in terms of its intake and its power for converting the flow of intake into behavior. It becomes quite obvious that very simple explanations in terms of single stimuli and single response patterns cannot give us a picture of this enormously facile being. Perhaps we need a concept like entropy.

In the past we have taken cross-section slices of the activities of the growing person at successive ages. More recently we have followed children over long periods in longitudinal studies, in order to picture the various aspects of growth. We have also placed children in specific situations and studied their reactions to controlled stimulation. From these studies we have come to understand much about perception and learning. Sometimes we have selected a particular characteristic which manifests itself in various ways, such as love, anger, or anxiety, and generalized it into a construct which makes possible explanations of what are on the surface very diverse types of behavior. By studying the resemblances between parents and children we have also been able to reach conclusions about the interrelations of heredity and environment for many behavior areas. We have also learned much about building a good environment for children in community, home, and school.

In breaking into this manifold, scientists have put forth a wealth of theories for the various phases of the process: genetic, cognitive, behavior, psychoanalytic, and social learning theories. We have cut across the developing organism in many ways, and we have sliced it vertically in many ways. But what of the whole, and particularly what of the whole as it moves through long segments of time? This is essentially the domain of developmental theory which asks whether or not we can abstract from the manifold, as it is transformed in time, general principles that will aid our understanding. Because these principles are encumbered by a wealth of concrete observation and fact, it is sometimes desirable to construct a theoretical model in order to obtain a generalized picture by separating the principles from their supporting data.

Models, whether laid out verbally, diagrammed visually, or constructed out of metal and wood, force the scientist toward a very clear and definite picture of the interrelations involved in whatever process is symbolized. Next they bring out new research problems or programs, and permit generalization by analogy from one type of system to another. A unifying concept which runs through studies of biological and social phenomena is that of system which changes with time. Immediately we think of mechanical systems or machines which are so fabricated that certain functions are accomplished. We may ask then, What is a machine, an organization, a structure, and how does it come to have its present shape or form?

In the models of the types of machines which are everywhere about us, form is imposed from without by a designer, engineer, or fabricator; in the living machine, design evolves within limits from the materials and relations within the organism and their interaction with the environment. Limits are imposed by the number and variety of the genes and chromosomes, the mechanical structure of the organism, and the environment, which is restricted in comparison with all possible environments.

The model of the growing person is one of an ongoing manifold passing through time with a multidimensional head, much as a tunnel-making machine moves under the river. When viewed from the front this head consists of many irregularly shaped components of various sizes with a complex series of interrelations. At the boundaries there are complex relations with the external world which involve a continuous intake and outgo at both the physiological and psychological levels. Through this interchange a dynamic or ongoing equilibrium is maintained. The system is however open-end, that is, always in imbalance. Over and above the maintenance level there is a progressive enlargement or growth with time which reaches a maximum in mature form. In the ongoing movement of the system successive levels of integration are achieved, and new properties emerge which must be studied and analyzed in and for themselves. The model has a self-propelling character and as it cumulates from its own experience, it progressively and irreversibly builds up energy from the free energy of the environment. This denies any possibility of explaining human behavior on the analogy of a physical or mechanical model which is limited to intake and outgo. In exploring this model, our concern is with its changing total shapes as well as with the changing form of its parts. Analyses of these transformations in time are crucial.

Consideration of this model brings us immediately face to face with the inadequacy of the language by which we symbolize our findings. In considering the development of plants, Arber (1954) points out that the one-dimensional ongoing stream of verbal language is inadequate. She suggests visual representation as preferable, and perhaps visual representation by a motion picture of growth collapsed in time, but points out the difficulties of such representation in terms of manipulability. D'Arcy Thompson (1952) using Cartesian coordinates, devised a visual system which proved more effective for symbolizing species transformations than either verbal or mathematical symbols. Medawar (Clark and Medawar, 1945) has successfully written the equations for the stretch of body form from fetal life to maturity from the frontal or two-dimensional aspect as portrayed in the celebrated Stratz figure, but states that the solution of the problem for the three-dimensional figure is beyond present mathematical equipment. In fact, those most expert in the use of mathematical methods for the analysis of growth question whether or not it will ever be possible to represent the variety of shapes and forms—the outcomes of growth—that arise from differences between species and within species, and from differences between individuals.

If the problems of symbolization are so great for structures with forms and shapes that can be measured with precision, how much greater must they be for psychological functions? At present our mathematical and statistical procedures can handle cross-sectional relations better than longitudinal ones, the relations of two factors better than those of three, and those of three better than those of four, and so on. It would be interesting to analyze scientific problems in terms of the manner in which our analytic symbols force restrictions on the conceptualization of problems; for example, how much does the present vogue of hypothesis research force problems and analysis into the determination of two-variable relations?

Turning now from the difficulties of symbolization, we may delineate the general characteristics of the developing system. We can consider these under the general heads of openness, activation, growth, selection, learning, mechanization, cumulation, emergence, and symbolization. In my opinion these topics represent the major dimensions of the developmental process or, if not the dimensions, at least the framework within which discussions of development must be oriented. In a sense they define development. All are processes extended in time; all influence in some manner the shape or form of behavior. Despite the interaction which creates difficulties in discussing any one without bringing in the others, they seem to me to be distinct enough to constitute the parameters of evolving behavior form.

Openness

The developing organism is an open system in which irreversible changes occur as a result of the relations within the system and the interactions with the environment. Contrasting with the open system is the closed system of the machine manufactured by man which likewise carries on an exchange with the environment, but which involves a fixed input and output and a series of relations which do not change except for deterioration. While machines can be built which will store information and operate over substantial periods of time in much the same way as a living system, nevertheless both the controlling and holding mechanism are laid out in minute detail in the original wiring plan. If there is to be any substantial change in the machine's operation, a new wiring diagram has to be evolved by someone outside the machine.

In an open system the control element is progressively modified in time, the wiring diagram changes itself, and the system has some capacity for self-correction or repair. Hence an outcome of momentary transactions is a modification which is carried forward in time. As a result, the changes in an open system are not reversible, a fundamental characteristic of an open system and of living as compared with mechanical systems. For example, a child who has once learned to talk can never be returned to the state of one who never has talked, as is shown in the studies that compare congenitally deaf children with those who acquire complete deafness at 18 months or two years or after even a small amount of language has been acquired.

Activation

As early as we know it, the developing organism is already an active system achieving a high level of interchange with the environment and within itself. This basic activation state is maintained by internal and external stimulation. Upon this level with its own patterns, there are superimposed the patterns initiated by the particular stimuli to which the organism reacts specifically.

The human system has been described in terms of two levels: first as a physicochemical machine which takes in food and converts it into action and waste products, and second as a sensory-neuromuscular mechanism which takes in stimulation and converts it into behavior by channeling the energy developed in the physicochemical system into specific responses. As Sherrington (1906) in his concept of the final common path has shown, the muscular system is so organized that only one major activity can be conducted at a time, and that energy goes to specific points in space time.

One implication of this principle often made in the literature emphasizes the energy output of young organisms in various species and leads to the surplus energy theories of play which have been current for centuries. And a practical conclusion sometimes drawn is that rather than looking upon the energy of the child as harmful and therefore seeking to repress it, we should concern ourselves with the guidance and direction of this valuable possession.

Recent work by Hebb (1955) indicates the existence of an arousal or activation level which is a function of the amount of stimulation rather than of its specific character. To this sum of stimulation, all forms of sensitivity contribute. Above this level there is the cue level, or the precise stimulation from specific items, objects, or elements within the range of "its" sensory equipment to which the organism builds specific reactions. If the activation level is lowered by a drastic reduction in stimulation, severe disturbances of the perception and action of the person soon appear. Evidence is also piling up from other fields that any marked reduction in activity level over a substantial period of time realigns the functions of the organism and may be more disturbing than the disorder which brought about the reduction. Immobility, prolonged rest, nonuse of parts or functions, bring their own particular type of deterioration. There is, then, a psychological tonus like the physiological tonus which must be maintained in order to keep organ systems ready for functioning. To use Hebb's analogy, the human mechanism is more like a steam engine with some steam always in its boilers than like an automobile engine which starts from a complete stop. Electroencephalograms show activity in the nervous system even when the person is at rest and indicate that specific stimulation sets up waves which are superimposed upon a going pattern.

From the standpoint both of practice and theory, then, a fundamental problem arises. Are we talking about a machine into which we pour a specific stimulation and out of which comes a specific response, or are we talking about an organization already moving under its own power, with its own patterns, upon which we are superimposing new patterns? The first point of view makes of the growing person a passive recipient of forces from the outer world. The second makes him an active agent with something to say about the imposition. Psychologists develop drive theories on the assumption that we have to explain activity as something which arises from without; whereas the second point of view recognizes activity as a property of a living system, and makes the explanation of inactivity or lack of drive the important problem.

At any moment in time much behavior consists of transactions within and without the organism which are concerned with the steady state and have little or no significance in terms of forward and backward reference. Thus we may set apart moment-by-moment transactions which do not modify the system. For example, we take in oxygen and expel it, take in food and convert it into energy, tissue, and waste. By analogy we can develop the same concept of behavior for social systems or for phylogenetic development and separate the inflow and output on a transactional level from the cumulative changes which affect later functioning.

Some learning is also transactional in that it occurs in a relatively brief time and involves acts or products which are not carried forward. Hence there are quotation marks about "permanence" because of the difficulty of drawing a hard and fast line between a transactional and a cumulative process.

Growth

As time passes the living organism increases in size and in complexity in a manner that can be described accurately and in detail. Growth begins with the fertilized egg and continues until maturity or a terminal point is reached. Psychologists have devoted some attention to a hypothetical zero point at which behavior originates, but while some study fetal behavior, for most the study of behavior begins with an organism already far along the developmental course.

With growth the shape of the organism changes in such a way that it moves progressively away from its initial state and progressively nearer its final state. As a result the correlations between the primal shape of the organism or any part and the shape at any later period decrease with growth, and the correlations with the final shape increase (Anderson, 1939). This holds for mental as well as physical functions.

With increase in physical size, significant changes take place in bodily structure and function. Since volume increase as the cube, surface as the square, and linear dimensions in the power of 1, changes are forced within the differentiated structures in order to make possible the carrying on of functions. For example, the cortex, which is a surface, must be folded to accommodate the number of neurons necessary for functioning in the complex system.

Some physical size relations clearly limit or modify functions. Because of differences in the length of the arms and legs relative to the total body length, the adult cannot put his foot in his mouth as can the infant. This instance, however facetious, covers a wide territory, since back of it there lie quite substantial variations in motor patterns and adjustments with age.

For behavior it is difficult to separate increase in size from differentiation. Nevertheless, we describe the change in the range and level of activities as an enlargement of the life space. With age the organism reacts to more and varied stimuli; has available many more skills and more complex patterns of reactions; reacts to objects which are more remote in space and in time; and adjusts to situations and solves problems of greater altitude or complexity. Several years ago, Thorndike, in a study of the dimensions of intellect (1926), found very high, almost perfect correlation between range, speed, and altitude, which point to the existence of a general underlying factor in mental growth that manifests itself in different ways. More recently an entropy theory of intelligence has appeared (Eysenck, 1953) that distinguishes between levels of intelligence, not so much by quality of performance as by the rate at which information moves through the nervous system. Theoretically, the higher the rate the more completely explored is the situation, and the greater is the likelihood of obtaining a good response. This suggests a developmental relation between size and function within broad limits.

Selection

With development, the growing person purchases efficiency at the cost of versatility. Facing time limitations and choice points, the organism loses the multipotentiality of the infant and gains the efficiency of the adult. Involved in this modification are whole series of processes which we divide among growth, learning, and cumulation.

Various observers have described the very young organism as fluid, polymorphous perverse, multipotential, and plastic; all are terms which indicate that subsequent development can go in any one of many directions. But once a choice is made and direction is set, cumulative and irreversible changes take place which determine the major aspects of subsequent form. These choice points are not points in the mathematical sense as they may be extended in time. Much more typical than the dramatic episode or trauma, which may be momentary, is oscillation for a week, a month, or a year before direction appears.

In his life the growing individual meets successively many choice points, at which decisions between alternatives are made. Modern computers are based upon a similar principle in reverse; complex problems are broken down into a series of binary choices for purposes of analysis. In ongoing behavior such choices are made in a forward sequence with complex form or behavior as the outcome. Needham (1936) gives as an example of this phenomenon the switching yard which makes up railroad trains by means of electrically controlled switches. At each switch the car can go in either of two directions. From a line of incoming cars that are jumbled or randomized many trains may be put together, each of which will differ in make-up from every other. Viewed from the air these assembled trains on their tracks have a distinct shape which varies from day to day and which is produced by a relatively simple process of consecutive binary choices. The choice point may be illustrated in human terms by two fourteen-year-old boys who receive chemistry sets for Christmas and are equally interested for several weeks. One boy then tosses the set aside, whereas the other over the next eight years receives professional training, and devotes the remainder of his life to chemistry. Although he begins with an interest in chemistry, he soon chooses between chemical engineering and chemistry; then between inorganic and organic chemistry; and then, within organic chemistry, he makes further choices. Considering the developmental significance of these choices for the subsequent behavior of the person, we know astonishingly little about the mechanism involved.

Similarly the pattern of social relations may operate to determine the final forms of behavior in what is known as the vicious circle. A child is delinquent in a minor fashion; his escapade becomes known in his neighborhood; the neighbors expect him to be delinquent and take precautions against him. Over a period of time, a combination of personal and social expectancies is built up and culminates in a form of behavior which we call delinquency. Later, when society steps in, it usually attempts correction or punishment for a time, but then returns the person to the same environment in which the cumulative expectancies have created an atmosphere that makes good adjustment unlikely. Some students of social problems point out how difficult it is to undercut or eliminate the context of expectancies which determine the shape of such behavior. Similar expectancies and atmospheres operate at successive choice points to shape desirable behavior.

It should be noted that there are difficulties in predicting the outcome of choice points at the moment the choice is made. Retrospectively we can describe the factors involved in great detail and can develop a number of explanatory systems, as is well illustrated by the variety of theories advanced for the causes of particular wars by historians. But because experience is assimilated, and because successive choices bring greater involvement of the system, it is difficult, if not impossible, to make firm predictions at the moment of choice. This principle has sometimes been generalized into a distinction between significant and nonsignificant events. The significant event for the individual is the one which modifies the subsequent amount of energy or practice devoted to the activity, or which modifies the subsequent current of social relations and their feedback into the individual.

As the growing person meets choice points, his behavior becomes more structured and multipotentiality is lost. Supporting evidence comes from studies of children's interests, which have the widest range at eight years and progressively narrow until the age of twenty-two. But the study of particular interest patterns shows greater involvement on the part of the adolescent and young adult, and indicates a shift from breadth of interest to depth of concern. In this shift, the fact that the time available to the growing person is limited forces choice. The boy cannot practice the piano while he plays baseball, as many parents know. Since the person lives only so many hours a day and is a creature that can do only one thing at a time, selection becomes necessary and shape evolves. Moreover, the more complex and advanced the organism, the more limited becomes the time available for realizing potentialities.

This progressive specialization of function within the person parallels the specific roles that characterize our complex society, which is composed of a series of interlocking roles giving individuals specific functions upon which other persons depend. In a sense, the capacity of the person to structure his available energy makes possible this differentiation of roles. In turn, the specialized roles offer him the possibilities upon which subsequent behavior depends. As a result of specialization and role playing, the growing person becomes more sensitive to some aspects and less sensitive to other aspects of his environment. Choice comes to be made more and more in terms of the availability in the environment of materials and events which resemble the system already operating. From this principle, which has been called the principle of congruence, emerge the projective methods for evaluating personality structure which theoretically should operate better in the older and more differentiated person.

An important question arises for all the developmental phenomena. Why do they not continue indefinitely? If left to themselves the factors we describe would make for an indefinite progression. Among the various explanations for the stoppage of development the best for our purpose seems to be the principle that the limiting factors are not within growth itself, but comprise an inhibition of growth.

If the concept of inhibitors of growth is broken down we find, first of all, competition among growth processes within the person which limits each component. Since the person is a very complex collection of structures and functions within a single whole, one subsystem will limit or circumscribe another. A second type of inhibitor is external, and arises from relations with other persons and the demands made by them. A third inhibitor arises out of restrictions in opportunity, and constitutes what is now known as deprivation. The deprivation studies imply limitations which have effect over a very long period of time; in some degree they fail to recognize the capability of the organism for self-repair or readjustment when appropriate facilities are made available. A fourth and very important inhibitor already mentioned is that all psychological processes take time, and time is limited.

If we turn to learning we find so-called task limits—those inherent in the tools or machines operated by the person—and physiological limits—those intrinsic to the fact that the person moves toward but never quite reaches the highest possible performance of which he is capable in terms of his genetic make-up and bodily mechanism. For very complex verbal skills involving vocabulary and language acquirements, there are limits in terms of the sheer mass of material which is to be grasped or handled within a given time.

In the past the students of development have more frequently concerned themselves with the factors that produce behavior change rather than with those that limit or restrict the organism. But the study of the limits of behavior which is implied in such phrases as "growth potential" and "developmental potential" would seem to be very important.

Learning

By learning we mean the modification of behavior as the result of stimulation. It is a property of the living organism which is present to an extraordinary degree in the human being. Because it can be compressed or distributed in time at the will of the experimenter or teacher, learning can be regarded as time-free in the sense of organizational time. It is not my purpose to discuss the very powerful intellectual tools and theories which psychologists (Bush and Mosteller, 1955; Hull, 1952) have developed for the analysis of learning, except to point out that these are concerned with the mechanisms rather than with the products or the content of learning. Developmental theory on the other hand studies the way in which the products or results of learning are incorporated into evolving systems of behavior, and how they function as constituents in subsequent development. For example, developmental theory is not concerned with how long it takes a child to learn to read or to learn to dance, but with whether or not he can read or dance and what reading and dancing do for him in his subsequent behavior. The experimenter in learning loses interest in the child as soon as he reaches one or ten perfect trials, i.e., some defined level of proficiency. Learning theory is therefore not the equivalent of developmental theory, but deals with different phenomena.

Learning may also be viewed as a disturbance of the steady state of the organism in line with what was said earlier about the activational level. In this sense, the organism seeks to recover from the disturbance of learning by forgetting and thus to return to balance. It follows for the developmentalist that questions about the phenomena that keep what is learned active and alive within the developing system are important. Hence he speaks of reinforcement, reiteration, and overlearning.

In the life situation as distinct from the laboratory situation, research shows that reinforcements run into the thousands and millions, that they are aperiodic rather than periodic, and that they are not clean-cut and separate phenomena but appear as parts of complex patterns.

A developmental phenomenon growing out of learning becomes important for us. First, in moving from birth to maturity, the person meets a succession of situations and problems for which he builds appropriate skills. The growing person learns to handle a spoon, to lace his shoes, to read, to write, to ride a bicycle, to shave, to dance. By the time he is an adult the number of response patterns, even in a very ordinary individual, is so great as to defy enumeration. In building these response patterns, regardless of their type, we find essentially the same basic learning mechanisms operate and that the person, despite the travail of learning, moves on to competence.

Recently investigators have obtained evidence that the early life of the child, with its tremendous impulses to curiosity and the manipulation of objects, is given over to the building of hundreds of thousands of simple response patterns to hundreds of thousands of behavior objects. For example, with the appearance of the eye-hand coordinations, perception of objects advances. The movements of the body, limbs, and head, which are made almost endlessly while crawling and creeping, become the basic coordinations upon which upright locomotion is based. The vocalizings of the infant become shaped and integrated into the symbols used in speech.

In the later periods of development the growing person combines these response patterns into larger units, and we pass from the period in which learning consists of establishing what Hebb (1949) calls simple cell-assemblies to the building of complex from simple cell-assemblies. To some degree, success in meeting the involved problems of adolescence and adulthood is a function of the repertoire, or the number and variety of responses that have been established in infancy. With development the number of reaction potentials increases faster than the opportunities for using them. Thus a great file of simple and complex responses and response assemblies becomes available for meeting the demands of life.

These larger systems become in some degree independent of the units of which they are constructed. This process is similar to the equipotentiality of which the biologists speak. Because one unit can be substituted for another, the behavior of a complex assembly becomes in some degree independent of its parts, and in a sense exists in its own right. Such substitution is, however, limited by the fact that if too large a section of the whole becomes inoperative, the remainder ceases to function. For behavior integration and equipotentiality are enormously facilitated by transfer, and by the mediation mechanisms of the symbolic process.

Mechanization

In development the growing person mechanizes or ritualizes much of his behavior. Achieving maturity is largely a matter of successively ritualizing one series of responses after another and incorporating them into the system of behavior. Mechanization and ritualization have counterparts in many biological and social systems. From this process there are several developmental results. First, less of the available time is needed for the particular response; second, energy is freed for responding to other aspects of the situation; third, higher-level integrations become possible, as units or subsystems already organized become available for incorporating into large systems. For example, the four-year-old lacing his shoes for the first time takes thirty minutes; a year later he takes one minute for the same operation, and has thereby gained twenty-nine minutes and a residue of energy for other response patterns. He also has added a pattern to his repertoire which can be elicited upon demand by the particular situation of shoes and laces and which has partial components such as tying a knot that can be transferred to other situations. Thus the way is cleared for complex patterns and the solution of more complex problems. As the person moves toward competence his affective involvement decreases, and the stability of expertness appears. Recent work on aging persons indicates the extent to which there is emotional disorganization upon retirement, when time-filling activities and routines are lost. As a new schedule of activities is worked out, the emotional difficulties of the older person are reduced or disappear.

Mechanization, or ritualization as it is sometimes called when social systems are discussed, has a very important function in stabilizing behavior. But the process has limitations. In a real sense these products of learning, like the constituents in biology, can be viewed as residuals or debris which ultimately will choke the system and cause infirmity and death. Out of this comes a very interesting and appropriate question for the study of development at all levels, namely, that of the optimal balance between stabilizing and adjusting processes. For behavior this is related to our earlier discussion of activation level, in which the question was raised of how stimulation and activity can preserve or maintain the adjusting capacity as long as possible.

Cumulation

Some events moving into the developmental stream modify either the internal relations or the external relations of the system. A product appears which is then carried forward in the total system as a permanent change to affect all subsequent relations. Some products are stable, as in the case of a specific skill or bit of knowledge, and are accreted. In many functions cumulation continues in such a manner that it changes the shape of the total system. Whether the product is *stable* or the function is *cumulative* depends upon its effects on the developmental stream.

In the earlier section on choice points it was said that decisions or events which modify the stream of practice or which affect the relations of individuals with other persons may exercise an influence out of all proportion to their extent and thus determine whole areas of subsequent behavior. What begins as a minor phase of the total system comes to include more and more of the system. This process resembles the cone concept which is used by Woodger (1952) to describe the fact that within the embryo some physical and functional phenomena increase out of proportion to the total system. A similar phenomenon occurs in history. Consider the example so frequently given that contrasts the discovery of America by Lief Ericson with that by Columbus. In the former the feat was greater but without consequence, whereas in the latter the series of social and economic changes which followed made it one of the outstanding events of history.

A corollary of the principle of cumulation is that of the recession of the cause. Any system moving forward in time which has a holding or memory mechanism will retain some of the effects of past action. When cumulation is possible, the origins of present behavior go far back into the life history, and it becomes difficult if not impossible to explain present behavior in terms of present stimulation even with the best of assessments. Thus arises the necessity of both an historical approach to the developing person and an approach in terms of multiple factors or causes, calling to mind the classical distinction between precipitating events and an underlying complex of causal relations.

Emergence

As the person moves forward in time, changes occur in the altitude of behavior because of the emergence of new properties and new relations. The enlargement of the life space proceeds by stages, each of which may involve a varying period of time for acquisition followed by a period varying in length during which the growing person adapts to his new found functions and properties. Thus there are sudden as well as gradual transformations of behavior with each change followed by a period of gradual adaptation.

In child development controversy has raged and still does as to whether development is a continuous process with small increments out of which higher levels of behavior gradually emerge, or whether there are distinct stages through which the organism must pass in order to move on to a new level of development. That the controversy is partly a matter of how data are obtained and interpreted may be illustrated by the Shirley (1931) studies of walking behavior. By supporting the baby under the armpits and recording stepping behavior, a continuous and gradual curve which shows increasing strength and better coordination is obtained. If, however, locomotion is viewed from the standpoint of what the child is doing, a four-stage picture emerges: (a) lying prone; (b) a crawling stage in which the abdomen is on the supporting surface; (c) a creeping stage in which the abdomen is lifted clear of the surface and pacing movements of the arms and the legs occur; (d) a walking stage in which the upright position is assumed and forward motion on two feet occurs. While the appearance of these stages spreads over several months, in terms of the whole life cycle the transitions are completed within very short periods. Criteria for describing these separate levels can be developed, and our popular language clearly recognizes their existence. While some investigators concern themselves only with the development of behavior itself, it is clear that immediately following the transition and permanently thereafter, the orientation of the child to his environment is changed. A child who walks and runs has different properties from one who crawls or one who lies in his crib.

Comparable instances of other relatively quick transformations can be given. Thus the transition from reflex grasping to thumb opposition not only modifies all responses but all perceptions as well. The most dramatic transition, and also the most difficult one to explain, is that from vocalizing without meaning to the use of sounds as symbols, or the appearance of language and speech. Although more questions are raised about the existence of later stages, the evidence suggests that the separation out of the qualities of objects or surrogates from the objects in which they are embedded, which occurs roughly between five and nine years, is a somewhat similar process, and that the reorganization in the forward reference of thinking during puberty is a comparable stage.

A fruitful way to attack the question of stages is to ask whether or not properties have emerged which cannot be described in terms of earlier behavior, irrespective of whether it takes a few days or a few years for the transformation to take place. Furthermore, the transformation which we describe in terms of properties may originate either in a process of maturation or a process of learning. Not only is the child who talks very different from the one who doesn't, but the child who reads in a world of reading material is very different from one that does not.

In any complex learning process it is clear that the expert performer not only can equal all that the inexperienced performer can do—he can also accomplish many things which are beyond the beginner's reach. While it is difficult to describe these properties of the growing organism which flow out of increased complexity, the speech of the common man is full of descriptions of behavior that are felt to be relevant. From the very beginning in developmental psychology there have been many controversies over whether or not the child's behavior is best described in terms of his bodily movements, like the number of steps taken toward another child, or in more generalized descriptive terms, like aggression and cooperation, that pertain to his social orientation. When we say a person is angry we describe a total property of a system, even though we have some difficulty in defining the term precisely. All such terms, like all others in science, need to be examined with great care in order to determine their appropriateness and their limits. In general, we describe the properties of simple organizations or systems more readily than we do those of the complex ones.

It may well be that the human being differs markedly from the animal in his capacity to build up integrations of behavior that give new properties. Learning may function with respect to behavior much as do the organizers in embryonic development that establish the gradients out of which new structures function and properties evolve. A human being is then a combination of a built-in series of stages, which he has in common with his biological ancestry, and a process of learning by means of which new stages with new properties can be built up within a single lifetime. If the stimulation for learning is set by the culture in such a way that all growing individuals meet similar stimulation at the same ages, an orderly progression is produced which, from the developmental point of view, is superimposed upon the development that is in the person by virtue of his biology. The classical instance is the imposition of the first grade upon children at six years.

Symbolization

In one fundamental respect the human system differs from other living systems as we know them. Through the symbolic process man is able to represent his experience in units which can be dissociated from their contexts and manipulated in and for themselves. Although symbols are often identified with their primary source in speech and language, they nevertheless exist in virtually every type of stimulation and response. It is the process of symbolization rather than the particular form taken that is important for our purpose.

Symbols not only function as the means of communication with other persons and thus make possible our social life, but they also act within the individual as triggers to set off responses, as tracking devices to control the direction of responses, as holding devices to store past experiences and incorporate them into the system, and as the manipulable carriers of meaning which make thought and problem solving possible. Within the person they function as reverberating mechanisms, and in social groups as feedback mechanisms. Through their storage and carrier functions they become the devices by means of which we anticipate the future, become aware of purposes and goals, and are able to free ourselves from the immediate demands of the present.

On the one hand, the symbolic system is a system in and for itself, set within the larger system of the person, but following its own rules and developing in its own way; on the other hand, it functions parallel to the other phases of the developmental process and in time becomes so pervasive that virtually every phase of stimulation and response becomes tied in with symbols in some fashion. In terms of properties, a comparison of a deaf and dumb child of ten years with a normal child of the same age would reveal even to the most casual observer such enormous differences that he would hesitate to place them in the same category. In their nature and extent the differences would be like but greater than those between a society with and one without a written language.

Symbols appear first between twelve and eighteen months, and are to be distinguished from the responses to signals which many animals make by the fact that the child, by deliberately using them as means for controlling others and himself, frees them from their attachments and comes to manipulate them in and for themselves. Thereafter symbolic development follows its own pattern by starting very slowly, then proceeding at a very rapid or accelerated rate. For many symbolic processes curves of progress are positively accelerated, which means the more you have the greater is your gain within the next period of time. Contrary to many other growth processes, some aspects of symbolization seem to be without limits. Symbolic processes also seem to be the behavior processes which are most resistant to the effects of aging.

When what has already been said in this paper is examined, it becomes clear that throughout I have assumed a human being with symbolic capacity. The symbolic system in itself is an open system with a high activation level; it increases in size and complexity, it is selective, it facilitates learning, it is subject to mechanization, it increases by accretion and cumulation, and new properties continually emerge. A word in particular should be said about its relation to learning, since it furnishes a ready method of advance which is a great gain over the slow and progressive simplification of behavior in the animal. By substituting symbols for overt action the rate at which information flows through the system is enormously increased, the rates of selection and of mechanization are increased, and the person is freed for attacking the more complex problems which could not be attacked in any other way.

Through symbols we are able to build up the systems of thought reflected in our sciences and philosophies, to establish the records which make our histories, to make the inventions and discoveries which transform our lives, to establish and operate the societies which make up our civilization. I am not interested in creating striking images; I am concerned with calling attention to the symbolic mechanism as a basic determiner of the shape or form of the human being wherever we find him.

Summary

We may now bring together into a final brief picture the factors in the development of the growing person that seem to be generalized into principles of form or shape organization. We began by pointing out that we deal with an open system with a very high rate of interchange within the system and with the environment in which it exists. The processes which evolve out of this interchange are irreversible. When we first study it the system is already a going and active concern upon which events and stimulation are superimposed. This implies that the system is itself in some degree a determiner of its experience. With age this system grows in size and differentiates. Because of limitations of time and the fact that basically only one thing can be done at a time, the entire process of development is inherently selective, and movement is in general from versatility to efficiency. In succession choice points are encountered at which the decisions are made which determine the ensuing developmental process.

Through the basic mechanism of learning which follows a describable and similar course for a wide variety of behaviors, the system acquires patterns of skill and knowledge. In the early period with its curiosity and manipulation, tens of thousands of specific reactions to stimuli are established; later, and to some degree simultaneously, these are integrated and organized into larger units. Experience may be merely transactional. Or it may result in stable products which are incorporated into the system ready for functioning. Or it may result in cumulations which acquire significance for later development by affecting the extent of practice or the system of social relations. As the system moves along it passes through successive stages out of which come new properties and accomplishments to which subsequent periods of orientation are devoted. Pervading the whole and facilitating all the factors involved in development is the symbolic process by means of which the range of communication, control, and direction, as well as the emergence of new properties, is enormously increased.

It should now be clear that development is a multi-faced process of enormous complexity. To describe it and explain it we need the efforts and thought of many investigators in many disciplines. The person interested in developmental theory hopes that in our concern with the transactional aspect of the living system, we will not lose sight of the long-term changes which will add to our understanding of the life process.

THE CONCEPT OF DEVELOPMENT IN COMPARATIVE PSYCHOLOGY*

T. C. Schneirla

The concept of development, connoting a pattern of changes occurring in a system through time, is fundamental to the psychological study of animals. But a comparative psychology as a scientific discipline does not arise merely through the general recognition that some kind of relationship exists between the phenomena of phylogenetic and ontogenetic change (Schneirla, 1952). Nor is it to be established by the same methods as a comparative anatomy or a comparative embryology.

---

* T. C. Schneirla, "The Concept of Development in Comparative Psychology," from Harris, Dale B. (ed.), *The Concept of Development: An Issue in the Study of Human Behavior*. University of Minnesota Press, Minneapolis.

Although, as investigation progresses, the molecular basis of behavior and of psychological capacities is revealed with increasing clarity in organic processes, it is becoming evident that the principles holding for the molecular and molar levels of organic reality cannot be the same (Needham, 1929; Schneirla, 1949). Although the principles for the *molecular* are basic to the molar level, the molar requires operations in investigation and theory beyond the specific terms of the molecular. No matter how thorough-going the studies of the embryologist and his colleagues in cognate biological fields may be, in themselves they cannot be expected to give an inclusive, comprehensive basis for dealing adequately with questions concerning the behavior and psychology of animals (Needham, 1929; Waddington, 1940).

It may not be altogether clear why a science of behavioral development should be responsible for deriving its own theories and validating its own principles, rather than accepting them fully formed from other fields in biology. But the probability exists that in all types of organisms, in ways depending on the characteristics of each respective phyletic level, behavior has differing degrees of indirectness in its relation to structure (Schneirla, 1956). The issue is often complicated by a confusion of abstractions, or preliminary working attempts to represent phenomena, with a positive or finalistic view of reality itself. To emphasize these considerations, a conceptual distinction which is both empirically grounded and heuristically recommended is suggested here, to differentiate what is connoted by the terms *growth, differentiation,* and *development.* In *growth,* the emphasis is on change by tissue accretion; in *differentiation,* on variation in the changing of structural aspects with age in an organism. In *development* the emphasis is on progressive changes in the organization of an individual considered as a functional, adaptive system throughout its life history. Growth processes, as well as those of tissue differentiation, are subsumed by the term development, which further stresses the occurrence of progressive changes in the inclusive, organized function of the individual.

This distinction has great theoretical importance for the study of behavioral and psychological capacities in the animal series. For it stresses the fact—increasingly apparent with the advance of morphology on the one hand and of psychology on the other—that although tissue and organ, local movement and inclusive behavior are closely and inseparably related in functions, the represented phenomena are not fully describable in the same terms.

The Problem of Behavioral Development

For an adequate perspective in the methodology of research and theory, we cannot accept an a priori definition of behavioral development either as an unfolding of the innate, with gains through learning presumably superimposed in superior phyla, or as a continuum expanding mainly through the pressure of environmental forces, with the genes merely contributing an initial push to the process. Rather, a defensible generalization is that species' genetic constitution contributes in some manner to the development of *all* behavior in *all* organisms, as does milieu, developmental context, or environment. The "instinct" question, then, is regarded as merely a traditionally favored way of posing the question of behavioral development in a general, preliminary way, with emphasis upon the role of genetic constitution (Schneirla, 1956).

The question of behavioral development is posed from the above standpoint not to be eclectic, but first of all to encourage behavioral analysis, an emphasis badly needed in all developmental psychology. In the history of this subject, studies of psychological development have been too predominantly descriptive, too discontinuous as to stages investigated, and too preoccupied with certain stages, particularly the adult, to promote the rise and use of adequate analytical methods.

The critical problem of behavioral development should be stated as follows: (1) to study the organization of behavior in terms of its properties at each stage, from the time of egg formation and fertilization through individual life history; and (2) to work out the changing relationships of the organic mechanisms underlying behavior, (3) always in terms of the contributions of earlier stages in the developmental sequence, (4) and in consideration of the properties of the prevailing developmental context at each stage. In attacking this problem, an adequate programmatic combination of cross-sectional and longitudinal studies must be carried out through an analytical methodology suited to the phyletic level under study, with the results then synthesized to the end of valid phyletic comparisons.

Development Considered Phyletically

Behavioral Relationships in the Animal Series

The phyletic study of development naturally presupposes some kind of series or order among contemporary animals as to their respective evolutionary backgrounds. However, no clear and supportable series, except in the broadest terms, has been established with respect to the behavioral properties of the respective animal phyla (Hempelmann, 1926; Maier and Schneirla, 1935). The phyletic tree has numerous long branches, with large sections now extinct, and present-day animals fall into a very irregular pattern on it. Although the order of evolutionary antiquity and recency is usually paralleled by the degree of specialization in the sense of behavioral plasticity, exceptions are numerous. For instance, the horse, which as a specific form of organism has had about the same evolutionary time as man to reach its present adaptive status, compares favorably with man only in specializations other than psychological. Our comparative behavioral study therefore must include all principal existing animal forms, and not just those that form some type of series (Schneirla, 1945 and 1952).

It is sometimes argued that a comparative psychology can be founded directly on comparative anatomy (Lorenz, 1950). This may presuppose or favor a preformistic answer to the instinct problem (Lehrman, 1953). But the concept of homology, connoting a significant evolutionary relationship between comparable mechanisms among species has not been validated as yet for behavior and its organization. To the extent that this may be possible in the future, what homologies may exist in the behavioral properties of different animals must be worked out essentially by psychological and behavioral methods; always, of course, in view of evidence from comparative anatomy and other biological sciences. For the chief difficulty, as we shall see, is that behavior seems to have very different relations to structure in different types of animals.

Psychological Levels

For this task of phyletic comparisons, a theory of psychological levels promises to be useful, in which the capacities and organizations in behavior are compared as *high* or *low* in respect to psychological status (Needham, 1929; Schneirla, 1949). If one knew only the reactive repertoire of the sea anemone's tentacle, the insect's antenna, and the cat's forepaw in their normal situations of use by the respective organisms, one would conclude without much doubt that among these three, the tentacle indicates by far the lowest psychological status in total behavior, the paw the highest. What we know about the respective owner organisms would not change this judgment.

Psychological level cannot depend upon adaptive success, for by hypothesis, all existing animal forms must be ranked more or less on a par in this respect. The principle of homeostasis, which concerns the physiological readjustment of an organism to biologically nonoptimal conditions, applies well to all animals considered against their respective environments. But at the same time, the mechanisms underlying homeostasis, along with other capacities crucial for behavioral organization, show progressive changes in their internal organization and in their relation to other individual functions, from psychologically low-status to high-status animals.

Complexity is often advanced as a criterion for distinguishing psychological standing. This, however, is a doubtful procedure until we know *what* it is that is complex. For instance, the compound eye is very complex in many insects, but in action it lacks associated perceptual capacities present in a mammal through experience. Insect visual discriminations through the compound eye are based on flicker vision rather than on pattern perception in the mammalian sense (Schneirla, 1949). The significant feature of behavior for psychological appraisal is not specifically dependent upon a multiplicity of molecular units, many sensory endings, and the like; it is rather a difference in *kind,* depending upon the qualitative aspects of behavioral organization and capacities. From what might be termed the additive point of view, the capacity for lasting neural trace effects gained through experience often is regarded as crucial for the psychological appraisal of an animal. But it is probable that higher mammals do not necessarily retain behavioral modifications better than do many birds, or even insects; it is in the versatile use of the traces under new conditions of emergency that mammals typically excel (Schneirla, 1945). Also, on higher levels, the nature of variability in behavior becomes much more than the capacity for some kind of change. To say that higher levels are characterized by plasticity means that capacities and organizations appear which admit the possibility of systematic variations appropriate to new environmental conditions. Adjustments to changing conditions are no longer made merely through the random shifting of items in a fixed repertoire, but through the appearance of opportune new patterns.

The developmental process characterized as behavioral evolution, therefore, cannot be viewed as a process of accretion, by the addition of further mechanisms similar to the old. If it were such a process, phyletic comparison would be easier than it seems to be. Rather, development evidently has taken place in evolution through the progressive transformation of old mechanisms—e.g., homeostasis (Pick, 1954)—as well as through the appearance of new mechanisms—e.g., neural patterns underlying reasoning. Through organic evolution higher functional levels appear, which, as individual adaptive systems and as behaving wholes, may be considered qualitatively new.

Ontogeny and Phyletic Comparisons

Phyletic Differences in Behavioral Attainments

Accordingly, on each further psychological level, the contribution of individual ontogeny is a characteristically different total behavior pattern arising in a different total context. This fact cannot be appreciated unless the comparison of different animal types is made in terms of their entire ontogenetic ranges, and not confined to certain stages such as the adult. This is a principle to which Orton (1955) recently has called attention from the standpoint of biological procedures in systematics. The "waggle dance" as described for the honey bee by von Frisch (1950) is an efficient communicative device and an impressive event, but when considered in the light of its stereotyped ontogenetic basis, it must be viewed—contrary to a frequent conclusion from general similarity—as fundamentally inferior in its qualitative organization to symbolic processes in higher mammals and language in man. The appearance of behavioral stereotype through ontogeny, if found characteristic of a species, indicates a lower psychological level, whereas a systematic plasticity in the organization of behavioral cues made broadly representative through experience indicates a high level.

Homologous mechanisms are transformed functionally in a new setting on each higher level, as through ontogeny a characteristically different total behavior pattern arises in a different total context. Consider, for example, how differently the vertebrate forelimb is involved in the feeding operations of a toad, an adult monkey, and a five-year-old human child. The same principle may hold roughly for analogous adaptive mechanisms in very different phyletic settings. For instance, the group unity and organization of an insect colony is held (Schneirla, 1946) as an extension of Wheeler's (1928) concept of trophallaxis, to derive from reciprocal stimulative processes reinforced by a variety of sensitive zones and special secretory effects. In different but comparative ways, parallel intragroup processes of this type may be held basic to social behavior on *all* levels from invertebrates to man. The fundamental mechanisms are similar, in that specialized sensitivity and responses affecting the formation of individual affiliations and reactivities in the group situations are always involved; however, the various resulting patterns of collective behavior are distinctively different. I have characterized that of social insects as *biosocial,* in that it is evidently dominated by specific mechanisms of sensitivity, physiological processes and response; that of mammals as *psychosocial,* in that although trophallactic mechanisms are still basic and indispensable to them, group behavior is now characterized by plastic processes arising through learning, anticipation, and the like (Schneirla, 1946 and 1949). The principle is: that similar biological processes can lead to very different behavior patterns through ontogeny, depending upon what developmental capacities in species and what limitations are effective in the prevailing context.

Within each level there is a range of behavioral capacities differing in underlying organization—as in carnivores, from random emotionalized escape to a roundabout food-getting response. The different patterns in the behavioral repertoire of any level may be termed *functional orders,* and ranked high or low in organizational status. In a chimpanzee, for example, casually flicking away a fly represents a pattern of low functional order, gesturing to another chimpanzee a pattern of high functional order, although the actual movements are similar. The highest functional orders characterizing any psychological level are those utilizing its maximal possible gains through ontogeny. With respect to behavior and its organization, the ontogeny of any level is not so much a retracing through the stages and functional orders of successive ancestral forms as a new composite leading to a new total pattern characteristic of the level. Functional orders characteristically attained through ontogeny on any phyletic level are not replicas of those appearing through ontogeny at other levels, either with respect to their order of appearance, their individual make-up, or their properties of organization as part-processes in the inclusive behavior pattern.

Genes and Behavior

It would be misleading to say that the behavioral patterns of lower psychological levels are inherited or innate, and those of higher levels learned. The traditional heredity-environment dilemma stands out more and more clearly as a pseudoproblem, as further evidence indicates that in all animals intrinsic and extrinsic factors are closely related throughout ontogeny (Schneirla, 1956). As a guide to investigation, it is no wiser to regard development as a natively determined unfolding of characters and integrations than as the result of the molding effect of extrinsic forces. We may therefore hold that genetic constitution contributes to *all* behavioral development at *all* phyletic levels, just as modern geneticists hold that without a participating environmental context at all stages there could be no development in any animal (Snyder, 1950; Waddington, 1940). The question is *how* development occurs in the particular animal under prevailing conditions, not *what* heredity or environment specifically contributes, or how much either contributes proportionally to the process. Genes are complex biochemical systems, integrated from the beginning of ontogeny into processes of increasing complexity and scope, the ensuing progressive processes always intimately influenced by forces acting from the developmental context (Waddington, 1940; Weiss, 1954).

In each species genetic factors contribute indirectly to advances in stabilizing organismic functions, and impose characteristic limits at which they tend to resist change. But as Dobzhansky (1950) has emphasized, it is undesirable to confuse this natural state of affairs with the impression of heredity *directly* determining development. Nor does a correspondence of the theoretical genotype with strain-specific behavioral characteristics, appearing in any animal in the standard environment, demonstrate a *more direct* relationship of genetic constitution and behavior in highly selected than in relatively unselected strains (Hall, 1951; Schneirla, 1956).

The fact that in all phyletic types modified or even radically different phenotypic alternatives may arise through ontogeny, according to variations in the context of development, militates against a doctrine of genic *Anlagen* directly determining the rise of behavior. Gesell's (1945) conception of a human ontogeny, in which early behavior expands through "the innate processes of growth called maturation," seems more adequate for insects, although not really suitable in their case (Schneirla, 1956). Harnly (1941) found, in experiments with the fruit fly, *Drosophila melanogaster*, that the same gene may influence the development of different wing size and structure according to what temperature prevails during the development of the phenotype. The types produced from the same genotypic strain were capable of full normal flight, erratic flight, or no flight, depending upon conditions of development. Not only wing size, but also wing articulation and neuromuscular control were believed differently affected according to what temperature prevailed during specific early stages of ontogeny. It is impossible to account adequately for such facts in terms of a preformistic theory of behavioral development.

Intervening Variables in Development

The nervous system is often represented as the carrier of an innate determination for behavior patterns presumed to arise solely through its passive development. For this type of conclusion, Coghill's investigation (1929) of larval salamanders is often cited as evidence. However, it cannot be presumed that the full significance of this research is understood as yet, even for amphibian development (Fromme, 1941). On any phyletic level, the nervous system develops as *one* part of the organism, closely interrelated with other mechanisms throughout, and like them, it is indirectly influenced by the genes (Lehrman, 1953; Schneirla, 1956). Nativistic views of behavioral development (Hunter, 1947) may often seem to furnish the sole available theoretical resolution of a problem in the appearance of individual behavior, but only when the developmental system is so incompletely understood that the burden can be placed upon neural growth alone. To carry weight, investigations on the ontogenetic causation of species behavior patterns must involve a thoroughgoing analysis of behavioral development as related to other processes at successive stages. Conclusions not based on such analyses are premature.

The problem of ontogeny is not one of instinctive patterns directly determined by the genes, but one of understanding what changing integrations of development underlie successive functional stages characteristic of the species. Two concepts may be used to represent the factorial complexes essential to the entire flow of events in development. The first is *maturation*, which connotes growth and differentiation together with all their influences upon development. Maturation is neither the direct, specific representative of genic determination in development, nor is it synonymous with structural growth. Much as an environmental context is now recognized as indispensable to any development (Dobzhansky, 1950; Snyder, 1950; Waddington, 1940; Weiss, 1954), students of behavioral development (Kuo, 1924; McGraw, 1946) emphasize the roles of structure and function as inseparable in development. The second of these two fundamental concepts is *experience*, which connotes *all* stimulative influences upon the organism through its life history.

As an abstract operation, for heuristic purposes, these factorial complexes may be conceptualized disjunctively in their effects upon development, and there are indications that this logical procedure may be carried further as experimental methodology advances. But realistically, the two concepts must be considered as standing for complex systems of intervening variables closely integrated at all stages of development.

Circular Functions and Self-Stimulation in Ontogeny

An indispensable feature of development is that of circular relationships of self-stimulation in the organism (Schneirla, 1956). The individual seems to be interactive with itself throughout development, as the processes of each given stage open the way for further stimulus—reaction relationships depending on the scope of the intrinsic and extrinsic conditions then prevalent. At early embryonic stages, circular biochemical effects of expanding range, in a sense *spiral*, may be invoked in these terms; somewhat later, proprioceptive processes capable of integrating neurally with the early cutaneous stimulative effects (Carmichael, 1936; Kuo, 1932b). From the time of fertilization, development is continuous and progressive (Gluecksohn-Waelsch, 1954; Willier, 1954), its processes and stages being closely interrelated in their intrinsic and extrinsic aspects (Schneirla, 1956).

The potentialities of redundant stimulative processes for development have been incompletely investigated. For instance, Fromme (1941) found, in replicating Carmichael's test (1926) with early amphibian embryos immobilized by chloretone, that specimens held inactive during stages 17 and 20, in particular, in later tests swam more slowly than normal specimens and with deficient coordination. The need of further work along the lines of the Carmichael experiment is indicated, with reference to its bearing on the interpretation of Coghill's findings (1929). The effects, particularly in the earliest stages, may not be those of obvious action or readily recorded impulses, so much as electromotive changes and action twitches of minute magnitude, slight afferent pulses, or other subtle energy releases capable of contributing in some way to the changing of patterns in development. Kuo (1932a, b, and c) demonstrated for the chick embryo ways in which stimulative effects from early functions, such as heartbeat and rhythmical pulsations of the amnion membrane, could influence the development of a head-lunge response. Early self-stimulative relationships, influencing the rise of a basic generalized response, contribute (with the entrance of extrinsic, differentiated conditions) to the specialization of this act into pre-hatching responses such as egg-chipping or into post-hatching patterns such as preening, drinking, and food-pecking.

Hatching or birth represents only one of the stages at which the young individual influences a notable change in its own environment made possible by stimulative gains through its own prior development. Although opening the way for a wider scope of action in a more heterogeneous environment, neither involves an absolute change for the stimulus-response progressions of development.

At all stages, the individual possesses the maturational properties of its species, which inevitably exert species-typical feedback effects upon its development. That the relationship of such effects to the rise of behavior at later stages may be elusive is suggested by Nissen's statement (1954): "Among animals at all levels in the phyletic scale there can be observed highly motivated, almost compulsive forms of behavior which have no relation to homeostasis in the usual sense of the term." He cites the nest-building and brooding behavior of birds as examples of activities that "have no obvious effect on maintaining the physiological equilibrium of the individual engaging in these activities." But actually, in many birds maturational processes affecting temperature change and causing irritability of the skin at the hypersensitive brood patches, lead to active, variable readjustments and appear to be key functions in the appearance of nesting and incubation. The outcome may be physiological adaptations of the homeostatic type. For mammals, Kinder's research (1927) indicated that temperature conditions at the skin must somehow provide a basis for nest-building in rats. But such factors are likely to influence behavioral development only as part processes in a complex of interrelated organic events. Lehrman (1955) found that feeding of the young by the parent ring dove did not occur except through the development of a behavior-influencing integration of (a) hormonal factors; (b) afferent effects, in particular, processes centering in the delicately sensitive brood patches; (c) neural factors, including certain processes bearing on a generalized "drive" excitation; (d) closely related circular processes in individual behavior; and (e) parent-young reciprocal stimulative processes occurring in a complex sequence. Without the last—an involved succession of stimulative exchanges between the two participants—there could be no feeding response by the parent to the perceived squab. Such findings suggest that the developmental pattern underlying behavior is a global one, including a range of circular stimulative and reactive processes based upon the gains of earlier stages.

Early self-stimulative relationships may provide a significant basis for later adjustments to new conditions involving members of the same species. Birch (unpublished manuscript) found that female rats, provided from early youth with wide collars which prevented stimulative access to the posterior body and genital zones, in a later test at primiparous parturition lost their young through neglect or cannibalism. Controls indicated that it was not the weight or general disturbing effect of the collar but the effect of the wide flange in preventing stimulative relations with the posterior body that accounted for the abnormal outcome in experimental animals. Thus, self-stimulative experiences normally inevitable in everyday behavior during youth seem essential to efficient delivery and care of the young in this animal. One normal gain denied to the experimental rats may have been a general perceptual orientation to the posterior bodily areas as familiar objects; another, the capacity of responding to the genital zones—and to objects such as neonate young, bearing equivalent olfactory properties of attraction—by licking and not biting.

This point of view, of the maturation and experience complexes always intimately related in behavioral development, may be adapted to the study of turning points and intervening stages in mammalian development. The term *stage* has been extrapolated from embryology to the study of behavioral ontogeny and we must accept embryological evidence as integral, but we must advance beyond this to higher functional orders in development to which organic maturation is *one* contributor. The accomplishment of given maturational events often seems to influence behavioral changes rather promptly, particularly at lower phyletic levels, as when an amphibian larva metamorphoses into an adult with rather different properties of sensitivity and response. Higher in the psychological scale, the relation of one stage to following stages seems to be more involved. But the principle is much the same for all. Investigations of domestic cats in our laboratory (Rosenblatt, *et al.*, unpublished manuscript) indicate how behavioral advances centering in birth, eye-opening, and weaning occur through complex intrinsic-extrinsic integrations. For a time after eye-opening, the kitten's approach to the mother evidently proceeds much as before, on a proximal sensory basis. Within a few days changes appear which seem attributable both to new gains through maturation—e.g., an increasing visual acuity (Warkentin and Smith, 1937)—and to an integration of previously established, nonvisual perceptual processes with new visual cues, which control a somewhat different adjustment to the mother. Transfer to solid food involves nutritive and behavioral adjustments that are accomplished by degrees and through the participation of various factors. Stages in behavioral development at higher psychological levels thus may be regarded as overlapping; their turning points occur in a more gradual manner than may be true of the events in maturation that contribute to them.

Ontogenetic Progress and the Role of Experience

Under normal conditions, *experience*—the effect of stimulation on the organism, including especially stimulative effects characteristic of the species' ontogenetic milieu—is indispensable to development at all stages. The growing organism takes more from its environment than the energy essential for nutrition; it is affected from the first by energy changes in its environs and responds according to its developmental stage. *Trace effects,* organismic changes affecting further development, result.

Experience may contribute to ontogeny in subtle ways. From insects to man, not only the general developmental progression, but also events critical for the pattern changes of later stages, may involve factors of experience. We may expect to find that many physiological and behavioral rhythms, now often considered purely endogenous and innate, may result from an interrelation between intrinsic and *essential* extrinsic factors. For insects Harker (1953) has found that the normal day-night activity rhythm of the adult mayfly does not appear unless the egg has been subjected to at least one 24-hour light-dark cycle. In some manner the organic effect of a specific physical condition, acting for a limited time at a very early stage, becomes deeply ingrained and influences complex daily behavioral variations in the adult stage. Comparable results were obtained by Brett (1955) for the rhythm of pupal emergence in *Drosophila.*

In birds and mammals, organic periodicities basic to the development of adult drives may be stimulatively influenced by periodic experiences, as through changes in the behavior of the incubating mother. Comparably, tidal, diurnal, and other rhythms in the physical environment, as well as nonrhythmic changes, may affect the developing stages of animals from lower invertebrates to mammals, influencing the basis for both periodic and aperiodic processes underlying adult behavior. A striking case has been demonstrated by Hasler and Wisby (1951) in Pacific salmon. When adult salmon ascend a river system to spawn, turns at stream branches are made toward the tributary carrying the chemical essence of the headwaters in which the individual developed from the egg before descending to the ocean.

The effects of early experience may influence the later stages of development in a nonparticulate manner. Moore (1944) found that neonate rats, raised in an environment of 95° F. temperature, as adults had longer, more slender bodies and tails, and performed less well in maze-learning tests than did others kept at an environmental temperature of 55° F. It is not surprising that such extrinsic-intrinsic relationships are often overlooked or their effects misinterpreted in behavior studies.

The nature of gains through experience is both canalized and limited by the relative maturity of species-typical afferent, neural, and efferent mechanisms, in dependence upon the developmental stage attained. A definite conditioned response to a combination of contact and shock stimuli could not be demonstrated in the chick embryo before about 16 days (Gos, 1935); in the neonate dog, before about 20 days (Fuller, Easler, and Banks, 1950). According to phyletic capacities, the influence of experiential effects upon later behavior depends critically upon the stage at which they act. Thus in ungulates the mother is a prominent feature of the environment for the neonate, who, susceptible to specific exteroceptive effects such as odors, and with the eyes open, soon establishes a strong bond with the parent (Blauvelt, 1955). In the phenomenon first described for birds by Lorenz (1935) as "imprinting," the primacy of some available and conspicuous attractive object after hatching seems crucial, for the normal bond with the parent first of all and, secondly, with species mates. Newly hatched goslings first exposed to stimulation from a person are later unresponsive to stimuli from species mates as against the human object. In normal hatching, parent and species mates of course have the advantage of primacy (Fabricus, 1951). The prepotent stimulus for eliciting the approach seems to be highly generalized and definable in quantitative terms such as size and rate of movement. Stimuli specific to the initially adequate object soon become prepotent and canalize later response tendencies rather fundamentally. Such early adjustments are often virtually irreversible.

The trace effects of early experience seem to be characteristically generalized, diffuse, and variable, their influence at later stages nonparticulate and difficult to identify. Even so, they must be reckoned with as important influences in the later behavior pattern. In tests of conditioning in chick embryos, Gos (1935) found that specific conditioned responses to a touch-shock pairing could not be accomplished before approximately 16 days. But subjects stimulated repeatedly, after about the 10-day stage changed in their general responsiveness to the extrinsic stimuli used, indicating some trace effect. Hypothetically, the extent to which a vertebrate embryo is mechanically disturbed in the egg or uterus might be expected to affect the later levels of emotional excitability. Experiences common to the usual species environment, such as knocking about of the eggs as they are turned by the incubating parent or as the parent leaves the nest and returns to it, might have trace effects influencing similarly the development of all normally raised young. Therefore, from early biochemical effects to later experiences promoting learning, the characteristic properties of the species-standard environment are never to be taken for granted or minimized, but always admitted to the field of possible indispensable factors in normal development.

Relative maturity at birth differs widely among the mammals, and is to be considered a cardinal factor in development, determining which environmental effects will be influential. In ungulates such as goats, from a time shortly after birth the social bond seems to advance with acceleration to the stage of visual perception, greatly facilitated by a precocious quadrupedal motility as well as by specific olfactory, auditory, and tactual effects (Blauvelt, 1955). In contrast, the advance of the neonate in carnivores such as cats and dogs seems to be much slower, against stricter early postnatal limitations along afferent, neural, and motor lines. Reciprocal stimulative relationships of the neonate with species mates seem based upon strong organic gratification from such stimulative processes, and must be considered fundamental to the wider social integration of the individual (Schneirla, 1946). Differences during and subsequent to the establishment of the primary bond with the mother, as well as in other group experiences bearing upon the litter period, may contribute basically to certain striking contrasts in group behavior tendencies characteristic of the various lower vertebrate orders. In carnivores, the typical disruption of such associations at the time of weaning would seem to be significantly related to organic and behavioral changes in the participants (e.g., maternal hormone effects) not common in ungulates, in which group associations typically persist with only secondary changes in the pattern of intragroup relationships.

The receptivity of the sexually mature mammal to a species mate may represent the rearousal of a perceptual responsiveness established earlier in life, as during the litter period. The appearance of this typical susceptibility, even in relatively nonsocial animals, would seem also to involve the lowering (through a hormonal priming of certain organic mechanisms) of afferent and other functional thresholds which serve as key items underlying the mating reaction. But the adjustments are perceptual, and for them, early experiences may well be critical. Thus, individual carnivores raised apart from their species mates are likely to display disturbance reactions upon encountering them for the first time in adult tests, and mate only after long delays and much disturbance, if at all.

In each species, the specificity and the nature of the effects of early experience upon later behavioral development seems to differ more or less characteristically according to developmental stage. Such trace effects may be major factors in ontogeny, but thus far have been difficult to identify as originating in the embryonic and early neonate stage of carnivores and other mammals with a protracted period of early development. One striking contrast is opened by the experiment of Fuller *et al.* (1950), who found that puppies, presented with a combination of electric shock and a visual, auditory, or other exteroceptive stimulus, exhibited no specific signs of conditioning before the twentieth postnatal day. This result was confirmed by James and Cannon (1952). It seems probable, however, that a less specific type of conditioning may occur at earlier postnatal stages in young carnivores, evidently mainly on a somesthetic and chemoceptive basis, and in terms of lower functional orders of neural integration. Studies in our laboratory (Wodinsky, *et al.*, unpublished manuscript) on the behavior of neonate kittens show that specialized feeding adjustments to the mother begin soon after birth and have taken on an individually characteristic pattern within the first day or two. The adaptations concern both a direct approach to the mother and the attachment of a given kitten to a particular nipple with increasing frequency, posted at the front, center, or rear in the nipple series. These adjustments, which usually are individually stabilized within the first four days after birth, clearly involve learning, and form a basis for wider social adaptations in the litter situation.

Species behavior patterns depend not only on characteristic maturation but also upon experiences normal to the species (Kuo, 1924; Maier and Schneirla, 1935). Kuo (1930) manipulated the early experiential context of kittens to control their later reactions to rodents; and Patrick and Laughlin (Patrick and Laughlin, 1934), by raising young rats in open areas as on table tops, obtained adults inclined to range more and to follow walls less than is usual for the species. Through tests with an isolated rhesus monkey, Foley (1934) emphasized early associations with species mates as a factor in normal grooming reactions. Such results prepare us for Hebb's proposition (1949, 1953) that through early experience basic perceptual organizations are built up, and that the gains from experience are likely to be greater at opportune early phases in ontogeny than at later stages. With rats (Forgays, 1952; Forgus, 1954) and with dogs (Thompson, 1954) early environments of greater scope and heterogeneity provide a more adequate basis for perceptual adjustment to new situations than do more restricted, simpler environments, and a more adequate basis for performance in maze and problem-solving tests at later stages (Bingham and Griffiths, 1952; Hymovitch, 1952). In mammals (Beach and Jaynes, 1954; Christie, 1951) a wide variety of conditions in early postnatal experiences, from trauma (Hall and Whiteman, 1951) and deprivation (Holland, 1954; Hunt, 1941) to litter conditions (Seitz, 1954) and maternal behavior (Kahn, 1954) can influence behavioral development. Age normally brings a variety of accretions, substitutions and changes, attributable to progressive conditions in the organism and in its experience. For example, Seitz (1950) found that young foxes, although they had been well tamed from an early age, from the age of about 36 days began to defend themselves against handling. Taming evidently serves to raise the excitation threshold for certain familiar perceptual patterns, but its inhibitive potency suffers at sexual maturity. Species properties like learning capacity, excitation threshold, tension adjustment, and a host of other characteristics must be considered among potential determinants of what experience may contribute at any stage.

Early generalized training often is basic to later more specialized adjustments. In Maier's tests (1932) of reasoning in rodents, which required the combination of previously separate perceptual experiences, chance scores were made by rats younger than six months, scores of 80 per cent or more by adults. Young chimpanzees given early experience in handling of sticks in play later proved superior to chimpanzees lacking such experiences, when given reasoning tests requiring the combination of sticks into tools for reaching food (Birch, 1954).

The range and nature of developmental gains through early experience naturally depend first of all upon species capacities for learning. In insects and lower vertebrates gains through experience are less pervasive in later life than is typical for mammals (Schneirla, 1945). But for any species capable of learning, what is acquired must depend in part upon the conditions of exposure. Whether or not a mammal's response to an attractive or disturbing situation serves to change its motivational adjustment to that situation may determine whether a simple conditioned reaction or a selective-learning pattern is acquired (Maier and Schneirla, 1935; Maier, 1942). Also, the tension level of the animal at the time of an experimentally introduced experience may determine the functional order of what is learned.

Finally, any generalization about stereotypy or variability in behavior through development must depend upon the phyletic level as well as on the conditions of individual experience and individual background. For instance, shock or punishment, as one well-studied condition, may promote variability (Stone, 1946) or stereotypy (Maier and Klee, 1943) in habit formation, according to the intensity of the shock and the manner in which it is administered. Many other conditions may determine whether experience in a mammalian species contributes to the individual repertoire a stereotyped habit of restricted scope or a more versatile pattern useful in new situations (Maier and Schneirla, 1935; Nissen, 1951).

Original Nature and Development

Rather than assume an "original nature," more or less modifiable secondarily, it seems preferable to regard behavior patterns on all levels as results of a developmental process representative of the species. In this process, genetic constitution is regarded as indirectly influential in the rise of structural-functional mechanisms through progressive intrinsic-extrinsic relationships characteristic of the level. In higher animals, developmental processes are progressively more labile in relation to capacities for profiting by experience and for behavioral plasticity. In the higher mammals, early behavior, although not formless, is most generalized of all, and the individual is most radically influenced by experience in the course of later development. But in no case can experience be considered a latecomer in development, limited to learning at later stages, for its earliest influences seem indispensable in the ontogeny of all animals. This is the case despite the fact that the trace effects of embryonic experiences seem qualitatively inferior and limited in comparison with higher functional orders of learning entering later on.

The concept of an original nature has found its main support in the widely used isolation method, purported to separate innate behavior from acquired (Tinbergen, 1951). Typically, in an animal separated at birth and raised apart from its species mates, the principal responses appearing are considered innate or inborn (Schneider, 1950; Seitz, 1950; Zippellius and Goethe, 1951). As a rule, this assumption places far too much weight on the experimenter's knowledge of the typical developmental setting and its effect in ontogeny (Riess, 1950). Actually, isolation tests unaccompanied by an appropriate analytical methodology show only what behavior can or cannot develop in either of the two situations—usual and unusual—through causes not demonstrated. For one thing, the contexts may be equivalent in certain important but unidentified respects, as, for example, when the animal through its *own* organic and behavioral properties has introduced a sequence of stimulative effects normal to the species (Schneirla, 1956).

Levels of Attainment through Ontogeny

Perception

Adaptive behavior may be achieved at different phyletic levels through capacities low or high in functional order, in relation to the species niche. Through ontogeny, irritability or sensitivity arises in all animals in very different manifestations which are basic to a variety of adaptive patterns. This fact does not justify generalizing the capacity of *perception* through all phyletic levels, as is often done by reducing it to any stimulative reaction whatever (Nissen, 1951). It is imperative to investigate whether a given species really can progress, through development, beyond the lowest functional order of mere reaction to stimulation. Actually, critical phyletic differences in capacities for organizing sensory data seem evident, not as differences in degree but as differences in kind (Hebb, 1949; Schneirla, 1949). Stimulative irritability, in its simpler forms, need entail no more than simple sensory effects, far inferior to and qualitatively very different from the specialized perceptual processes of which higher mammals are capable.

Experimental psychologists have been unable to analyze out the raw-sensory effect element from other components in a perceptual adjustment (Hall, 1951), possibly because adequate investigations have yet to be made at early ontogenetic stages and with different phyla. Protozoa assuredly possess stimulative irritability, but it remains to be demonstrated that they perceive, i.e., *sense with meaning*. Recognizable perceptual adjustments of a low organizational order appear in lower vertebrates in a form which resists later ontogenetic modification. Ehrenhardt (1937) could train lizards only to a limited extent to snap at the member of a stimulus pair with the more irregular outline, in opposition to the characteristic naïve response of this animal, i.e., the response most easily established early in ontogeny. In mammals, specialized and more plastic perceptual adjustments are mastered more slowly in ontogeny, but are modifiable to far greater extents as a rule than is possible in lower vertebrates.

Perceptual habits seem very different on lower and higher vertebrate levels. Within a few trials after hatching, the chick begins to discriminate grains from small, bright, inedible objects, improving on the nonparticulate stimuli which initially evoked the embryonic head-lunge (Schneirla, 1935). In contrast, young carnivores require more time to integrate visual cues with nonvisual perceptual adjustments established to mother and litter mates earlier in the neonate period (Rosenblatt, *et al.*, unpublished manuscript). Methods must be devised adequate to the task of tracing ontogenetic progress from sensory naïvete to perceptual adequacy in animals at different phyletic levels.

Claims for "innate perceptual schemata" have not been validated in adequate experimental investigations. Presumed innate "sign stimuli" (Tinbergen, 1951; Tinbergen and Kuenen, 1939) seem to resolve themselves, not into initially meaningful patterns, but into very generalized stimulus effects (Ginsberg, 1952; Schneirla, 1949). For example, the "hawk reaction" of nestling birds in certain species, which had been attributed to an innate meaning content, seems due rather to a diffuse stimulative shock effect than to an inborn pattern significance (Ginsberg, 1952; McNiven, 1954). For perceptual meaning and organization, experience seems necessary, although its gains are much more restricted in lower vertebrates than in mammals. Through experiments in which, by means of translucent hoods, ring doves were deprived of early experience in visual definition, Siegel (1953) found that even in this psychologically inferior bird, such experience normally must contribute an essential perceptual basis for learning pattern discriminations in adulthood. In a similar study of the chimpanzee, Chow and Nissen (1955) found that experience is very essential for adult interocular equivalence in pattern discrimination. Experiments with the rat led Teas and Bitterman (1952) to conclude that initial visual adjustments are "loosely organized wholes out of which the perception of objects and relations is subsequently differentiated."

The sensing of the new and strange is increasingly striking in ontogeny, from rodents to primates (Hebb, 1949), and invites further study as indicating the integration of a wider scope of environmental cues. Correspondingly, from the fact that the chimpanzee's curiosity seems dominantly destructive in its motor outlets, and man's on the whole constructive, Nissen (1954) refers to the greater command of object and situation relationships in the latter. In a comparative study of early development in the chimpanzee and the human infant, Jacobsen *et al.* (1932) found the chimpanzee well ahead in all aspects of sensitivity and reactivity during the first nine months. However, as the infant began to accelerate in perceptual accomplishments and in early concept formation, as indicated by the mastery of his first words, his superiority in psychological development became increasingly apparent.

Motivation

The manner in which differentiated drives appear through ontogeny is still unclear for any animal. Comparative investigations are needed as to the ontogeny of motivation, and I wish here to offer some theoretical considerations with respect to this problem.

It is probable that homeostasis affords a comparable basis for adaptive behavior in all vertebrates: a generalized excitation set up by some marked deviation from the theoretical "steady state" forces a readjustment to the pattern of metabolic processes normal to the species. But, as Pick (1954) has pointed out, homeostatic mechanisms elaborate and change from fishes to man. It is not only in the organic mechanisms of homeostasis themselves that the evolutionary elaboration occurs, but also in their potentialities for widened behavioral relationships through ontogeny. A primary question is, How do these reactions become differentiated appropriately as adjustments of approach or of withdrawal?

Theoretically, in all organisms, evolution has involved the selection of mechanisms favoring approach to *mild* stimulative effects and avoidance of *intense* stimulative effects (Schneirla, 1939 and 1949). This set of relationships between individual and environment may be proposed as a primary condition in the ontogeny of all animals, accounting for basic tendencies of the types traditionally contrasted as "appetite" and "aversion" (Craig, 1911; Holt, 1931). Species violating this condition as basic in ontogeny would risk extinction. Other things equal, the pseudopod of an amoeba shrinks with intense local stimulation and extends with weak stimulation; correspondingly, the forelimb of a neonate mammal flexes or adducts, extends or abducts, according to the energy of local stimulation (Carmichael, 1946; Carmichael and Smith, 1939).

The selective process postulated as operative in these terms throughout the phyletic series presumably has included all types of functional thresholds, from afferent and neural to those of visceral and motor systems. Fundamentally, before ontogeny advances very far, these mechanisms are specifically dominated by the intensity of stimulative energy presented in any given experience. The response here depends critically upon the intensity of the neurophysiological energy discharge produced, and not upon what *kind* of object or situation produced it, nor upon its potentialities for eventual benefit or harm to the organism. In the earthworm (Hess, 1924; Maier and Schneirla, 1935), the response is typically a whole-body approach or withdrawal, stereotyped and appearing early in ontogeny, a forced type of reaction not much changed throughout life in its typical relationship to the intensity range of stimulation (other things—e.g., temporary sensory adaptations—being equal). Once ontogenetically established, as with the typical lunge reaction of lower vertebrates to small moving objects and withdrawal from larger moving objects (Eibl-Eibesfeldt, 1952; Honigmann, 1945), the stereotyped patterns resist modification (Ehrenhardt, 1937). On higher levels, motivation must be thought of as having an increasingly complex ontogeny. New rules are possible; the primary naïve relation of response to objective stimulus intensity may then be modified and, indeed, reversed, according to species capacities and the external circumstances encountered in individual experience.

In mammals, a more generalized condition of drive is postulated in early ontogeny, although still paralleling the described biphasic basis for response according to stimulus energy effect, viewed as a central feature of development in all animals. Theoretically, the ontogeny of mammalian *withdrawal* adjustments is primarily centered in high-threshold mechanisms in afferent, neural, flexor-adductor muscles and in the sympathetic autonomic and related visceral systems; that of *approach* adjustments is centered in low-threshold, afferent, neural, extensor-abductor muscles and the parasympathetic and related visceral systems. It is presumed that the neural changes underlying learning are maximally affected through conditions activating either one of these systems in a synergic manner, as when intense stimulation is reduced or the source evaded through withdrawal, weak stimulation held or the source gained through approach. The ontogenetic basis of individual learning is thus viewed as essentially biphasic in these terms.

In contrast to the stereotypy characteristic of lower levels, higher mammals attain increasingly plastic and individually specialized drive, incentive, and consummatory relationships. "Searching for . . . ," as a term applied to lower animals, is a loose generalization usually indicating random behavior under drive impulsion; but through ontogeny in a mammal with advanced capacities for learning, the expression can mean the specific, drive-impelled anticipation of an incentive. Infant rats, subjected to regular experiences of food and water deprivation, as adults learn to locate these objects with greater facility than do normal subjects (Christie, 1952). Such results would not be expected in reptiles; but in chimpanzees, capable of more involved differentiations in motivation, assemblage of appropriate symbolic tokens could occur under comparable conditions. Mammalian motivation is a superior attainment through higher functional orders in behavioral development than are possible in lower phyla.

Conditions Widening Ontogenetic Attainment

The potentialities of ontogeny depend most significantly upon the extent to which the central nervous system has developed a basis for the processes underlying learning. Notwithstanding arguments for "learning" in the radially symmetrical invertebrates below flatworms (Warden, Jenkins, and Warner, 1940), a general distinction of qualitative differences seems necessary here (Maier and Schneirla, 1935). The trace effects of experience in organisms such as protozoa and coelenterates seem to be diffuse or peripheral, and not a basis for more lasting changes in the individual behavior pattern. For example, in paramecia subject to the combined effects of light and temperature, temporary behavior changes may be produced which suggests conditioned responses (Bramstedt, 1935; Warden, Jenkins, and Warner, 1940). However, these changes depend upon some evanescent condition in metabolism, due solely to thermal effects (Best, 1954; Grabowski, 1939) altering sensitivity to light for a limited time. In contrast, invertebrates such as worms and insects are capable of neural trace effects through experience (Maier and Schneirla, 1935), as a basis for more lasting changes in the individual behavior pattern.

In addition to neural trace effects, or *fixation,* there is the capacity of advanced neural systems for *correlation,* or the organizations of trace effects. Phyletic differences in the capacity of the nervous system to correlate the effects of experience must account for the fact that although many insects are capable of forming rather complex habits, these patterns seem much more limited in plasticity (i.e., in properties for change and transfer) and far more situation-bound than is typically the case with mammals (Maier and Schneirla, 1935; Schneirla, 1945).

Simple conditioned responses established when the animal's reaction alters the situation in some critically adaptive way have different properties from the simpler conditioned-reflex types of change learned by more passive subjects. The former pattern is more similar to selective learning in its properties (Maier and Schneirla, 1935; Maier, 1942). The conditioned response of the flatworm, in its general make-up, vaguely resembles that typical of decerebrate carnivores (e.g., as generalized), but seems far inferior to the gains possible through learning in normal mammals. The cerebral cortex, evolving as a special modification of the vertebrate forebrain (Herrick, 1924; Jerison, 1955; Kappers, Huber, and Crosby, 1946), introduces new properties and higher functional orders of attainment in species ontogeny, freeing the individual from the domination of specific afferent or motor mechanisms.

With the rapid acceleration of this correlative asset to predominance in the mammals, there appear striking advances in species attainments related to experience in ontogeny. Penfield (1954) attributes man's superior intellectual attainments to the frontal and parietal lobes of the cerebral cortex, increasing through the mammals to man but without a specific counterpart in lower vertebrates. This superiority is also, although less obviously, due to evolutionary advances in the lower-center neural mechanisms (Pick, 1954) which specifically favor the acquisition of more extensive organizations of central and peripheral processes—as in motivation (Schneirla, 1956). Improved neural correlation promotes the attainment of a wider, more intricate set of environmental adjustments, and of new behavioral patterns as individual accomplishments, provided that such acquisitions are favored by the circumstances of ontogeny.

From these considerations, one would anticipate significant improvements in animal intelligence tests through measuring plasticity in modifying learned habits, rather than through measuring acquisition or retention. This seems to be the case after equivalent experience. Rodents are on the whole less gifted in the roundabout type of adjustment than are dogs, but carnivores fall short of the primates in this respect. Accordingly, in a test requiring the selection of the alternative learned response appropriate to whichever of two experienced incentives was presented, Fischel (1950) found that although turtles and birds failed and dogs succeeded, the latter were distinctly inferior to human subjects. In higher animals, responses to young, food, and other goal objects are not necessarily more intense in energy output, but are richer in meaning content and more versatile in their appropriateness to changing environmental situations than responses in lower forms. With increasing cortical correlation capacity influencing ontogeny, species attainments in behavior are marked increasingly by the characteristics of appropriate variation and newness.

The existence of qualitative differences between species in capacities for behavioral organization cannot be doubted, and more extensive comparative studies are overdue (Schneirla, 1945 and 1952). Learning clearly furnishes the ontogenetic basis for higher attainments such as reasoning; but reasoning does not therefore reduce to learning, any more than these capacities may be considered fully distinct from each other. Harlow (1949) has shown for the monkey how cumulative experience in "learning to learn" object discriminations may progress to a high degree of complexity and skill. However, visual-discrimination methods may not be well suited for the analysis of higher mental functions based on learning (Maier and Schneirla, 1935). If systematic tests adequate for this purpose can be devised for a variety of animal forms, phyletic differences in potentialities for psychological development may prove more striking than has been suspected.

The advantages of the comparative method have been scarcely explored in the study of behavioral development (Schneirla, 1945; Werner, 1940). A child's attainment of sentences marks a new advance from the stage of unitary verbal symbols, and contrasts sharply with a monkey's inability to master symbolic relationships beyond the simplest abstractions. In a far wider sense, man's capacity for repatterning verbal symbols serially, or for attaining such symbols at all, is qualitatively far above the functional order represented by the gestural symbolic processes to which the chimpanzee seems developmentally limited, although not altogether dissimilar in its ontogenetic basis.

Summary and Conclusions

As a recapitulation doctrine has not been validated for behavior, the concept of *psychological levels* is advanced to express the phyletic range of behavioral organization and psychological capacities, and the concept of *functional orders* is advanced to express the ontogenetic range on any one level.

The term *development* with respect to individual behavior stresses progressive changes in organized adaptive function through ontogeny. Behavioral development on any phyletic level is not so much a retracing through the stages and levels of successive ancestral forms as a new composite leading to a new pattern distinctive of the level.

Genetic constitution and developmental setting influence *all* behavioral development in *all* organisms, operating jointly through a complex formula of intervening maturation-experience variables in a casual nexus typical of the level. *Maturation* connotes processes contributed through growth and differentiation, and *experience* connotes the effects of stimulation on the organism.

The nature and specificity of the developmental effects of early experience depend upon species capacities, developmental stage, and setting. The trace effects of early experience, although mainly generalized and diffuse in comparison with those of later stages at which specific conditioning and learning enter, may be fundamental to behavioral development. At higher levels, simple conditioning must be considered a possible developmental factor even in prenatal stages.

Within the unusual developmental setting of the species, typical patterns of successive self-stimulative effects, as well as of other feedback effects and of inevitable experiences, offer key factors for the progression of stages characteristic of species ontogeny. In many species, ontogeny may also entail characteristic reciprocal stimulative ("trophallactic") processes between individuals, essential to the rise of group behavioral affiliations as well as of mating.

In lower phyla, behavior patterns are stereotyped within a relatively short ontogeny in terms of lesser functional orders, and thereafter resist change; in higher phyla, after generalized early stages, through a longer ontogeny they involve functional orders of increasing scope and plasticity. But for no species is ontogeny based upon an initial innate pattern ("original nature") modified to different extents at later stages.

Plasticity, a superior functional order in variable behavior, depends both upon the species capacities for the fixation and correlation of neural trace effects and upon the range and character of individual experience. From a simple sensory irritability prevalent in the lowest phyla, animals in higher phyla advance through ontogeny to higher orders of relationships and meaningful organization in perception. In mammals, early naïve sensory responses are basic in ontogeny to progressively higher perceptual orders attained according to species capacities and opportunity through experience.

A wide parallelism exists among phyla in the mechanisms underlying adaptive behavior, susceptible to specializations of approach and withdrawal adjustments according to ontogenetic capacities and opportunities on the respective levels. On this basis, increasingly elaborate drive processes underlie the attainment of successively higher orders of anticipative motivational adjustments.

Superior behavioral attainments at higher psychological levels depend not only upon capacities for neural correlation and systematic organization, but also upon advances in phyletically old mechanisms (e.g., those of homeostasis). Species attainments in behavioral plasticity broaden correspondingly, with greater situation-appropriate variation and newness, when extrinsic conditions are favorable in ontogeny.

QUESTIONS FOR A THEORY OF COGNITIVE DEVELOPMENT*

William Kessen

I.

Modern dictionaries say that the word "cognition" is obsolete and many psychologists agree. You may go searching, if you like, in textbooks of psychology, and you will find that the place in the index between "cochlea" and "cold spots," or between "coefficient of correlation" and "coition," is usually unmarked by "cognition." Herbert Spencer (1899, p. 368), with his fondness for words at the edge of the language and for intricate schemes of classification, was able to put "cognition" to heavy work but, between his time and our yesterday, the word died. Even now, with all the permission given to the open discussion of cognition, there are some of us who feel uneasy with the notion and speak it out with the same mixture of fear and daring that we felt when we first said "damn" in front of Mother. There is need for the daring and reason for the fear but, rather than deciding at the outset whether we should resurrect or reinter "cognition," I propose that we consider two related issues: What lies behind the new ferment over cognition? What questions should be asked of a theory of cognitive development?

Frequently used pejoratively, frequently used shallowly, "cognition" sometimes seems to be, in this time of rapid and wide-ranging change in psychology, a protest word, a sign that we have come on new problems or on old problems newly seen, that we have new models for human behavior, and that we are beginning to wonder whether we are as free of our entanglements with philosophical presupposition as we once thought. The talk about cognition is also a sign that psychologists are returning willy-nilly to an examination of one of our most ancient psychological puzzles—the problem of knowledge. The concern of a psychology of cognition is the relation between reality and man's representation of reality; the concern of a psychology of cognitive development is the way in which the child comes to know the world. There are a number of reasons why we may be uneasy about such talk: chief among them are the strong scent of philosophy and the absence of a prototypical observa-

* William Kessen, "Questions for a Theory of Cognitive Development," from Stevenson, ed., *Concept of Development*. Monographs of the Society for Research in Child Development, vol. 31 (1966), pp. 55–70.

tion that stands for "cognition" in the way that "pigeon-in-box" stands for "learning" and "taking the Stanford-Binet" stands for "intelligence." Many of my comments here will bear on the issue of philosophical odor: my conclusion, to give you anticipatory warning, is that the musty smell is not the smell of dead issues but of issues buried alive. The absence of a prototypical case for cognitive study marks the diversity, agitation, and openness of the new psychology of knowledge. I recognize that these characteristics may be renamed ambiguity, confusion, and fuzziness, but there is much to be said for not prematurely reducing the notion of "cognition" to a specific and confining representative observation. Even without a formal definition, one may say that a psychology of cognition must include (at least) perception, thinking, play, aesthetics, dreams, and language.

What lies behind the recent revival of interest in these matters? What are the sources of our concern, particularly in child psychology, with new conceptions of cognition? It is tempting, after the fact, to see the current excitement about human knowledge as an inevitable consequence of some particular event (publication of Hebb's *The Organization of Behavior,* Hull's death, the Americanization of Piaget, to name a few), but human history, even the history of ideas, does not accept such neatness. Any serious account of the revisionism of recent American psychology must include considerations of the context of general culture—even psychologists look anew at root premises when the air is full of talk about anomie, alienation, and existential anxiety—and of the erosion of our hope that simple learning would provide a model for all psychology, of new data, of new metaphors for man, and of new doubts about the security of our first principles for the construction of knowledge. The remarkable synchrony of the appearance of new observations and the decay of old ideology has provided the freedom and uneasiness that seem to characterize contemporary psychology. We have a moment or two, before a new Boulanger rides up to capture psychology, to examine anew our fundamental postulates, I will say no more about our alienated, anomic, and anxious culture, and only a word in passing about learning as a model for cognition. Rather, permit me to make just an observation or two about the new data and new metaphors that have forced our attention again to the psychology of knowledge. Then, I would like to examine at somewhat greater length new doubts about first principles.

Our reasonable desire for simplicity in accounts of the sources of knowledge has been shaken by findings widely scattered through psychology—animal exploration, the play of children, the response to beauty—but the center of recent research in cognition (especially research with implication for the child) has been in the fields of perception, thinking, and language. To be sure, the tradition of research and speculation in perception has been a continuous one and has maintained the virtue of great sensitivity to epistemological presuppositions. But, even in this respected area of psychological study, the last years have seen novelties that are relevant to the development of knowledge. One need only point to recent studies on the informational analysis of perception, to discoveries of infantile perceptual structure, to studies of perceptual search and scanning, to indicate the diversity of empirical outcome that has led us to search for changed views of man's knowing. The corresponding continuing stream of research in thinking has broadened to a river; Berlyne's (1965) splendid new book outlines the remarkable demands that must be met by a comprehensive theory of directed thinking. As for studies of the development of thinking, it is no exaggeration to say that the turn toward Piaget over the last decade has transformed developmental psychology; his observations are the aliment that any theory of cognitive development must assimilate. In studies of perception and of thinking, the phenomena outrun our power to provide explanation and they support the reach for new categories that is represented by our hesitant talk of cognition. But it is in the area of language study that the new wine has tested the old wineskins. The work of linguists and psycholinguists so overtaxes our simple linear models of human behavior that, taken alone, it might have forced us to the reconsideration of cognitive development.

But the pressure of data is rarely enough to force a revision of our prejudice. We need new models—or better, new metaphors of mind—before we will leave the security of the old ways. In one of those apparent coincidences that leaves a deposit for the historian of psychology, two apparently unrelated developments over the last decades have provided metaphors for the new interest in cognition. For one, investigations of neuroanatomy and neurophysiology over the last several years have revealed—to put the matter almost in the brevity of parody—that we need not trim our psychological theories to physiological constraints; there is enough variety in the central nervous system to encompass our wildest psychological speculation. For another, the computer as a metaphor of mind has dispelled, even for the most dedicated mechanist, the illusion that simple machinery limits our conception of man. Recent enthusiasm for the computer as model of man gives some credibility to the disturbing proposition that some of us will assign to man only that variability and range that the best current machine permits but, even for the uncommitted, it is clear that neurophysiology and computer technology have provided conceptual possibilities that investigators of human behavior have scarcely touched.

New data, new metaphors—perhaps enough new data and enough new metaphors to comprehend the agitation about cognition that is our present concern. But I submit, with some trepidation, that there are issues more complicated and even perhaps more profound that are raised by our re-examination of the sources of human knowledge. The theme I would like to whistle in the dark of my own confusion has to do with epistemological first principles; it has been hummed, sung, and orchestrated by psychologists and by philosophers for centuries. To put the issue in capsule form, we may speak of the problem of *constrained analysis* or, more precisely, of the problem of the limits imposed on psychological analysis by premises that are not fully explicit and often not testable. A philosopher or a historian of ideas might call this theme "the epistemology of the psychologist" and describe its development from Aristotle to Hume to Kant; such a survey is outside my competence and my intention. I dare to raise the issue of constrained analysis only because of my conviction that a serious examination of the relation between development and cognition requires that we look anew at some of our basic preconceptions about psychology.

As an opening into this range of problems, consider the way in which James Mark Baldwin (1915) summarized the complex relations that exist between social requirements for knowledge and individual variation. He wrote as follows:

> The assumption that the individual's results are corrected and made reliable by social tests rests upon the further assumption that the social results by which they are tested are themselves correct and reliable . . . the best that society can do for the individual is to bring him into agreement with itself; but the result may be right and it may be wrong (p. 17).

In these words, Baldwin argues that the forms of our knowledge—our system of cognitions, if you like—are determined in some degree at least by social agreement. This may appear at first merely to be a restatement of the principle of intersubjective verification—that we can say we know only what several of us agree to—but the issue goes deeper than that. It is difficult enough to decide about the occurrence of an event—whether the truck ran the red light, whether the child blinked at the loud sound, and so on—and it is a monument to man's ingenuity that we have devised techniques to resolve some issues of fact. But the touchier problem, and the one that must make us all amateur epistemologists, is that there must be social agreement, as well, about the structure of the matrix within which decisions about facts are made. Space, time, motion, the independence of objects—to name just a few entries in one possible list—may be seen as widely shared constructions of human knowledge. It is a measure of the generality of social agreement that we so seldom have occasion to discuss—even less, to question—such fundamental constructions; but I will maintain that the psychologist concerned with the development of cognition cannot stand free of the epistemological morass. A case can be made in defense of the proposition that even the freedom of other psychologists is illusory, but my concern today is to suggest to you that, if we want to understand how the child comes to his theories of physics, mathematics, and psychology, we will have to dissect not only our psychological principles but our epistemological principles as well. Before we can establish a psychology of cognition, we must state the basis of our belief in stable characteristics of reality. By and large, this requirement will lead us to the recognition that reality is constructed, not immanent in man or in stimulus, and that contained in every particular psychological theory is a particular epistemological system. The form of epistemology typical of American psychologists has been naïve realism, and it has been a profitable prejudice. I do not propose that it must be abandoned; however, I most strongly maintain that our epistemological preconceptions, whatever they be, infiltrate our view of the developing child. The child who is confronted by a stable reality that can be described adequately in the language of contemporary physics is a child very different from the one who is seen facing phenomenal disorder from which he must construct a coherent view of reality. It is a most unhappy conclusion for an empirical child psychology, and I welcome the psychological critique that will liberate me from its implications. But consider the alternatives to a critical analysis of the relation between our understanding of the child's widening knowledge and our root suppositions about the nature of reality. Either object, number, time, causality, and their cousins are given immediately in experience, and therefore their development requires no explanation; or they are part of the organism's instinct capacity, and therefore their development requires no explanation; or else one rejects the whole notion of categories of knowledge as useless or misleading in studying the development of the child. Neither of the first two alternatives has attracted wide adherence among psychologists although both have had at least occasional defense. The third solution—rejection of epistemological considerations as psychologically irrelevant—commands wider support among psychologists and must be addressed before we again hail philosophers as brothers. When we contemplate rejection of the epistemological puzzle, we meet again the problem of constrained analysis. Let me be as clear as I can about the core of my argument. I am not proposing that psychologists interested in the development of thinking, play, perception, and language adopt a particular set of epistemological doctrines—I am not an agent for the *Kantsverein.* Rather, the burden of my argument is that each of us, who approaches the developing child to ask questions about the way his constructions of or responses to the world change, carries a full kit of untested presuppositions about the nature of man and about the nature of reality. Our apparent freedom as scientists is only the freedom to ignore the primitive elements of our psychology, to say with another popular model of man, "What—me worry about epistemology?"

In part, the proposals made here are another sermon on the theme, beloved of psychologists, that we be self-conscious about the logical apparatus of our discipline. But the proposals are more than that. Even the sophisticated psychologist digs back toward his presuppositions a very short way indeed; he often finds his primitive terms in refinements of the common language of the educated adult—for example, in words like *event, property of an event, intensity, duration, position,* and so on. For most problems having to do with perception, motivation, learning, and problem solving in the adult of the same epistemological community as the researcher, such a limited analysis of first principles works. For the study of the child who is busily *constructing* the notions of event, property classes, intensity, duration, and position, we must go further. The psychologist's elaborated, systematic, and schematic view of reality depends on a particular language, a particular place in the development of philosophy and science, and a particular place in psychological theory. To use this view as a model of the child's reality is to miss the most interesting problems of developmental psychology: the problems that turn precisely on the issue of how the child comes to build the adult world out of his encounters with phenomena. To be sure, the psychologist interested in cognitive development must start somewhere; it is not necessary for him to regress to an infantile conception of reality in order to understand the development of perception and thought. It is incumbent on him, however, to recognize that his preliminary guesses about the world of the child are themselves constructions of great complexity and, further, that, for the very young infant especially the psychologist's epistemological constructions are likely to make a bad fit.

Unhappily, the issue of the psychologist's preconceptions goes beyond considerations of general agreement among adults of a common culture. To be sure, we move through the world like dentists and dockworkers and probably share with many human beings a set of principles about the way the world is. But this is not usually the set of principles that we bring to bear on our understanding of the child's developing comprehension. Analysis of the child's cognitive development is constrained not merely by the fact that we see it through the working reality of a Western adult; our analysis is also constrained by the dimensions of psychological theory. Consider the following example: Piaget is concerned with the development of the child's conception of space. To approach this problem, he designates a conception of space toward which the child will develop. Piaget chooses for this purpose, and for reasons that are not quite perfectly clear, the conception of space that is held by the educated adult. Incidentally, Piaget's educated adults have more the characteristics of graduate physicists, but that point is not critical. It is critical that an endpoint for this development has been stated or assumed and that observations of the child and their interpretation are organized around this specification. Observations, however available and reliable, which are not congruent with and structurable in terms of the specified end-point, cannot be considered a relevant part of development. The example of Piaget's procedure is a particularly telling one because, of all psychologists, he has made the best attempt to free the developing child from the pull of his ultimate knowledge.

Consider an even simpler case. A 5-year-old child is given a set of cards that vary two values of three dimensions, and he is asked to sort them into four piles. When he cannot do so according to a predetermined scheme, the experimenter concludes that the child does not have the concept of intersecting dimensions. Neither case represents bad science or improper procedure, but both illustrate that our conclusions about the child's cognitive apparatus may be based as much on the construction of reality we imposed on him in the first place as on the reliability, stability, and generality of our observations. "Yes, yes," our no-nonsense psychologist says, "we know all that; theories vary in their primitive terms, theories vary in their rules of manipulation, different theories have different domains of application, all theories are incomplete, and God's in his heaven. What have you told us about cognitive development?" Just this: The defining problem of cognitive development is to comprehend how an organism of a particular kind, in encounters with phenomena defined in a particular way, constructs the world. For a task of this range, it is not possible to duck the specification of philosophical—particularly epistemological—underpinnings for a psychological theory. The danger that our conclusions about the development of human knowledge may derive in large measure from the preconceptions of the nature of man and the nature of reality that we stuffed—or worse, let slip—into our initial construction of the psychological task (a danger that I believe to be clear and present in all current attempts to understand cognitive development) requires that we take a long uncomfortable look at our governing presuppositions. Perhaps we will be as satisfied with our epistemology as we are with our psychology, but I know of no way to determine our satisfactions—and our disagreements—without open and continuing discussion of the principles behind psychological theory.

It is worth noting here, although the point deserves more detailed discussion than there is time for, that the constraint of our first presuppositions is most obvious in those studies of the child that are, in effect, "discovery games." Under such procedures, and they are common in child psychology, the investigator determines the correct solution to the problem he presents the child—whether in a study of perception, language development, directed thinking, or concept selection—and it is the role of the child to discover what the experimenter has set as the problem.

Generally speaking, such a procedure only permits pass-fail scores on the child's ability to match the experimenter's conception of what the child's perception, language, thought, or concept attainment ought to be or might be at that moment. A subtle and productive examination of the nature of the child's mistakes, or an examination of the possibility that his correct solutions may be attained in a fashion different from the one proposed by the researcher, demands some attention to a theory of errors or a theory of alternative cognitive procedures. And, to make my point in yet another way, we cannot be sensitive to the implications of error or responsive to the possibility of alternative explanation unless we have a reasonable idea of the constraints that we have imposed on our analysis in its very first statement.

Even if one sees some merit in the argument just outlined, the question immediately arises of what the child psychologist interested in problems of knowledge should do. Let me address this operational question in two ways: first, by speaking of several conclusions that do not follow from renewed concern with the premises of cognition, and second, by proposing a series of questions that any developmental theory of cognition must address.

It should be emphasized again that I am not proposing any doctrine of a single truth. Because we are so ignorant of the development of the child's cognitive ability, and because there is such a wide range of presuppositions available to us, it is, on the contrary, almost compulsory that a number of different theories of cognitive development lie side by side for a long time. More than that—the presence of epistemological variation supports the evidence of neurophysiological variation and of the variation of computers—any number of conceptual models may rattle around in our heads or in our theories, whichever locus the theorist prefers.

Nor is it a solution to the problem of constrained analysis to call for a return to a taxonomy of the child's cognitive development, to propose that we are so enmeshed in our philosophical and psychological confusions that it would be better if we just went out and looked at the child as he perceives, speaks, and plays. It is, of course, valuable to know something about the behavior that must be explained by our elaborated theories, but it is only an evasion of the problem of our prejudices to maintain that there is anything like uncluttered and premise-free observation. Taxonomy has its own epistemological basis.

It should be clear, too, that the broadening of our concern about human knowledge does not represent a call for the abandonment of analytic and empirical procedures for the understanding of the child's development. We cannot discard the splendid work of the last decades and race for the nineteenth-century collection in our historical libraries to determine what the structures of human knowledge really are. If anything, the demand for careful analysis and careful empirical specification of problems is greater in studies of cognitive development than elsewhere in psychology because it is necessary for the psychologist of cognitive development to chart for us very carefully the route he follows in going from the problem of knowledge to the laboratory. Let me, in this connection, make a stipulation that may save some unnecessary argumentation. Like all other psychologists, I believe in S's and R's if they are interpreted in their most general sense to state that psychology is about behavior on one side and characteristics of the world on the other and, further, I believe that we can test our best propositions only by observations that are available to intersubjective agreement. The difficulty with the notions of stimulus and response as they are used in current psychological speech is that they often function as screens for complex epistemological presupposition. If we knew what went into making a stimulus class for a newborn, a 2-year-old, and an adolescent—if we knew what constituted response classes across their development—we would already have solved most of the problems involved in the child's construction of reality. The language of learning theory is the best systematic language we have in psychology, but its comparative excellence should not blind us to the epistemological skeletons it protects. A theory may have empirical relevance even if it does not define the environment physicalistically or the behavior topographically.

Our new theories of the development of cognition, then, will be plural, systematic, analytic, and empirical. The wish I have just expressed is that they will also better reveal their relation to the matrix of social and philosophical presuppositions from which they are derived.

II.

Let me turn at last to the proper subject of my commentary—a set of questions that may properly be addressed to any theory of cognitive development. It is not meant to be a complete list, and, in the light of what I have said earlier, it is unlikely that a complete list of such questions can be made. Moreover, the questions do not fall in formal deduction from considerations of the problem of constrained analysis and its ancillary difficulties, although a weaker linkage can be established. Specifically, several of the questions raised here derive from a consideration of the places in theories of development where major (and frequently unstated) presuppositions enter. Because these presuppositions—or, better, predecisions—constrain all that follows after them, and because the predecisions are usually consequences of a particular (but, again, frequently unstated) epistemological position, the questions that follow are but one degree removed from the issues discussed earlier.

Nonetheless, one may reject all that has gone before on the topic of epistemological wariness as a packet of philosophical frills and fancy talk and still consider the following group of questions as relevant to theories of cognitive development.

The Developmental Goal

There is no dragon of entelechy here; to understand a developmental progression, one must usually state the developmental end point. However obvious this issue may seem, it is a revealing exercise for the developmental theorist to state as carefully as he can what he is about to explain with his theory. The constraints of presupposition make their appearance at once. If you are going to predict the language of the 4-year-old child in terms of kernels and syntactic transformations, then clearly the range of variability in your analysis of his development is severely restricted. If you are going to explain the behavior of adolescents in terms of their structures of defense, you have predetermined a substantial part of the investigative endeavor. This question goes beyond the necessary and appropriate statement of the range of observation in an empirical study. If you set as the developmental end point in a study of mother attachment the child as a system of habit-family hierarchies (or as coordinated schemata, or as a set of interrelated ego functions—the particular example does not matter), you thereby determine the observations that are relevant to your conclusion. You will not be able to observe factors that may be important for our understanding of mother attachment, not because you "interpret the data differently," as the phrase goes, but because your presuppositions about the nature of man and the nature of reality lock out of your inquiry those possible observations that do not lie on the course that runs toward your predetermined end point. To say it briefly, the specification of the dimensions of the developmental goal limit the relevant evidence; only some parts of the observational range are seen as part of appropriate inquiry; even before a single datum is collected or a single measure taken, there is reduction of the possibilities of answer. The reduction is unavoidable, certainly, but it need not be unconscious. It is no scientific triumph to conclude that certain factors are not relevant to your findings when your presuppositions do not permit you to observe such factors; the triumph is ambiguous when you conclude that certain other factors *are* relevant to your findings if your presuppositions permit you to observe only such factors. In an ultimate developmental theory, there will be no need for a specification of developmental end points; the exchange between child and phenomena will be so well understood that each developmental change will be predictable from a statement of cognitive structure and environmental conditions.

Until that glorious day, however, we must define the domain of most developmental sequences in part by a statement of their end points, and we must beware the constriction of any such statement.

Cognitive Function[1]

In what follows, I am assuming that there is some general agreement that human beings do not respond to each aspect of the environment as particulate but, rather, that an environmental event is classified or coded, and that there are collections or classes of responses that are appropriate to the division on the environmental side. Further, I am assuming that these classes themselves are related in certain ways. The reduction of environmental variability and of behavioral variability that follows from coding and classifying can be discussed in terms of "operators" or "concepts," and the relations among "concepts" can be called "rules" without predisposing us toward pleading a single psychological position. Other names are given these compressors of variety—"generalized habits," "cognitive structures," "schemata." For the moment, the name is not critical, and I will speak about rules shortly.

The question I pose for theories of cognitive development under the present heading is prior to a specification of the nature of rules. What functions does the theory assign man that govern the development or invention or growth of operators for the reduction of variability and rules for relating operators? What primitive cognitive functions do we ascribe to man? For Kant, the functions are the categories of quantity, quality, relation, and modality; for Piaget, they are adaptation and organization; for recent learning theory, they are (chiefly) learning and generalization. It is important to note that, in the usual language of psychological analysis (that so recently seemed so neutral), cognitive functions are not responses or habits or even rules; they are animals of a higher and perhaps incommensurable order. Because the assignment of cognitive functions to man is prior to the elaboration of psychological theory, we rarely have occa-

[1] *Cognitive Functions* is the third unsatisfactory heading given to this section. It was preceded by *Rule-making Functions* and *Categories of Cognition* and shares with them the burden of widely diverging and confusing use in philosophy, psychology, and mathematics. An attempt to corral the idea behind these names appears in the next several paragraphs of the text. It may help to note that *cognitive functions* are closely akin to Piaget's *functional invariants* and that, had I the courage, I would propose a term such as *cognitive structors* for the notion. An additional complication arises because some theories of cognition, Kant's and Piaget's among them, introduce a conceptual layer between cognitive functions and rules. This is the layer of epistemological pigeonholes—*number, degree, causality,* and so on. Theories in the associationistic tradition ignore this layer altogether. The place of such superordinate schemata in theories of cognitive development will not be discussed further in the present paper.

sion to re-examine our assignment. These are the abilities we assign man that organize the way he organizes the world. If this formulation seems flighty, consider the following simple but nontrivial example. We do not assign to rocks the ability to learn; we do assign such a cognitive function to men. Some theorists of behavior use as fundamental cognitive functions instinct and maturation; others are less willing to do so. Some theorists aim for understanding within a single function (most often, learning); others (for example, the writers of textbooks) draw on cognitive functions avidly, reconstructing man at the beginning of each new chapter.

Two points deserve emphasis—first, there is a wide variety of cognitive functions that psychological theorists put into their subjects; and second, these functions are at a level of analysis different from habits, schemata, program subroutines, or rules. Incidentally, if cognitive functions sound to you like faculties in the older usage, I can only confess that they sound so to me as well. Perhaps we did not escape the implications of *senses, intellect, feeling,* and *will* by redesignating the cognitive functions *information processing, learning, motivation,* and *decision-making.*

Do new cognitive functions appear with development? Do children, as they grow in wisdom and strength, take on new procedures for the construction of rules to process environmental input? Older theorists, from Aristotle to Bühler, have suggested that there is a developmental progression from a dominance of instinct to a dominance of training to a dominance of intelligence. Spencer saw the problem in a phylogenetic form that can be easily translated into its developmental implication.

> For scientific and artistic progress is due not simply to the accumulation of knowledge and of appliances: The impressibilities and the activities have themselves grown to higher complication (1899, p. 368).

Most modern theorists have preferred stable assignment of cognitive functions, but, no matter how the question of developmental progression is answered, we must remember that every theorist of cognition is under some responsibility to designate these governing procedures for the elaboration of rules. Once the designation is made, all subsequent analysis is thereby constrained.

The Nature of Rules

We are not yet to the safe ground of the intersubjectively reliable because, even with some statement of cognitive functions (e.g., whether learning or adaptation), there remains an important decision to be made about the nature of the rules (or principles, or schemata, or generalized stimulus-response hierarchies) that are produced by the operation of the cognitive functions. Why do we need to consider rules at all? Why can we not consider all responses (or response-analogous concepts) to be of equivalent value? Why do we have, in a sense, responses of a higher order? Whether we like it or not, some responses are, like Orwell's pigs, more equal than others. It is not necessary to ascribe to the school-aged child an enormous number of particular responses or habits to manage $1 + 1 = 2$, $1 + 2 = 3$, $111{,}111{,}112 + 3 = 111{,}111{,}115$. We say (in every responsible theory of man) that the child can add, and, in so saying, we give a higher value to the response called *adding* because of its governance over other responses, such as saying "three." The mathematicians and the psychologists influenced by mathematicians speak of "operators," the computer simulators speak of "search routines" or "do loops," Piaget speaks of "schemata." In each case, and there are others, there is a recognized hierarchization of process in which some procedures dominate, determine, or include others. Stimulus-response theories have been deficient in their explicit recognition of hierarchization of psychological process, partly to maintain simplicity of theory, partly because many of the typical empirical interests of the theorist did not obviously require hierarchization, partly because of the peculiar restraints of a unidirectional linear model for psychology. Berlyne (1965) has recently proposed, within the language of stimulus-response theory, an ingenious more equal response called the "transformational response" that works much like an operator in set theory and that may open the way to a more subtle handling of cognitive issues by classical learning theories. But do not mistake this characteristic of theories of cognitive development: explicitly or not, they contain the notion of rule, and the theory will be limited in its range of application and its relevance to cognitive development by the character of its rules.

Throughout these paragraphs, I am speeding over issues of great scope with only a few words; nowhere is the discrepancy more apparent than on the issue of the nature of rules. Once the notion of hierarchization is introduced into our theories of cognition, it is clear that there are likely to be far more than two levels of rules in most sensitive theories; even more disconcerting is the need for different kinds of rules—rules encompassing simple relations, search procedures, recursive maneuvers, response-selection procedures, and so unboundedly on. It is easy to understand why psychologists have been reluctant to open the door to this disarray, but we will not understand the development of knowledge until we have theories that make a place for the generation, selection, and modification of rules.

Conditions for Rule Change

I raised earlier the question of whether cognitive functions change. No doubt obtains in considering changes in rules. There is a time when the child cannot add and a later time when he can; there is a time when he cannot perform concept-sorting tasks of a certain kind and a time when he can; there is a time when he cannot generate sentences in the language and a time when he can. These facts, more or less common property of all theories of development, force our attention to a fourth general question: What are the conditions under which rules change? This question is easily as complicated and difficult to answer for existing theories as are the earlier questions about developmental goals, cognitive functions, and rules. It pertains to two subsidiary and included problems: What are the occasions for change, or, in another language, what is the motivation for change? and what events control change, or, in the more common mode, what reinforces cognitive change? You know of the wide diversity of opinion of this central issue not only for cognitive development but for all psychology. It is interesting that, over the last decade, the conditions of cognitive change have come to sound more Apollonian; there has been a decay in the models based on physiological deficit and pain and a corresponding increase in the use of models based on discrepancy in the environmental presentation, curiosity, and uncertainly reduction. Most of these revisions, by moving the locus of "motivation" and the locus of "reinforcement" beyond the easy observation of the psychologist, have made the task of the theorist of cognitive development both more difficult and susceptible to more flexible solution. Perhaps because of the long-lasting interest of American psychologists in moving behavior around—the revived interest in cognition has not diminished the cultural commitment to manipulation—there has been far more attention to this question in theories of cognitive development than to the preceding, perhaps somewhat more general, questions about goals, functions, and rules. For that reason, I hope it is not unduly pessimistic to suggest that the problem of the conditions of cognitive change is not yet solved. Fortunately, the pessimism is bounded by the intensity of interest among psychologists in these questions. A reminder may be in order, nonetheless, that the question of the conditions of cognitive change is not independent of decisions on the issues raised by the preceding questions.

These four questions—the goals of developmental sequences, cognitive functions, rules, and the conditions of change in rules—are an irreducible minimum for any psychological theory that purports to deal with the development of human knowledge. Whenever a decision (and it is a decision almost always, an induction from nature almost never) is made about each of these dimensions of a theory of cognition, constraints are put on the choice of observation, the choice of method, the choice of analysis, and the choice of conclusion. In order to obtain even a dim view of what the child contributes to our ultimate statement and what is inherent in our prior decisions about theory, it is fundamental that these decisions be as openly and explicitly exposed as is humanly reasonable. When we do so, we may find that the governing principles of our theory are less obvious and less securely defensible than we had thought. It may even turn out that a philosophical defense of some ways of proceeding to the study of cognition will be useful and attractive; philosophical justifications may even be preferable to justification by availability of apparatus and justification by the prejudice of the cult.

In addition to these abstract (although universal) questions, there are three more technical issues in the study of cognition that warrant some attention. Again, they are questions that can be addressed to all theories of cognitive development, and they are questions with implication beyond the thought of the child.

Construction and Selection

In most of our perceptual and problem-solving tasks for adults, we clearly do not call on the subject to invent or develop a rule on the spot to handle the problem posed. Rather, he is required (to put the matter somewhat loosely) to search through available operators or rules and to select a relevant, possible, or correct one. Rule selection is an important area of research for the study of human thought, but it should not be confused with the original invention, discovery, learning, or construction of the rule, the concept, or the transformational response. It is, presumably, in the domain of a psychology of cognitive development to describe and explain how the child takes these primitive and fateful steps. How does the child construct the notion of object, or the concept of numerosity, or the rules for search in problem solving? With an occasional striking exception, most of the studies called "concept-formation" with children have, like the studies with adults, been studies of rule search and rule selection; we have very little evidence on the equally interesting and certainly more fundamental step of rule construction. Perhaps the issue can be sharpened by a somewhat exaggerated statement: There is little evidence to suggest that the psychological principles appropriate to an understanding of rule selections are relevant to an understanding of rule construction. Theories of cognitive development appear to need a distinction between the two processes (whatever the theoretical language), and child psychologists need a great deal more research on rule construction or true concept formation. In passing, it should be noted that the idea of "selection" is a peculiar one in psychological theories strongly based on deterministic metaphysics; it is by no means clear in current theories of perception and thought how this process of selection can be given nonmetaphorical expression.

Availability and Use

The linguists and the psycholinguists have been critics and tutors to child psychologists for some time now, and it is a defensible proposition that the linguist's observations about language and his questions about the adequacy of available psychological analysis to encompass his observations, taken together with the responsive attempts of psychologists to modify their theories to make room for language, have been the major source of new thinking in the study of cognition during the last several years. Out of this dialogue has come a reminder of a distinction that it would be wise for our prospective theories of cognitive development to keep in view. It is the distinction between studying whether the child *can* use a concept or rule and studying the occasions on which he *does* use it. We may, for example, study the occasions upon which the child says a particular phrase, and we may also ask whether the child is able to speak the phrase even if we cannot specify the usual occasions of its utterance. We may know about availability without knowing much about the occasions of use. A similar distinction is valuable in studies of thought and of concept formation as well, and can probably be adapted for investigations of perception and play. We may, in presenting the child with materials like the Vygotsky blocks, raise questions about his use of concepts; we may also (with somewhat different procedures) raise questions about the availability of different categories or methods of sorting. Some of our inquiries about the development of knowledge will concern the study of the occasions and circumstances under which the child uses a particular rule or generalizes a particular habit; some of our inquiries will be probes for rules, attempts (like the linguist's "Is the following sentence a proper sentence in your language?") to determine what operators and rules the child has available even though we are not always able to answer the more obviously psychological question of the occasions of use.

Conforming, Transferring, Generating, and Formulating

This last point is not the goal of the present paper but another somewhat prosaic matter that bears on the issue of the proper measure of the child's cognitive ability. Under what tests will we be willing to say that a child possesses a concept (another puzzling phrase), or uses a rule, or selects a particular transformational response? By the way, here as elsewhere, I do not mean to suggest that the phrases I put in disjunctive chains are psychologically or epistemologically equivalent; it is an attempt to indicate that the problems raised go beyond any particular view of cognitive development. As you know, the standard tests have been of four kinds, and I would like to propose that we continue to be careful in maintaining a distinction among the four and that we consider the possibility that different kinds of sub-theories or subsets of principles may be necessary to account for the results of the different tests or probes. Conforming (bringing the behavior of the subject under the control of the dimensions that the experimenter sees as relevant), transferring (subject's making an appropriate response in a novel setting of the problem or concept), generating (subject's producing an instance of the problem solution or concept not previously encountered), and formulating (an explicit, complete, and correct statement of the principle of sorting or of the principle of problem solution), are very different tests. The behavior of the child in the face of these four tests of cognitive skill may show different developmental courses and require somewhat different theoretical expression. The statement of this problem is a fitting last question—for an issue like this; where the apparatus of careful and reliable observation can be brought to bear and where one can hope to make at least preliminary specifications of the nature of the setting and the nature of the behavior, one can see the beginning of a solution to the problems of theory invention that have threaded through this paper.

My closing plea reflects the theme of my first remarks. For a number of reasons, child psychology—particularly the psychology of cognitive development—is in an era of doubt, of openness, and of variation. These times are probably rare in a science, and we should make the most of them—by examining the footings of our psychology, however painfully indefinite the examination seems—by continuing to explore empirical questions of relevance to the child's developing construction of the world, even if we must invent new conceptual dimensions and try untested methods of study—and finally, by avoiding the premature precision (and the premature constraints of domain) that increase the size of the frog by decreasing the size of the pool.

INFANT LEARNING AND DEVELOPMENT: RETROSPECT AND PROSPECT*

Frances Degen Horowitz

Occupying the last position in a program of substantive papers can reflect either the assumption that the area of infant learning represents the rear guard of infant research or the avant-garde of infant research. As the avant-garde, infant-learning research is being carried on by a very active, rigorous, though relatively small, group who are at the beginning of some long delayed but significant advances. If on the other hand you consider infant-learning work to represent the rear guard of infant research, then I leave the characterization and the adjectives to your own predilections.

In his 1963 review of infant learning in the first year of life (Lipsitt, 1963) and in his 1965 address to this conference, Lewis Lipsitt (1966) noted the scarcity of infant-learning studies as reflected in both the sheer number of studies and in the almost total lack of systematic programs of research which seek to identify the variables and parameters of the learning processes in the young human organism. The concluding optimistic tone of his 1963 review would appear to have been somewhat predictive of the present state of infant research. It is interesting to note that in the 68-year interval between 1896 and 1963 Lipsitt mentions, and our own check indicates, only some 35 to 40 studies of infant learning. But in the recent 4-year interval from 1963 to 1966, give or take a few overlapping studies (between published and unpublished status), there have been some 15–20 reports of research on infant learning.

These recent reports have been imbedded in the geometric expansion of general infant research—a body of data about the nature and development of the human organism that is raising questions of significance and resurrecting the excitement that infant researchers in the 20's and 30's

* From Frances Degen Horowitz, "Infant Learning and Development," the *Merrill-Palmer Quarterly*, vol. 14 (1968), pp. 101–120. Reprinted with permission of the author and publisher. Presented at The Merrill-Palmer Institute Conference on Research and Teaching of Infant Development, February 9-11, 1967, directed by Irving E. Sigel, chairman of research. The conferences was financially supported in part by the National Institute of Child Health and Human Development. The author would like to acknowledge the bibliographical aid of Lucile Paden, Patricia Self, Saunny Silverman, and Robert Aitchison, and the suggestions of Kastoor Bhana in the preparation of this paper. Barbara Etzel's critical reading of the manuscript was very helpful. Partial support of the author while writing this paper was derived from Grant HD-00870 from the National Institute of Child Health and Human Development to the Bureau of Child Research at the University of Kansas.

must have felt. Indeed, there is the distinct feeling that we are all involved in unravelling the complex fabrics of the developmental process. Those fabrics are, however, still tightly woven; locked into those weaves are the old questions of learning, maturation and development, of acquisition and innate capacity. Some of the key tactics that will unlock the process involve clearing the semantic overgrowth of terms like "maturation." Others will become available only through the ingenuity of the laboratory researcher and the sometimes hard to come by cooperation of his naïve subjects. Still other advances will come by way of relevant experiments on non-human primates—experiments which are, for strong ethical reasons, impossible to conduct with humans in our society. Such comments about the complexity of human development have, however, almost reached the point of being professional clichés and in so being reflect on the one side the terribly small gains in knowledge that have been made. But, from another point of view, the reiteration of these clichés reflects a return to the research problems previously laid aside but which are now being attacked at higher levels of methodological and theoretical sophistication.

Historically, with the early demonstrations from Pavlov's laboratory and from the animal-learning research in this country, it was a natural next step to demonstrate the relevance of learning research procedures and the relevance of identified learning variables to the study of human development—in particular, to the development of the very young child. The results of those attempts could be regarded as strongly supporting learning as the major phenomenon accounting for early development, as casting serious doubt on how much of early development could be accounted for by learning, or as leaving the basic problem unsolved. To declare yourself primarily *for* the learning position you had to take the following position: Enough studies demonstrated the success of conditioning to suggest that, indeed, these processes are ongoing and can be made to account for behavioral acquisition—assuming the liberty to stretch the story here—and add the logical steps there.

On the other hand, to declare yourself as primarily *against* the learning proposition you had to say: The phenomenon of conditioning while sketchily established, is clearly of the unstable sort and requires such a contortion of laboratory arrangements—arrangements which are hardly ever found in the "real world"—that only a preference for fantasy could induce one to put his scientific bets on a conditioning model to account for behavioral development. And, even if conditioning accounts for some acquisition, infant development is so rapid that there is not enough time for the slow trials, association and error procedures, to be taking place. Logically, then, the development of the organism must, in large part, be the function of maturational processes inherent in the organism—which, if they need a benevolent environment in which to develop, certainly do not depend solely upon the carefully juxtaposed arrangements of the learning model for their operation. Or, to declare a draw between the two alternatives you had to say: The problems of methodology are such that one is unable to say to what extent a conditioning paradigm can be used to account for behavioral development and therefore no assumptions can be made.

I'm afraid that hardly any of us is the skeptic enough to have taken the last position. By the tradition of our graduate upbringing our lots have for the most part been cast with either the "for" or "against" camp. It is almost paradoxical that the importance and serious implications of these conclusions should have become the cornerstone of all the theoretical and semi-theoretical systems of the world of child psychology without generating a frenzy of frontier research to check the validity of these positions. In part this has rested upon the problems of adequate methodology—problems which have needed for their solution some of the recent technological advances in observation and measurement of behavior. A review of the some 50 studies which one would clearly classify as studies of infant learning demonstrates this. From the crude attempts of investigators such as Valentine in 1914 to apply reward contingencies, or from the successful and unsuccessful infant-conditioning studies of Jones (1931), Bregman (1934), Kasatkin (1935), Wenger (1936), and others (Marquis, 1931; Jones, 1930; Kasatkin and Levikova, 1935; Kantrow, 1937) in the 1930's, we come finally to the elegance of the recent conditioning studies being reported by Papoušek (1966) from the Prague laboratories and by Lipsitt, Siqueland, Kaye and others (Lipsitt, Pederson and DeLucia, 1966; Siqueland and Lipsitt, 1966; Lipsitt, Kaye and Bosack, 1966; Kaye, 1965) from the Brown University infant laboratories. While the early 1950's are sprinkled with reports of successful infant conditioning in Russia, in this country Rheingold's study, published in 1956, launched a new era of experimental infant-learning research. Using gross variables in a complex setting, she demonstrated the experimental modification of social responsiveness in institutionalized infants. Likewise Brackbill, in 1958, using a more delimited setting showed the smiling response to be controllable by complex contingent social stimulation.

During the ten-year period between 1956 and 1966 both the volume and sophistication of infant research in general, and of learning research in particular, has grown. It is instructive to trace the infant research in this period. First I should like to mention those studies which are clearly studies of learning. From 1957 to 1966, a quick review of infant learning goes something like the following, year by year.

1957—Polikanina and Probatova (1957) report the establishment of conditioned motor alimentary reflex to an intermittent light in prematurely born children.

1958—Brackbill's (1958) study, already mentioned, appears. Usol'tsev and Terekhova (1958) report a failure to condition the eye-blink response to skin temperature change in infants 1 to 4 months old.

1959—From Prague, Papoušek (1959), using a combination of classical and instrumental procedures, offers evidence that conditioned head-turning responses can be reliably established at 4 to 6 weeks of age; Janoš (1959) suggests that prematures show later conditioning than full-term infants but that if an age correction is made no difference is found. Rheingold, Gewirtz and Ross (1959) demonstrate that vocal responses of 3-month-old infants can be increased by the occurrence of a contingent complex stimulus.

1960—An English language publication appears of a 1953 study by Kasatkin, Mirzoiants and Khokhitva (1960) which demonstrated successful conditioning of the orienting reflex to sound in infants at about 2½ months.

1961—Kol'tsova (1961) reports the establishment of what she calls temporary associative connections, without consequent or reinforcing stimulation, in 20- to 24-month-old infants between signal-type stimuli, such as a bell, and a reflex such as the eye blink elicited by an air puff. Polikanina (1961) indicates she has obtained conditioned autonomic and motor responses using an aversive odor with a tone at 2 weeks of age in infants born 1 to 1½ months prematurely, and at 3 months in infants born 2½ to 3 months prematurely. Rendle-Short (1961) fails to obtain a conditioned eye blink in infants younger than 6 months but is successful with infants over 6 months of age.

1962—Bartoshuk (1962) observes rapid habituation of tonic arousal reaction and suggests the possibility that learning may account for the response. Papoušek (1962) reports further success in studies of conditioned head turning for food.

1963—Weisberg (1963) concludes that social contingencies can account for vocalization increases as opposed to certain non-social contingencies, or to social and non-social stimulation presented non-contingently.

1964—Gullickson and Crowell (1964) report permanent increase in electroactual threshold in neonates from Day 1 to Day 3 and interpret their results as suggesting that habituation phenomena may be thought of as learning. Lipsitt and Kaye (1964) demonstrate classical conditioning of sucking to a tone in neonates. Simmons (1964) publishes research results indicating 12-month-old infants show operant discrimination learning. Siqueland (1964) reports successful operant conditioning of head turning in 4-month-old infants.

1965—Kaye (1965) indicates successful conditioning of the Babkin reflex to hand transport in newborns. Keen, Chase and Graham (1965) suggest that neonates show 24-hour retention of an habituated heart rate. Sameroff (1965) claims that modified suction responses as a function of consequential events does not indicate learning, because changes are not persistent from feeding to feeding. Stern and Jeffery (1965) present positive results of operant conditioning of non-nutritive sucking in the neonate.

1966—Lipsitt, Pederson, and De Lucia (1966) report the use of visual conjugate reinforcement to control operant behavior in one year old infants. Siqueland and Lipsitt (1966) give the results of three experiments which demonstrate the successful conditioning of head-turning responses in newborns through reinforcement contingencies, including positive results in cue reversal procedures. Weisberg and Fink (1966) demonstrate that 14½- to 19½-month-old infant performance on a fixed ratio schedule bears some resemblance to animal performance under similar conditions. Lipsitt, Kaye and Bosack (1966) report neonatal tube sucking as a function of reinforcement. Papoušek (1966a, 1966b, 1967) provides more complete information about his conditioning research with human infants. He suggests four stages in the conditioning process as illustrated in his work on conditioned head turning: First, initial indifference to the conditioned stimuli; then, inhibition of general activity; followed by the unstable emission of the conditioned response; and, finally, achievement of a stable conditioned response. His work illustrates early conditioning and the gradual process of acquisition with stable, conditioned responding finally elicited even when infants are food satiated. Fitzgerald, Lintz, Brackbill, and Adams (1966) indicate successful pupillary conditioning in infants 3½ to 12 weeks of age.

From such a stacatto recitation it should be clear that the tentative indications of the infant conditioning pioneers of the 1930's now appear to be more clearly substantiated and important elaborations are being demonstrated. As many of the cited reports conclude so we *must* also agree the young human infant is a *learning* organism, subject to a variety of environmental contingencies, able to behave reliably in relation to associated events, and, within a range of response capabilities, his behavior is modifiable, trial by trial, over a somewhat generally predictable conditioning course. I should point out that while this conclusion implies strong support for the learning position, it would be foolhardy to assume that the nature of the organism—leaving the "nature of the organism" undefined for now—does not impose some limitations. Further, our saying, "within a range of response capabilities behavior is modifiable" is a purposely vague reference. And, it must remain vague until certain essential information about the organism is known.

Such information is in the process of being discovered. For, during this same approximate ten-year period the human infant has been yielding up a variety of information about itself in laboratories dotted about the country. This information may be described as delineating characteristics of the human infant as well as capabilities of the organism—demonstrated in a variety of ingenious experimental and observational procedures, but differing from the research just cited is the fact that the infant is demonstrated as being a certain way or as having a certain characteristic without the process of acqusition being involved or investigated. If you will forbear another historical trip, a selective trail of this descriptive research would look something like this when summarized. (I might add that this is *very* selective—a sampling procedure, if you will.)

1958—Fantz (1958) reports his now historic findings of the presence of selective attending to patterns in the human infant.

1960—Bell (1960), through the analysis of film records of 96-hour-old infants, suggests that the neonate exhibits well-organized behavior systems which are complex and interrelated. Gibson and Walk (1960) report apparent depth perception is present as soon as mobility is present in infants and other animals.

1961—Fantz (1961b) reports further on infant visual attention, indicating that stimulus complexity is a relevant variable and suggesting infants may have an innate preference for the human face as a visual stimulus. Kessen and his co-workers (1961) point to individual variation and day-to-day changes in the movements of newborns (Kessen, Hendry and Leutzendorff, 1961). They also single out the stability of individual differences in measurements of newborn behavior (Kessen, Williams and Williams, 1961), and report that hand-mouth contact in newborns is not related to activity level (Williams and Kessen, 1961).

1962—Birch, Thomas, Chess, and Hertzig (1962) indicate early stable individual differences among infants in measures of activity levels, rhythmicity, adaptability, threshold of responsiveness, attention span, etc. Fantz and his colleagues (Fantz, Ordy and Udelf, 1962) report that the young human infant appears to have more extensive ability to make visual accommodations than previously supposed. Walk and Dodge (1962) demonstrate depth perception in a 10-month-old monocular human infant.

1963—Engen, Lipsitt, and Kaye (1963) describe the newborn's ability to exhibit discrimination and adaptation to olfactory stimuli. Fantz (1963) proposes that the infant has innate ability to perceive form, when he demonstrates selective attending in newborns under 5 days of age. Kessen and Leutzendorff (1963) report that a non-nutritive sucking stimulus elicits mouthing and inhibits crying and movement, while stimulation to the forehead has none of these effects. Lewis and his colleagues (Lewis, Meyer, and Grossberg, 1963) report cardiac deceleration accompanying visual attending behavior. Lipsitt, Engen, and Kaye (1963) report increasing responsivity to chemical olfactory stimuli over the first days of life and also note strong individual differences. Salzen (1963), on the basis of observations of one infant, suggests that stimulus factors controlling the elicitation of infant smiling may involve contrast and change of brightness.

1964—Bartoshuk and Tennant (1964) report reliable correlations of EEG records with sleep-awake episodes in newborns. In another article Bartoshuk (1964) describes neonatal cardiac responses to auditory stimuli as a linear relation of the log scale of cardiac measures to auditory intensity. He says the power function resembles the power law relation between stimulus and response shown by Stevens to hold for a large number of sensory systems. Bower (1964) uses an operant analysis to demonstrate that infants 70–85 days of age exhibit depth discrimination. Fantz (1964) publishes data which show decreasing visual fixation to familiar visual stimuli and increasing fixation to novel stimuli in infants 2 to 6 months of age suggesting that visual responsiveness habituates and is a mode of environmental exploration for the infant. Hershenson (1964) indicates human newborns show reliable visual fixation preferences for stimuli intermediate in brightness and low in complexity. Kaye (1964) demonstrates increasing skin conductance in human newborns over the first 4 days of life. Keen (1964) reports that a 10-second auditory stimulus initially produces greater suppression of non-nutritive sucking than an auditory stimulus of 2 seconds duration, but that then the response shows habituation. Leventhal and Lipsitt (1964) demonstrate that newborns exhibit adaptation to auditory stimuli and sound localization ability. Levin and Kaye (1964) note highly consistent individual sucking rates in newborns. Saayman, Ames, and Moffett (1964) propose visual fixation as an indicator of visual discrimination and report that 3-month-old infants can discriminate circles from crosses. Spears (1964) notes some color pattern preferences in 4-month-old infants, with shape dominating color in visual preferences. Steinschneider, Lipton, and Richmond (1964) present evidence for discriminable individual differences in cardiac responsivity among newborns. White, Castle and Held (1964) describe eight 2-week stages leading up to the development of visually directed reaching just prior to reaching at 5 months of age.

1965—Bell and Darling (1965) report male neonates show higher muscle strength than females, but that this is not clearly correlated to lower tactile sensitivity. Birns (1965) demonstrates the presence of individual differences among neonates in response to controlled stimulation. Birns and her co-workers (Birns, Blank, Bridger, and Escalona, 1965) suggest that auditory stimulation may have an arousing or soothing effect on neonates and demonstrate that aroused infants are more effectively soothed by a low-frequency tone as opposed to a high-frequency tone (a scientific validation of the lullaby). Haynes, White, and Held (1963) propose that 19 cm. is the necessary distance for neonatal focus, but that visual system undergoes rapid changes and visual accommodation approximates adult performance by 4 months of age. Hershenson, Munsinger and Kessen (1965) report neonatal preference for shapes of intermediate variability. Kistjokovskaya (1965) is reported to suggest that smiling is an innate response appearing at the 4th and 5th week of life to stimuli having certain auditory and visual characteristics—that these stimuli need not be, but often are, human. Lenneberg, Rebelsky, and Nichols (1965) publish evidence that infants of both deaf and hearing parents produce similar vocalizations and therefore conclude that crying and cooing responses through 3 months of age are independent of environmental stimulation. McGrade and Kessen (1965) report significant individual differences in neonatal activity. Weitzman and Graziani (1965) indicate sleeping newborns show complex EEG patterns of electrical activity under auditory stimulation. Weller and Bell (1965) demonstrate a positive correlation between activity level and basal skin conductance.

1966—Bower (1966a) reports that infants 50 to 63 days of age show shape constancy and interprets this as indicating that the human infant is "set" to respond to a high-order invariant of shape constancy in the perceptual world. Brackbill and her colleagues (Brackbill, Adams, Crowell, and Gray, 1966) further substantiate the quieting effect of continuous auditory stimulation, and fail to replicate Salk's heartbeat phenomenon. Brennan, Ames, and Moore (1966) report infants at 14 weeks fixate more complex patterns, while at 3 weeks they prefer the least complex and at 8 weeks are intermediate. Charlesworth (1966a) shows that orienting and attending behavior are most persistent when the locus of stimulation is maintained at a high degree of uncertainty. Doris and Cooper (1966) indicate brightness sensitivity undergoes rapid development in the first 3 months of life. Gottlieb and Simner (1966) report that seemingly anticipatory heart rate increases precede spontaneous non-nutritive sucking in infants 25–36 days of age. Haith (1966) presents evidence that newborn sucking rate is reduced in the presence of a moving stimulus, that no habituation is seen, and confirms other findings of stable individual differences in sucking rate in neonates. Kaye (1966) reports that feeding affects non-nutritive sucking, but that except for the cumulative effects of a loud tone auditory stimulation does not. Lipton, Steinschneider, and Richmond (1966) report significant changes in cardiac control mechanisms over the first 4 months of life. Roffward, Muzio, and Dement (1966) describe the course of sleep in normal newborns. Rovee and Levin (1966) indicate a relationship between pre-stimulus activity level and the change resulting from oral stimulation. Salapatek and Kessen (1966) report that ocular fixations in the newborn are not equally distributed over a triangle but tend to be directed toward the vertex. Stechler, Bradford and Levy (1966) show that attending behavior in the newborn is accompanied by increased skin potential reactivity. Steinschneider, Lipton, and Richmond (1966) describe motor responsivity and cardiac changes in neonates under conditions of auditory stimulation, noting individual differences. Tauber and Koffler (1966) report the occurrence of optokinetic nystagmus in response to apparent motion. Watson (1966) suggests that the zero degree of orientation of an object initially elicits the greatest amount of smiling. Korner and Grobstein (1966) find that holding neonates at the shoulder of an adult appears to induce scanning and visual alerting behavior.

From this sketchy recital of a variety of research findings descriptive of the human infant one cannot fail to be impressed at the characteristics and capabilities of the human young. One can systematize the findings in a variety of ways. Studies tracing out the workings of any one characteristic are not numerous and, at the risk of sounding like an encyclopedia, I chose the chronological presentation for the purpose of demonstrating the increasing number of interesting and often isolated descriptive facts being accumulated about the human infant. With replication and extension some of the facts will be modified or discarded; some will remain stable. What is very clear is that gone forever is the notion that the human infant is at birth set adrift in a big, blooming, confusing world to be pitied for his helplessness. Rather we have come to admire him as a complex, fascinating and revealing organism, the result of a phylogenetic history embarked on a course of ontogenetic development.

Now, let us return to some of the basic issues—which primarily involve interpretation and synthesis of data—and to some extent touch upon research strategy. I am always surprised that the impressive characteristics of the human infant (the descriptive data cited in the second list) are used to support the notion that, after all, the course of human development is almost totally pre-programmed, that the infant is "set" to respond to predetermined stimulus conditions, and that while a benevolent stimulus-endowed environment is necessary for development it is certainly not primarily instrumental in development. Ignoring such a logical contradiction for the moment, this position is analogous to saying that oxygen is set to combine with hydrogen ("See? It does!") and we have water—so it is all pre-programmed and, isn't the physical world a wonderfully complex system? To take a descriptive sequence as a causal sequence is to return to the old logical fallacy of using norms as explanations, only this time with increased sophistication. It is obvious that the young human organism *at* birth exhibits a variety of response characteristics, is physiologically complex, is in some ways quite different from the eventual adult organism, and is reliably discriminated on an individual difference continuum on many measures. It is further clear that charting and describing these characteristics and capabilities has only just begun. But it should also be obvious that the demonstration that the organism can do something or exhibits certain capabilities does not necessarily identify the process responsible for the result. While descriptive data is no more or no less important than process data, one does not substitute for the other.

Let me give an example. Kaye (1964), who has been involved in significant research on infant learning, has also reported that infants show increasing skin conductance over the first 4 days of life. No psychologist doubts that a physiologist can properly take such a descriptive statement and ask, "What is the process by which this phenomenon occurs? Under what conditions will it not occur?" Assuming the data are reliable and replicable no one doubts it does occur. But we would hardly expect the scientific community to come to rest with the fact of its occurrence. The question of "how it works" is clearly a legitimate one. However, in the realm of behavioral development there is often the tendency to regard the appearance of some *responses* as behavioral prime numbers, phenomena which cannot be understood as part of a process for which the significant conditions of acquisition can be specified.

If the argument revolves around the term "learning" then let's drop the term and talk about process of development—or process of behavioral acquisition—and ask ourselves the question, "How much of the process of behavioral acquisition has been worked out?" By process I mean mechanisms, the identification of the relevant variables, their parameters, and their interactions. From such an inquiring vantage point let us, then, evaluate where we are.

It could be maintained that in the last ten years research on infant learning has been largely involved in demonstrating, on methodologically sounder grounds, the claims of the '20's and '30's that conditioning can be made to occur in neonates and very young human infants. A minimum of 15 carefully controlled studies report successful conditioning; most of these used neonates. Only two to three studies fail to establish conditioning. It would appear that, among other things, what has been proved is that in a well-equipped laboratory using methodologically sophisticated techniques results can be obtained which clearly establish the human infant as conditionable. The results further identify some of the variables which must be taken into account. The recent work of Siqueland and Lipsitt (1966), for example, provides some exciting evidence of how much we can learn about infant learning from well-designed and carefully controlled experimental procedures. These impressive facts gave rise to an earlier statement that in this research we are now on the threshold of making significant new advances in working out the processes of behavioral acquisition, aided and abetted by the mounting evidence as to the variables which must be considered. It is these promising research directions to which I should now like to turn.

What are some of the investigatory paths which, if successfully traversed, might provide significant advances in our understanding of the process of behavioral acquisition?

One area that touches upon a variety of problems concerns the sensory modality of external stimulation involved in conditioning. First, it is clear that much information is yet to be gathered on auditory, visual, tactual, and kinesthetic sensitivity. Among other things, it would be helpful to know about differences in habituation and adaptation in the different sensory modalities. For instance, how much monotony is necessary to accelerate habituation? How much stimulus variability is necessary to retard habituation? Under what circumstances do habituated responses become disinhibited, and under what circumstances are they maintained in a state of habituation? Knowing these things for instance has important implications for our presentation of effective stimulation in conditioning experiments. Is there generally greater resistance to adaptation with visual stimuli as opposed to auditory stimuli? If so, we should expect some important parametric differences between conditions for learning involving the visual mode and effective conditions for learning in the auditory mode. If there is greater resistance to habituation in the visual mode than in the auditory mode, one would predict more rapid development in situations which involve visual input and discriminated visual responses, as compared to auditory input and the development of auditory discrimination.

In light of these considerations, it is simply not valid for Bower (1966a) to conclude that because at 50 days of age human infants show shape constancy (assuming, for the moment, that Bower's findings will be replicated) such perceptual facts are innately available to the organism. If one generously figures 20 hours of sleep a day for an infant—an assumption many mothers would spiritedly dispute—one arrives at a rockbottom minimum of 200 hours of distributed visual experience in a perceptually stable environment during the first 50 days. If even only half of that time involves associative conditioning, the amount of time is still considerable. Think of *how much* differential sensory stimulation and feedback is involved, perhaps in a modality which involves little habituation and for which effective stimulation is, relatively, quite plentiful. We know little about the time required to establish associative connections which do not require gross motor responses, and it is possible that the speed is much faster than we are usually accustomed to considering. Such a consideration does not necessarily dismiss the interpretation of Bower's findings. Indeed, the organism may, as a function of phylogenetic processes, have such information available to it under stable stimulus conditions. Lorenz's (1965) recent volume, *Evolution and the Modification of Behavior*, argues cogently for considering such a point of view.

The point is that at the present time we do not know and must await more definitive research before such conclusions can be drawn. This relates to the whole problem of assuming something cannot be so because our present knowledge of a process is so limited. I refer here to the proposition that because of the rapidity of early development it is therefore unlikely that the traditionally slow trial-by-trial acquisition under carefully arranged conditions will turn out to account for behavioral acquisition. This argument has gained renewed vigor in recent years, spurred on by all the wonderous things infants are able to do in the first year of life. Again, we know little of the speeds involved in conditioning. We are only just beginning to investigate systematically those properties of stimulus events which make them act as reinforcements, or the role which stimulus change plays in conditioning. It is possible that, in the complexity of many natural environments, the frequency of correctly juxtaposed events is in excess of that necessary for conditioning to occur; that, in contrast to the picture of the random and relatively infrequent chance occurrence of reinforced responses, the normal environment is actually chock-full of reinforcing events, in abundance as it were. Such a possibility suggests that the identification of the process which determines how it is that an event will act as a reinforcer would be a significant path to pursue.

Reference to the important role of stimulus change in determining reinforcing phenomenon is not a novel suggestion. It is to be found in the writings of Berlyne (1960) and in discussion sections of numerous articles (e.g., Brackbill, et al., 1966). Indeed, the current prescriptions of stimulus enrichment for institutionalized and low-income child populations involve the underlying assumption that the natural environments of these children do not provide sufficient stimulation—often forgetting, I am afraid, the qualification that perhaps what the environments lack is not necessarily stimulation per se but effective *contingent* stimulation. I defy anyone to seriously maintain that a child growing up in a four-room flat housing eight people is in a setting that can be correctly described as one of "stimulus deprivation."

The role of stimulus feedback and effective contingencies which make events act as reinforcers should provide evidence which will make possible some significant advances. Again, at our present state of knowledge, it is not valid for Lenneberg, Rebelsky, and Nichols (1965) to conclude that crying and cooing up to the age of 3 months are independent of environmental stimulation because infants of deaf and hearing parents have similar vocalizations up to that age. They report that deaf parents usually have visual signal light systems arranged to indicate to them when the infant is vocalizing and crying. One assumes that these lights act to make the parent approach the infant's crib. As such, their presence may act as an effective contingency for vocalizing and indeed, because the deaf parents may thus respond in a non-discriminatory or non-differential fashion to many vocalizations, may in fact eventually help retard language development. Thus the presence or lack of parent verbal stimulation per se may not be the primary factor for vocalizations. Indeed, some interesting questions may be asked about the sensory modality of the contingent event, relative to the sensory modality of an eliciting stimulus or of the response. Is there an interaction, for instance, in the auditory and visual stimuli relative to the sensory modality of the required response and the most effective reinforcer?

There is another path of inquiry which I think has great promise, and that is the one related to individual differences. You may have noted the frequency with which findings of studies cited earlier include some mention of the stability of individual differences.

And if such stable differences do not hit the light of print, every infant researcher knows and reads into all reports the fact of large individual variability both between and within subjects. Traditionally, an individual difference approach has involved testing the stability of characteristics over periods of time—quite often, years via correlational methods. These efforts have not really yielded us much except in the way of some gross group statistical predictions. Experimentally, individual differences have sometimes provided for some levels controls in an analysis-of-variance design and again, the yield has not been exceptional. However, the persistence of something like individual variation, so obvious and often annoyingly intrusive, might profitably be capitalized upon. It is possible that an accurate assessment of the variables and parameters of individual differences relative to stimulus dimensions of the environment will eventually provide greater control over the conditioning process. We would predict that when environmental stimulation matches responsiveness characteristics of the organism the number of conditioning trials required should be roughly similar for all subjects.

For instance, suppose it can be shown that individuals reliably differ in the duration of visual fixations and one is involved in conditioning visual discrimination. Using ordinary procedures of a standard stimulus-exposure time, the number of trials necessary to establish that discrimination would vary greatly in a sample of infants. However, if the experimenter can assess individual fixation durations and match the time of stimulus exposure to the infant's fixation duration, then a number of possible results can be envisioned, each suggesting further experimentation relating to the identification of the variables involved in the process of visual discrimination. One possibility, of course, is that the number of necessary stimulus presentations is roughly the same for all subjects. Under these circumstances then at the physiological level the question becomes, How is visual information processed and what determines the duration of a fixation? What determines the ability of the organism to process a given amount of information or the differences in the time necessary to load up X amount of information?

We are presently involved in testing some of these ideas in our infant laboratory at the University of Kansas. Should they prove fruitful and extendable into more complex situations, the use of individual differences in this way has, of course, some rather exciting implications about manipulating developmental rate—a topic beyond the scope of this paper (see Horowitz, 1965). All that I mean to suggest here is that the demonstration and appreciation of stable individual differences is not enough if we are ever going to understand the processes involved in behavioral acquisition. We need to identify the dimensions of individual differences and then plug them in as variables in experimental work. To paraphrase what I originally thought was a well-known saying but have since discovered is not so well known, It is better to use them than to curse them.

The proposal made by John Watson to this conference in 1966 is relevant here and has promise of opening up some exciting paths in infant learning research. Perhaps you will remember (see Watson, 1967) that he has suggested that the occurrence of a stimulus following a response—that is, a contingent event—must be "remembered" by the organism for as long as it takes the organism to repeat the response. If the duration of the awareness of the contingency is shorter than the recovery time for response repetition, one should find that learning does not occur. And conversely, only when contingency awareness equals or exceeds response recovery time should associative processes be demonstrated. If research shows Watson's hypothesis to be correct, I think we will have identified some major variables that determine when an infant does and does not learn.

From this point of view, it seems to me that the study of how the organism learns provides the central focus of understanding developmental process. The issue of "maturation vs. learning" is, hopefully, dead but I find no comfort in the supposed compromise of "learning *and* maturation." If by "maturation" one means physiological changes that occur over time as a function of chemically and electrically induced structural modifications, that these modifications result in changes in response capabilities or, relative to Watson's proposal, changes in speed of response recovery—then on the one hand maturation so defined probably occurs. On the other hand, the job is to discover the variables and mechanisms that account for these changes. To the extent that these changes are phylogenetically determined to follow a somewhat orderly sequence, then to that extent the results of process are determined. But maturation in this sense then becomes amenable to scientific investigation. Whether this be classified physiology or psychology, it must be subject to some systematic process which can be described as a function of quantifiable variables. To the extent that those variables involve external stimulation and opportunities for structural changes to occur and permit the emission of responses as a function of associative experience, then maturation can perhaps be thought of as the physiology of learning. But in no case is it useful to think of maturation as a mystical term to explain what we do not understand.

In this regard, recent results reported by Sackett (1966) provide provocative pause. Sackett, as you may know, reared infant monkeys in total isolation except for some controlled visual stimulus input. He observed the differential occurrence of responses to some of the visual stimuli. That these responses occurred differentially in the absence of any obvious contingencies or differential stimulus feedback—except as the differences were inherent in the stimuli themselves—suggests the possibility of a natural response sequencing to certain kinds of stimulus conditions and raises questions concerning the nature of stimulus input, the role of time both for exposure to the stimulus and elapsed age of the organism, and the differential conditions required for eliciting and maintaining responses. It would be wrong to interpret Sackett's findings as supporting "maturation." They imply that processes are involved which do not include the kinds of contingent stimulus feedback of the conditioning model. Now the questions become, What is that process, how does it work, and how and to what extent is it intertwined in the conditioning model? How does it operate in the maintenance and development of the increasingly complex repertoire of responses? These problems are not of the X versus Y type. Ultimately the workings of all the processes of development will be known; or if that is too grandiose, then a great deal more will be understood than is presently in our ken.

In conclusion, it is hopefully not the nether side of optimism which suggests that the current success of the research in infant learning holds the promise of providing some of the significant answers to how the developmental fabric is woven. Investigators such as Lipsitt, Siqueland, Kaye, and Papoušek have provided the firm foundation on which they and others can now build the models of the basic processes. Should that promise be delivered, a summation of evidence ten years from now will tell a more substantial story.

CONTRIBUTIONS OF LEARNING TO HUMAN DEVELOPMENT*

Robert M. Gagné

One of the most prominent characteristics of human behavior is the quality of change. Among those who use the methods of science to account for human behavior are many whose interest centers upon the phenomena of behavioral change, and more specifically, on change in behavior capabilities. Sometimes, changes in behavior capabilities are studied with respect to relatively specific forms of behavior, usually over relatively limited periods of time—hours, days, or weeks. In such instances, the investigator names the processes he studies *learning* and *memory*. Another major class of phenomena of capability change comprises general classes of behavior observed over long periods of time—months and years. The latter set of events is usually attributed to a process called *development*.

The reality of these two kinds of capability change is obvious in every day experience, and requires no special experimentation to verify. The capabilities of the young child, for example, change before our eyes every day, as he learns new names for things, new motor skills, new facts. In addition, his more general capabilities develop, over the months, as he becomes able to express his wants by means of word phrases, and later to communicate in terms of entire sentences and even longer sequences of ideas, both in oral and printed form. From these common observations one can distinguish in at least an approximate sense between the specific short-term change called learning, and the more general and long-term change called development.

To distinguish learning and development is surely a practically useful thing, for many purposes. At the same time, the two kinds of processes must be related to each other in some way. The accumulation of new names for things that the child learns is quite evidently related to the capability he develops for formulating longer and more complex sentences. The specific printed letters he learns to discriminate are obviously

* Robert M. Gagné, "Contributions of Learning to Human Development," *Psychological Review*, vol. 75, (1968). Copyright 1968 by the American Psychological Association, and reproduced by permission. This article is a slightly modified version of the Address of the Vice President, Section I (Psychology), American Association for the Advancement of Science, Annual Meeting of the Association, Washington, D. C., December 1966.

related to the development of his skill in reading. The question is, how? What is the nature of the relation between the change called learning, on the one hand, and the change called development, on the other?

Over a period of many years, several different answers have been proposed for this question of the relation between learning and development. Investigators in this field have in general been concerned with accumulating evidence which they interpret as being consonant or dissonant with certain theories, or models. Usually the model they have in mind is fairly clear, even though it may not be explicitly represented in their writings.

Models of Human Behavioral Development

It is my purpose here to consider what certain of these models are, and what their implications are for continuing research on human learning and development. Specifically, I am interested in contrasting certain features of models which appear to be of commanding interest in present-day research. I hope by this means to clarify some issues, so that they may, perhaps, be subjected to experimental testing in a manner that will allow us to sharpen and strengthen our inferences about the nature of human behavioral development.

It is inevitable that the theme of genetically determined growth, or maturation, as opposed to influences of the environment, will run through any discussion of the nature of behavioral development. Everyone will agree, surely, that development is the result of an interaction of growth and learning. There are enormous practical consequences associated with this issue—for example, in designing education for the young. If growth is the dominant theme, educational events are designed to wait until the child is ready for learning. In contrast, if learning is a dominant emphasis, the years are to be filled with systematically planned events of learning, and there is virtually no waiting except for the time required to bring about such changes.

It will be clear enough that my own views emphasize the influence of learning, rather than growth, on human behavioral development. But this is not because I deny the importance of growth. Rather it is because I wish to come to grips with the problem of what specific contributions learning can make to development, and by inference, what kinds of learned capabilities enter into the process of development. I want particularly to contrast a model of development which attempts to account specifically for learning effects with certain other models that do not do so. When I describe this model, you will perhaps agree that it can be conveyed briefly by means of the statement: Within limitations imposed by growth, *behavioral development results from the cumulative effects of learning.*

To set the stage for a model of this sort, it seems desirable first to mention two other models that are more or less in current use, and which have been in existence for some time. The first of these may be called the *growth–readiness model*, which has been associated in previous times with such theorists as G. Stanley Hall (1921) and Arnold Gesell (1928), among others. Briefly, it states that certain organized patterns of growth must occur before learning can effectively contribute to development. Major evidence for this theory comes primarily from studies of the development of physical and motor functions in young children. A prototype study in this field (Gesell & Thompson, 1929) involved special training in stair-climbing for one of a pair of identical twins at the age of 46 weeks, no special training for the other twin. At 53 weeks, the untrained twin did not climb as well as the trained twin. But after 2 weeks of training, one-third as much as the total given to the trained twin, she actually surpassed the trained twin in performance. What this and many similar studies are usually interpreted to mean is that training for a motor performance might as well wait, in fact had better wait, until the child is maturationally "ready," before beginning the specific regime of training leading to the desired proficiency. The findings are consistent with this model. Other writers have pointed out that giving the untrained twin no special training doesn't mean that the child is learning nothing during this period. Unfortunately, the study is not therefore a truly critical one for testing predictions from the theory. Actually, it must be said that much other evidence bearing upon this model suffers from this kind of defect.

A second model of considerable importance, particularly because it has attracted much attention, is that of Piaget. Although the interaction of the child with his environment is given a specific role in this theory, it is well to recognize that it is in some fundamental sense a theory which assigns only a contributory importance to the factor of learning (Flavell, 1963, p. 46; Sonstroem, 1966, p. 214). The model may be summarized, briefly and therefore not without some injustice, in the following statements:

1. Intellectual development is a matter of progressive internalization of the forms of logic. The sequence of development manifests itself at first through motor action, later through concrete mediation of ideas, and still later through complete symbolic representation.
2. Progress in development is affected by the interaction of the child with his environment. New experiences are *assimilated* into

existing cognitive structures, and newly acquired structures in turn make possible *accommodation* to the demands imposed by the environment. The total process, as Flavell (1963, p. 47) points out, may be considered one of *cognitive adaptation.*

This theory has been accompanied by a great mass of observational evidence, gathered over a period of many years, by Piaget and his colleagues in Geneva. They have observed children's performance of a variety of tasks, including those having to do with number, quantity, time, movement, velocity, spatial and geometrical relations, the operations of chance, and reasoning, among others. Generally speaking, the method has been to present the child with a concrete situation, say, two arrays of beads differing in spatial arrangement, and to ask probing questions in the attempt to determine the nature of the child's understanding of the situation. The behavior of the same child may then be tested again at a later age; or his behavior may be compared with that of older children on the same task.

There have been a number of confirming studies of Piaget's findings carried out by several investigators in various countries of the world (Dodwell, 1961a; Elkind, 1961; Lovell, 1961; Peel, 1959). More important for present purposes, however, are the several studies which have attempted to induce particular kinds of intellectual development by means of specific instruction (or learning). Many of these are described by Flavell (1963, p. 370 ff.), and need not be reviewed here. One prototype investigation, by Wohlwill and Lowe (1962), took the following form: Kindergarten children were tested on a task dealing with "conservation of number," requiring them to recognize that the rearrangement of a set of objects in space does not alter their number. Three different groups of the children were given three different varieties of training, each designed to provide them with a mediational way of arriving at conservation of number. A fourth group served as a control, and was given no training. The results were that no effects could be shown of any of the kinds of training. The group improved their performance somewhat, but the experimental groups gained no more than the control group. Other experiments by Smedslund (1961a, 1961b, 1961c, 1961d, 1961e) lead to much the same conclusion.

Another example is provided by a recent experiment reported by Roeper and Sigel (1966), this time concerned with the tasks of conservation of quantity, using standard situations described by Piaget and Inhelder (1964) for conservation of substance, liquid substance, weight, and volume. In this case the trained groups of 5-year-old children were given fairly extensive general training in classifying, in reversibility, in seriation, three mental operations identified by Piaget as involved in the development of ideas of conservation in children. To summarize individual results very briefly, it was found that some trained children *did* improve on some tasks, but not on all of them. In contrast, the untrained control children showed no improvement. But the effectiveness of training was by no means general—one child might achieve a success in conservation of weight, but not in conservation of volume.

There have been quite a number of experiments using conservation-type tasks, and I have only mentioned here what seem to me a couple of representative examples. Generally speaking, the results seem to be summarizable as follows. Tasks which require young children to respond to situations reflecting conservation of substance, volume, weight, and number do not appear to be readily modifiable by means of instruction and training which is aimed rather directly at overcoming the typical deficiencies exhibited by children. Where such training has been shown to have some effect, it is usually a very specific one, tied closely to the situation presented in training, and not highly generalizable. On the whole, any impartial review of these studies would doubtless be forced to conclude that they do not contradict Piaget's notions of cognitive adaptation, and in fact appear to lend some support to the importance of maturational factors in development.

It is my belief that there is an alternative theory of intellectual development to which many students of child behavior would subscribe. In particular, it is one which would be favored by those whose scientific interest centers upon the process of learning. Naturally enough, it is one which emphasizes learning as a major causal factor in development, rather than as a factor merely involved in adaptation, as is true in Piaget's theory; or rather than a strictly incidental factor, as in the theory of maturational readiness. It is easy enough to identify the philosophical roots of such a theory in American psychology. Perhaps the proponents who most readily come to mind are John B. Watson (1924) and B. F. Skinner (1953), both of whom have given great weight to the importance of environmental forces, of learning, in the determination of development.

But philosophy is not enough. As Kessen (1965, p. 271) points out, for some reason not entirely clear, those theorists who have generally emphasized the influences of environment, as opposed to growth, have also generally espoused a rather radical type of associationism. Thus, they have maintained not only that learning is a primary determinant of intellectual development, but also that what is learned takes the form of simple "connections" or "associations." To account for how a child progresses from a stage in which he fails to equate the volume of a liquid poured from one container into a taller narrower container, to a stage in which he succeeds in judging these volumes equal, seems to me quite impossible to accomplish on the basis of learned "connections." At the least, it must be said that there is no model which really does this. Furthermore, the experiments which have tried to bring about such a change, largely on the basis of "associationistic" kinds of training, have not succeeded in doing so.

In contrast to a weak and virtually empty "associationistic" model, it is not surprising that a theory like Piaget's has considerable appeal to students of development. It tells us that there are complex intellectual operations, which proceed generally from stages of motor interaction through progressive internal representation to symbolic thought. As an alternative, we may choose a theory like Bruner's (1965), which conceives the developmental sequence to be one in which the child represents the world first enactively (through direct motor action), then ikonically (through images), and finally symbolically. These are models with a great deal of substance to them, beside which the bare idea of acquiring "associations" appears highly inadequate to account for the observed complexities of behavior.

The Cumulative Learning Model

The point of view I wish to describe here states that learning contributes to the intellectual development of the human being because it is *cumulative* in its effects. The child progresses from one point to the next in his development, not because he acquires one or a dozen new associations, but because he learns an ordered set of capabilities which build upon each other in progressive fashion through the processes of differentiation, recall, and transfer of learning. Investigators of learning know these three processes well in their simplest and purest forms, and spend much time studying them. But the cumulative effects that result from discrimination, retention, and transfer over a period of time within the nervous system of a given individual, have not been much studied. Accordingly, if there is a theory of *cumulative learning*, it is rudimentary at present.

If one cannot, as I believe, put together a model of cumulative learning whose elements are associations, what will these entities be? What is it that is learned, in such a way that it can function as a building block in cumulative learning? Elsewhere (Gagné, 1965) I have outlined what I believe to be the answer to this question, by defining a set of learned capabilities which are distinguishable from each other, first, as classes of human performance, and second, by their requirements of different conditions for their acquisition. These are summarized in Figure 1.

The basic notion is that much of what is learned by adults and by children takes the form of complex rules. An example of such a rule is, "Stimulation of a neural fiber changes the electrical potential of the outer surface of the neural membrane relative to its inner surface." I need to emphasize that "rule" refers to what might be called the "meaning" of such a statement, and not to its verbal utterance. These ideas are learned

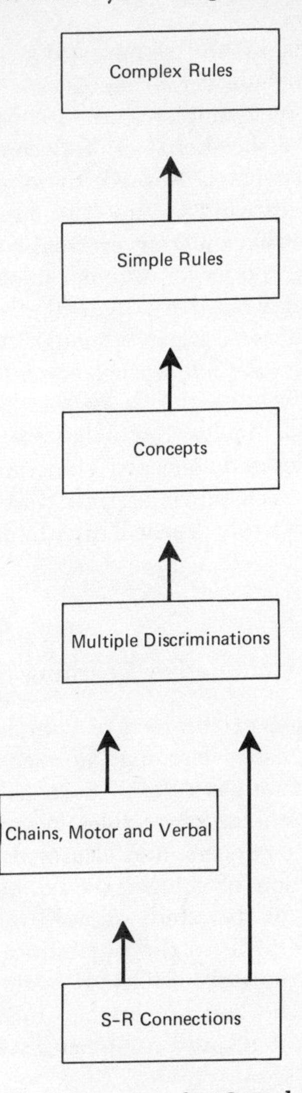

Figure 1 A General Sequence for Cumulative Learning.

by individuals who have already learned, and can recall, certain simpler rules; in this instance, for example, one of these simpler rules would be a definition of electrical potential. Simple rules, in their turn, are learned when other capabilities, usually called concepts, have been previously learned. Again, in this instance, one can identify the presence of concepts like "stimulation," "fiber," "electric," "surface," and "membrane," among others. In their turn, the learning of concepts depends upon the availability of certain discriminations; for, example, the idea of surface has been based in part on prior learning of discriminations of extent, direction, and texture of a variety of actual objects. In the human being, multiple discriminations usually require prior learning of chains, particularly those which include verbal mediators. And finally, these chains are put together from even simpler learned capabilities which have traditionally been called "associations" or "S-R connections."

The identification of what is learned, therefore, results in the notion that all these kinds of capabilities are learned, and that each of them is acquired under somewhat different external conditions. By hypothesis, each of them is also learned under different *internal* conditions, the most important of these being what the individual already has available in his memory. It is clear that associations, although they occupy a very basic position in this scheme, are not learned very frequently by adults, or even by 10-year-olds. Mainly, this is because they have already been learned a long time ago. In contrast, what the 10-year-old learns with great frequency are rules and concepts. The crucial theoretical statement is that the learning of such things as rules and concepts depends upon the recallability of previously learned discriminations, chains, and connections.

Examples of Cumulative Learning

Some verification of the idea of cumulative learning has come from studies of mathematics learning, an example of which is Gagné, Mayor, Garstens, and Paradise (1962). Seventh-grade students acquired a progressively more complex set of rules in order to learn the ultimate performances of adding integers, and also of demonstrating in a logical fashion how the addition of integers could be derived from number properties. The results of this study showed that, with few exceptions, learners who were able to learn the capabilities higher in the hierarchy also knew how to do the tasks reflected by the simpler rules lower in the hierarchy. Those who had not learned to accomplish a lower-level task generally could not acquire a higher-level capability to which it was related.

These results illustrate the effects of cumulative learning. They do so, however, in a very restricted manner, since they deal with a development period of only 2 weeks. Another form of restriction arises from the fact that only rules were being learned in this study, rather than all of the varieties of learned capabilities, such as concepts, discriminations, chains, and connections. In another place (Gagné, 1965, p. 181) I have attempted to spell out in an approximate manner a more complete developmental sequence, applicable to a younger age, pertaining to the final task of ordering numbers. In this case it is proposed that rules pertaining to the forming of number sets depend upon concepts such as joining, adding, and separation; that these in turn are dependent upon simpler capabilities like multiple discriminations in distinguishing numerals;

while these depend upon such verbal chains as naming numerals and giving their sequence. Following this developmental sequence to even earlier kinds of learning, it is recognized that children learn to draw the numerals themselves, and that at an even earlier stage they learn the simplest kinds of connections such as orally saying the names of numerals and marking with a pencil.

It should be quite clear that this cumulative learning sequence is only a suggested, possible one, and not one which has received verification, as was true of the previous example. I doubt that it is at all complete. It attempts to show that it is possible to conceive that all of the various forms of learned capabilities are involved in a cumulative sense in the first-grade task of ordering numbers—not only the specific rules that are directly connected with the task, but also a particular set of concepts, discriminations, chains, and connections which have been previously learned. Normally, such prior learning has taken place over a period of several years, of course. And this means that it would be quite difficult to establish and verify a cumulative learning sequence of this sort in its totality. If such verification is to be obtained, it must be done portion by portion.

A Cumulative Learning Sequence in Conservation

Can a cumulative learning sequence be described for a task like the conservation of liquid, as studied by Piaget (cf. Piaget & Inhelder, 1964)? Suppose we consider as a task the matching of volumes of liquids in rectangular containers like those shown in Figure 2. When the liquid in A is poured into Container B, many children (at some particular age levels) say that the taller Container B has more liquid. Similarly, in the second line of the figure, children of particular ages have been found to say that the volume in the shallower Container B, exhibiting a larger surface area, is the greater.

What is it these children need to have learned, in order to respond correctly to such situations as these? From the standpoint of the cumulative learning model, they need to have learned a great many things, as illustrated in Figure 3.

First of all, you may want to note that "conservation of liquid" is not a behaviorally defined task; accordingly I have attempted to state one that is, namely, "judging equalities and inequalities of volumes of liquids in rectangular containers." However, such behavior is considered to be rule-based, and could be restated in that form.

"Nonmetric" is also a word requiring comment. What this diagram attempts to describe is a cumulative learning sequence (in other words, a developmental sequence), that obtains approximate volume matchings without the use of numbers, multiplication, or a quantitative rule. I

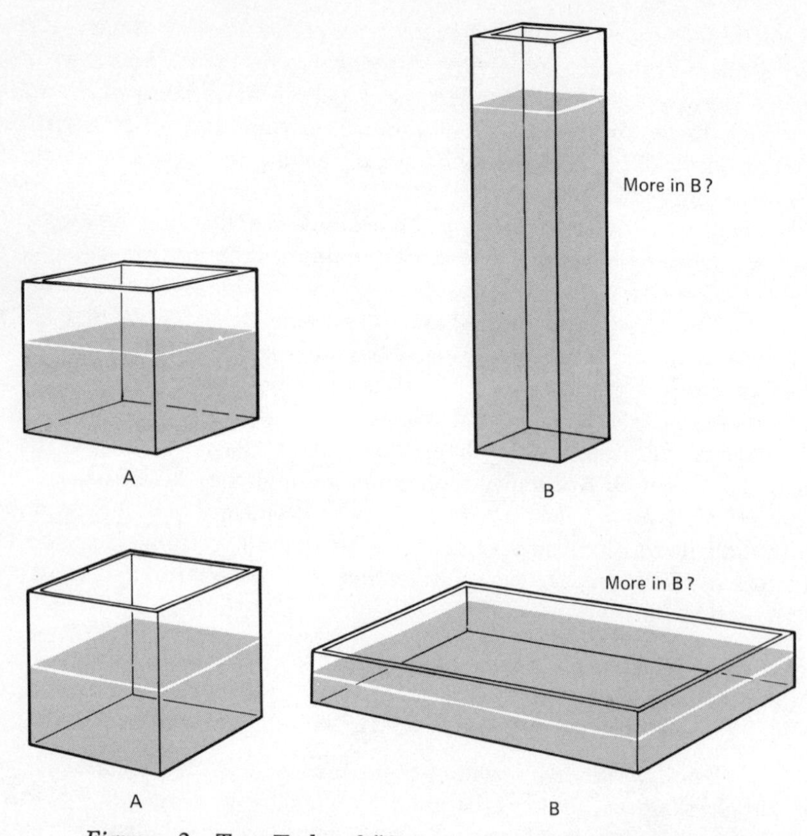

Figure 2 Two Tasks of "Conservation of Liquid" of the Sort Used by Piaget and Other Investigators.

believe such a learning sequence can occur, and perhaps sometimes does occur, in children uninstructed in mathematical concepts of volume. Choosing this particular sequence, then, has the advantage of application to children who are more like those on whom Piaget and others have tried the task. But let it be clear that it is by no means the *only* learning route to the performance of this task. There must be at least several such sequences, and obviously, one of them is that which *does* approach the final performance through the multiplication of measured quantities.

The first subordinate learning that the child needs to have learned is the rule that volume of a liquid (in rectangular containers) is determined by length, width, and height. A change in any of these will change volume. This means that the child knows that any perceived change in any of these dimensions means a different volume. Going down one step in the learnings required, we find three rules about compensatory changes in two dimensions when another dimension remains constant. That is, if the height of a liquid remains the same in two different

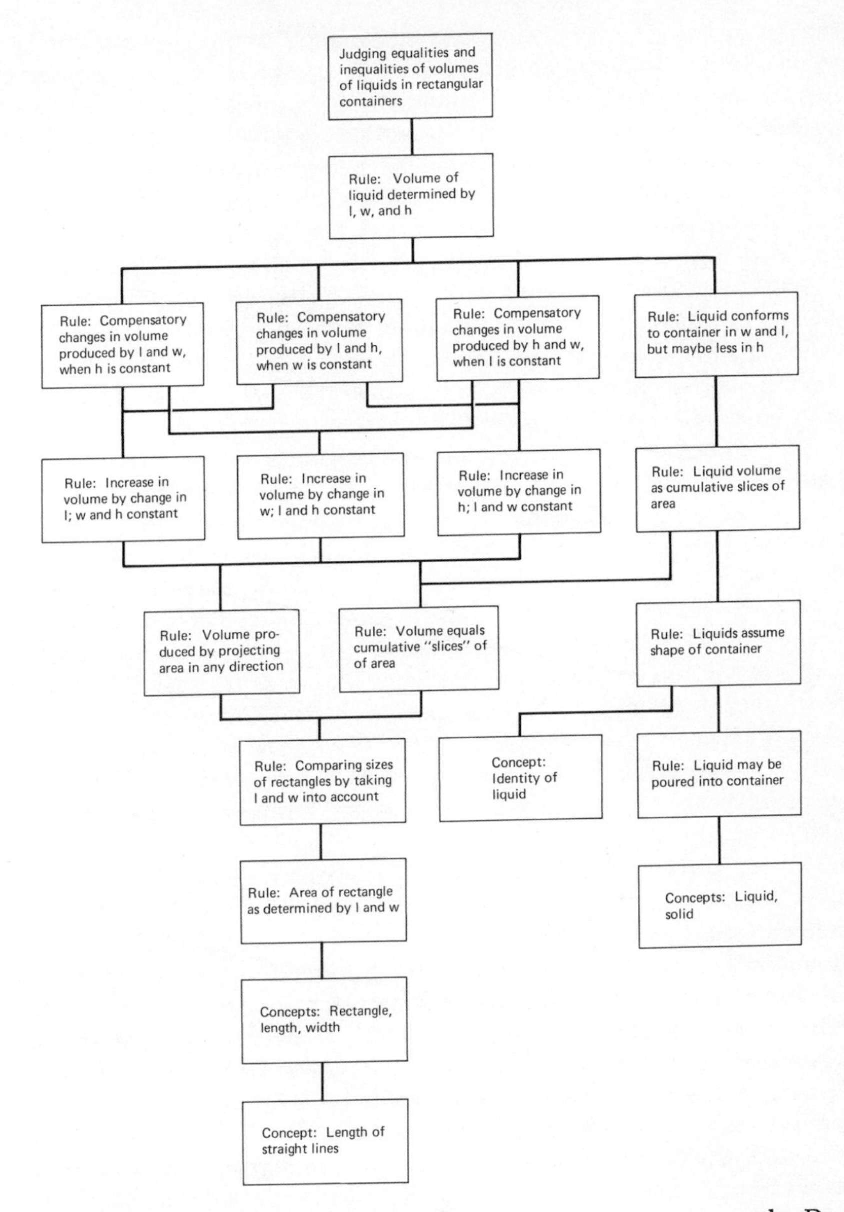

Figure 3 A Cumulative Learning Sequence Pertaining to the Development of Nonmetric Judgments of Liquid Volume.

containers one can have the same volume if a change in width is compensated by a change in length. Similarly for the other instances of compensatory change.

Now, in order for a child to learn these compensatory rules, the model says, he must have previously learned three other rules, relating to change in only one dimension at a time. For example, if length is increased while width and height remain constant, volume increases. Again, similarly for the other single dimensions. These rules in turn presuppose the learning of still other rules. One is that volume of a container is produced by accumulating "slices" of the same shape and area; and a second is that volume can be projected from area in any direction, particularly, up, to the front or back, and to the right or left. Finally, one can work down to considerably simpler rules, such as those of comparing areas of rectangles by compensatory action of length and width; and the dependence of area upon the dimensions of length and width. If one traces the development sequence still farther, he comes to the even simpler learned entities, concepts, including rectangle, length, width, and an even simpler one, the concept of length of a straight line.

Just to complete the picture, the model includes another branch which has to do with liquids in containers, rather than with the containers themselves, and which deals on simpler levels with rules about liquids and the concept of a liquid itself. This branch is necessary because at the level of more complex rules, the child must distinguish between the volume of the liquid and the volume of the container. Of particular interest also is the concept of liquid identity, the recognition by the child that a given liquid poured into another container is still the same liquid. Such a concept may fairly be called a "logical" one, as Piaget does. Bruner (1966c) presents evidence tending to show that identity of this primitive sort occurs very early in the child's development, although its communication through verbal questions and answers may be subject to ambiguities.

Having traced through the "stages" in learning which the model depicts, let me summarize its characteristics as a whole, and some of their implications.

1. First, it should be pointed out that this model, or any other derived in this manner, represents the hypothesis-forming part of a scientific effort, not the verification part. This specific model has not been verified, although it would seem possible to do so. In the process of verification, it is entirely possible that some gaps would be discovered, and this would not be upsetting to the general notion of cumulative learning.

2. According to this way of looking at development, a child has to learn a number of subordinate capabilities before he will be able to learn to judge equalities of volume in rectangular containers. Investigators who have tried to train this final task have often approached the job by teaching one or two, or perhaps a few, of these subordinate capabilities, but not all of them in a sequential manner. Alternatively, they may have given direct practice on the final task. According to the model, the incompleteness of the learning programs employed accounts for the lack of success in having children achieve the final task.

3. In contrast to other developmental models, some of them seemingly based on Piaget's, the cumulative learning model proposes that what is lacking in children who cannot match liquid volumes is not simply logical processes such as "conservation," "reversibility," or "seriation," but concrete knowledge of containers, volumes, areas, lengths, widths, heights, and liquids.

Generalization and Transfer

There is still another important characteristic of a cumulative learning model remaining to be dealt with. This is the fact that any learned capability, at any stage of a learning sequence, may operate to mediate other learning which was not deliberately taught. Generalization or transfer to new tasks, and even to quite unanticipated ones, is an inevitable bonus of learning. Thus the child who has been specifically instructed via the learning sequence shown in the previous figure has actually acquired a much greater learning potential than is represented by the depicted sequence itself.

Suppose, for example, we were to try to get a child who had already learned this sequence to learn another requiring the matching of volumes in cylindrical containers. Could he learn this second task immediately? Probably not, because he hasn't yet learned enough about cylinders, volumes of cylinders, and areas of circles. But if we look for useful knowledge that he *has* acquired, we find such things as the rule about liquids assuming the shapes of their containers, and the one about volumes being generated by cumulative "slices" of areas. The fact that these have been previously learned means that they do not have to be learned all over again with respect to cylinders, but simply recalled. Thus a cumulative learning sequence for volumes of liquids in cylinders could start at a higher "stage" or "level" than did the original learning sequence for rectangular containers. Cumulative learning thus assumes a built-in capacity for transfer. Transfer occurs because of the occurrence of specific identical (or highly similar) elements within developmental sequences. Of course, "elements" here means rules, concepts, or any of the other learned capabilities I have described.

It will be noted that the final tasks of the developmental sequences I have described are very specific. They are performances like "matching volumes in rectangular containers." Does the existence of transfer imply that if enough of these specific tasks are learned, the child will thereby attain a highly general principle which might be called *substance conservation?* The answer to this question is "no." The model implies that an additional hierarchy of higher-order principles would have to be acquired before the individual might be said to have a principle of substance conservation. Transferability among a collection of such specific principles will not, by itself, produce a capability which could be called the principle of substance conservation, or the principle of conservation.

What *is* possible with a collection of specific principles regarding conservation, together with the transfer of learning they imply, is illustrated in Figure 4.

Suppose the learner, making use of transfer of learning where available, has acquired all four of the specific conservation principles shown in the bottom row—dealing with conservation of number, conservation of liquid volumes in both rectangular and cylindrical containers, and conservation of solid volumes. Others could be added, such as conservation of weight, but these will do for present purposes. The property of learning transfer makes possible the ready acquisition of still more complex principles, such as the example given here—judging the volumes of liquids in irregularly shaped containers. It is easy to see that by *combining* the principles applicable to volume of rectangular containers, and others applicable to cylindrical containers, a learner could easily acquire a capability of estimating volumes of irregularly shaped containers. Other kinds of combinations of previously acquired knowledge are surely possible. As I have pointed out, this is the kind of generalizing capability made possible by the existence of learning transfer.

In contrast to this new entity in the developmental sequence, an external observer may, if he wishes, look at the *collection* of what the individual has learned about conservation, and decide he will call this collection the principle of conservation. An external observer is perfectly capable of doing this, and he may have legitimate reasons for doing so. But what he achieves by so doing is still an abstraction which exists in his mind, and not in the mind of the learner. If the external observer assumes that because he can make this classification of such an entity as a principle of conservation, the same entity must therefore exist as a part of the learner's capabilities, he is very likely making a serious mistake. The learner has only the specific principles he has learned, along with their potentialities for transfer.

I believe that many of the principles mentioned by Piaget, including such things as reversibility, seriation, and the groupings of logical operations, are abstractions of this sort. They are useful descriptions of intellectual processes, and they are obviously in Piaget's mind. But they are not in the child's mind.

Another example of how such abstractions may be useful for planning instructional sequences, but not as integral components of intellectual development, may be seen in exercises in science for elementary school children, titled *Science—A Process Approach*, developed by the Commission on Science Education of the American Association for the Advancement of Science (1965). One of the processes these exercises intend

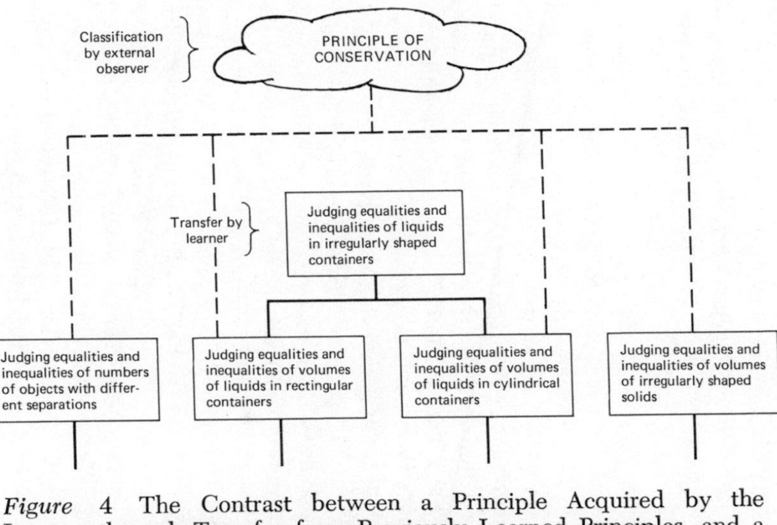

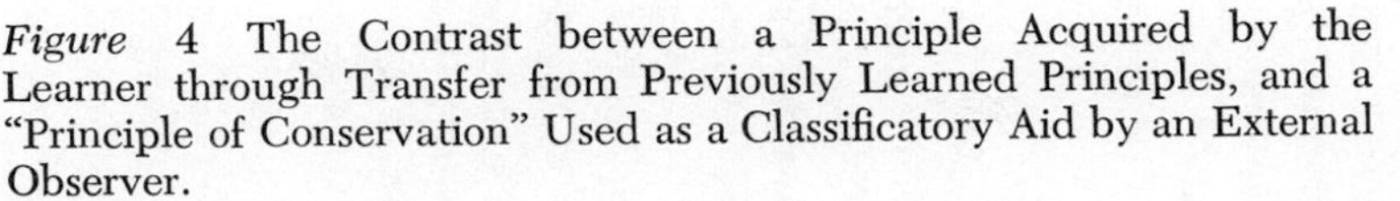

Figure 4 The Contrast between a Principle Acquired by the Learner through Transfer from Previously Learned Principles, and a "Principle of Conservation" Used as a Classificatory Aid by an External Observer.

that young school children learn is called Observation. But it would be incorrect to think that the designers of this material believe that something like the Principle of Observation is to be directly taught to children as an intellectual entity. Observation in this case is an abstraction, which exists in the minds of the designers, but not in the minds of children. What the children *do* learn is a rather comprehensive collection of specific capabilities, which enable them to identify several fundamental properties of the world of objects—tastes, odors, sounds, the solid-liquid distinction, color, size, shape, texture, as well as changes in these. Each is a fairly specific capability, applying to a class of properties only one step removed in abstraction from the objects themselves. At the same time, transfer of learning makes it possible for the child to build upon these things he has learned, and to learn to identify objects or changes in them in a manner which requires the use of several senses at once.

These instructional materials make it clear that the specific capabilities of observation are considered to have transfer value to other kinds of things which are learned later on—to classifying and measuring and predicting and inferring, as well as to other activities involved in scientific experimentation. Transfer of these specific capabilities takes place in many ways and in many directions. But the processes themselves are not acquired as a part of the child's mental constitution. They are merely external names for a collection of capabilities, as well as for the developmental sequences on which these are built.

Returning to the general theme, it should be clear that the various kinds of capabilities that children learn cumulatively, despite their relative specificity, provide a totality of transferable knowledge that is rich in potentialities for further learning. New combinations are possible at any time between principles acquired, let us say, in a context of containers of water, on the one hand, and in the very different context of exchanges of money, on the other. Furthermore, it is recognized that such generalizations can readily occur when the individual himself initiates the intellectual activity; the new learning does not have to be guided by external instruction. The process of cumulative learning can involve and be contributed to by the operations of inductive and deductive thinking. The cumulative learning model obviously does not provide a theory of thinking; but it suggests the elements with which such a theory might deal.

Summary

What I have attempted to describe is a model of human intellectual development based upon the notion of cumulative learning, which contrasts in a number of respects with developmental theories whose central theme is maturational readiness, as well as with those (of which the best known is Piaget's) of cognitive adaptation. It is a model which proposes that new learning depends primarily upon the combining of previously acquired and recalled learned entities, as well as upon their potentialities for transfer of learning.

As for the entities which are learned, the model assumes that complex principles are formed from combinations of simpler principles, which are formed by combining concepts, which require prior learning of discriminations, and which in turn are acquired on the basis of previously learned chains and connections. The "stage" in which any individual learner finds himself with respect to the learning of any given new capability can be specified by describing (*a*) the relevant capabilities he now has; and (*b*) any of a number of hierarchies of capabilities he must acquire in order to make possible the ultimate combination of subordinate entities which will achieve the to-be-learned task. In an oversimplified way, it may be said that the stage of intellectual development depends upon what the learner knows already and how much he has yet to learn in order to achieve some particular goal. Stages of development are not related to age, except in the sense that learning takes time. They are not related to logical structures, except in the sense that the combining of prior capabilities into new ones carries its own inherent logic.

The entities which are acquired in a cumulative learning sequence are relatively specific. They are specific enough so that one must specify them by naming the class of properties of external objects or events to which they will apply. At the same time, they possess great potential for generalization, through combination with other learned entities by means of a little understood, but nevertheless dependable, mechanism of learning transfer.

This kind of generalization through learning transfer is internal to the learner, and thus constitutes a genuine and measurable aspect of the learner's intellectual capability. Another kind of generalization is not necessarily a part of the learner. This is the classification an external observer may make of a collection of learned capabilities. While the observer naturally has the capability of making such a generalization (and often does so), the learner may not have such a capability. Thus, an external observer may classify a collection of learner capabilities as "the conservation principle," or "the principle of reversibility." Such abstractions have a number of uses in describing intellectual capabilities. Because they are so described, however, does not mean that the learner possesses them, in the same sense that the external observer does.

Intellectual development may be conceived as the building of increasingly complex and interacting structures of learned capabilities. The entities which are learned build upon each other in a cumulative fashion, and transfer of learning occurs among them. The structures of capability so developed can interact with each other in patterns of great complexity, and thus generate an ever-increasing intellectual competence. Each structure may also build upon itself through self-initiated thinking activity. There is no magic key to this structure—it is simply developed piece by piece. The magic is in learning and memory and transfer.

part two

PERCEPTION

N. L. Gage observes that the term *perception* is often used to refer to a number of different processes which vary according to several continua. For example, the amount of complexity within a perception varies from simple dimensional responses to brightness or pitch, to the perception of a class of events or objects, to the perception of complex meanings and values. Our speed of perceptual response also may vary from a slow estimation of an indistinct object to the rapid discernment of a well-defined object. Further, our individual differences in perception may be small or very great; we all respond to illusions in roughly the same way but respond very differently to such projective tests as the Rorschach. As a final example, some of our perceptual experiences may be regarded as part of an ongoing process, while others may be seen as results of those same processes. Inevitably, controversies arise among theorists over the use of the term *perception* when it is used to describe any of these phenomena in terms of these or similar continua.

William O'Neil's essay, "Basic Issues in Perceptual Theory" defines perception in terms of five philosophical issues, and then classifies several theories of perception with respect to these issues. Although this essay was written more than a decade ago, his five issues continue to divide theorists in all three categories of cognition. For example, the issues as to whether perceiving is an active or passive process has its parallel in the category of thinking. To paraphrase Kessen, the essentially passive child, encompassed by a stable reality, differs greatly from the child seen facing a world of chaos from which he must construct a coherent view of reality.

The second essay in Part Two, Al Dick's "Perception as Information Processing: a Stage-Analysis," is an attempt to analyze the complex behaviors involved in the act of perceiving. Dick relates the study of information processing to O'Neil's framework of issues. He then describes a three-stage conceptual model of perception and elaborates upon the characteristics and functions of each stage.

William Schiff, in his essay "The Comparative Study of Sensory and Perceptual Processes," points out that these processes are most fruitfully studied with both ecological and evolutionary factors clearly in mind. Schiff's essay indicates how theory and methodology together limit what we observe of behavior.

Complementary essays by Soltis and Wohlwill discuss difficulties which complicate our efforts to relate the study of perception to other cognitive processes. Soltis' essay, "The Language of Visual Perception" is a philosophical analysis of the difficulties involved in defining how we see and how we recognize and identify what we see. Wohlwill, in his "From Perception to Inference: A Dimension of Cognitive Development," presents a psychological comparison of four ways in which perception has been related to thinking. Wohlwill concludes his essay by postulating three dimensions along which perception and thinking can be related.

The fifth essay in Part Two is Eric Aronson and Edward Tronick's "Implications of Infant Research for Development Theory." These authors direct our attention to research with respect to infants' spatial organization of experience. They discuss and evaluate the theoretical implications of recent perceptual studies, especially with respect to Piaget's theory.

The final essay in Part Two is William Epstein's "Developmental Studies of Perception." Epstein lists criteria for well-designed developmental studies and then presents examples of research which met these criteria. He notes that since most research does not meet these criteria, we are not in a position to make many authoritative statements about perceptual learning or development.

Some Topical Readings

SURVEY OF THEORETICAL POSITIONS

Bevan, William, "Perception, Evolution of a Concept," *Psychol. Rev.*, vol. 65 (1958), pp. 34–55.

Haber, Ralph, "Introduction" to *Contemporary Theory and Research in Visual Perception.* New York: Holt, Rinehart and Winston, Inc., 1969.

Hochberg, Julian, "Nativism and Empiricism in Perception" in L. Postman, *Psychology in the Making*. New York: Knopf, 1961.

COMPLEXITY OF PROCESSES

Garner, W. R., "The Stimulus in Information Processing," *American Psychologist*, vol. 65 (1970), pp. 350–358.

Gibson, James, *The Senses Considered as Perceptual Systems.* New York: Houghton Mifflin, 1966.

COMPARATIVE STATUS

Gibson, Eleanor, *Principles of Perceptual Learning and Development.* New York: Appleton, 1969.

INTERRELATIONSHIPS

Neisser, Ulrich, *Cognitive Processes.* New York: Appleton, 1967.

INFANT RESEARCH

Cratty, B., *Perceptual and Motor Development in Infants and Children.* New York: Macmillan, 1970.

LATER RESEARCH

Epstein, W., *Varieties of Perceptual Learning.* New York: McGraw-Hill, 1967.

Additional Readings in Perception and Perceptual Learning

Armstrong, D. M., *Perception and the Physical World.* London: Routledge 1961.

Arnheim, R., *Visual Thinking.* Berkeley, Calif.: University of California Press, 1969.

Attneave, F., *Applications of Information Theory to Psychology.* New York: Holt, Rinehart and Winston, Inc., 1959.

Baltes, P., and L. R. Goulet, *Theory and Research in Life-Span Developmental Psychology.* New York: Academic Press, 1970.

Bevan, W., "Perceptual Learning, An Overview," *J. Gen. Psychology,* 64 (1961), pp. 69–99.

Broadbent, D. E., *Perception and Communication.* New York: Pergamon, 1958.

Bruner, J., *On Going Beyond the Information Given.* Cambridge, Mass.: Harvard University Press, 1957.

Bruner, J., "On Perceptual Readiness," *Psycholog. Rev.,* 64 (1957), pp. 123–157.

Bruner, J., *Processes of Cognitive Growth: Infancy.* Worcester, Mass.: Clark University Press, 1969.

Brunswick, E., *Perception and Representative Design of Psychological Experiments.* Berkeley, Calif.: University of California Press, 1956.

Chisholm, R., *Perceiving, A Philosophical Study.* Ithaca, New York: Cornell University Press, 1957.

Deutsch, C., "Environment and Perception" in *Social Class, Race and Psychological Development.* New York: Holt, Rinehart and Winston, Inc., 1968.

Drever, J. D., "Perceptual Learning," *Annual Review of Psychology,* Palo Alto (1960).

Forgus, R., *Perception: Basic Process in Cognitive Development.* New York: McGraw-Hill, 1966.

Gibson, E., "Perceptual Development," *NSSE Yearbook,* 1963.

Gibson, E., *Principles of Perceptual Learning and Development.* New York: Appleton, 1969.

Gibson, E., and V. Olum, *Methods of Studying Perception in Children.* P. Mussen, ed., Handbook of Research Methods in Child Development. New York: Wiley 1960.

Gibson, J., and E. Gibson, "Perceptual Learning, Differentiation or Enrichment," *Psycho. Rev.,* (1955), pp. 32–41.

Haber R., *Contemporary Theory and Research in Visual Perception.* New York: Holt, Rinehart and Winston, Inc., 1968.

Haber, R., *Information-Processing Approaches to Visual Perception.* New York: Holt, Rinehart and Winston, Inc., 1969.

Hanson, N. R., *Patterns of Discovery.* New York: Cambridge University Press, 1958.

Hebb, D. O., *The Organization of Behavior.* New York: Wiley, 1969.

Hershenson, M., "Development of the Perception of Form," *Psychological Bulletin,* 67, no. 5 (1967).

McCleary, R., *Genetic and Experiental Factors in Perception.* Skokie, Ill.: Scott, Foresman, 1970.

Merleau-Ponty, M., *Phenomenology of Perception.* London: Routledge, 1962.

Murphy, S., and J. E. Hockberg, "Perceptual Development: Some Tentative Hypotheses," *Psychol. Rev.,* 58 (1951), pp. 332–347.

Pastore, N., "An Examination of the Thesis that Perception is Learned," *Psychol. Rev.,* 63 (1956), pp. 309–316.

Phaup, M., and W. Caldwell, "Perceptual Learning: Differentiation and Enrichment of Past Experience," *Journal of General Psychology,* 60 (1959), pp. 137–147.

Piavio, A., "Mental Imagery in Associative Learning and Memory," *Psychol. Rev.,* 76 (1970), pp. 241–263.

Popper, K., *Conjectures and Refutations.* New York: Basic Books, 1967.

Postman, L., "Perception and Learning," in Koch, S., *Psychology, A Study of a Science,* Vol. V.

Rock, I., *Nature of Perceptual Adaptation.* New York: Basic Books, 1966.

Segall, M. H., *et al., Influence of Culture on Visual Perception.* Indianapolis: Bobbs-Merrill, 1966.

Smith, K. U., and W. M. Smith, *Perception and Motion.* Philadelphia: Saunders, 1962.

Taylor, J. G., *The Behavioral Basis of Perception.* New Haven: Yale University Press, 1962.

Tighe, L. S., "Discrimination Learning: Two Views in Historical Perspective," *Psychological Bulletin,* 6 (1966), pp. 353–370.

Trabasso, T., and G. Bower, *Attention in Learning: Theory and Research.* New York: Wiley, 1968.

Uhr, L., *Pattern Recognition.* New York: Wiley, 1965.

Vernon, M. D., *Psychology of Perception.* Baltimore: Penguin Books, 1962.

von Senden, M., *Space and Sight.* New York: Free Press, 1960.

Wapner, S., and H. Werner, *Perceptual Development.* Worcester, Mass.: Clark University Press, 1957.

White, B., "Research on Attending Processes in Learning," Progress Report, USPHS Grant, M3689, 1962.

White, S., "Evidence for a Hierarchical Arrangement of Learning Processes," in L. Lipsitt and C. Spiker, eds., *Advances in Child Development and Behavior.* New York: Academic Press, 1965.

Wohlwill, J., "Definition and Analysis of Perceptual Learning," *Psychol. Rev.,* 65 (1958), pp. 383–395.

Wohlwill, J., "Developmental Studies of Perception," *Psychological Bulletin,* 57 (1960), pp. 249–288.

Wohlwill, J., "Perceptual Learning," *Annual Review of Psychology,* Palo Alto, (1966).

BASIC ISSUES IN PERCEPTUAL THEORY*

W. M. O'Neil

An attempt to classify modern theories of perception will be used to bring out some major points at issue between them. Some of these issues are epistemological but, as the psychologists' attempts to shelve such questions as nonpsychological have proved unavailing, it is better to face them without too much apology.

Five bases of classification are proposed. These interpenetrate yielding cells, as it were, in a five-dimensional solid. Though the five bases are logically independent, there is an historic tendency for the position taken on one to be associated with that taken on another. Thus some of the cells are recipients of a null class. This fact enables a simpler classification, which nevertheless is best approached through the more elaborate one. Incidentally, the scheme being proposed readily accommodates some other distinctions between types of perceptual theory offered by Koffka (1935), Brunswik (1939), Bruner and Postman (1949), and Boring (1952a).

Perceiving as an Active or as a Passive Process. The issue is whether perceiving is something done by or something done to the organism. Those who think of it as apprehension, or as the attainment of knowledge in the course of the organism's coping with its surroundings, take an activist position, whereas those who think of it in terms of the reception of stimulation, or as the input of information, are passivist. It is improbable that any contemporary theorist is as thoroughgoing a passivist as was, say, James Mill (1878). With the common agreement that perception is selective, there is an admission that the organism is at work on the available stimuli or information. Nevertheless,

---

* W. M. O'Neil, "Basic Issues in Perceptual Theory," *Psychological Review*, vol. 65, no. 6 (1958).

there is a strong passivist tendency amongst the more behavioristically minded of the contemporary theorists. After Titchener there has not been any strong tendency of this sort amongst the phenomenalists. Within the activist theories there are some important differences. Primary among them is the treatment of perception as reactive or as proactive, to use Murray's terms (1951). Those who regard perception as a discriminative response and who stress the impact of needs, motives and the like upon this response are reactivists. The more thoroughgoing purposivists, rarer in perceptual theory than in personality theory, it must be admitted, are proactivists.

The Perceived as a Real or as a Phenomenal Object. The issue, an essentially epistemological one, is whether we perceive a real or externally existing object or a representation (or some corresponding impression) of it. The man in the street, who is a naive realist, believes that when he sees a door, he does in fact see an object existing in the external world. A sophisticated phenomenalist such as Scheerer (1954) maintains that a door existing in the external world reflects patterns of light energies which stimulate the retina, thus giving rise to neural processes which in turn produce a unit in the phenomenal field which represents the door. Though Scheerer and others of similar persuasion differ from Johannes Müller in many important ways, they agree to the principle in his statements: "Sensation consists in the sensorium receiving through the medium of the nerves, and as the result of the action of an external cause, a knowledge of certain qualities or conditions, not of external bodies, but of the nerves of sense themselves" (1842, p. 1065) and "The immediate objects of the perception of our senses are merely particular states induced in the nerves, and felt as sensations" (1842, p. 1073).

Early experimental psychology inherited a clear phenomenalist tradition from both sides of its family tree. Indeed, the analytic phenomenalism of British associationism blended quite readily with the analytic phenomenalism of the sense-physiologists such as Bell, Müller, and Helmholtz. The rise of realism in British philosophy in the early years of this century gave some special twists to the tradition, but it was American realism, especially working through behaviorism, which rejected it. However, the behaviorists largely vacated the field of perception. Their prime interest was in effector-based responses, and any interest they had in sensory processes was limited in the main to questions of sensitivity. There was an early promise, made first by the philosopher Perry (1912), to give a full treatment of perception in terms of discriminative response, but only recently have serious efforts at fulfilment been evident. Thus it has been left largely to the phenomenalists to cultivate the greater part of this field.

The problem of error in perception or, put the other way, of veridical perception is a genuine one for psychology as well as for epistemology. All too often the psychologist with phenomenalist convictions regards it as not his concern to make out how the organism attains knowledge of the external world; this includes, of course, the scientist attaining knowledge about his subject matter. As a psychologist he is ready enough to confine the organism to an acquaintance with phenomena, yet as an organism theorizing about perception he presumes an acquaintance with both appearance and reality.

Two phenomenalists as different as Titchener and Koffka used the fact of error in perception to distinguish the arena of psychology from that of physical science. Titchener (1909) analyzed visual and other illusions in order to illustrate his distinction between experience viewed as independent of and as dependent upon an experiencer. Koffka (1935) made his distinction between the geographical and the behavioral environment the more convincing by his description of the dog bounding into a snow-filled crevasse which he took to be solid snow-covered earth. In general the phenomenalists make out a persuasive case that we perceive not real things but appearances, and that we respond to things not as they are but as they appear to be. These propositions, however, make it impossible to deal with or to recognize the fact of error which had so large a part in leading to them. If it were the case that we know only phenomena, there would be no way for us to get to know about error. Only if we truly know the real on some occasions can we tell that on others we have fallen into error. Within the realist position it is not difficult to show how the fact of error in perception can be recognised, though precisely which piece of knowledge is fallacious may be quite difficult to decide. Were realism to make out that such a decision was easy, it would be embarrassed by the fact that scientific progress is so difficult. In accordance with the formula about the whole truth and nothing but the truth, we may distinguish errors of omission and errors of commission. The former is quite easy for the realist to handle. For instance, in terms of perception as a discriminative response, it may be said that the organism has failed to respond to the object or that it has failed to respond differently to different objects. The error of commission is illustrated by the situation where we seem to have an object of such and such properties before us and yet where other, especially later, experience convinces us that there was no such thing confronting us. The realist, content with the discriminative response view as it is usually formulated, has no answer to the phenomenalist's question: "Where and what was the apparent object?"

The Perceived as a Term or as a Proposition. The distinction being made is a logical and not a verbal one. A term is any property, such as what the word "red" means, or any set of properties, such as what the words "the red piece of paper" mean. A proposition is the predication of one term to another. The words "the red piece of paper is on a gray background" express or symbolize a proposition. In the behaviorist theories, as well as in most phenomenalist theories of perception, the perceived object is treated as a term. Thus the organism discriminates the red paper from its gray background or the perceiver experiences the field as a red paper figure on a gray ground. But if we consider perception as attaining knowledge, at any rate what James (1890) meant by *knowledge about*, we need to think of what is known as having propositional form. Facts are not in the ordinary sense things or terms; they are not what can be expressed by means of nouns, adjectives, verbs, or adverbs alone. Facts are what we need sentences to talk about, and knowledge of a fact is judgmental. Those adopting the propositional view have been in recent times few in number and almost always no more than implicit adherents. There being so many who have treated the perceived object as a term, a much greater diversity amongst them can be expected.

Three major conceptions of the percept can be distinguished amongst the phenomenalists. At one end are to be placed those whom Boring (1952a) calls the *Leipsigers*, and at the other are his *Berliners*. The *Grazers*, kept company by Spearman and a large band of functionalists, are located somewhere in between. The Leipsigers, especially in the purified form of Titchener, agree with the radical behaviorists in being satisfied with simple discrete substantial terms. Titchener's sensations, despite his talk of context, are just as simple and discrete as Watson's stimuli, despite his talk of situations. Titchener had no way of including relationships within the experienced, and a behaviorist who conceives stimuli either as energy sources or as energies is equally handicapped. The earnest attempts of the Hullians to explain transposition phenomena in terms of stimulus generalization show the distaste for the inclusion of relationships among the stimuli. The Grazers found a place in the percept for relations, but did so by regarding them as imported into or imposed upon the situation by the perceiver. Both they and the Berliners agree in admitting relations as well as qualities in the percept. But the two disagree upon how relations get there. For the Berliners they are not the result of some "production process," but are as primary as the qualities related.

There was a distinct inclination amongst the *Akt* psychologists to adopt a propositional view. Brentano's distinction between the "lower" act of ideation and the "higher" act of judgment included a recognition that the object of the latter was propositional. Stout was quite explicit when he said:

> In general, it would seem that a complete object [of consciousness] can be adequately described in language not by isolated words but by prop-

ositions capable of being asserted, denied, doubted or assumed. . . . When we are said to perceive or know something this may be taken to express the fact that we perceive or know that something is the case (1929, p. 100).

More recent formulations have not been as explicit. But the recent analyses by MacCorquodale and Meehl (1953, 1955) and Meehl and MacCorquodale (1951) reveal that expectancy in Tolman's theorizing is propositional in form. Insofar as expectancy is equivalent to hypothesis, as used by Tolman and Brunswik (1933), and as later adopted by Bruner (1951) and Postman (1951), all four theorists may be claimed to adopt a propositional view. Helson (1951), in stating his adaptation-level theory as a general theory of perception, gives some signs of subscribing to this view too.

Descriptive Versus Abstractive Modes of Analysis. In offering a characterization of something in a way which distinguishes it from other things, we may resort either to descriptive or to abstractive terms. The former are features of the object actually before us when we inspect it, features that stand up for inspection, whereas the latter are not literally present in the form in which our analysis might suggest they are. The Wundtians recognized most of the time that the elementary sensations with which they characterized a given percept were in fact abstractions. A pure sensation, as they conceived it, is not something that we ever experience. Likewise, the dimensions of experience which have been derived by Boring (1933) from the later Titchenerian doctrine of sensory attributes (1929) are not components of experience, but the dimensions of a reference frame within which some particular experience may be located. On the other hand, features like figure-and-ground, contour, and gradient of texture are deemed by those referring to them to be the very material or structure of our experience as we have it. They are descriptive terms. Wundt, the early Titchener, and the Leipsigers generally preferred elemental abstractive analysis; the later Titchener, followed by Boring and Stevens and their associates, have adopted a dimensional abstractive analysis; David Katz, Rubin, the Berliners, Gibson and the phenomenologists generally have employed a descriptive and usually structural analysis.

Preferred Location of Causal Conditions. In seeking to give a causal account of perception, the early theorists stressed proximal stimulus conditions. Later there was a considerable swing to distal stimulus conditions, as Brunswik (1939, 1952) pointed out. But both place the emphasis upon what Bruner and Postman (1949) call autochthonous factors, whereas these two, in concert with many others in the last decade or two, have emphasized needs and other related "central" factors. There is, of course, another kind of "central" factor represented by the brain fields of Köhler and Wallach (1944) and of Krech (1950a and b), and by the cell assemblies of Hebb (1949). The classification of perceptual theories contained in Allport's treatise (1955) is based primarily upon the determinants stressed in the theory.

It is appropriate at this point to comment upon the unfortunate tendency of psychologists to identify causes with their effects, or to assimilate them. Werner and Wapner (1952), in arguing that if both sensory and motor (tonic) processes interact they must be of the one kind, provide an example of assimilation. The widespread resort to dispositional concepts such as attitudes and sets usually involves an identification. No more is said about them than that they have specified observed effects and that they differ from one another in the effects they each have. Operational or dispositional statements are well enough, but the trouble arises when the need for anything more is not felt. It is well enough to identify opiates by means of their soporific effects, but it is misleading to identify them with those effects. An interesting example of this sort of identification may be found in the instructive controversy between Boring (1952a and b) and Gibson (1952). Boring, supposing for argument's sake our planet to be more liberally endowed with satellites, asks what we would be perceiving when we perceived the size of a moon. He is not a naive realist, and so he is not satisfied by the reply "Why, the size of the moon we happen to be looking at!" His own studies of the apparent size of our one and only pre-Sputnik moon led him to conclude that major determinants of such a perception are the actual distance, size, and elevation of the moon, the perceived distance of the moon, the actual elevation of the perceiver's regard, and the perceiver's attitude. He goes on, quite reasonably, to suppose that these values may be used as the parameters of some function, ø, which would be invariant when the perceived size of the moon is invariant. So far, to adopt a Boring phrase, this is good science. But it is probably only a confusion of cause and effect when he says ". . . then, in judging size, you are perceiving not object size, not retinal size, but ø. To discover the object of perception, you have to discover what function of the stimulus is invariant when the perception is invariant" (1952a, p. 146).

A Threefold Classification

The historical association of views on these five basic issues has been such that a workable division into discrimination theories, phenomenalist theories, and judgmental theories is possible. No one of these is a compact cluster, so that in any general discussion of them some exceptions have to be made. The second cluster is indeed a clear doublet, the rather tough-minded abstractive theorists standing somewhat apart from the more tender-minded phenomenologists. In the sections below each of these types of theory will be characterized in general terms. Attention will be paid to the relative strengths and weaknesses of the discrimination and the phenomenalist theories. Finally, as part of an attempt to bring out what has usually been left implicit in judgmental theories, it will be argued that this type of theory affords a hope of our being able to retain the best of the other two worlds.

Discrimination Theories

Discrimination theorists are ordinarily slightly on the activist side of center. They are ordinarily realists in that the stimulus (wherever it is to be located, distally or proximally, and however it is to be analyzed) is real in some sense and not phenomenal. They are ordinarily given to abstractive analysis. The stimulus to which the response is made is a term rather than a proposition. There is a marked tendency to pack as much as possible into the stimulus, which is scarcely distinguished conceptually from the cause of the response. Graham (1950), indeed, disposes of determinants such as sets and attitudes by referring them, at least in experimental settings, to instruction stimuli. Boring, whose theorizing is best thought of as moving from the tougher-minded phenomenalism towards discrimination theory, speaks of the perceiver's attitude as one of the stimulus properties which is a parameter of ϕ.

Apart from any disagreements on specific perceptual problems, a few general distinctions can be made amongst the proponents of discrimination theories. Those with the greater physiological interests tend to be more proximally focused, and to look for perceptual determinants largely in the sensory mechanism. Those who have been most influenced by modern positivism tend to have the more sophisticated views about the stimulus, to have a greater "focal length," and to favor dimensional analysis through their concern with stimulus continua. Those who are led to their problems by practical considerations—such as the need to establish the perceptibility of signs, signals, displays, and the like—tend to be empiricist in the rule-of-thumb sense.

This type of theory has many attractions. Foremost amongst these is the scope provided for rigorous theorizing and rigorous experimentation. Another, of less certain value, is the avoidance of the bothersome problem of introspection. As the aim of a perceptual experiment within this theoretical context is the testing of limits of discrimination and the seeking of invariant relations between stimuli and discriminative responses, it does not matter what the responses be, whether verbalization, key-pressing, or making matches; and it does not matter how they be established, whether by conditioning or by instruction stimuli. When a verbal response is used, it does not have to be treated as a report. To treat it as a report the investigator must concern himself with what it alleges and not with its own distinctive features as an event. Such a concern raises the nagging doubts about intersubjective corroboration to which the discrimination theorist seems prone. Their avoidance in this way does not, however, preclude the discrimination theorist from dealing with many problems that have often been deemed to be subjective. For instance, it can be argued that, in establishing a subjective or phenomenal scale of brightness or loudness, the subject is being calibrated as a different kind of instrument for application to some physical variable (*vide* Bergmann and Spence, 1944).

The discrimination theorist also has the advantage of the phenomenalist in that he can offer some coherent account of error, as Perry (1912) and other American realists have demonstrated. However, saying that when in error the organism has made the "wrong" response does not go far enough. It needs to be shown in what way that response is wrong. When the outcome of an animal's having made some response is the survival or the death of that animal, or even when the outcome is a shorter or longer period of delay in obtaining food or a mate, it may be said that the response made was right or wrong. However, there are occasions when an illusion contributes to survival or to a maximization of benefits, but is nonetheless an illusion. What distinguishes it as an illusion is that the seeming knowledge in it is false. This same limitation obtains when this type of theorist attempts an account of knowledge in the sense of our store of knowledge. Apart from the limitation of the dispositional terms employed, for instance, by Hull (1930) and later Berlyne (1954), no way is provided for a distinction between true knowledge and false knowledge.

The advantages gained by the eschewal of introspection must have offset against them some handicaps accruing at the same time. It would appear helpful to know that chromatic colors disappear in scotopia, and that it is not merely a matter of some one chromatic color replacing all other colors. A purely discriminative experiment could tell us no more than that, apart from any intensity differences, stimuli of one wave length are not discriminable from those of another including a mixture of all wave lengths. We could have no indication whether the central processes following different stimuli were the same or whether these processes were functionally equivalent though qualitatively different. The pattern of verbal responses made with the addition of appropriate instruction stimuli would not afford a clue to this unless we regard such responses as alleging what is the case, i.e., as being introspective reports.

Finally, a discrimination theory usually involves a renunciation of what seems to some the genuine and important psychological question expressed by Koffka (1935) as: "Why do things look as they do?" This question seems possible only if we are willing to consider introspective data and make some form of distinction between the real and the phenomenal. A distinction between tape-measured and subject-measured distance, for instance, does not go far enough. But as soon as the discrimination theorist begins to make the necessary admissions he is in danger of losing many of the advantages he had otherwise gained for himself. Berlyne suggests: "If we think of the perceptions which constitute the 'phenomenal world,' etc., as reactions to stimulation from physical energy-changes, and overt behavior in its turn as a reaction to them, we may be able to enjoy some of the benefits of both points of view" (1951, p. 145). However, if perception is taken as a response in the ordinary Hullian sense, there are difficulties arising through there being no effectors to perform it; and if it is taken as a stimulus-producing response or purely theoretical or intervening variable, it is puzzling how we come to know that it has the properties attributed in other theories to the "phenomenal" world.

Phenomenalist Theories

Phenomenalists have a broader spread over the active-passive variable. Contemporary representatives more often than not come down on the activist side, sometimes well out towards the extreme. They are never realists in predominant flavor, although many struggle, as it were, towards realism. Generally speaking, the more distally focused have the stronger realist tendency. In the past they preferred elemental analysis, but now the predominant preference is for descriptive structural analysis. Once the basic classifying and counting had been done by the elementarists, the remaining part of the program appeared tedious and sterile. Phenomenology has afforded a much richer alternative. In both varieties, phenomena are conceived as terms—sensations when the focus is proximal, and objects when it is distal. The favored determinants are stimulus properties (though played down as compared with the discrimination theorists), other phenomena, brain states, and, with the more markedly activist, motives. There is some dispute, however, amongst them as to the legitimacy or profit in using one or other of these. For instance, Gibson (1948) seems mainly concerned to find stimulus correlates and little impressed with the need to locate neural determinants, whereas Werner and Wapner (1952) center their attention upon the total organismic state. The rocks that have seriously split phenomenalism have been the issue of abstractive versus descriptive analysis and the issue of proximal versus distal focus. Though independent issues, those choosing abstractive analysis have usually chosen a proximal focus, and those choosing descriptive analysis have usually chosen a distal focus.

Though an epistemological history of phenomenalism should go back at least as far as Locke's distinction between primary and secondary qualities, a psychological history can appropriately begin with that part of Müller's doctrine of the specific energies of the sensory nerves which stated that we are aware not of external objects, but of the states they produce in our sensory nerves. Within this doctrine a veridical knowledge of external objects would be attained to the extent that there is some parallelism or isomorphism between the external objects and the states they produce. When that parallelism breaks down we have error or illusory perception. But how anything about the parallelism or its absence can be discovered is a puzzle. It is not enough to say, as Müller did, that vision is ordinarily elicited by light, and that when it is elicited by pressure on the eyeball there is an incongruity of stimulus and sensation. To say this assumes that we may know light and pressure on the eyeball as other than phenomena. But even were there two sorts of knowledge—the sort we have in experiencing color and the sort we have in knowing that light is impinging on the eye—it is difficult to show how we would relate one to the other and to say in which realm of knowledge this relating occurred (*vide* Anderson, 1927).

There is an interesting paradox in the phenomenalism of Müller's sort. Our knowledge of our sensory states must in one sense be veridical, i.e., in terms of the doctrine, we know only these states and can not know other than what is present in them. Yet it almost always seems to be knowledge of external objects, which, according to the doctrine, is a nonveridical impression. Further, what is said to be knowledge of the states of our sensory nerves gives us the knowledge we have of external objects and very little, if any, information about the states of our sensory nerves. The way we are able to get such knowledge is to have those states as the external objects of other sensory states. Müller, in effect, told us that it is not the moon that we see but a state of affairs in the visual mechanism, whereas Boring tells us that it is not the moon that we see but a mathematical function. Both seem to be confusing what we see with the conditions of our seeing.

Though the later phenomenologists reject sensory states located at the organismic surface, their objects are not literally *out there*. Their focus, as Brunswik (1952) shows, is not fully distal. The term "intermediate object," which Brunswik used earlier in his work on object constancy, might well be applied with a slightly different meaning to the phenomenal object in these theories. There remains, nevertheless, the impassable gap between the phenomenal and the real. However, the phenomenological approach has the virtue of producing an orderly account of how things look. The now classical work of Katz, Rubin, and Wertheimer and the more recent work of Michotte (1946) and Gibson (1950) is evidence enough of this. Furthermore, such work has produced data which seem amenable to corroboration and which set what has already been said to be a genuine psychological problem.

Judgmental Theories

Perhaps some form of discrimination theory has greatest appeal to the physiologically minded, and some form of phenomenalism to the psychologically minded, especially when in the ranks of the professional physiologists. Judgmental theories have certainly been propounded most often by the philosophically minded. However, if, as will be asserted here, such a theory can accommodate some distinction of the sort attempted in the contrast between the real and the phenomenal, and at the same time preserve some of the advantages of objective reference attained by discrimination theories, it may well be the type of theory the psychologist will find best adapted to his needs.

Unfortunately, we have no contemporary judgmental theory in sufficiently elaborated form to let us prove these hopes. Bruner (1951) and Postman (1951) gave separate statements of their joint thinking which constitutes one promising outline. However, much that is relevant to the claims being made is left implicit in their theorizing. An explication is rendered the more difficult by a more recent paper by Postman (1953) which reveals a distinct swing away towards a discrimination theory. A later paper by Bruner (1957b) extends some aspects of the judgmental treatment but does not relate its terminology to that of the earlier statements. Without asking these theorists to accept any responsibility for what is to be maintained, many things which perhaps they thought best to leave implicit will be made explicit. Then, in the light of that interpretation, some indication will be given of where their theorizing seems to have gone off (or not far enough along) the track favored here. Finally, some promissory notes will be proffered in the absence of hard cash to capitalize the expansion of this type of theory. Bruner, Postman, and their associates have already a great deal of hard cash to the credit of other aspects of their theorizing.

The full course of cognition was said in 1951 to be an hypothesis-information-testing cycle. This conception applied not only to perception but also to thinking such as remembering, imagining, and reasoning. This is an immediate attraction. All too often we are asked to conceive perception in one way and thinking, with all it entails, in so unrelated a way that we remain puzzled as to how what we acquire through perception can be incorporated in our thought.

The central notion is "hypothesis," which in Bruner's later statement seems to be replaced by "category." Hypotheses "serve to select, organize and transform the stimulus information" (Postman, 1951, p. 249); or "stimulus inputs may be sorted, given identity, and given more elaborated, connotative meaning" in terms of categories (Bruner, 1957b, p. 148). Hypotheses are deemed to vary in strength, or categories to vary in accessibility. But what hypotheses or categories are is a matter left a little obscure. Bruner in general was and is content with a dispositional conception—these are we-know-not-whats recognizable only by their determinants or by their effects. Postman went a little further in saying that an hypothesis is an expectation that such and such is so and so, or that such and such does so and so. He relates this to Tolman's notion of expectancy. MacCorquodale and Meehl have argued convincingly that an expectancy in Tolmanian theory has the form "if (S_1 and R), then S_2"; e.g., "if, when the light flashes, the lever is pressed, then a pellet of food falls into the cup." This, of course, is an action hypothesis which is the sort required in a theory of performance. Where the account puts the emphasis upon cognition, the "response component" may be ignored; thus "if S_1, then S_2." Cognition of whatever type is to be regarded then as a predicating (asserting, denying, supposing, doubting and so on) S_2 of S_1; or, if no imputation of conscious deliberation is made, as a judging of S_1 to be S_2. It is important to stress that S_1 and S_2 may be quite complex, and that a distinction between a judgment and the verbal or other expression of it is intended.

The judgment or predication may be true or false, depending upon whether, in fact, occasions of S_1 are occasions of S_2. In illusion there is a false predication. This arises in several ways, one of which is fallacious inference, though the latter is more prevalent in thinking than in perceiving. In an error of omission S_1 is judged to be S_2 alone when in fact S_1 is also S_3, S_4, etc. Thus we must recognise that there are, first of all and independently of our knowledge of them, the facts. These are the empirical content of the term "real" in the traditional distinction. When we judge that S_1 is S_2 and in fact it is, we have attained a knowledge of that fact. When we have attained a knowledge of some of the infinite facts confronting us and have failed, by omission and by commission, to attain the knowledge of others, we have the state of affairs emphasized by the term "phenomenal." To say that we know only the phenomenal is true in the trivial sense that we know only what we know. But it should not be interpreted as having more significance than that. It is demonstrable that, in a rough manner of speaking, the moon looks smaller than it really is. Further, if we assume, as some of our forebears did, that the law of visual angle is sufficient for the explanation of size perception, it may be said, equally roughly, that the moon looks bigger than it ought to. To ask, "What is this too small and too large thing that we see?" is to ask the wrong question. It assumes that there is such a too small and too large thing somewhere. As it is not *out there*, it is then concluded that in some sense it must be *in here* or somewhere in between. The question, "What do we see when we see the size of the moon?" is answerable properly not with "the real moon" (for that gives us no way of coping with the anomaly we have noted), nor with "the retinal image of the moon" (for that is contradicted by experience), nor with "an intermediate or a phenomenal object" (for that is a piece of fabrication which leads us into confusion), but with "that the moon is size X," or "that the moon is the size of X." That this visual judgment is false, and that it varies with the position of the eyes in the head, are matters of great interest to the psychologist—but not because he studies phenomenal moons, leaving the astronomer to take care of the real ones.

Frequently sense perception is regarded as having the particular and the concrete as its content whereas thought, apart from its purely reproductive forms, is regarded as being characteristically concerned with the universal, the conceptual, and the abstract. The psychologists prone to make this distinction seem to have forgotten the long-standing arguments about the nature of universals, and not to have learned that the scholastic positions known as realism, nomalism, and conceptualism do not exhaust the possibilities. Discrimination theorists, when dealing with cognitive matters other than perception, are prone to resort to a sort of nominalism: knowledge of universals is merely making the same response, a symbolic response, to several stimuli. Phenomenalists of the elementarist kind tend to follow a similar line, substituting such phenomena as composite images. Phenomenologists in psychology turn regularly to conceptualism, Goldstein and Scheerer (1941) affording an example singularly rich with scholastic vestiges. Bruner, in his theorizing about categories, reveals an inclination towards conceptualism. Nevertheless, his discussion has the merit of showing that perception is not confined to the particular, and that a concern with the universal is as characteristic of perception as it is of thought. A clearer recognition that all cognition is judgmental would lead to a rejection of both nominalism and conceptualism without forcing one into the mystification of medieval realism. What distinguishes the universal is that it is predicable of many things, a point recognized by Bruner in respect of his categories. However, this is the case with anything that is predicable, even when it is a so-called concrete particular. Some psychologists seem to think that there are some logical considerations requiring a special class of "things" called concepts. The nominalists know that there are no such considerations, but the psychologists among them fail to see that speaking of stimulus and response generalization does not take them far enough in accounting for knowledge. It is not that we confuse one member with another of its class, but that we know each of them to be things of the same sort.

The explication given here of the term "hypothesis" (or "category") reveals a difficulty in giving a qualitative account of hypothesis-strength (or category accessibility) which has an important place in the theory. Though the operational criteria provided especially by Postman suffice for the mounting of an extensive and fruitful experimental program, they are not adequate for a full explanatory use of the term. An hypothesis itself cannot vary in strength; S_1 is simply S_2 but never more or less S_2. No ground is gained by saying that the hypothesis is held more strongly unless it can be said what holding an hypothesis is. To say that the hypothesis is used more or less frequently in certain situations does not explain what hypothesis-strength was meant to explain. The same strictures may be made with respect to habit-strength in Hullian theory. Should habits, now defined dispositionally, turn out to be neural structures, then habit-strength could be some variable property of those structures. A judgment may have a similar neural determinant, in which case hypothesis-strength could be given positive meaning. This is a hope, however, with little factual backing at the moment.

The information and test phases in the cycle also need some clarification. At first sight the information phase is a good deal like what we have all been calling perception. Were further consideration to confirm this impression, then this theory would be not so much an account of perception itself as of the wide context of processes in which it is embedded. Further, too much stress is placed by Bruner and by Postman on stimulation as such, that is, the impact of stimulus energy upon the receptor mechanism. It appears important to recall at this point the distinction made by Harper and Boring (1948) between cues and clues. What is referred to as information should on some occasions be recognized as a nonperceived but sensorily mediated determinant of perception, and on others as a hypothesis (judged state of affairs) relevant through inferential processes to the hypothesis under test. The former type of information is a cue (which is all that is considered in cybernetic theory), whereas the latter is a clue which may be a premise as well as a mere determinant. It may well be that a similar distinction should be made in respect of the test phase of the cycle, yielding something like reinforcement in that brand of learning theory with which discrimination theories are most readily affiliated, and something like inference in Tolman's theory, which is at least implicitly judgmental. The strengthening of an hypothesis is akin to acquisition in learning, and the weakening of an hypothesis is akin to inhibition and extinction.

The distinction between cue and clue may well help with the resolution of the present paradox of size and shape perception. The bidimensional retina can of itself encompass only visual angle and projected shape. Perception of size should thus be dependent upon access to information about distance, and perception of shape upon access to information about inclination. There is a fair amount of evidence to support the former of these suggestions, though little available that is relevant to the latter. Several theorists, e.g., Koffka (1935) and Boring (1952a), have been led to suppose that there must be an invariant relation, to put it in Boring's way, between, in the one case, perceived size, perceived distance, and retinal image size. Now an experiment like that of Holway and Boring (1941) does not bear on this alleged relation, for no distance judgment was obtained. An experiment like that of Gruber (1958) does bear on it; that it falsifies the alleged relationship is ameliorated only by its inherent design defects. Stavrianos (1945) failed, despite a commendable persistence, to show that errors in perceived shape were related to perceived inclination. Her experiment, too, has some basic design defects. Nixon (1958), in a series of studies free of these limitations, has falsified the hypothesized relationship. The puzzle here would be resolved were it to turn out that this is an "overdetermined" situation in the sense that there is more information, not all of which is congruent, than the minimum required for a given judgment of size or shape. Some of this information is in the form of a judgment of distance or of inclination. This is a clue to size or to shape when it is conjoined with information such as retinal size or retinal shape. It may operate as a clue through some inferential process, as Helmholtz would have maintained. In addition, there may be cues to distance (inclination is a matter of relative distances) which on this occasion do not influence the judgment of distance or of inclination, but which may influence the judgment of size or of shape. If ocular torsion is relevant to the moon illusion, it would be just such a cue: the paradox that Gruber leaves was left also by Boring in respect of that illusion, namely, that the changed apparent size of the moon is due to a changed cue to distance, even though the apparent distance has been inversely affected. The way a cue affects a judgment would not be through some inferential process. Inference seems peculiar to discourse or to knowledge. It is not merely a matter of cause and effect. When we speak of cues, no more than a causal relation is being referred to. The nearness of the horizon moon is possibly the result (through inference) of its appearing so large.

Were a considerably modified notion of response developed, one able to embrace what Woodworth (1915) called a "mental" response as well as what he called a "motor" response, the proposal made by Berlyne (1951) would have greater merit. As was pointed out above, a pure stimulus act such as he had in mind will not suffice. Nor, of course, is it enough to say that the perceptual response is to be characterized as discriminative. Little responding, whether cognitive or otherwise, is indiscriminate. The kind of response needed is judging that S_1 is S_2. To fit such a concept on to Hullian theory requires rather more general renovations than Berlyne suggests. However, if it can be done, we should have, as he is hoping to have, the best of the two worlds of discrimination theory and phenomenalism in coping with perceptual problems.

PERCEPTION AS INFORMATION PROCESSING: A STAGE ANALYSIS*

A. O. Dick

> *I can read words of one letter. Isn't that grand? However, don't be discouraged. You'll come to it in time.*
> —*White Queen to Alice.*

So will we come to it in time, but first let us start by considering the role of perception in experimental psychology. For many observers, it has become clear that the traditional topics such as learning, perception, and memory have so many common elements that it is difficult to study one topic without also studying the others. Indeed, behavior is a function of so many variables that, of necessity, each experimental psychologist must delimit or define a relatively narrow view of behavior. The study of perception is no exception. It is more of an attitude toward the study of behavior than a discrete discipline. As will be noted in this discussion, the particular position which a theorist adopts has a strong influence on the sorts of questions he asks and the way in which he attempts to answer them.

Approaches to Perception

A number of years ago, O'Neil (1958) described three general classes of perceptual theory as discrimination, phenomenal, and judgmental theories. Research under any one of these classes is determined in large part by the definition of that class. The discrimination theorists, for example, have focused upon the stimulus, with the stimulus implicitly

* This selection was written expressly for this volume. Preparation of this paper was supported by National Science Foundation Grant GB-7848. The critical comments of Thomas Schroeder and Dr. Bruno Kohn are gratefully acknowledged.

thought to be the cause of the response. By contrast, the phenomenalists have been more concerned with sensation and the way in which sensory qualities give rise to perceptions. Phenomenalists would argue that knowledge is a direct result of the sensations and not of the stimulus itself. Sensations may not, however, be faithful representations of the stimulus and consequently differences between the stimulus and sensation may give rise to error. Because sensation leads to perception, a distinction is often made between sensation and perception; this distinction, however, has frequently been ignored in recent years because it is not clear where sensation ceases and perception begins. In the third classification, the judgmentalists have concerned themselves with the question of what the perceiver does to the stimulus in relation to what he already knows. They emphasize how a stimulus is used in conjunction with other internal states rather than how a stimulus is perceived. All three of these classes of theory fall under the rubric of perceptual theory, although such a variety of perceptual positions indicates simultaneously that perception is both complex and difficult to define.

Since O'Neil wrote his essay, there has been a sizable outpouring of data about perceptual behavior which has overtaxed O'Neil's classification and which has caused psychologists to search for more encompassing frames of theoretical reference. Developments in computer science have produced such a frame of reference in the study of information processing.

Information theorists have concerned themselves with the ways stimuli are encoded, analyzed, transformed, stored, and retrieved. The study of information processing has several advantages over earlier theoretical approaches. First, it parsimoniously incorporates the three positions into one. Such integration is necessary, as Haber (1969) points out, if we are to understand how various aspects of the perceptual system both interact among themselves and also work in harmony. Second, the study of information processing cuts across many topics of psychology. For example, the approach has been adopted by a number of investigators in perception, (for example, Haber, 1969), in memory (for example, Atkinson and Shiffrin, 1968), in cognition (for example, Neisser, 1967), and in thinking (for example, Reitman, 1965). Thus, the study of information processing facilitates the theoretical integration of a variety of behaviors in one enterprise. As a result, our understanding of behavior is increased on a number of points; at the same time, however, the apparent complexity of perception is also increased because we are forced to consider the complex interrelationships between various kinds of behaviors.

Rather than elaborate upon possible rough classifications within information processing, it might be more instructive to consider the issues described by O'Neil in relation to the way theorists have developed information processing. Although not often discussed, the issues which O'Neil employed as the basis for his classification are still with us. By considering information processing within O'Neil's context, perhaps we can see some of the ways in which the study of perception has changed, as well as appreciate some of the complexities which the newer approach has introduced.

It should be understood, of course, that there are limitations to discussing the study of information processing in terms of O'Neil's issues. First, a number of stages are often postulated in information processing, and different properties or functions are frequently assigned to different stages. Second, it is difficult to talk about information theorists because there are diverse positions within this approach, many of which are the consequence of procedural complications.

O'Neil's Issues

1. Active or Passive Process. Despite the passive implications in the phrase "stimulus input," perception is regarded by most information processors as an active process. The observer *does* something to the stimulus. There are degrees of activity, of course. The amount of activity can vary both with the stimulus and with the task. For example, the observer expends less effort in perceiving differences in brightness than in perceiving form. Forgus (1966), for example, has argued that the perception of brightness is a more basic process than the perception of form because less learning is involved in the perception of brightness. Since learning is frequently a result of a stimulus input, it is also possible to argue that a stimulus may do something to an observer. However, for the most part, the position is mostly active—sometimes the perceiver dominates and sometimes the stimulus dominates, but at all times there is a little of each. The domination of one or the other depends upon the perceptual task posed.

2. Location of Perception. There is no single answer to the question as to where perception occurs. The locus of perception depends upon several factors. Two important factors are the number of stimulus dimensions and the kind of instructions (implicit or explicit) given the subject. A stimulus dimension, as used here, refers to a physical continuum along which stimuli can be ordered according to physical characteristics. Thus, hue constitutes a single dimension because it can be ordered along the physical dimension of wavelength. Similarly, brightness can be ordered along the physical dimension of intensity or energy. Many psychological studies have shown that different dimensions are often being analyzed at different points within the same perceptual system. Indeed, sometimes investigators overlook the fact that they are dealing with two or more dimensions within the same experiment. For example, the perception of a letter depends not only upon the lines which make up the letter and their specific composition, but also upon the location of the letter (spatial location) and the context in which the letter occurs.

This point can be further demonstrated by studies of pattern recognition. Neisser (1967), for example, has divided the global dimension of form into features. A number of features, such as concavity upward or downward, closed perimeter, and angularity may be analyzed separately. Neisser argues that analysis need not be confined to parts of a form but may also occur in terms of stimulus features. He argues that feature processing suggests we detect form in some hierarchical manner —a point which he has supported by data (Neisser, 1967, pp. 79 ff.). Thus, whether we talk about stimulus dimensions or about the features of a stimulus, some sort of processing hierarchy emerges. This hierarchical processing, however, makes it difficult to identify a locus of perception because analysis may occur at many points. It is often thought more convenient, therefore, to talk about *stimulus analysis* rather than about *perception* because the latter carries too many connotations to be a useful term.

3. Descriptive or Abstract Modes of Analysis. As O'Neil suggests, the descriptive mode refers to the physical features of the stimulus which distinguish it from other stimulus. By contrast, the abstractive mode refers to features of the object that are not physically present in the object. Information processors postulate the existence of two interrelated mechanisms—one dealing with descriptive information and another dealing with abstract information. Thus, information processing can be both descriptive and abstractive, depending on the level of analysis. For example, Posner and Mitchell (1967) asked subjects to compare two objects either in terms of physical features or in terms of abstract features (for example, names and vowel-consonant categories). They found large differences between the descriptive and abstractive modes in their experiments. Unfortunately, it is not clear from their data or from other data where the descriptive mode ends and the abstractive mode begins.

4. Real or Phenomenal Object. Many psychologists argue that perception cannot begin until a stimulus excites a receptor. Others argue that objects are perceived as representations. Many investigators have proceeded as though the representation were identical with the object. Those who study information processing also have a strong phenomenal tradition. However, not many investigators have examined phenomenal characteristics within the context of information processing—although Haber (1967) and Standing, Sales, and Haber (1968) are exceptions to this statement. Specifically, Haber (1967) examined the increasing clarity of a visual stimulus as the stimulus was repeated. He found that clarity increased beyond the point at which the subject could identify the stimulus, and that repetition enhanced clarity. Although there is obviously a relation between the dimensions of one physical stimulus and the response to that stimulus, his data clearly showed that the relation is not perfect.

In summary, it should be noted that information-processing approaches do not resolve the issues discussed by O'Neil. Perception is neither active nor passive: it is both. There is no locus of perception; perception occurs at many points within the system and employs both descriptive and abstractive modes of analysis. Furthermore, both the real object and the phenomenal representation are involved in perception; perception and behavior will break down if either are denied for very long. These issues, however, underscore the complexity of perceptual behavior and provide a background for describing an information-processing model.

An Information-Processing Model of Perception

The particular model to be described is a three-staged construct originally conceived by Atkinson and Schiffrin (1968) to account for data from memory experiments. The model is a conceptual one which cannot entirely be supported by physiological data because such data are not yet available. Visual perception is emphasized, although it would be possible to generalize the model to other perceptual modalities such as audition (Neisser, 1967).

A diagram of the model is shown in Figure 1.

The three major stages are indicated by boxes in the figure. These stages presumably correspond to different treatments of stimulus information both simultaneously and over a course of time. Within each stage are posited a number of operations which are carried out upon the stimulus. For example, as I look out the window, stimulus information enters the sensory register, where it is held for a very short period of time while certain analyses are carried out upon it. These analyses appear primarily concerned with the physical attributes of the window scene—attributes such as object location, color, orientation of lines, and so forth. As will be shown later, these analyses occur sequentially on the dimensions and features of the scene.

After these preliminary analyses, stimulus information is then transferred into short-term storage, where another set of analyses occurs. The

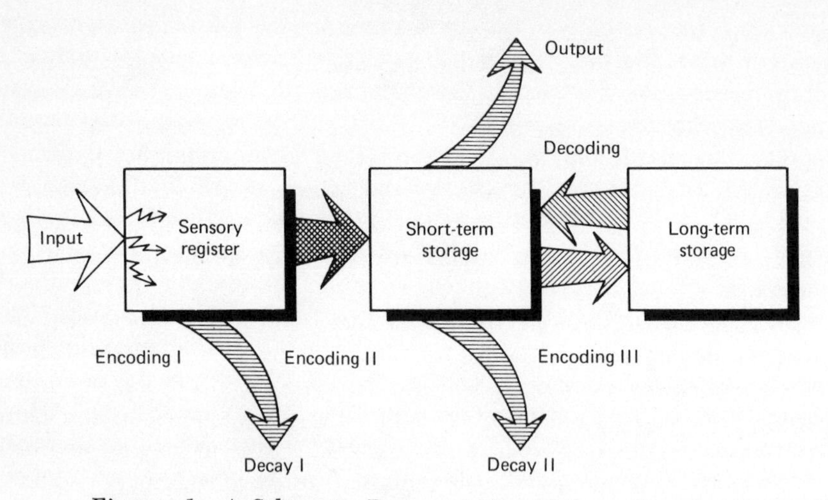

Figure 1 A Schematic Representation of Some Processes Assumed to Occur in Perception and Memory.

major differences between the first two stages of analysis are the amount of time involved and the modes of analysis. Where the sensory-register analysis is essentially descriptive, the short-term analysis is abstractive in character.

From short-term storage, information is converted into long-term storage, where it is retained permanently. It should be possible for information to be in two places at once and therefore shifting information would involve "copying" information rather than removing it from one place to put it elsewhere. Under these conditions it would be possible to have two types of analyses occurring simultaneously. A response might occur after any analysis, but it would seem that output could not occur directly from long-term storage because of the way in which material is stored there.

Many things happen to a stimulus between input and output or between input and storage. Stimulus information may be torn apart, transformed, jostled, forced to wait, lost, and even pushed all the way through the system. Since Cattell's work in 1885 (cited in Woodworth and Schlosberg, 1954, p. 101) it has been known experimentally that the system is limited in capacity and is incapable of dealing with all available stimulus information. A modern experimental example of this is found in Sperling's (1960) work. He examined some of the parameters of this effect using a rather ingenious procedure. He presented briefly several rows of letters but asked the subject to report only part of the display, for example, one of the rows. After the visual stimulus had been exposed, Sperling presented coded auditory cues to indicate to the subject which row to report. By requesting a report for one of the rows each time, he was able to request a report of the entire display over a series of trials. Further, Sperling occasionally asked a subject to report all the information in a display. By comparing the accuracy of single-row reporting with that of total-display reporting, Sperling found that partial reports were consistently more accurate than full reports. He concluded that with total reporting, more information is absorbed than can be reported. Indeed, he estimated that approximately 2^{64} alternatives may register, but only $8(2^3)$ alternates can be reported (Miller, 1956).

Because of the large discrepancy between the amount of input and output, there is no question that an informational bottleneck exists. The locus of this bottleneck, however, has been difficult to identify. One reason for limited capacity seems to be due to the transfer from the sensory register to short-term storage. As will be demonstrated later, this transfer seems to occur for one item at a time. Thus, some items must sit and wait; when their turn comes some of them will have faded away and cannot be transferred. In Figure 1 this is represented as Decay I.

Information is lost because of the processes in short-term storage. New information can apparently bump out older or previous information. If this previous information has not yet been entered into long-term storage, it is lost. This loss is represented in the figure as Decay II. Both Decay I and Decay II seem to occur relatively automatically. At present, unfortunately, we know of no way to break this informational bottleneck. (Miller, 1956).

The Sensory Register

The sensory register has been defined largely through the use of very brief exposures of a stimulus display. Such brief exposures (by means of a tachistoscope) reduce the confounding effects of voluntary eye movement. Although subjects often continue to analyze a stimulus display after it has physically disappeared, they must carry out this analysis without the benefit of self-correction. This imagery or relatively undefined representation constitutes a transient form of memory. Indeed, following a brief tachistoscopic exposure, the stimulus apparently is held for a brief period of time before being transferred into short-term memory.

Sperling's experiments have also produced data which bear upon the problem of informational decay in the sensory register. When he used his partial-report procedure, he was also able to vary systematically the time between visual display and the auditory signal to report. In general, the longer the signal was delayed, the less accurate the report. When the delays were as long as ¼ of a second, accuracy of reporting levelled off and remained relatively constant. These data have been interpreted as indicating that a portion of the neural representation decays away or is lost before it can be processed and transferred into short-term storage.

In a more recent experiment, Dick (1970) has shown that informational decay is directly tied to the physical dimensions of the stimulus. He used a partial-report procedure similar to Sperling's. However, Dick varied the stimulus dimension which the subject was to report. He made displays consisting of four letters and four numbers which were arranged in two rows of four. Moreover, he colored half the items red and half the items black. One group of subjects replicated Sperling's report procedures by reporting stimuli in rows; a second group reported according to color of the items; and a third group reported by class of items, letters, or numbers. Dick found that, as the interval between the exposure and the signal increased, the accuracy decreased for both report by rows and by color but *did not* decrease for report by class. He explained these results as differences in the mode of analysis which subjects used. For example, both spatial and color information entailed a descriptive dimension and were analyzed in the sensory register. By contrast, classification by letter-number required abstractive behavior, and thus occurred as short-term analysis. The fact that spatial and color information were lost over time, but that class information was retained, supports this interpretation.

These studies of information decay suggest that the descriptive analysis in the sensory register might occur in a sequence of substages. The physiological work of Hubel and Wiesel can be interpreted to support such a proposal. These experimenters placed very fine electrodes into the visual cortex of both cat and monkey and then recorded the neural activity of a single cell when different visual stimuli were displayed. They identified several categories of neural cells, each category being maximally sensitive to a particular kind of stimulus. For example, they reported that some cells were sensitive to both the orientation and the location of a line on the retina. A cell with a vertically oriented field responded maximally if the stimulus was oriented vertically in a particular portion of the visual field. A weak response was obtained from the cell if the line was oriented horizontally. No response was obtained from that particular cell if the line was located outside the receptive field. Hubel (1963) concluded that these "simple" cells represent the fifth-order neurons, that is, the fifth cell in the visual system. The simple cells, when they receive input from four cells preceding them, are the first cell in the cortex of the visual system and thus are responsive to both location and orientation of a particular kind of stimulus.

These same investigators reported other kinds of cell activity (for example, some cells were sensitive to movement and some sensitive to edges). Overall, these data suggest that the visual system is organized hierarchially, with different types of stimulus information being analyzed at different points in the system—orientation and location at one level, edge and movement at another level. Other investigators have described corresponding effects in other parts of the visual system (Hartline, 1969). The conclusion to be drawn from these studies is that different kinds of stimulus information are analyzed at different points within the visual system. These analyses are descriptive and would seem to occur in the sensory register.

Once the physical attributes have been analyzed in the sensory register, this information is transferred into short-term storage. The act of transfer itself, however, represents an interesting and complex process. Within the sensory register, it appears that several inputs can be analyzed simultaneously. Data show that transfer from the sensory register to short-term storage, however, can occur for only one item at a time.

This item-by-item transfer has been demonstrated in two experiments by Mewhort, *et al.* (1969). Subjects were presented tachistoscopically two rows of eight letters. The experimenters varied the exposure time, the familiarity of each row, and the presenting of a masking or confusing stimuli. The results indicate that a conflicting or masking stimuli stops further analysis in the sensory register. Moreover, accuracy of reporting increases as more time is made available—a fact which supports the argument for sequential transfer. Mewhort, *et al.* argue that if all information is transferred simultaneously, performance differences would have been minimal when exposure time was varied.

In sum, humans can absorb more information than they can report. Moreover, humans first analyze the physical dimensions of stimulus information. Apparently, there is a difference between the mode of analysis used in this initial stage and subsequent stages. Finally, one item of information at a time is transferred from the initial to subsequent stages of analysis.

Short-Term Analysis

The second stage of the model is characterized by its processing time, its mode of information processing, and by its apparent reduction of information to some sort of coded form. Although our emphasis will continue to be upon data related to vision, it is assumed that descriptive analysis for other modalities has occurred in the first stage of the model, and that these are also entered into the second stage. In this respect, the short-term stage of our model should be regarded as a common meeting ground for stimuli from a number of perceptual modalities.

In contrast to the very brief life span of information processing in the sensory register, the life span for short-term processing has been estimated to be about 30 seconds. This estimate is derived from two sources. First, Milner (1966) showed that human patients with lesions in the hippocampus (a brain structure located in the temporal lobe) have great difficulty remembering information for as long as 30 seconds even without distractions unless verbal labels are readily available. Second, Feigenbaum and Simon (1962), in a brief review of the literature, concluded that regardless of how fast a serial list of words was presented, it takes about 30 seconds for a subject to learn each word. In other words, each item of information must sit in short-term storage for about 30 seconds before it can be stored for reliable retrieval. Although these data do not provide direct evidence for the life span of the short-term storage stage, they suggest that the short-term stage requires 120 times more processing time than the sensory register stage and, furthermore, that the two stages are quantitatively different.

In addition to differences in life spans, the two stages differ in the way information is processed. Whereas information in the sensory register was analyzed in terms of its physical dimensions, information in the short-term stage is analyzed in a much more abstractive mode. Evidence for this distinction comes from studies of categorization and naming, both of which probably involve linguistic as well as perceptual activity.

Certainly the naming or labeling of objects is a linguistic activity. There is nothing inherent in an object that requires a name but, as *Alice in Wonderland* says about naming, "It's useful to the people that name them, I suppose. If not, why do things have names at all?" Indeed, if the young child learns language through hearing and speaking it, then, in an overly simplified way, the naming or labeling of objects or events enhances the establishment of connections or associations between different stimuli. For example, it is likely that children learn the word *ball* by associating a round, red visual stimulus with an auditory stimulus "ball." As he grows older, the perception of a ball comes to represent a complex interplay of cognitive processes—entailing both visual activity and the activation of existing auditory and visual memories. The amalgamation of these activities results in the verbal response *ball.* If short-term storage is to be involved in this amalgamation, one would expect to find some sort of linguistic activity as part of its functioning. The finding that acoustically similar items are more likely to be confused in recall than items that are acoustically dissimilar supports this argument to some extent (Conrad, 1964).

The problem of interference and decay in short-term processing has received more attention by psychologists than any other related phenomena. For example, in an experiment on short-term memory, Peterson and Peterson (1959) demonstrated that accuracy of recall will decline rapidly if the subject is distracted between original learning and testing. They presented subjects with a three-letter nonsense syllable and then demanded that the subjects immediately count backwards from a number by threes. Peterson and Peterson varied the amount of backward counting before asking their subjects to recall the original nonsense syllable. The more backward counting, the poorer the recall.

In a more recent study, Posner and Rossman (1965) showed that two variables may reduce accuracy of recall. The first variable was the difficulty of the intervening task (for example, naming numbers versus adding numbers). The more difficult the tasks, the more subjects forget. The second variable was the length of time that the material was in the memory.

Some investigators have decided to ascribe two kinds of operations to short-term storage: one which holds information passively, and the other which holds and renews information in a more active manner. The second operation has been variously described as rehearsal, the rehearsal buffer, or the buffer. Loss of information from the buffer seems qualitatively different from loss of information which is held passively. Even the loss within the buffer may occur for different reasons. The buffer, according to Atkinson and Shiffrin (1968), maintains information in much the same way as I often talk to myself. For example, when I look up a telephone number, I often repeat the number to myself until I have dialed it, and then forget it. The buffer, in other words, renews information and makes it available for a longer period of time.

Because the rate of rehearsal is slower (Landauer, 1962) than the rate at which information can get into short-term storage (Mewhort, *et al.*, 1969), not all stimulus information can be maintained in the buffer. Thus, one reason why information is lost is that information gets "kicked out" of the buffer by newer, incoming information. In fact, if subjects are asked to recall a list of nine items, they will be more accurate if they recall the last three items first, and then recall the others. (Howe, 1965).

A second reason why information is lost involves the buffer itself. For example, if you were shown the letter sequence D G A C B H F E, it is likely that you will tend to remember the items in alphabetic sequence because that sequence is easier to remember than the original. Reordering the item requires time between presentation and recall. Thus, the more time between presentation and recall, the more likely that you will reorder the sequence. Such reorderings occur with increasing frequency as more time for rehearsal is allowed (Murdock and vom Saal, 1967). Of course, it is possible that you could correctly recall all of the names of the items but recall nothing about their order, as would be the case in the response sequence A B C D E F G H. In this case, transfer from sensory register to short-term storage would be sequential, and name information would be converted into order information. Whereas name or spatial information is apparently represented topographically in the sensory register, order information is represented relationally in short-term storage; in other words, the order in which the items are maintained represents order information. Thus, loss of spatial information occurs before transfer to short-term storage, while loss of order information occurs after transfer or a shuffling of items in short-term storage. In summary, information can be lost from short-term storage for two reasons: (1) because it is not maintained in the buffer, and therefore decays and is no longer available; and (2) because items are shuffled within the buffer in an attempt to preserve one kind of information at the expense of another kind.

Long-Term Analysis

A third stage of our model is needed to account for the influence of past perceptions upon behavior. Although long-term storage is usually considered to be outside the province of perceptual problems for the most part, there are a number of observations about long-term storage which are relevant to a discussion of information processing. For one, information seems to be processed differently from the other two stages. It may be that information undergoes some biochemical transformation. Exactly how this occurs is not understood. Although many analogies have been made between the storage of genetic information in DNA (deoxyribonucleic acid) and the storage of environmental (learned) information in RNA (ribonucleic acid), the data seem to suggest that RNA is involved, but is not the whole explanation. For this reason, some psychologists argue that processing entails alterations in protein structure.

The theoretical limits of long-term storage capacity may be defined in terms of the number of possible combinations in the memory molecule—a number so large that it is unlikely that anyone would use all of them in his lifetime. The restrictions on human information processing, therefore, are not due to long-term capacity. Furthermore, Penfield (1956) reported that patients undergoing brain surgery can, with the aid of direct electrical stimulation, recall events that they had not remembered for years. Thus, these data suggest that once information reaches long-term storage it is there permanently.

The fact that we cannot recall everything we have experienced is related to the problem of retrieval. An analogy may help explain the difficulties entailed in retrieving information from permanent memory. If we place a black ball in an urn, the probability of retrieving the black ball from the urn by chance is perfect, inasmuch as it is the only ball in the urn. However, if we add nine white balls, the probability of retrieving the one black ball is only one in ten. If ninety more white balls are added, the probability of finding the black one becomes $1/100$; and as we continue to add white balls, the probability of finding the black one decreases steadily. Note we do not have to assume that the black ball has been changed in any way; the probability of locating it changes simply as a function of the number of other balls which are in that urn.

Fortunately, storage and retrieval from long-term memory is not as random as the urn analogy would imply. Information appears to be organized into categories, and retrieval is tied to these categories quite closely. Mandler (1967) has shown this relationship experimentally in a series of studies. He gave each subject a deck of cards with a highly familiar word printed on each card. The subject was asked to sort these cards into anywhere from two to seven categories using any system he chose. Each subject repeatedly sorted the deck until he came up with two identical sorts in succession. Following the final sort, the subjects were asked to recall as many of the words as possible that they had just sorted.

In one experiment, the results indicated that subjects will recall an average of 5.6 words from each of the categories they use in sorting. Thus, if they used seven categories, they will recall about 5.6 words from each of the categories; if they used but two categories, they will also recall about 5.6 words from each of those two categories. Throughout these experiments, Mandler found a definite relationship between the number of words recalled and the number of categories used. As he points out, the words most frequently used actually required little learning. His data, however, support the argument for categorical processing of information in long-term storage.

How one searches through different kinds of information in permanent memory, however, is a subject which cannot be discussed adequately in this paper. The interested reader might consult Shiffrin and Atkinson (1969) for a theoretical discussion of these problems. An important point to note, however, is that although specific information cannot always be retrieved from long-term memory, it may nonetheless influence our behavior. This influence might be called implicit retrieval, and it is important in a variety of perceptual situations. One of these situations will be discussed in a later section of this paper.

Interplay of Stages

Thus far, we have concerned ourselves with a three-staged, conceptual model which serves to organize data about perception and the perceptual act. Although the data discussed indicate boundaries between stages, perceptual behavior normally does not involve just one of the stages—it entails an interplay of information processing from all three stages. A perceptual response to a window scene, in other words, rarely entails a simple one to one relation to a stimulus.

Experimenters have used two relatively simple ways to examine use of stimulus information: either change instructions or stimulus materials, while holding other conditions constant. In most of the studies we have discussed, the type of materials and the instructions remained the same throughout the experiment for a given subject. By contrast, in the experiments which will be described in this section, the investigator either varied the instruction or the materials. These variations in procedure affect the observations made about the use of information. In particular, different questions are asked, and different approaches are used. One approach is to consider the ways information is extracted at different stages in the system. Just because a stimulus may have been presented does not always mean that a complete analysis of the stimulus will occur before a response is made. The second approach considers the ways in which information gets processed over time. Even though stimuli may be presented in a particular order or arrangement, an experimenter has no assurance that the subject will respond to the stimuli in that order. The order which a subject imposes upon stimuli has practical implications with respect to such tasks as reading.

Interstage Analysis: Categorization

Categorization, according to Bruner, Goodnow, and Austin (1956) allows one to treat discriminable objects as though they were equivalent. The effect of categorization is to reduce the amount of information which a person must process. Categorization also reduces precision and detail. For example, consider the classification *chair*. Many different stimulus objects are lumped into this classification—hard chairs, soft chairs, arm chairs, folding chairs, kitchen chairs, classroom chairs, and so forth. Through categorization, the compression of information about a specific chair is more than compensated for by the use of a category for a number of functionally equivalent objects. The utility of categorization, of course, depends upon the availability of categories, or what Bruner (1957b) calls perceptual readiness. A person's development of categories or his concept attainment will not be discussed here, however see Bruner, *et al.*, 1966. Instead, categorization will be considered in terms of the three-stage model.

Recent experimentation has shown that categorization can occur at different stages within the model. Posner and Mitchell (1967), (for example), presented subjects with a pair of objects simultaneously and asked the subjects to categorize these items as being "same" or "different." In several experiments reported, they varied the combinations of pairs of stimuli presented. For example, the stimuli were physically identical on some trials (for example, two capital letters, *AA*); on other trials the stimuli were of the same name but physically different (for example, one capital and one lower-case letter, *Aa*); and on still other trials several stimuli were presented which involved combinations of vowels and consonants in capitals and lowercase (for example, *Ae*). Subjects were told to respond "same" if the elements were both vowels or consonants, and "different" if one vowel and one consonant were presented. (In the examples given, the correct response would have been "same" for all three; of course, appropriate trials were run on which the correct response were "different.") The results of this experiment suggest that categorization is an economical form of processing. Although subjects were asked in this experiment to make responses based on the categories of vowels or consonants, more efficient analyses of the stimuli were sometimes made. The results are shown in Figure 2.

Posner & Mitchells' data suggest that the subjects went through the

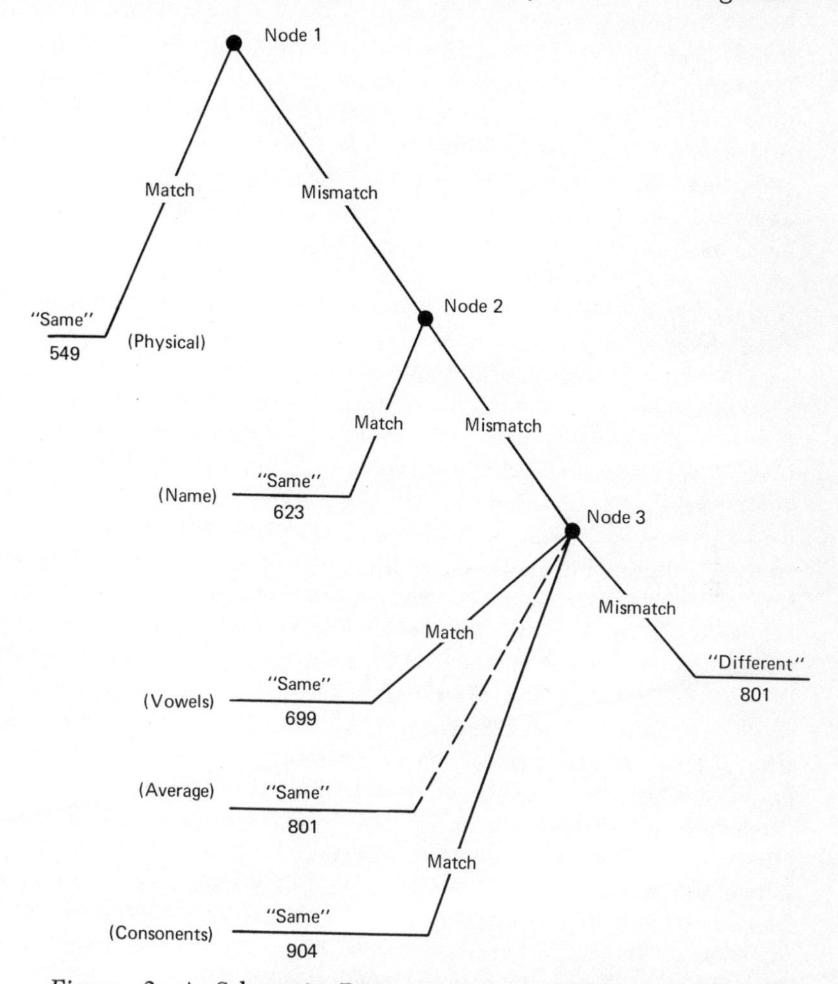

Figure 2 A Schematic Representation of Categorization in the Posner and Mitchell (1967) Experiment. The NODES Represent the Points at Which Decisions Seem To Be Made: Physical Similarity at Node 1; Names at Node 2; and Class at Node 3. The Numbers at Each Response Point Represent the Average Reaction Time of Subjects Making That Type of Comparison. (Data from M. I. Posner and R. F. Mitchell, "Chronometric Analysis of Classification," *Psychological Review,* vol. 74 (1967). Copyright 1967 by the American Psychological Association, and Reproduced by Permission.)

following steps to make their response: First they compare objects in terms of physical similarity. If the two stimuli are physically similar, it usually follows that both stimuli will have the same name and that both will belong to the same category; and a "same" response can be made purely on the basis of physical similarity. If the two stimuli are not the same physically, a second comparison must be made, this time comparing the names of the items at Node 2. Physical similarity or a match here also implies membership in the same category and therefore a response about categorizing can be made purely on the basis of name. If a mismatch occurs, analysis proceeds to Node 3. Here, according to the data of Posner and Mitchell (1967), the sequence of analysis at Node 3 is first checked for vowels; if "no," then it is checked for a vowel and a consonant; if "no," then it is checked for two consonants. There exists a general bias toward first checking for matches and then examining mismatches. This is reasonable because a match at Node 1, for example, implies matches at Nodes 2 and 3, whereas a mismatch at Node 1 implies nothing about matches or mismatches occurring later.

In terms of our model, it would appear that Node 1 is a sensory-register activity while Nodes 2 and 3 involve short-term storage activity. This statement is based on data from other experiments reported by Posner and Mitchell (1967) in which the reaction time or speed of Node 1 activity is unaffected by experience, although Node 2 is affected. Subsequently, Posner, Bois, Eichelmann, and Taylor (1969) asked subjects to respond "same" or "different," this time on the basis of name. In these experiments, however, the two letters were not always presented simultaneously. In some cases, the first letter was presented and remained available for up to 2 seconds before the second letter was presented. By forcing the subject to wait and hold his response, it was thought that the first information had been fully analyzed, and was being held in short-term storage; the longer the delay, the more likely this would be true. If an item were held in short-term storage on the basis of its name only but not in terms of its physical characteristics, the entire matching procedure would have to be on the basis of name, even if the two stimuli were physically identical. Indeed, the results indicate that the longer the delay between the two stimuli, the longer the reaction time for a Node 1 match. The delay probably forces the subject to make all matches at Node 2 in this experiment, even those that would occur at Node 1 under more optimal circumstances.

These data of Posner and his associates suggest again that stepwise analysis occurs both in the sensory-register and in short-term storage. More important, these data show that the occurrence of a response does not depend upon the amount of information available in the stimulus but rather on the amount of processing it requires. A response can occur on the basis of sensory-register analysis alone and need not involve either short-term or long-term storage or analysis. The number of subsequent analyses performed after the sensory register will depend upon the availability of previously learned schemes for classification. Given this well-developed system, the data clearly show that not all of the classifications will be used for each and every stimulus.

Intrastage Analysis: Order of Processing and Report

Just as one can examine relations between dimensions, one can also examine relations within a dimension. By and large, experimenters studying this problem have used multiple-item tachistoscopic displays, and have varied both the types of items presented and the instructions for recall. Of interest is the way in which a series of display items is integrated. It was pointed out earlier that transfer from the sensory register to short-term storage occurs in a serial manner (Mewhort, *et al.*, 1969). This means that spatial information must be converted to temporal-order information; several studies show how this takes place with different materials.

Bryden (1960) varied both the familiarity of the material and the instructions for recall. On a given trial, he presented a single row of letters or geometric forms tachistoscopically. He asked subjects to report the material from left to right or from right to left. On half of the trials, subjects were told which order to use after they had been shown the material; these trials constitute the data of major interest. Bryden found that subjects could use either order of report with equal facility when reporting the geometric forms. In contrast, the subjects were more accurate when reporting letters from left to right than when reporting from right to left. In addition, they were always more accurate on the left side of the display when reporting letters. Bryden suggested that the order of report is more flexible with unfamiliar than with familiar material, a suggestion supported by Winnick and Dornbush (1965). Other experiments, however, show that the inflexibility of report is not related to the familiarity of the material per se but rather is a function of the material and of the contextual-spatial properties of the stimulus (Dick and Mewhort, 1967).

To test this idea, Bryden, *et al.* (1968) asked subjects to report number sequences from left to right or from right to left. The rationale of the experiment was that although numbers are highly familiar, spatial information may not be necessary for integration. Consequently, in addition to manipulating the order of reporting, the numbers in the tachistoscopic display were either spaced normally or were spread out across the visual display. Spreading numbers out, however, had no effect on accuracy. Analysis of order of reporting indicated that subjects could report from right to left with good accuracy, although this order was not quite as good as left to right. A comparison of the right-to-left accuracy of letters and numbers indicated that it is much easier to report numbers in this direction than letters. Thus, the authors concluded that familiarity does not have an important influence on order of reporting. They argued that numbers are almost as familiar as letters and, consequently, if familiarity were the crucial variable, there should be little difference between letters and numbers. The results suggested instead that accuracy is influenced by the specific material and the experience of the subject with that material, especially integrative reading habits specific to letters.

The important variable here seems to be the order of internal operations carried out on the stimulus. Since transfer of material from the sensory register to short-term storage is sequential, and since printed English is also sequential, a particular order must be imposed on the stimuli to obtain maximal information from the display. The most convenient strategy for most subjects is to employ the order normally used in reading. In fact, in the absence of the other instructions, that is the order which most subjects do impose. For example, Hebrew readers who normally read from right to left will report nonalphabetic binary symbols from right to left (Harcum and Friedman, 1963). The use of a particular order is strongly ingrained; some of Bryden's (1960) subjects reported that they had to visualize the letters from left to right so they could report from right to left. With numbers, there is also a tendency to stick to the left-to-right order, but subjects also must locate the decimal under normal reading conditions and then read the number backwards to locate thousands, millions, and so forth. Experience with numbers is bidirectional but with letters experience is unidirectional.

Spatial information seems to play an important role in the integration of a series of particular types of items. Although many of the data obtained have been from adult subjects, it would seem that the use of spatial information is learned because use varies with different materials. Whatever the mechanism responsible, spatial information must be represented in long-term storage in such a way that it can influence operations being carried out in the sensory register and short-term storage. How this mechanism works or is established, however, is not known.

For the most part, our concern in this paper has been to describe a three-stage model, and to offer some hypotheses and data about interstage and intrastage analysis. To conclude, we need to consider additional influences upon information processing. In particular, we need to consider such influences as information reduction and selection, directing conditions, and the problem of response bias.

Normally, each of us is bombarded with a variety of stimuli every moment. Many experimental results lead one to the conclusion that all of this information is received by the organism. At the same time, other data show that the organism does not make use or even show evidence of receiving the information. There are a number of factors involved in producing differences in the amount of information received and the amount processed. Of the information received, some is lost because the rate of input is too fast for complete processing. This was discussed earlier with the sensory register. A second portion of the information received is not processed because the organism apparently chooses to ignore it. This second type of loss comes about because stimulation does not impinge upon a passive organism. Rather, the organism is differentially disposed to respond to particular kinds of stimulation. Such a disposition has the effect of reducing the amount of information which is processed.

Reduction in Processing

Many of the stimuli that impinge on an organism do not need to be processed. It is only when stimuli change that information is gained; thus if a stimulus is repetitive, little is gained from the second and subsequent presentations. The ticking of a clock, for example, soon passes out of awareness. If, however, the ticking changes in some way—if it becomes louder, softer, or ceases—then you are again aware of it. Often we respond overtly to such a change (for example, you might turn your head to see what has happened). This orienting reflex, as it is called, can be measured physiologically, moreover, even in the absence of overt behavior. By using measures of skin potential, heart rate, and skin conductance (Raskin, Kotses, and Bever, 1969), it can be shown that the nervous system is constantly monitering repetitive stimulation (Sokolov, 1969). When a stimulus such as an auditory tone is first or initially presented, physiological changes can often be observed in heart rate and skin conductance. However, when the stimulus is continually repeated these physiological activities return to normal and are not observed as long as the stimulus is constantly presented. If the stimulus is increased or decreased in intensity or even omitted, the physiological indicators reappear but will again disappear if the stimulus remains constant at its new level. Occurrence of the orienting reflex is dependent upon changes in stimulation.

Several important points can be made concerning the orienting reflex. First, it is apparently an innate mechanism and common to many organisms. Second, there seems to be some sort of transient memory involved. According to Sokolov (1969) the organism compares new, incoming stimulation to that which it has previously experienced. If a new stimulus matches the representation of the old one, no processing occurs because the information in the stimulus has already been analyzed. Some form of memory is thus obviously involved, for if the organism cannot recreate recent stimuli in some way, then there cannot be a way to match incoming stimulation. The orienting reflex, then, enables the organism to monitor incoming stimuli; if stimuli have been previously processed, further processing is unnecessary. Thus, the organism is biased against repetitive stimuli and therefore has more time available to process stimuli that carry new information.

Past Experience: Increased Efficiency

There are other ways in which the organism may be differentially prepared for stimulation. Whereas the orienting reflex may be an innate mechanism which relies in part upon memory, the study of illusions suggests some striking examples of the influence of past experience upon perception (Allport and Pettigrew, 1957; Pollack, 1969). Allport and Pettigrew, for example, examined the effects of the rotating trapezoidal illusion among Zulu children. When tested on this illusion, rural Zulu children were likely to report that the window was rotating (no illusion). In contrast, both urbanized Zulu and European children were likely to report the window as oscillating. Because there is an almost complete lack of angularity in the rural Zulu environment, the authors conclude that experience with angularity is an important influence on the perception of illusions. Similarly, other types of experience play a role in preparing the organism for stimuli.

Although the mechanisms are not completely understood, Bruner (1957b) has suggested that perception depends on the construction of categories into which stimulus inputs may be sorted. In this context, he defined perceptual readiness in terms of the relative accessibility of these categories. With increasing accessibility of the category, less stimulus information is required for the input to be categorized. A reduction in the amount of stimulus information available is equivalent to a reduction in the amount of processing needed. Therefore, the amount of processing necessary to categorize the stimulus is a function of the accessibility of the category. The existence and strength of a category are, of course, dependent upon past experience. This experience helps to establish categories which in turn influence processing of current input. Although not shown in the figure of the model, these effects imply a "feedback" arrow from long-term storage to the sensory register. Accessibility can be stated in terms of the efficiency of the operation of this feedback loop.

Directing Conditions

Motivational factors are also reported to have an influence on the processing of stimuli. For example, Wispe and Drambarean (1953; cited in Dember, 1960, p. 310) deprived some subjects of food and water for 24 hours. At the end of this 24-hour period, these subjects were tested on identification thresholds for words relevant or neutral with respect to food and water. The deprived subjects showed lower thresholds for the motive-related words than did a nondeprived control group; there was no difference between the groups for the neutral words. Thus, these data support the notion that motivation will affect perceptual processing. Dember (1960), however, is very guarded in his discussion of this type of effect. Many of the experiments conducted on this type of problem are open to alternative interpretations, the most important being response bias.

Response bias refers to the fact that the subject will frequently report not what he actually sees but something else instead. Consider, as an example, an experiment in which subjects are presented both neutral and "dirty" words. The words will first be presented at a very brief exposure, and then gradually the duration of exposure will increase until the subject correctly identifies the word. Results from experiments of this kind are quite consistent. The duration of exposure must be longer for identification of the dirty word than for the identification of the neutral words. Does this indicate that subjects do not see the dirty words? Probably not. As one rather unihibited subject put it, "*If I didn't know better*, I would have sworn that I saw #%*&." The word that he thought he saw was exactly what he was shown. The point of this example is that the subject may have seen clearly what was shown to him but does not want to say it lest the experimenter think him foul minded. Instead, the subject will say he saw *puck* or *shot* instead of the word actually presented. Although the effects of response bias are more dramatic in dirty-word experiments, this effect has been observed in many experimental situations. When a subject "guesses" about a stimulus, he is usually correct more often than would be expected by pure chance. Because these effects are so general, it is fortunate that procedures have been developed to measure this bias (Galanter, 1962).

Conclusions

Much like Alice, the reader has been taken on an excursion through the looking glass, this time to see how the person in the glass perceives. The picture that has been presented differs from earlier efforts and continues to be incomplete. We began by placing the study of information processing in the context of four issues which continue to divide theorists concerned with perception. We then described a three-stage model of information analysis and storage. Although it continues to be a conceptual model, it nonetheless serves two important functions. First, it organizes a wide spectrum of research from studies of single-cell activity to studies of the entire organism performing complex tasks. Second, the model suggests areas for further research.

The three-stage model is based upon three different categories of information processing which occurs as perceptual-memorial behavior. The major properties of each stage are summarized in Table 1. Recent experimentation indicates that substages exist within each stage.

Table 1

SOME DISTINCTIVE ASPECTS OF EACH OF THE STAGES

	Sensory Register	Short-Term Storage	Long-Term Storage
Psychological form of representation (mode of analysis, O'Neil, 1958)	Descriptive	Abstractive	
Form of physiological representation	Neural (modality specific)	Neural (largely auditory and linguistic)	Probably biochemical
Capacity of stage	Relatively unlimited	Limited by rehearsal	Unlimited
Approximate life-span of input within stage	¼ second	30 seconds	Equal to life-span of organism
Interference	Time dependent	Time dependent and related to difficulty of task	Time independent
Mode of operations on the memory representation	Parallel	Serial	(Not known)

The emphasis of this paper has been upon data which show how humans process stimulus information. The paper and the model were organized around questions about the analysis, storage, encoding, and the use of information. The weakest aspect of this model is our understanding of the intersections of the stages. In many instances, for example, visual activity occurs in the absence of an appropriate visual stimulus, as in dreams, hallucinations, and in just everyday activity. Also, an auditory stimulus of the printed word is often sufficient to evoke a visual image: the waves at the seashore, a red rubber ball, or a lobster. Such visual activity must emanate in long-term memory and activate sensory register mechanisms. Two questions then immediately arise: (1) What is the function of imagery? and (2) What are the mechanisms responsible? In answer to the first question, Paivio (1969) has shown that concrete, image-evoking words are more easily learned than abstract words. Furthermore, he has argued that while it is obvious that imagery and verbal processes are functionally related, they constitute two alternative memory-coding systems. In this way, it would seem that images may facilitate information processing, perhaps by allowing more complete processing in addition to the double-coding system. There is no answer to the second question, although Hebb (1968) has offered some physiological and theoretical suggestions concerning the general role and mechanisms of imagery.[1] Although not well understood, the implications of imagery for understanding the interplay of the various stages seem clear.

THE COMPARATIVE STUDY OF SENSORY AND PERCEPTUAL PROCESSES*

William Schiff

Man has always been interested in the animal worlds about him. Throughout history, he has developed an extensive animal lore from casual observations of animal life. More technical forms of formal naturalistic observation (Schiller, 1957) were devised by European ethologists during the nineteenth and twentieth centuries and, more recently, carefully controlled experimental methods have been employed by ethologist and psychologist alike in the effort to unravel some of the mysteries surrounding animal behavior.

Among the vast array of questions stemming from man's curiosity about animals are those concerned with their sensory-perceptual abilities.

[1] For a different interpretation of the mechanisms and role of imagery, see Neisser (1967).

* This selection was written expressly for this volume.

Man's curiosity about the other inhabitants of the world in which he finds himself has stimulated, but not totally determined, his study of animals' sensory-perceptual processes. The formal requirements of science have imposed constraints upon both the methods and language of modern comparative perceptual psychology. Few twentieth-century scientists ask the question: "How does the world *appear* to a cat, fish, or insect?" With certain exceptions, for example von Uexküll's (1934) classic phenomenology of animal worlds, behavioristics has altered the way of posing questions about animal perception so that psychologists and ethologists studying animal perception talk about their subject matter differently, but still behave as if they are interested in perception, not merely discrimination.

Questions about how and why animals act the way they do lead naturally to questions about how they gather information with their sensory systems (Schiller, 1957). For only with information from their environments about objects to be avoided, approached, mated with, put together, dug out, climbed, devoured, and so on, can animals survive. Many of the apparently "superhuman" feats of the subhuman animal kingdom are related to specific sensory-perceptual abilities—the keen sense of smell in the dog (Kalmus, 1964), the visual acuity of certain birds (Donner, 1951), and the auditory "radar" of bats (Griffen, 1958) are but a few examples.

Yet, despite such concern with questions of animal behavior, comparative psychology has not proceeded on a direct course in unraveling the nature of animal perception. Indeed, comparative treatments of sensory-perceptual processes have run a somewhat convoluted course in the past 75 years. Around the turn of the century, psychologists such as McDougall were attempting to compare sensory-perceptual processes along the phylogenetic scale within the context of theories of instinct. Naturalists in the zoological disciplines were operating on a more species-specific basis but were also primarily concerned with instinctive behaviors. McDougall's concept of instinct included sensory-perceptual components, cognitive-emotional components, and motor components; it also included the notion of decreased stereotypic nature and increased modifiability of instinctive animal behaviors as one ascended the phylogenetic scale. With the exception of the cognitive-emotional component, his view was remarkably like those found in modern ethology (Schiller, 1957), although it was far less specific. McDougall was interested in the common features shared by animal perception and behavior throughout the phylogenetic scale, despite species-specific differences. Thus the exhibition of fear might be elicited by different stimuli in different species (the cry of a predator, or a verbal threat) and might lead to quite different motor behaviors (flight, or a phone call to the police), but the underlying functional similarity was thought as important as species-specific differences. McDougall wrote:

> We may, then, define an instinct as an inherited or innate psychophysical disposition which determines its possessor to perceive, and to pay attention to, objects of a certain class, to experience an emotional excitement of a particular quality upon perceiving such an object, and to act in regard to it in a particular manner, or, at least, to experience an impulse to such action. . . . Now, the psychophysical process that issues in an instinctive action is initiated by a sense-impression which, usually, is but one of many sense-impressions received at the same time; and the fact that this one impression plays an altogether dominant part in determining the animal's behaviour shows that its effects are peculiarly favoured, that the nervous system is peculiarly fitted to respond to just that kind of impression. The impression must be supposed to excite, not merely detailed changes in the animal's field of sensation, but a sensation of complex sensations that has significance or meaning for the animal; hence we must regard the instinctive process in its cognitive aspect as distinctly of the nature of perception, however rudimentary. (1908)

McDougall's treatment of animal sensory-perceptual processes was not only comparative and remarkably modern in conception, but it insisted upon the selectivity of stimuli based upon species characteristics. Moreover, it was thoroughly functional in its treatment of sensory-perceptual factors.

Because McDougall's treatment stressed the *instinctual* aspect of behavior, it fell before the anti-instinct movement which pervaded American psychology after the second decade of this century. Although there have been some notable exceptions—such as Maier and Schneirla's classic treatment of animal behavior (1935)—with the advent of behavioristic psychologies stressing learning and behavior modification, functional comparative treatments of sensory-perceptual processes in animals were rarely to be found in the mainstream of American psychology for years. Psychologists tended to ask questions about the limits of sensitivity of sensory receptors under specialized conditions, using conditioned discriminations of stimuli in such apparatuses as the Y maze, the Lashley jumping stand, and the Skinner box. With such techniques, it was and is possible to determine if an animal (formerly a rat; more recently, a pigeon) is able to learn to discriminate between certain wavelengths of light (Cohen, 1967; Ehrlich and Calvin, 1967), intensities and frequencies of sounds (Wendt, 1934), shapes or patterns (Lashley, 1930), or other "discriminanda." Such research is generally concerned with some issue in psychology other than how the animal in question uses its sensory-perceptual apparatus in guiding its normal behavior, or how it extracts information from its environment (Forgus, 1966). The experimenter may be interested in generalization gradients (Baron and Vacek, 1967; Butter, 1963), signal-detection theory (Blough, 1967), or some other aspect of how behaviors may be *controlled* by experimenter-manipulated stimulation (Mentzer, 1968). But the sensory discrimination made by the animal is relevant to the animal only in that it enables it to obtain the experimenter's food, avoid the experimenter's shock, or answer his question; such behaviors usually have little to do with the animal's perceptually guided behaviors outside the laboratory.

> Consider Pavlov's dog isolated in a cubicle containing a food tray and a bell. The rule of this special environment was that whenever the bell sounded, food appeared. . . . As long as Pavlov chose to make this improbable sequence a law of the cubicle (and only so long as he did), the dog might be expected to detect it.
>
> What about the instrumental conditioning of responses? We must now consider Skinner's rat isolated in a box containing a lever and a food cup. . . . Skinner had created this little world (perhaps in six days, resting on the seventh) so that depression of the lever caused delivery of a food pellet. In order to detect this strange invariant, the rat had to behave before he could perceive, but in the course of exploration the utility of the lever became evident: it afforded food. (J. J. Gibson, 1966)

To the extent that data provided by such techniques determine whether an animal has discriminated a reward-relevant stimulus from a reward-irrelevant stimulus (or set of stimuli), we may classify the findings under the rubric "the sensory-sensitivity approach." These techniques are especially useful in determining the limits of sensitivity of an animal's sense organs. For example, if an animal can distinguish a patch of light differing from another only in wavelength of light, we can be certain he can "see color," providing other sources of discrimination have been properly eliminated by control procedures (Maier and Schneirla, 1935).

For many years psychologists have been gathering sensitivity data in many species of animals. Such data provide knowledge about different animals' abilities to discriminate energy levels within the physical continua described by classical physics (light and sound) and chemistry (chemical composition of substances detected by taste, olfaction, and other "chemical senses"). Further, some of these studies (Michels and Schumacher, 1968) are concerned not only with the relative ability to discriminate color, sound, and the like, but are also concerned with the evolutionary status of different sensory and neurological structures and processes. Only relatively recently however, have comparative psychologists again come to realize that although sensory-perceptual processes *may* be studied in isolation from the animal's natural environment and behavioral context, they are most fruitfully studied with both ecological and evolutionary factors clearly in mind. A few recent examples may help clarify this point.

Much of the existing animal research has focused upon the process of learning (and perceptual learning) and has often used electric shock as a stimulus. Electric shock is convenient because we can control and measure its intensity, and because it can quickly "motivate" an animal in a reliable fashion. Unhappily, it also has its serious drawback. In a recent publication dealing with the use of such a stimulus, for example, Morrow and Smithson stated:

> In the study of learning capacities of various invertebrate species, it may be particularly important to manipulate variables most relevant to their modes of adjustment. For example, it is unlikely that many invertebrates encounter electric shock in their environment. It is also unlikely that such stimuli played any great role in their phylogeny. Hence the effect of such stimuli on behavior may be too disruptive to allow demonstration of full adjustment capacities. (1969)

The same argument may well hold for vertebrate as well as invertebrate species. For example, electric fishes may encounter electrical stimuli in their natural environment. Psychologists have studied these animals rather extensively, with special reference to the emission and controllability of their electrical discharges (Mandriota, Thompson, and Bennett, 1968). One recent investigator of less intense electric "signals" has concluded, however, that no biologically generated signals were detectable in the natural environment of these animals, although it has been well established that such signals are detectable and can be brought under the experimenter's control in the laboratory aquarium.

> What role, if any, these signals play in the lives of fishes remains unanswered, although one would suspect that evolutionary intermediates between "nonelectric fishes" and the highly specialized electric forms do exist, and that they have put such an electric sense to good use. (Barham, *et al.*, 1969)

In other words, while the basic data regarding the control of behavior by a stimulus and the emission of a potential stimulus by an animal have been available for some time, only very recently are they being utilized to study the natural functions of the sensory system in question (Mandriota, Siegel, and Gallon, 1969). Only recently are the limitations of artificial stimulus being fully realized (Morrow and Smithson, 1969).

Although psychologists have been interested in problems of a comparative sensory-perceptual psychology for years, their treatments of these problems have drifted from early attempts at comprehensive examination of animal perception and action (such as McDougall) to more recent studies of conditioned discrimination and sensory sensitivity. In contrast ethological investigators such as Tinbergen and Lorenz have persisted in trying to attain a comprehensive understanding of species-specific behaviors (especially of submammallian species). Fairly recently, an integrative approach has emerged (see E. J. Gibson and Walk, 1960 and 1961; and J. J. Gibson, 1966).

The work by psychologists and ethologists, when integrated, have helped to evolve a more truly functional comparative perceptual psychology. Beach's lament (1950) regarding the narrowness of comparative psychology, which by its very nature should be both broad and precise, is apparently being heeded after years of floundering. What has emerged —from a lamenting comparative psychology dealing with a few processes of a few species, from the stimulating example of naturalistic European (and American) ethology, and from a functionalistic perceptual psychology emphasizing the importance of stimulus information (not just central processes)—is a comparative approach to sensory-perceptual processes which might be called "the sensory-utility approach." The emphasis is upon how animals of various species *use* their sensory equipment when they extract behaviorally relevant information from their environments, rather than upon what sensory discriminations an animal can be taught to make under idealized conditions. The sensory-utility theorist typically concerns himself with stimulus properties of the sort an animal is likely to use in his basic perceptual adjustment to *his* world, rather than those properties which represent the classical or sensitivity dimensions used for measurement in the physical sciences (J. J. Gibson, 1962). The sensory-utility theorist is concerned with the sensory equipment as perceptual systems (J. J. Gibson, 1966), rather than the sensory-sensitivity view of sensory equipment as general detection apparatus, isolated from the rest of the organism. He is interested in the animal's sensory preferences, its ecologically geared selectivities and sensitivities, and its response-guided successive information pickup systems.

A growing number of provocative hypotheses and theoretical views about perception have helped to turn comparative perceptual psychology toward this sensory-utility approach (J. J. Gibson, 1966). Studies of "looming" and "visual cliff" phenomena have provided empirical support for these views. Moreover, since these views of perception tend to stress the *active* aspect of perception, rather than the passive receptivity of stimulus energies it is not surprising that several animal studies having to do with their "mapping" of perceptual space have appeared (Held and Freedman, 1963; Held and Hein, 1963). Indeed, movement-produced stimulation has begun to receive the emphasis which stimulus-response psychology placed upon stimulus-produced movement.

Despite empirical advances, the establishment of a strong sensory-utility approach to comparative sensory-perceptual psychology has been hindered or constrained by its traditions and its procedural biases. In order to understand the nature of these hindrances, we need to consider particular issues and procedures used by comparative psychologists.

Problems in Animal Studies of Sensory-Perceptual Functions

The Organization of Research around Controversial Issues: toward an Understanding of an Umwelt

> The *Umwelt* of any animal that we wish to investigate is only a section carved out of the environment which we see spread around it—and this environment is nothing but our own human world. The first task of *Umwelt* research is to identify each animal's perceptual cues among all the stimuli in its environment and to build up the animal's specific world with them. (von Uexküll, 1934)

One way in which animal studies have entered strongly into psychology is in the service of such issues as the "nature-nurture" issue. A large body of animal literature has been generated by this controversial issue, but—with the exception of knowledge about the ill effects upon sensory apparatus when stimulation is prevented from reaching it, especially patterned light to the eyes (Riesen, 1950; Riesen and Aarons, 1959)—little has been learned about animals' sensory processes or, for that matter, about human perception and ideas, which was the source of the issue (Hochberg, 1963).

The nature-nurture issue has long been thought testable by a variety of perceptual studies. Theorists have sought to determine whether or not depth perception, size perception, shape perception, localization of objects in space, and the like, are sufficiently present or accurate in animals without visual experience (or association between vision and another modality) to support either a theory of innate perceptual ability or a theory that all perception must be learned. Animals have been the chosen subjects for most of these studies because of the relative ease of controlling their early experience and their rapid maturation. Moreover, experiments with animals provide freedom for dark-rearing experimental manipulations, or the use of such physiological techniques as ablations and neural severing. Human data, with the exception of such philosophical speculations as "What would happen if a man born blind could suddenly see?" (Berkeley, 1709), or a few accounts of cataract patients recovering their sight after operations (Hochberg, 1963), have been rare. Animals, in other words, have provided most of our data about the nature-nurture issue, and for this very reason we need to raise questions again about inferring whether animals can tell us much about "rational" or "experiential" bases of the human mind.

Generally speaking, although the findings and interpretations of studies involved in the nature-nurture issue remain controversial, they generally support the notion that animals exhibit rather accurate distance, size, and shape perception (as well as brightness and color perception) without learning, although learning may improve the accuracy of the perception involved. The young children, and probably the rat and cat, perceive—or act as if they perceive—depth at an edge (E. J. Gibson and Walk, 1960 and 1961), and the rapid approach of an object (Fishman and Tallarico, 1961a and 1961b; Schiff, 1965) without opportunity for associational learning. If "unlearned" may be loosely considered as equivalent to "innate" (neither term *explaining* much of anything), one can conclude that at least some species walk on the "nature" side of the issue.

The findings are less clear regarding the localization of objects in space. One oft-quoted finding is that chicks wearing vision-distorting prisms do not improve their pecking accuracy. Presumably, this finding indicates not only a built-in perceptual localization mechanism, but a rather inflexible perceptual-motor system as well (Hess, 1956). However, a more recent study (Moray and Jordan, 1969) has produced results which expose the earlier findings to some searching questions.

Rather than pursue the debate in terms of *either* nature *or* nurture, Held and his co-workers have been attempting to study the entire sensorimotor "mapping" process in various species (1963). Sensorimotor "mapping" processes entail questions concerning the necessary and sufficient conditions for such behaviors as reaching, grasping, manipulating, and alighting on a surface. Their studies involve testing animals at various stages of development, a procedure which has yielded important information for comparative perceptual psychology.

Whenever the nature-nurture issue, or similar issues, have prompted animal perception studies, there has been a tendency to focus upon *specific* effects of early experience on later, perceptually guided behavior, rather than explore the range of experiences ordinarily occurring in the normal course of development. Since a rigorous curtailment of normal developmental experiences is required to support the "nature" side of the issue, the constrictive role of the experimental manipulation mitigates against discovering the usual causes of sensorimotor development. Yet, the simple "nurture" answer that a perceptual behavior is learned is also a dead end, since it is falsely considered a sufficient answer when it is at best a categorical statement containing little in the way of useful information.

At the present time, we know something about the comparative aspects of certain perceptual-motor behaviors (such as cliff avoidance and approaching object avoidance), but we lack a phylogenetic perspective with regard to most sensorily guided behaviors. Indeed, predictions about behavior along the phylogenetic continuum are rare, with differences between species often emerging as something of a surprise. When psychologists are concerned with issues other than the ecological and developmental sources of perceptual-motor differences, knowledge is found the hard way.

An Excursion on the "Visual Cliff." A case in point is the recent set of studies using the "visual cliff." The visual cliff is a device used for testing the perception of depth at an edge, and consequent locomotor behavior. An animal is placed on a board on either side of which is a glass surface, below which are stimulus patterns simulating a "deep" and "shallow" drop-off. Preferences in descending to the visually shallow side of the platform are regarded as evidence of depth perception. Gibson and Walk's pioneer comparative studies (1961) of the cliff-avoidance behaviors of hooded and albino rats, chicks, lambs, pigs, turtles, cats, dogs, monkeys, and human infants, were concerned with the nature-nurture issue (since some studies included dark-rearing animals later tested on the cliff). But these investigators were primarily interested in determining what stimulus information animals actually *use* to avoid sharp drop-offs. Although several possible depth cues were available to the animals, it was found that motion parallax—the differential rate of displacement of nearer and farther objects across the visual field during locomotion—was the information which animals most often used. The aquatic turtle was found to be relatively poor in choosing the shallow side over the deep side, although it more often moved toward the shallow side than toward the deep side. Later studies (Routtenberg and Glickman, 1964) showed that although land turtles preferred to descend to the shallow side on a similar apparatus, aquatic turtles descended to shallow and deep sides about equally. Although these authors attributed the difference in behavior to differences in ability to perceive depth in land and aquatic turtles, when one considers the ecology-related behavioral differences of the two, it becomes clear that the conclusion is not justified without independent behavioral indexes of depth perception. As Gibson and Walk stated:

> It is necessary, therefore, to take a lesson from the ethologists and consider how the kind of discrimination to be compared fits with the ecology and biological requirements of the species its method of reproduction, defense, food-getting and territorial adjustment. A real comparative psychology will only be written in such a context. (1961)

Aquatic turtles typically *dive into* visually "deep" places when startled by a sound or sudden movement in their environment. They possess, then, an approach tendency toward deep places as when in their natural habitat they slip off a rock, island, or log into the protective embrace of deep water. This tendency, combined with a tendency to explore visually deep as well as shallow places (Routtenberg and Glickman, 1964), cancels the validity of the conclusion drawn from a minimal or nonexistant preference for the shallow side of a visual cliff. Such animals may, in fact, *be* inferior in their ability to discriminate depth, as Yerkes (1904) indicated over 60 years ago. That is not the critical point! The point is that the particular *test* of depth perception, while appropriate for many species, is a noncritical test for this species because of the animal's characteristic behavior patterns. To use descent from a platform within the context of a large-scale comparative study of depth perception where the method must be kept as constant a factor as possible is understandable. But to continue to use the technique when interested in the depth perception of this particular species reflects the myopia of the psychologist more than that of the turtle!

Had the visual cliff been used simply to test the nature-nurture issue, little knowledge would have accrued to comparative perceptual psychology. But the early finding that some animals manifested depth perception without opportunity for learning led the way for further study within and without the context of controversy. Since early experience is usually invoked to account for otherwise "innate" processes, many investigators attempted to demonstrate that depth perception is related to experiential factors. It was found that early visual, tactile, and habitat-related factors do play a role in some species' cliff avoidance (Carr and McGuigan, 1965; Krames and Carr, 1968; Tallarico and Farrell, 1964). Herring-gull chicks were studied as well as domesticated species (Emlen, 1963). And, in a more comparative vein, it has been discovered that different species rely on different sensory information, which affects their cliff behavior. Chicks, with their reliance on visual cues, are responsive to a lack of optical support (standing over the chasm), while rats, which ordinarily use nonvisual sensory information to guide their behaviors, are indifferent to a lack of optical support (Schiffman, 1968). In fact, "errors" in descent in the rat—noted by Gibson and Walk in their studies—are apparently related to the rat's reliance on tactile information from its forepaws and vibrissae when it is available (Schiffman, *et al.*, 1967). Only when forced by the circumstances of the experimental apparatus to use visual information, does the rat do so.

Thus, while many species avoid sharp drop-offs, probably for a variety of reasons, some do not. The Mongolian gerbil, a burrowing animal ordinarily found in desert areas, shows no significant preference for either side of a visual cliff (Thiessen, *et al.*, 1968), while the closely related Egyptian gerbil initially goes to the shallow side, then explores the deep side (Routtenberg and Glickman, 1964). However, the conclusion drawn about the Mongolian gerbil's "deficient" depth perception (Thiessen, *et al.*, 1968) is again possibly an artifact of a single test of depth perception. I have found that another burrowing species, the fiddler crab, shows excellent depth perception when the stimulus used to test it is that provided by a rapidly approaching object (Schiff, 1964 and 1965). This test of depth perception was also used with a variety of animals, including frogs, toads, pigeons, kittens, monkeys, and humans (Schiff, 1964 and 1965), and was later used with turtles (Hayes and Saiff, 1967), with the finding that they reliably withdrew their heads as the object apparently approached them. Even in the quest for rather simple answers regarding animals' perceptual abilities, it is not always sufficient to use a single indicator of that ability. One must consider the nature of the stimulus being presented in relation to the animal's ecology and its typical response characteristics. Fiddler crabs often run when approached, but then flatten out on the surface supporting them, raising their large claw before their eyes. All of this implies depth perception, yet the opposite conclusion might be reached if one tested them on a visual cliff. Kittens flinch in response to the stimulus of a rapidly approaching object, monkeys flee and "huddle," humans blink, frogs jump or "freeze" (Schiff, 1964 and 1965).

The ontogeny of the organism must also be considered before predictions can be made with any assurance. I found that nestling pigeons *gaped* in response to the "looming" stimulus, whereas mature pigeons manifested avoidant behavior to the same stimulus (Schiff, 1964 and 1965). In this case, two radically different behaviors in the same species pointed to the same percept (that of an approaching object) at different stages of development.

A related developmental difference was noted in young and mature fiddler crabs. Simply because one behavior may indicate a percept at one stage of development does not necessarily imply it will and has always been irrevocably linked with that percept. Simple one-to-one relationships between stimuli and responses are often not to be found when many species are employed in sensory-perceptual studies. While some species of birds gape in response to visual stimuli, others do so in response to auditory stimuli provided by the parent or a suitable surrogate (Tinbergen and Kuenen, 1939). Although the emphasis has been on visual stimuli in nestling behaviors (Tinbergen and Kuenen, 1939), the role of auditory and vibratory stimuli has also been noted.

Two related points emerge from the visual cliff and approaching object examples. First, if all the comparative psychologist concerns himself with is a particular psychological issue, such as the nature-nurture issue, or the presence of absence of depth perception at a particular stage of development, little will be learned about the animal's utilization or sensory-perceptual information. Second, unless the animal's ecology and stage of development are considered, even those conclusions reached may be in error. These points have been broached in the past, but are only gradually being taken seriously, as I have indicated here.

> Finally, we can suggest that the nativism-empiricism controversy be abandoned as such, with the aim of restating the problems of development more specifically. They should be stated in terms of the species under consideration, its environment and means of adjusting to it, and especially in terms of the information provided by the environment for this adjustment. (E. J. Gibson and Walk, 1961)
>
> It is well known that the "sign" of the taxis responses of many insects and other animals varies according to the age or excitability, developmental stage, and external conditions. . . . These phenomena appear to have in common an effect on metabolism, decreasing or increasing functional thresholds according to their direction. There are many physiological means of changing an animal's readiness to respond, which may require more or less involved relationships in the motivational pattern according to species and ontogeny. (Schneirla, 1964)

In summary the organization of animal sensory-perceptual research around controversial issues in psychology has tended to constrict research about how animals use their sensory apparatus. Even though psychologists have been seeking evidence to support theories about issues such as nature-nurture, and even though we have amassed considerable data about the sensory sensitivity of animal species, we need to integrate theory and this growing body of empirical evidence into a more functionally oriented approach to comparative sensory-perceptual psychology. We need to avoid the tendency, moreover, to ignore the ecology and the response capabilities of different species, while attempting to be broadly comparative when approaching perceptual questions.

The Further Organization of Research in the Psychologist's *Umwelt*: Apparatus and Procedure Biases

> The whole rich world around the tick shrinks and changes into a scanty framework consisting, in essence, of three receptor cues and three effector cues—her *Umwelt*. But the very poverty of this world guarantees the unfailing certainty of her actions, and security is more important than wealth. (von Uexküll, 1934)

A contemporary ethologist investigating the *Umwelt* of a contemporary comparative perceptual psychologist would doubtless identify certain "sign stimuli" involved in the "innate releasing mechanism" producing research (Tinbergen, 1951). As mentioned previously, one set of stimuli focusing the psychologist's attention has to do with controversial issues in psychology; another has to do with the physical layout of his laboratory, and habitual procedures he uses in measuring, recording, and presenting stimuli.

Many ethologists have attempted to devise stimulus displays pertinent to the animal's perceptual world. Typically these displays consist of patterns of stimuli which more or less resemble those found in actual encounters in mating, courtship, feeding, fighting, construction of living quarters, migratory excursions, and the like. The purpose of these displays is to discover what the critical stimulus is that the animal is responding to. Unlike McDougall, who spoke of response to *objects,* it is now realized that animals often respond to stimuli present in an object that have no necessary connection with the object at all. For example, if one paints a piece of cardboard properly, makes a "dummy" properly, it is not necessary that the object resemble the original source at all. In fact, a model quite unlike the original may elicit a response more readily, or more strongly than the original object did, even when the model may be a completely improbable one—too large, made of wood, and so on. Certain insects approach and climb vertical dark bands which apparently resemble trees; proper spotting of models may elicit egg retrieval, feeding behavior, courtship and sexual behaviors, or aggressive responses, depending on the colors used, their spatial arrangements, and similar stimulus factors (Tinbergen, 1951; see Kalmus, H. in Ratner and Denny, 1964). Other animals (ducklings and chicks, for example) flee when appropriate sounds or silhouettes are presented (Tinbergen, 1951). The origin of such behaviors has been questioned (Lehrman, 1953) as has the exact nature of the stimulus eliciting them (Hirsch, Lindley, and Tolman, 1955). But the fact remains that such research attempts to discover the *critical stimulus feature* the animal responds to. Such stimuli are thus functionally related to the animal's natural behavior, and the basis of their discrimination is thus behaviorally relevant.

In contrast to the ethologist's studies, quite another picture emerges when one looks into the usual comparative laboratory. Not only is the terrain dotted with programming equipment, but the typical behavior spaces for the animals being studied are the cubicles for operant studies, an occasional choice box or Y maze, or, in the case of monkeys, the ubiquitous Wisconsin General Test Apparatus. There is nothing wrong with these devices. They are useful. They provide a basis for cross-laboratory comparison, a precise control of stimulation, and the measurement of responses. Some problems do arise, however, when these devices become so figural in the investigator's *Umwelt* that they dictate the kinds of research he does, and the kinds of *stimuli* he incorporates in his perceptual investigations. These devices, particularly the operant devices, are remarkably useful in assessing the sensory sensitivity of some species, at any rate. But the stimuli are most often the meaningless patterns of transmitted light, tones, geometric figures, stripes, and gratings which have evolved from classical psychophysical research with humans. They often reflect (or necessitate) the sensory-sensitivity interests of the investigator, and are well suited for that purpose. But they are, for the most part, stimuli for animal "sensation," not animal "perception," as J. J. Gibson has noted:

> When one studies the evolution of the "senses" in animals, about which a great deal is becoming known, a puzzle appears in that they seem to have evolved not to yield sensations, but perceptions. For example, there is no survival value in being able to distinguish one spectral wavelength from another (pure color), but there is a great value in being able to distinguish one pigmented surface from another in variable illumination. Similarly, the distinguishing of pure tones was useless in evolution, although the distinguishing of sounds like those which specify running water, or animal cries, or the vowels of speech were very useful. There was no utility for the chemical senses to identify chemicals, but great utility in identifying the taste or odor of food, species, mate, or young. . . . In short, the survival value of the "senses" is found in the ability of animals to register objects, places, events, and other animals; that is, to perceive. (1962)

Thus, while there is nothing wrong with particular apparatus, such as the Skinner box, there is a danger that they will be used too generally in the study of animal perception, simply because, like Mt. Everest, "they are there." Some animal behaviors—for example: direction and distance of locomotion; seasonal behaviors, such as flock migrations; interindividual-linked and once-a-year behaviors, such as "queuing" (forming trains or columns of animals during mass migration) of lobsters (Herrnkind, 1969); and so forth—are not relevant to the "rate of response" measure used in most operant procedures, and some of the stimulus configurations appropriate for animal perception are not conveniently programmed into the *Umwelts* of the Skinner box, or the WGTA. If we are primarily interested in animals' use of sensory information, not simply in controlling their behavior, a flexible perceptual-behavioral system is required of the comparative sensory-perceptual psychologist.

In conclusion, comparative approaches to sensory-perceptual processes have evolved from early attempts to interrelate sensory functioning throughout the animal kingdom, through a sensory-sensitivity phase, to a sensory-utility phase. The rise of the behaviorist in this country led to a great use of animal subjects, but contributed relatively little to our understanding of how animals use their sensory systems to extract the information for survival from their environments.

Interest in certain controversial issues in psychology, as well as in broader comparative questions, has focused attention upon stimuli which are more appropriate to perception than to sensation.

The sensory-sensitivity approach (animal psychophysics) has yielded information of limited value in explaining how animals actually use their senses. Its influence may be prolonged and unnecessarily generalized by the presence of expensive and habit-forming equipment in the typical animal laboratory. The attainment of a knowledge of animals' *Umwelt(s)* is thus related to the *Umwelt(s)* of comparative psychologists, and one hopes that the psychologist's *Umwelt* will include the stimuli, response characteristics, ecology, and developmental level of the animal being studied.

THE LANGUAGE OF VISUAL PERCEPTION*

Jonas F. Soltis

Both philosophers and psychologists have been concerned with the phenomenon of visual perception, but it has been the psychologist who has concerned himself with an aspect of perception which has been slighted too frequently by the philosopher and which proves to be directly relevant to education. The traditional philosophical approach to perception has been through epistemology, and the primary focus has been on the development of an adequate theory to explain how we gain knowledge of the external world by means of the senses. Not so concerned with the *acquisition* of knowledge but more with the *way* the mind operates, the psychologist, in theory and experiment, has attempted to deal with the intimate relationship between the mind and the eye as knowledge *already possessed is utilized* in seeing. Currently, however, some contemporary analytic philosophers have turned from the traditional epistemological problems in which perception as a philosophical

* Reprinted with permission from Jonas F. Soltis, "The Language of Visual Perception," in Komisar and Macmillan (eds.), *Psychological Concepts of Education* published by Rand McNally Company. This essay is a distillation of some of the aspects of the language of visual perception dealt with at more length in the author's book *Seeing, Knowing, and Believing* (London: George Allen and Unwin, Ltd., 1966 *and* Reading, Massachusetts: Addison Wesley Publishing Company, 1966).

concern has been embedded, and have looked at the relationship between seeing, knowing and believing.[1]

If there is any area of agreement which can be taken as common to both the psychologist and these philosophers, it is the basic assumption that what we see and the way we see it is directly related to what we know and believe about the world and the things in it. As an example of this relationship, imagine an archaeologist and a school child in a museum both viewing the same Grecian urn. The knowledge which the archaeologist possesses and brings into play in his perception of the urn makes for a *different* visual-mental experience from that of the child. The claim that the archaeologist's experience is not only different, but also richer and fuller, is based precisely upon the assumption that what he knows affects the way he sees what he sees. In a thousand similar but perhaps less obvious ways, the knowledge we possess is constantly being utilized in the daily business of seeing. For the educator, recognition of this fact should make explicit at least one important avenue in which knowledge acquired in the schools is directly relevant to the total life of the individual. To be brought to realize that what a student learns will obviously color not only what he sees but also his appraisal of it and his action toward it, is to be made aware of an important aspect of the relationship between learning and living.

Common also to the philosopher, the psychologist, and the educator is their initial conceptual frame of reference: the ordinary language of perception in which are embedded certain other assumptions about seeing which enter into the framing of epistemological problems, experimental procedures, and pedagogical methods and purposes. To make explicit some of these assumptions is the purpose of this essay. Thus, in following the current analytic approach to perception, I hope that an unraveling of what we assume about seeing as indicated by the very way in which we speak about seeing will prove relevant to the work of philosophers, psychologists, and educators. Due to limitations of space, however, I will attempt to focus mainly upon those ideas and distinctions which are most relevant to education.

Broadly speaking, there are three basic ways in which the term "seeing" is literally used, and these will provide the framework for this analysis. First, there is the simple, literal sense of "seeing" used only when minimal requirements are met which allow that seeing has taken

[1] For a short basic statement of this new concern, see G. J. Warnock, "Seeing," *Proceedings of the Aristotelian Society,* New Series, Vol. LV. Among others, the following works deal with some aspects of this new approach to perception: Gilbert Ryle, *The Concept of Mind* (London: Hutchinson House, 1949); Roderick Chisholm, *Perceiving: A Philosophical Study* (Ithaca: Cornell University Press, 1957); D. M. Armstrong, *Perception and the Physical World* (London: Routledge & Kegan Paul, Ltd., 1961); N. R. Hanson, *Patterns of Discovery* (London: Cambridge University Press, 1958).

place, or, put in another way, used to separate cases of seeing from not seeing. A man who "sees" pink elephants is not literally seeing, whereas a man who sees grey elephants at the zoo is seeing, and it's not just the color of elephants which forces this distinction. Second, we use the term "seeing" to indicate not only that someone has literally seen something, but also that he is *right* about what he sees. A man who sees a snake and is right in taking what he sees to be a snake is in a different position than the person who sees a piece of wire but thinks he sees a snake. Thus, the third use of the term "seeing" is reserved for situations in which something is literally seen, but the one who is doing the seeing is *wrong* about what he is seeing. There is, of course, a fourth basic ordinary use of the term "seeing" which is not taken to be literal seeing. Thus, one is said to 'see' the solution to his problem, or to 'see' the point of the argument, but in no way do we take this use literally. This figurative usage is frequently involved in our literal talk about seeing, however, and therefore must be given at least minimal consideration in this analysis, even though the main goal of this essay will be to examine the three literal uses of "seeing" put forth above, in an attempt to provide a preliminary map of the way we use the ordinary language of visual perception.

I

Literally, in its simplest sense, our use of the term "seeing" seems to demand at least the presence of a physical object to be seen plus appropriate visual sensations had by the observer. Obviously, however, these requirements are not sufficient because within the peripheral limits of vision there are objects which we "overlook" or "fail to see," even though in a physical sense they produce visual sensations in us. Ryle, in treating "seeing" as an achievement verb, recognizes this deficiency and attempts to remedy it by requiring that the observer also apply a "perception recipe" to the sensations he has.[2] "Perception recipes," according to Ryle, are made up of knowledge of the looks of things under ordinary conditions and are learned by seeing things and being taught to talk about them.[3] For Ryle the term "seeing" is used to indicate recognition, an achievement, a situation in which an individual correctly uses the knowledge he possesses in application to his visual sensations. He gives the example of seeing a thimble:

> A person who espies a thimble is recognizing what he sees, and this certainly entails not only that he has a visual sensation, but also that he

[2] Ryle, *The Concept of Mind*, pp. 202, 217–228, 230–231, *et al.*
[3] Ryle, pp. 230–231.

has already learned and not forgotten what thimbles look like. He has learned enough of the recipe for the looks of thimbles to recognize thimbles, when he sees them. . . .[4]

He argues that we withhold use of the term "seeing" from situations in which we are wrong about what we've seen.[5] Thus we do not honor the claim that one has seen a snake when there is only a piece of wire in view. We withhold the term "seeing" in such an instance and reassure the individual that he did not see a snake, but merely something which looked like or appeared to be a snake. In such an instance Ryle brings in the figurative sense of seeing which he calls 'seeing.' Thus one can say that the individual *fancies he sees a snake*, but really he only 'sees' a snake. He couldn't see one because there is none to see. But even though the wire is not recognized, in our ordinary view of perception it should be clear that we do assume that the individual has literally seen something (the wire, in this instance). Ryle's formulation of seeing as *recognition* leaves little room for dealing with this very legitimate sense of "seeing" in which we would say something is seen though a mistake is made about what the thing is. Chisholm is more liberal in his statement of requirements for the use of the basic sense of "seeing" and overcomes this deficiency in Ryle. He argues:

Perhaps we would not want to say that a man sees an object x unless, in addition to sensing in the required way, the man also took the object x to be something. . . . There is no paradox involved in saying that a man sees a dog without taking what he sees to be a dog. It may be, however, that we would hesitate to say that he sees a dog if he didn't take it be anything at all. . . .[6]

Chisholm's "taking" requirement[7] is less stringent than Ryle's "perception recipe" requirement in that it agrees with our ordinary notion that something is seen even though a mistake is made, but we may also ask if this requirement is adequate to the task of accurately describing our assumptions about seeing in its simplest sense. Imagine yourself a hunter in a forest accompanied by a trusted and able guide who spots a deer not fifty yards away. You look, but don't see it. The guide exactly specifies where the deer is standing. You correctly focus on that spot and pay particular attention to it but still refuse to say that you can see the

[4] Ryle, p. 230.

[5] Ryle, pp. 150, 153, 234, 246, *et al.*

[6] Chisholm, *Perceiving*, p. 150. Copyright 1957, Cornell University. Used by permission of Cornell University Press.

[7] Chisholm presents a formal definition of taking as: "S *takes* something x to be f means: S believes (i) that x's being f is a causal condition of the way he is appeared to and (ii) that there are possible ways of varying x which would cause concomitant variations in the way he is appeared to." (Chisholm, *Perceiving*, p. 77.)

deer. However, you believe the guide (and it is in fact true that there is a deer there in your field of vision). You take the deer to have the characteristic of being the same color as the brush or the characteristic of indistinguishability from the background, or the characteristic of perfect natural camouflage (for you at least, though not for the guide). This situation oddly enough seems to meet Chisholm's requirements[8] in that it is a proper visual stimulus, you sense in a way functionally dependent upon it (if light is reflecting from it, it would indeed seem that you must), and you have taken it to have some characteristic (a correct one from your viewpoint at least)—but you don't see it, or do you? In Ryle's recognition-achievement sense, of course you don't, but in Chisholm's terms you do, even though you cannot *visually discriminate* the deer. This is not the seeing of something which you can't recognize nor mistaking something you could recognize, but rather it seems to be a case of having something in your field of vision which you could recognize *if you could discriminate it*; but you can't, and hence in a very legitimate sense of the term, we would ordinarily say that you can't see the deer.

Thus, it should be apparent that Chisholm's requirement is too lenient, but in testing it, we seem to have arrived at an appropriate criterion necessary to the basic sense of "seeing" as it is used in ordinary discourse to separate cases of seeing from not seeing. This is the "discrimination" requirement which, coupled with the demand of physical object and visual sensations, provides us with an explicit rendering of the assumptions we make when we use the term "seeing" in its simplest sense. Note, however, that when we use the term "seeing" in this basic way, we neither assume anything about the knowledge possessed and utilized by the observer, nor do we assume anything about the beliefs he may acquire in the seeing situation. It is only when we move on to more complex uses of the term "seeing" that the factors of knowledge and belief take on fundamental importance. Therefore, let us refer to this basic fundamental sense of seeing as "simple seeing," using this meaning as base upon which we can rely as we turn to what may be called "successes and/or failures in seeing." It is in these uses of the term, when knowledge and belief become relevant to our speaking about seeing, that seeing becomes most interesting as a phenomenon for the analytic philosopher, experimental psychologist and professional educator.

II

To speak of success in seeing, we must add to the three assumptions made with respect to simple seeing the requirements that (i) some knowledge possessed by the observer is utilized in (ii) arriving at a true

[8] Chisholm, *Perceiving*, pp. 76–85.

belief about what is being seen. Thus we could say that simple seeing logically (not temporally) precedes success in seeing. Though the ambiguous term "seeing" is often used to mean successful seeing, we do have some special terms in ordinary language to convey the sense of these additional requirements. Most frequently used are the terms "recognition" and "identification." To correctly visually recognize or identify something x, then, is to assume beyond the requirements for simple seeing that one must *know* what an x looks like (have a perception recipe for x) and *believe* truly that what he is seeing is an x.

This simple formulation of what more is assumed in cases where the idea of success is seeing is appropriate belies the variety of seeing situations covered by this sense of "seeing." The variable of knowledge utilized by the observer is the key both for making clearer the *range* of possible types of such situations and for providing some insights into the *ways* in which knowledge may be appropriately utilized in seeing.

We may begin with the question, What minimally must one know to be able to visually recognize or identify something? Obviously, one need not know the object's name or label or anything else about it. He need only know what it looks like, or, put another way, he must possess a perception recipe for the object. Passing by a shop window, one can *recognize* the odd-looking object there as the same one (or similar to the one) he saw as he passed the window yesterday. He need not know what its appropriate name or label is nor anything else about it, just as a witness to a murder could identify the murderer without knowing his name. Basically, then, recognition and identification involve the possession and proper utilization of knowledge of the looks of a thing, but need not involve any further knowledge about the thing seen (nor any verbal knowledge whatsoever). The reverse side of this coin is interesting to examine from the point of view of education. Suppose, for instance, that one did indeed know (has learned) a lot about combustion, but was unable to visually recognize an instance of combustion whenever he came upon one. The knowledge he does possess would be blocked from use because he fails to possess the basic knowledge of a perception recipe.

But let's reverse the coin again. It would seem that being able to recognize a thing in terms of having a perception recipe for it would open the gates for a direct application of other knowledge about the thing seen which the individual might possess. The moral for the educator here, of course, is that if he expects knowledge acquired in schools to be used in the life of the individual, some provision must be made to insure that the student will be able to recognize instances in which knowledge is appropriate. For the psychologist, one might add that in testing for the possession of knowledge by an individual, lack of utilization of appropriate knowledge in a specific instance does not necessarily indicate that the individual does not possess that knowledge.

One important point should be made before leaving this basic notion of perception recipe as a triggering device for bringing other knowledge into play in seeing, and that is that no claim is being made here that in order to be able to visually recognize a thing, a person must have seen it at least once before (thus perhaps suggesting a dominant role for visual aids in educational methods). Rather, it should be clear that one could come to possess a perception recipe through purely verbal acquisition. If one knows what a horse, black and white, and stripes look like, he could acquire the perception recipe for a zebra without ever having seen one. Essential, of course, to this discussion of perception recipes is the focus on the experience of the learner and the need to tie knowledge to experience in some way if one is to expect that it will be utilized in future appropriate situations.

Beyond knowing the looks of a thing, however, is even a vaster area of verbal knowledge which may be utilized in successful seeing situations. I would like to point out three types of such knowledge which, though there may be others, are certainly relevant to education. First, the ideas of recognition and identification are also frequently aligned with the idea of labeling or naming the thing seen. Thus we talk of recognizing an object as a chair, a table, a necktie, an urn. Depending on the context, we may be more or less precise in our labeling of what we see, and it should come as no shock to anyone that many objects have more that one appropriate label. Thus we may label an object seen as a "tree," a "maple tree," or a "sugar maple tree." Or we might identify an object as a "dog," an "animal," a "cocker spaniel," or as "Fido." It should be clear that, like the perception recipe, the label chosen is important as a triggering device to bring into play other knowledge possessed by the observer and associated with a particular label. The knowledge associated with "animal" may be quite different from that associated with "Fido;" hence an appropriate choice of label in a particular context may be most important both with respect to the quality of the visual experience had and in terms of future action called forth by the visual experience.

The "labeling" discussed thus far may be seen in terms of what Austin has called "capping" as opposed to "fitting the bill."[9] That is, the situations represented above are characterized by the presence of some visual object in need of a particular perception recipe and name or label to "cap off" an appropriate recognition or identification by someone. There is, however, another way to view the recognition situation in terms of labeling. There are contexts in which we already have a perception recipe and/or label and are looking for an object to fit it. Thus one might need something to fit the head of a screw and such a recipe may

[9] J. O. Urmson and G. J. Warnock, eds., *Philosophical Papers by the Late J. L. Austin* (London: Oxford University Press, 1961), p. 187.

not only be fitted by a screwdriver, but also by such objects as a coin, a letter opener, a shoe horn, etc. In fact, if we chanced upon an automobile accident and found the driver bleeding profusely, though we don't normally think of a necktie as a tourniquet, overlooking it and failing to see that it would "fit the bill" (relabeling it "tourniquet") would be disastrous in this emergency. So we see that labeling as a type of knowledge utilized in seeing is broad, varied, and flexible in range, besides being fundamentally important to our present appraisals and our subsequent actions.

The second type of knowledge utilized in seeing which I would like to briefly discuss may be called "expectation-producing" knowledge. A simple example should make clear how this type of knowledge operates in a visual situation: Before me on my desk is my Thermos bottle. I see it, recognize it as my Thermos bottle and know that it is filled with hot coffee. This bit of knowledge leads me to expect that if I remove the cork and pour the contents into a cup, I will have a cup of hot coffee. I also know that Thermos bottles have glass liners and therefore I know that if I were to drop it on the floor, the liner would probably break. Much of my knowledge about my Thermos bottle and Thermos bottles in general and, indeed, rigid physical objects leads me to a set of certain *expectations* with regard to future actions and/or perceptions of this selfsame bottle.

Now most philosophers of perception take it that knowledge about an object of perception which is utilized in perception is fundamentally of an *expectation-producing* type. This emphasis has resulted in a tendency to neglect or in a failure to recognize the possibility of another way in which knowledge enters into seeing which is not expectation-producing. For lack of a better term, we might call this third type "embellishment-producing" knowledge in that it adds to and/or ornaments the bare perception, giving it a fuller significance, and giving meaning to the one doing the seeing. If I know, for instance, that this Thermos bottle was a birthday present from my young daughters, this bit of knowledge would enhance my seeing by perhaps making the Thermos more valuable personally to me in this seeing of it as a generous gift but would not in any way be similar to the above example which led me to have certain expectations.

Thus I would argue that in ordinary visual experience, although some of the knowledge we utilize *is* expectations-producing, some in point of fact *is not*. One could multiply such examples of embellishment-producing uses of knowledge in seeing with such contrasts as that already given of the archaeologist viewing a Grecian urn in the museum, bringing his wealth of embellishing knowledge to this seeing, and the viewing of the same urn by the youngster ignorant of the Greeks and their culture. Both no doubt would see that the vase would break if dropped on the floor, but only one could see that the urn is approximately 2000 years old, the type in which the Greeks stored olive oil, was made by an accomplished artisan of the period, etc. Such a multiplication of obvious examples hardly seems necessary, since so much of what comes to consciousness in our seeing of objects is of this embellishing type.

Recognition of this distinction between these two types, however, offers important ramifications for education. Knowledge of the expectation-producing type is frequently found in a law-like statement, a generalization, a principle or the like. Hence, recognition of an object as having membership in a particular class allows one to apply in deductive fashion all those generalizations which apply to the class to which the particular object before him belongs, and this leads to certain expectations about that object. Psychological studies in transfer[10] indicate that transfer by means of the learning of general principles is an efficient vehicle in the utilization of knowledge, and the above discussion of expectation-producing knowledge shows quite clearly how learned general principles are applicable in the everyday business of seeing.

More interesting are the implications for education which come from the distinction of embellishment-producing types of knowledge utilized in seeing. Rather than being general and universal in scope, more often than not this type of knowledge is specific and narrow. The contrast might best be exemplified by the history teacher who sees himself on the horns of a dilemma about which to teach, the facts or the fundamental concepts of history. The tendency in recent educational practice has been to look down the nose at a factual approach to any subject and to applaud one which attempts to emphasize the general principle of the discipline being taught.

The realization of the way that such factual knowledge may embellish our perceptions, however, may force a better balance between these views. After all, it is the specific knowledge of the Greeks which makes more significant and meaningful the seeing of the urn by the archaeologist and not the general knowledge which might lead him to expect the urn to break if dropped. So too, it is the specific knowledge one has about van Gogh and his troubled life which may add zest to the viewing of his self-portrait minus one ear.

The upshot of these distinctions, then, in terms of their relevance for education, is that the learning of both expectation-producing and embellishment-producing types of knowledge is valuable in one's everyday interaction with his world. Neither ought to be neglected, and both offer power to the individual to cope with his environment. On the one hand

[10] For a concise summary of the recent discussion and findings of psychological investigations of transfer of learning, see McGeoch and Irion, *The Psychology of Human Learning* (New York: Longmans, Green and Co., 1956), pp. 228–230. Particular attention is called to Judd's theory of generalization.

is the power to predict and control, and on the other is the power to embellish and give meaning and significance to what he can predict and control.

III

We come finally to the third type of situation, failure in seeing, in which we also use the ambiguous term "seeing." The fact that we are sometimes deluded by our senses has been a thorn in the side of epistemology for centuries; but this same phenomenon has provided the psychologist with a fascinating area for investigation and experiment. The educator, however, also has a stake in the laying bare of assumptions made with respect to the human tendency to err, for few if any students are perfect and always get things right.

Philosophers have generally failed to make significant distinctions between the many ways in which we may fail at seeing and have tended to lump all such failures in a single class under the heading "illusion." Only a moment's reflection is needed, however, to realize the importance and need for a distinction to be made between hallucination and other types of failures. Hallucinations, to adopt Ryle's terminology, are pure cases of 'seeing,' fancying one sees something when there is nothing there to be seen. But in such cases as the wire-snake failure, there is something to see and in fact something is seen (in the sense of simple seeing), though we can also say that one merely thinks he sees a snake or that he only 'sees' a snake. Hallucination is really, then, a case of failure *to* see (no seeing takes place), whereas a wire-snake instance is a case of failure *in* seeing (some literal seeing takes place but goes awry). In what follows, I will limit my discussion to failures *in* seeing, where, in fact, what is assumed beyond the requirements of simple seeing being met, is the additional assumption that one has acquired a *false belief* about what he is seeing.

Immediately, a basic distinction is possible when we also consider the assumption that one of the factors making for the acquisition of a false belief in a seeing situation is the knowledge which the individual utilizes in the seeing situation. The knowledge of the looks of snakes makes possible the false belief that one is seeing a snake when in fact he is really seeing a wire. But it is important to note that one could not believe he sees a snake unless he knows something about snakes, nor, and perhaps more important, could one correct his mistaking of the wire for a snake unless he knew something about wires. These facts lead to the necessity for a distinction between what I will call "mistakes" and "errors."

In order to get a thing right after being wrong about it requires that one know how to get it right. The term "mistake," then, could be used to indicate just such a situation. However, if one got a thing wrong, but did not possess the appropriate knowledge for getting it right, we could differentiate this situation from "mistake" by calling it an "error." Furthermore, this distinction between "mistake" failures and "error" failures has fundamental relevance for education. Teachers, I would assume, are interested in their students' getting things right, but when a teacher attempts to achieve this end by indicating in a specific situation that a student is wrong, the teacher is only doing half the job unless the student has only made a mistake and not an error. If a student has made an error, it is important for the teacher to realize that the student needs to know more than that he is wrong; he needs some additional knowledge before he can be expected to get the thing right. In essence then, the distinction rests on the fact that some failures are corrigible without the need for new knowledge to be acquired by an individual (mistakes), while others are not corrigible unless new knowledge is acquired by the individual (errors).

There is another distinction implied by our assumptions about failures in seeing which I also feel has relevance to the philosopher, psychologist, and the educator. Once the distinction is made between failing *to* see and failures *in* seeing, it is important to note that there is also possible a basic distinction with regard to types of failures *in* seeing which has to do with the loose use of the term "illusion." The corrigibility factor is also basic to this distinction, but in a slightly different way.

There are, on the one hand, certain failures in seeing which we may characterize as "*ordinary* mistakes or errors" as opposed to what more strictly may be called "illusions."[11] Taking the wire-snake example as representative of a large number of similar kinds of failures in seeing, we should note that when and if the mistake or error is corrected, the look of the thing seen is changed at least to the extent that it now has a more wire-like look than a snake-like look. But in the case of illusion, there is, even after the individual ceases to be deluded by it, a relative constancy of looks not found in the ordinary cases of mistake or error. Thus, even when we know about, for instance, the principle of refraction, the stick-in-water still looks bent, or while we know railroad tracks are parallel, they still look like they converge in the distance. Thus illusions seem to have a relative constancy of look about them whether we are fooled by them or not, and in terms of our assumptions about the

[11] Of course, the notions of "mistake" and "error" can be applied to an illusory situation and the contrast implied here receives its force from the use of the term "ordinary" to indicate that there is a type of failure like the wire-snake example which is unlike the *extra*-ordinary quality of the "relative constancy of looks" of an illusion as discussed below.

use of the term illusion, this fact seems basic to any strict application of the term, especially if we are forced to distinguish illusion situations (in which we may either succeed or fail) from other types of seeing failures.

Utilizing this distinction would, I believe, make clearer both the philosopher's ambiguous general discussions of illusion and the psychologist's experiments and theories about the phenomenon. But for the educator, again, there seems to be a point of basic relevance between this distinction and what the educator attempts to do. One might say that the correcting of ordinary failures is easier in that the corrected mistake or error provides the individual with a different appearance once the appropriate knowledge is applied. But in terms of illusions or cases in which no real ostensible difference results between a correct and an incorrect belief, convincing a student that he ought to adopt or incorporate the correct view and the knowledge which makes the correct view possible may be much more difficult.

As an example of this point, recall the experiments of Piaget on concept formation in children. There is an experiment in which a tall, thin vessel is filled with a liquid and a shorter, wider vessel is filled with an equal amount. The first reaction of the child is to say that the taller contains more than the shorter. Even when the short vessel is emptied and the contents of the taller poured into the shorter, there still is a persistence for many children to hold that the taller container holds more because it "looks like it does," even though they know that both containers can be filled with exactly the same amount of liquid! If both containers looked alike to begin with, the problem would not arise; similarly, if after the demonstration of equivalent capacities there were a radical change of appearance of the two containers, the idea of equivalency would be easy to establish.

In this brief presentation of fundamental distinction in the consideration of failures in seeing, I have characterized failures in seeing by pointing to the acquisition of a false belief in a seeing situation. But, much as the consideration of successes in seeing led to the realization of the degree of variety possible from the simple recognition of something without any other knowledge about it coming into play to the more sophisticated seeings, which involve embellishing and expectation-producing knowledge, similarly, I would argue, there can be degrees of failure in seeing. From being all wrong, having only false belief(s) about what is being seen, to having many true beliefs and only one false belief presents a tremendous range of possibilities. Though this range and variability of failures in seeing cannot be dealt with here in this preliminary sketch of the language of visual perception, just to realize its existence should be important to the future philosophical, psychological and educational concerns with the phenomenon of visual perception and even relevant to the notions of success and failure as general categories.

In this brief essay, then, I have tried to sketch and make explicit some of the assumptions we hold when we talk in an ordinary way about visual perception. Rather than solve any problems or present any theories, I have merely tried to argue that "there is more to seeing than meets the eyeball,"[12] and what more there is, is quite relevant to the pursuits of the philosopher, the psychologist, and the educator.

FROM PERCEPTION TO INFERENCE: A DIMENSION OF COGNITIVE DEVELOPMENT*

Joachim F. Wohlwill

Introduction

How shall we conceptualize the changes which the child's mental processes undergo during the course of development? This question has been answered most frequently in terms that emphasize an increase in powers of abstraction or an increased intervention of symbolic processes. More generally, one might say that there is a decreasing dependence of behavior on information in the immediate stimulus field. For instance, in the delayed reaction experiment we find that the maximum delay that may intervene between the presentation of a stimulus and a discriminatory response increases with age (Munn, 1955, pp. 306ff.). Similarly, much of Piaget's work on the development of concepts—particularly that on the conservation of length, weight, volume, number, and so forth—is interpretable in terms of the increasing stability of concepts in the face of (irrelevant) changes in the stimulus field.

We have here, then, the makings of a significant dimension along which to analyze the course of cognitive development. The eventual aim of this paper is to suggest a more systematic approach for such an analysis, based on certain principles relating to the ways in which the organism utilizes sensory information. However, the realization of this aim

[12] Hanson, *Patterns of Discovery*, p. 7.

* Joachim F. Wohlwill, "From Perception to Inference: A Dimension of Cognitive Development," from Kessen and Kuhlman, eds., *Thought in the Young Child.* Monographs of the Society for Research in Child Development, vol. 27 (1962), pp. 87–112.

presupposes an adequate understanding of the interrelation between perception and thinking; it should therefore prove valuable to undertake a prior examination, in some detail, of the various ways in which this relation has been conceptualized, and more particularly of the developmental aspects of this problem.

A prefatory note of caution—given the notoriously elusive and ill-defined nature of such concepts as perception and thinking, no single, uniformly acceptable characterization of their relation is to be expected. For the same reason, the analysis of the developmental changes in the relationship between these two functions is beset with obvious difficulties. Nevertheless, we shall find that the alternative formulations that have been proposed to deal with this problem, and especially Piaget's illuminating comparison between perceptual and conceptual development, are not only of great interest in their own right, but contribute materially to the dimensional analysis of mental development.

Three Views of the Perception-Conception Relation

Let us start by reviewing three different ways in which theorists have conceptualized the relationship between perception and conception. These three clearly do not exhaust all of the different positions that have been taken on this question, but they probably represent the major trends of thought; of greater importance, they define three sharply differentiated foci from which this problem may be approached, so that their consideration should bring out some major theoretical issues. It should be noted at the outset that all three of these viewpoints are essentially nongenetic, at least insofar as any explicit treatment of development is concerned.

The Gestalt Position

One of the solutions to the problem at hand is to take a model of perception and to attempt to fit it intact to the area of thinking, thus reducing these two functions to a common set of basic processes. This appears to be in large measure the course followed by the Gestalt school in its efforts to interpret phenomena in the field of the thought processes—as seen in Köhler's classical work (1925) on the problem solving behavior of his chimpanzees or Wertheimer's analysis of "productive thinking" (1959) in the solution of mathematical and other conceptual problems. In these works we find a heavy emphasis on such quasiperceptual terms as "insight," "restructuring of the field," "closure," and the like, which seem to represent the sum, if not the substance, of the repertoire of concepts used by the Gestaltists to handle the processes of human reasoning. This point is expressed quite explicitly by Koffka in *The Growth of the Mind.*

After paying lip service to the increasing importance in the development of thinking of psychological processes affecting a delay between a stimulus and a consequent reaction of the individual, Koffka, states that

> . . . the ideational field depends most intimately upon the sensory, and any means that enable us to become independent of immediate perception are rooted in perception, and, in truth, only lead us from one perception to another (1954, p. 49).

This formulation, quite apart from its rather meager empirical yield, does not seem to have proved overly successful in its theoretical power. Not only has a major portion of problems in the field of thinking been left aside (e.g., concept formation, the nature of symbolic processes, and so forth), but even when applied to the situations with which the Gestaltists have concerned themselves, the explanatory worth of their concepts appears quite limited.[1] Thus, interpretations of problem solving in terms of restructuring of the field have a somewhat hollow ring in the absence of attention to the question of how a Gestalt may be restructured and of what keeps it from being appropriately structured at the outset. In fact, the whole problem of the ways in which conceptual activity may *transform* an immediate percept is ignored. Paraphrasing Guthrie's dictum about Tolman, whom he accused of "leaving the rat buried in thought," one might therefore be justified in criticizing the Gestaltists for leaving the organism too readily short-circuited in closure to permit him to think.

Last, but by no means least, the a prioristic and thus inherently non-genetic bias of the Gestalt school should be noted. In their work, even when it deals with the behavior of children, as in the books by Koffka and Wertheimer cited earlier, there is little interest in matters relating to developmental changes underlying such behavior—a limitation for which Piaget (1946, 1954), among others, has repeatedly taken them to task.

Bruner's Position

Let us examine next a point of view diametrically opposed to the Gestaltists, one which regards perception as basically an inferential process, in which the perceiver plays a maximal—and maximally idiosyncratic—role in interpreting, categorizing, or transforming the stimulus

[1] The work of such investigators as Duncker and Maier might be cited in refutation of this statement. But these psychologists really fall outside the classical Gestalt tradition, utilizing concepts that bear little direct relationship to the principles of this school of thought—cf. Maier's "functional fixedness" and the general attention given to problems of set.

input. This view is represented generally by the latter-day functionalist school of perception, particularly that of the transactionalist variety. Its most explicit statement has, however, come from Bruner (1957b), according to whom

> Perception involves an act of categorization . . . the nature of the inference from cue to identity in perception is . . . in no sense different from other kinds of categorical inferences based on defining attributes . . . there is no reason to assume that the laws governing inferences . . . are discontinuous as one moves from perceptual to more conceptual activities (1957b, pp. 123f.).

While Bruner claims neither that all perception processes can be encompassed in such a theory nor that it precludes a distinction between perceptual and conceptual inference, he does argue that the theory covers a wide variety of perceptual phenomena which conform in many essential respects to principles akin to those observed in the conceptual sphere.

Bruner's formulation raises a number of difficult questions. What is the implicit definition of perception on which it is based? What is the role assigned to structural aspects of the stimulus in such a model of perception? Most importantly, perhaps, to what extent does the operation of conceptual mechanisms in perception depend on conditions of inadequate or impoverished stimulation? Bruner has not ignored this latter problem, but he is inclined to dismiss its importance; for example, he reduces the difference between ordinary and tachistoscopic perception to a matter of degree—inferential mechanisms are always at work, but categorizations vary in the univocality of their coding of stimulus cues in proportion to the amount of stimulus information provided. Thus, for Bruner, veridical perception is a joint function of redundancy in the stimulus and the accessibility of appropriate categorizing systems, in the following sense:

> Where accessibility of categories reflects environmental probabilities, the organism is in the position of requiring less stimulus input, less redundancy of cues for the appropriate categorization of objects . . . the more inappropriate the readiness, the greater the input or redundancy of cues required for appropriate categorization to occur (1957b, p. 133).

We will find this notion of some interest in connection with one of the dimensions to be proposed later for tracing the development from a perceptual to an inferential level of cognitive functioning. For the present, it may suffice to point out, as Piaget and Morf (1958a) have, that Bruner's model of perception presupposes an adult perceiver; it would be difficult to apply it to the perceptions of a very young child, whose conceptual categories were still in the process of formation. Not surprisingly, under the circumstances, we find that Bruner has thus far failed, as much as the Gestaltists, to consider the developmental aspects of perception and thinking, either in the paper discussed here or in his monograph on thinking (Bruner, Goodnow, and Austin, 1956).

Brunswik's Position

The third viewpoint to be considered is that of Brunswik, who occupies a place somewhere between the two poles just discussed, emphasizing as he does the differences between perception and thinking, rather than attempting to explain one in terms of the other. While his untimely death kept him from pursuing this question beyond the sketchy treatment of it in his last work (Brunswik, 1956), his ideas still may contribute significantly to a workable distinction between perception and thinking—a point which we shall have occasion to acknowledge in the last portion of this paper.

Brunswik starts out by drawing a comparison—based on an actual empirical study—between the achievements of perceptual size judgments in a constancy situation and those of arithmetic reasoning where the equivalent task is presented in symbolic form. The perceptual task yielded the typical clustering of settings within a fairly narrow range of the point of objective equality; in contrast, a majority of the answers given to the arithmetic reasoning task coincided exactly with the correct value, but several subsidiary clusters of answers were found which were quite discretely separated from this mode and which corresponded to false solutions of the problem.

Generalizing from this example—the significance of which is obviously purely demonstrational—Brunswik contrasts the machinelike precision of the reasoning processes with the more approximate achievements of perception:

> The entire pattern of the reasoning solutions . . . resembles the switching of trains at a multiple junction, with each of the possible courses being well organized and of machinelike precision, yet leading to drastically different destinations . . . the combination of channelled mediation, on the one hand, with precision or else grotesquely scattered error in the results, on the other, may well be symptomatic of what appears to be the pure case of explicit intellectual fact-finding.
>
> On the other hand, . . . perception must simultaneously integrate many different avenues of approach, or cues. . . . The various rivalries and compromises that characterize the dynamics of check and balance in perception must be seen as chiefly responsible for the above noted relative infrequency

of precision. On the other hand, the organic multiplicity of factors entering the process constitutes an effective safeguard against drastic error (1956, pp. 91f.).

This conception of the difference between perception and thinking, while hardly exhaustive, is a fairly intriguing as well as plausible one. It has, moreover, definite implications for the analysis of the development of reasoning, although Brunswik has not given these explicit consideration. It is pertinent, however, to note his suggestion in regard to the developmental changes in color and shape constancy which he studied in his early work; he attributed the decline in constancy found in adolescence to the intervention of cognitive mechanisms which lessened the *need* for precise veridical perceptual achievements (cf. Brunswik, 1956, p. 91).

Developmental Approaches to the Interrelationship between Perception and Conception

The three contrasting positions just discussed serve to sketch out the boundaries within which one can trace the course of cognitive development from perception to thinking. As noted above, of the three positions, Brunswik's embodies the sharpest differentiation between these two functions and will be found the most useful for our purposes; in fact, we will presently see a striking similarity between Brunswik's view and Piaget's conception of this problem.

The Views of Piaget

The Two Piagets. Let us turn, then, to the work of Piaget, who has given us by far the most explicit and formalized comparison between perception and thought and between their respective developmental patterns. We should note at the outset that there appear to be at least two altogether different Piagets. On the one hand, we have Piaget, the psychologist of the development of intelligence, author of a long and impressive series of books covering an array of cognitive functions (language, reasoning, judgment) and of dimensions of experience (time, number, quantity, space, and so forth). On the other hand, there is Piaget, the psychologist of perception, author or sponsor of an equally impressive and even longer series of studies on a variety of perceptual phenomena, published in the *Archives de Psychologie*.

To these two divergent areas of interest correspond two sharply differentiated modes of approach to research. The "clinical" method which Piaget has followed in his study of the development of intelligence, with its deliberate avoidance of standardized procedures and quantitative analysis, stands in marked contrast to the more traditional experimental approach which he has favored in his perception research. Furthermore, while Piaget's aim in his work on thinking is essentially a genetic one, his purpose in tracing developmental changes in perception appears to be rather different. The developmental dimension in the perception research represents primarily an additional variable, coordinate with other situational, experimentally manipulated variables through which basic perceptual processes are exhibited. In this connection it is worth pointing to Piaget's view that developmental stages exist in the realm of intellectual, but not of perceptual, development (1965, p. 33). We will consider later the possible grounds for such a position.

In view of these various symptoms of a double personality, it is hardly surprising to find Piaget attempting to divorce thinking from perception and to minimize their mutual interrelatedness. Like East and West, "ne'er the twain shall meet"—or hardly ever. One of the very few instances where they do meet, i.e., where Piaget confronts perception and thinking in the context of the same experimental situation, provides an illuminating picture of his basic position. This is a study by Piaget and Taponier (1956), devoted in part to the investigation of a constant error arising in the comparison of the length of two parallel horizontal lines, drawn to form the top and bottom of a parallelogram (without the sides). In this situation the top line tends to be slightly overestimated; this illusion increases, however, from a zero-order effect at the age of 5 years to a maximum at about 8 years; for adults the extent of the error is intermediate. Piaget contrasts this developmental pattern with that obtained when the same judgment is made in the context of a cognitive task: The two equal lines are presented initially in direct visual superposition, so as to be perceived as equal; the top one is then displaced horizontally, the arrangement of the two lines corresponding to that of the previous problem. In this cognitive task, it is the 5-year-old children who show a pronounced bias in their judgment, which leads them to pronounce the two lines as unequal following the displacement. In other words, there is an absence of "conservation of length" in the face of configurational changes. By the age of 8, however, the equality of the lines is maintained fairly uniformly—conservation of length has been acquired. On the strength of these findings Piaget argues against a simple perceptual explanation for the young children's lack of conservation; since their error of perceptual judgment is at a minimum, their failure to maintain the equality of the two lines in the cognitive task must be due to other factors.

This example illustrates well the independence, in Piaget's thinking, between perception and conception or inference—even at the stage of "intuitive thought" where the child's responses appear to be governed by particular aspects of the stimulus field. In fact, as we shall note, Piaget has repeatedly stressed that these two functions follow very different paths and arrive at different ends during the course of developments (1946c, 1957b). With Brunswik, although on somewhat different grounds, he has been impressed by the statistical, probabilistic nature of perceptual judgments, as opposed to the precise, determinate, and phenomenologically certain results achieved through conceptual inference.

The Concept of "Partial Isomorphisms." Piaget's most recent and most systematic treatment of this question is contained in an article (Piaget and Morf, 1958a) the title of which states his position succinctly: "The partial isomorphisms between logical structures and perceptual structures." In spite of his characteristic reification of such concepts as "structures" and "schemata," Piaget is concerned here with the correspondence between the achievements or end products of perceptual as against conceptual mechanisms, the mechanisms themselves being left largely out of the picture.

In this paper Piaget and Morf discuss a number of phenomena which Werner (1957) has considered as illustrative of "analogous functions," i.e., functions serving similar ends but operating at different levels of cognitive organization. Like Werner, Piaget and Morf draw parallels between perceptual groupings and conceptual classes, between invariance in perception (the constancies) and in conception (the conservations); between the perception of stimulus relationships and the conceptual representation of relationships at the symbolic level. For these authors however, these analogies, or isomorphisms, are only partial; they emphasize, rather, the ways in which perceptual mechanisms differ from the corresponding inferential ones. They point out that perceptual phenomena generally do not meet the requirements of the fundamental operations of logic (reversibility, additivity, transitivity, inversion) except in a limited and approximate sense. For example, with respect to additivity, a line divided into a number of equal segments is actually perceived as slightly longer than its undivided counterpart (the Oppel-Kundt illusion); similarly, in the case of figure-ground reversals the perceptual inversion fails to satisfy the logical criterion of inversion insofar as the boundary line always remains part of the figure. To these examples relating to the logic of classes are added several others involving the logic of relationships. Thus, lack of additivity is illustrated in threshold phenomena, where two subthreshold differences when added together may yield a suprathreshold difference (i.e., $= + = \rightarrow \neq$ is possible in perception). Again, a person's difficulty in judging projective size is considered a case of lack of inversion of the relationship between retinal size, distance and perceived size ($r \times d = p$): given r and d jointly, the subject may "solve" for p, but he cannot obtain r by "dividing through" by d—i.e., by abstracting size from distance.

Finally, Piaget and Morf argue that there are "pre-inferences" in perception which are partially isomorphic to the inferential mechanisms of logical reasoning. Indeed, all perceptual judgment *qua* judgment is thought to involve a decision-process partaking to a greater or lesser extent of the character of an inference from the sensory information given. The extent to which it does so depends on the level of complexity (mediation?) of the judgment, ranging from the simple, direct judgments found in psychophysical thresholds to judgments dependent on "perceptual activity" as in size constancy. Here the difference between these perceptual pre-inferences and conceptual inferences can be found not only in the certainty or univocality of the outcome of the conceptual inference, but also in the subjects' lack of awareness of the separate steps in the inferential chain in the perceptual pre-inferences.[2]

The Perception-Conception Relationship in the Development of the Child. Despite the semblance of a link between perception and inference represented in Piaget's concept of "perceptual pre-inferences," the over-all impression one obtains from his treatment of partial isomorphisms, as well as from other discussions of the differences between these two functions, is of a parallelistic conception—perception and thinking represent two sharply differentiated processes which display certain structural similarities, but even more important differences. Developmentally, too, he considers perception and thinking as following two separate and independent courses, as may be seen in his comparison of the development of the "conservations" from the conceptual realm with that of the perceptual constancies (1957b).

Conservation may be exemplified by the invariance of the volume of a liquid under changes in its container, as when water is poured from a narrow glass into a shallow bowl. Piaget invokes here a gradual process of "equilibration," leading the child from an initial stage at which he focuses only on one biasing aspect of the stimulus (e.g., the height of the container) through an oscillatory stage where he shifts back and forth between this aspect and a competing one (here the width of the container), to a third stage in which the compensatory role of these two aspects begins to be suspected, and then to the final realization, with perfect certitude on the part of the child, of absolute, exact conservation, despite the perceptual changes. In the perceptual constancies, on the

[2] This specification of lack of awareness as a characteristic of pre-inferential processes in perception clearly brings to mind Helmholtz's "unconscious inference." Piaget is careful, however, to dissociate himself from those (e.g., Cassirer) who have read into this concept implications of a ratiomorphic process.

other hand, all aspects of the stimulus field, and notably the two stimuli to be compared, are always included in the individual's perceptual exploration of the situation, at least from a very early level of development. The only developmental change is in the extent and efficacy of this exploration or, conversely, in the potency of distorting factors present in this situation. These factors (e.g., a favored attention to the near object) bring about a relative lack of constancy in younger children, which is reduced in later childhood due to more intensive and complete perceptual exploration of the stimuli. But in the domain of perception the exact compensations achieved in the fourth stage of the development of conservation are not realized; instead, the compensations either fall short, as in most illusions, or actually lead to overcompensation, as in size constancy where overconstancy is the rule for adults.

We are now in a position to appreciate the reasons that probably motivated Piaget's denial of the existence of stages in perception, while affirming it for mental development. This distinction would be warranted, not in the sense that ontogenetic change in perception is necessarily more gradual, but rather in the sense that no meaningful structural criteria can be found in the area of quantitative perceptual judgments for distinguishing among different stages. The differences between successive perceptual achievements are necessarily only quantitative, whereas structural differences of a qualitative type, as in the above-mentioned sequence of stages, can be specified for conceptual development.

Some Critical Comments on Piaget's Views

The foregoing presentation is a highly condensed distillation of Piaget's ideas in which many and frequently subtle lines of reasoning—not to mention a number of obscure points—have been omitted. It would therefore be somewhat inappropriate to base an evaluation of the merits of his argument on the picture of it given here. Nevertheless there are several criticisms of Piaget which can safely be anticipated; let us consider three of these points in particular. This will lead us to a somewhat more general question regarding Piaget's approach and will pave the way to a reformulation in the final portion of this paper.

The first objection that is bound to be raised concerns the nonoperational, and at times frankly mentalistic, terms used by Piaget which may seem to leave his analysis devoid of empirical, and perhaps even of theoretical, significance. For example, the criteria which he proposes for a diagnosis of inferential and pre-inferential processes are anything but unambiguous; indeed, his whole conceptual apparatus of schemata, operations, centrations, and so forth appears to lack direct empirical reference. Admittedly, Piaget does little to dispel this impression; concrete illustrations or applications are at best sporadic, and rigorous, systematic efforts at tying the empirical phenomena to his constructs are generally eschewed in favor of ad-hoc and post-hoc arguments.

It is important to remember, however, that Piaget's ideas on the interrelation between perceptual and conceptual development are not in themselves intended as a theoretical system; they serve rather to explicate, in formal terms, the different models underlying Piaget's theories of perception and intelligence, respectively. Furthermore, a few empirical studies relevant to this discussion can actually be cited (e.g., Piaget and Lambercier, 1946; Piaget and Taponier, 1956; Piaget and Morf, 1958b), and, while the first two of these are mainly demonstrational in character, Piaget and Morf's investigation of "perceptual pre-inferences" represents a step toward a more systematic empirical approach in this area through the manipulation of stimulus cues which change the nature of the task from a perceptual to a more nearly judgmental one. Unfortunately, the experimental design of this study leaves much to be desired, and the rather elaborate interpretations of the results seem unconvincing, if not unwarranted.

A second criticism might well be directed at Piaget's highly idealized conception of adult thought and, at the same time, at his insistence on the distorting and probabilistic character of the processes of immediate perception. In regard to the first point, Piaget has of course been repeatedly taken to task for his inclination to see nothing but perfect logic and rationality in adult intelligence. His reliance on the principles of abstract logic as a model for human thinking has blinded him to the question of the breadth and stability of logic as *used* by the individual. In actual fact, of course, it is little more than a truism that logical principles understood in the abstract may not be applied in particular contexts (as in the atmosphere effects in syllogistic reasoning); likewise, even in the thinking of adults we find frequent instances of failures to apply or generalize a concept or principle when it is presented in unfamiliar ways or extended to novel situations. Differential generalization in the realm of thinking, furthermore, may have all the earmarks of the generalization *gradient* familiar from sensory phenomena.

Conversely, one may argue that Piaget overstates the case for the statistical, approximative, and generally biasing aspects of perceptual achievements. For quantitative judgments, to be sure, Piaget's probabilistic model seems quite appropriate and, indeed, seductively appealing in its simplicity and generality (cf. 1955a).[3] If we deal, on the other hand, with qualitative judgments and more particularly with judgments of identity

[3] An interesting feature of this model is its ability to account for the instances of nondeforming shape perception represented by the Gestaltists' "good figures" as a special case in which the relationships among the component parts are such as to yield, on the average, zero-order errors due to complete mutual compensation among the various possible distortions arising in such stimuli.

or difference among discrete categories of stimuli, we typically find something closely approaching the reliability and specificity of conceptual classifications. Parenthetically, it may be noted that for Bruner it is precisely this type of perceptual judgment which serves as his model of perception, a fact which presumably accounts for some of the ratiomorphic flavor of this model.

If Piaget, then, even more than Brunswik, overestimates the discrepancy between the respective achievements of perception and thinking, he seems also to exaggerate their functional independence. The very fact that the conceptual processes of adults can be characterized along such dimensions as concrete-abstract testifies to the continual interplay between these two functions in much of conceptual activity. Pointing in the same direction are the results from one of Piaget's own experiments (Piaget and Lambercier, 1946) involving size-at-a-distance judgments in which the correct matches could be arrived at inferentially by the intermediary of a reference stimulus. At a certain age level (in middle childhood) there is clear evidence of a "perceptual compromise," showing the mutual interaction, rather than absolute separation, between perception and thinking. We shall attempt to show below how the conceptualization of the development of the symbolic processes in general can be furthered by assigning to perception a differential role in conceptual tasks at different age levels.

Piaget's Structural Approach. If one examines Piaget's thinking further in order to account for his espousal of some of the views just discussed, as well as for the somewhat unsatisfactory explanatory status of the constructs of his system, one finds a ready answer in the structural approach which he has consistently favored in his theory of intelligence. What he seems in fact to have done is to specify the *formal* properties of the products of the thought processes at different stages of development. This has led him inescapably to a picture of successive metamorphoses in the mental development of the child. From this structural point of view, the difference between the reasoning processes of a child lacking "reversibility" and "observation" and an adult whose thinking does conform to these principles will in fact appear comparable to the differences between a caterpillar and a butterfly—or, to suggest a rather more pertinent analogy, between the pattern of locomotion of the 6-month infant and that of the child who has learned to walk. At the same time, this process of conceptual development will emerge as quite incommensurate with the much less dramatic and seemingly more continuous changes in the area of perception. However, it may be that the structural differences between the *products* of perceptual and conceptual processes obscure the continual interplay between the two in most, if not all, cognitive activity and therefore detract from a true appreciation of the differential involvement of perception in conceptual activity at varying developmental levels.

This interdependence between perception and thinking is the major premise for an alternative conception of intellectual development to be offered presently—a conception built around the person's dependence on various aspects of the information contained in the stimulus field. Such a conception, it is hoped, will contribute to a more truly experimental attack on the phenomena of mental development and their determinants, and thereby serve to supplement the structural analysis which Piaget has given us.

Three Dimensions of the Transition from Perception to Conception

If we ask ourselves how one might operationally distinguish between a purely perceptual and a purely inferential task, one criterion for inference would be the opportunity for the subject to supplement or replace the sensory data with information or knowledge not contained in the immediate stimulus field. As a matter of fact, this criterion differentiates the two portions of the study by Piaget and Taponier (1956) referred to earlier in which perception was contrasted to conception within the same stimulus context. The only difference between the two tasks was that in the conservation task the subjects were in effect informed beforehand of the equality of the two lines; this knowledge could take precedence over the lines themselves, and provide a basis for the subsequent judgment under altered stimulus conditions.

It seems possible, however, to formulate this criterion in quantitative, rather than all-or-none terms; that is, the relative amounts of information which the subject needs from the stimulus field in order to make the judgment may vary over a wide range. The precise sense in which this quantitative criterion permits us to place perception and conception at opposite ends of a single dimension will be more fully explained below. For the moment, let us simply grant the possibility of doing so and propose this dimension, along with two others that are closely related, as a skeleton for the construction of an experimentally useful conceptual framework within which the cognitive development of the child may be traced.

The three dimensions along which perception and conception can be related may be specified as follows:

1. *Redundancy:* As one proceeds from perception to conception, the amount of redundant information required decreases.
2. *Selectivity:* As one proceeds from perception to conception,

the amount of irrelevant information that can be tolerated without affecting the response increases.

3. *Contiguity:* As one proceeds from perception to conception, the spatial and temporal separation over which the total information contained in the stimulus field can be integrated increases.

It should be noted that these dimensions are stated in such a way as to be applicable either to intertask differences or to intersubject differences. Let us examine these three dimensions in some detail from the double standpoint of their relevance to the differentiation of perceptual from conceptual tasks on the one hand and to the analysis of changes during the course of development from a perceptual level of functioning to a conceptual level on the other hand—bearing in mind that these two terms are to be regarded as the poles of a continuum.

The Dimension of Redundancy

The dependence of perceptual functions on a high degree of redundancy in the stimulus input is rather easily demonstrated. Redundancy is basic to the differentiation of figure from ground; similarly it is a requisite for shape perception, the perception of speech, and to some extent for perceptual constancy (as shown in the multiplicity of overlapping cues involved in size constancy). In contrast, at the conceptual end we find redundancy reduced to an absolute minimum—typically zero—in the symbolic representation of mathematical or logical relationships. Whether the average adult is capable of operating consistently at this rarefied level is another question; the difficulty which most people experience in dealing with such nonredundant material, and the fact that a considerable amount of redundancy is built into our language, suggests that there are definite limitations in this respect. This conclusion is supported by work on concept formation, such as that of Bruner, Goodnow, and Austin (1956).

A developmental trend in the direction of decreasing reliance on redundant stimulation can be found in a variety of contexts. First of all, within the area of perception as such, the writer found considerable relevant evidence in a recent survey of the literature on perceptual development (Wohlwill, 1960). The clearest example of this point comes from studies on the identification of geometric or familiar-object stimuli on the basis of partial cues (e.g., Gollin, 1956), where the degree of completion of the figure necessary for its identification gradually decreases during the course of development. It seems justifiable, in fact, to regard such a task as becoming increasingly inferential as the amount of information which the subject has to "fill in" increases.[4] Indeed, this appears to be in part the import of Piaget and Morf's study of "perceptual pre-inference," in which the importance of continuity of lines serving as cues in a perceptual judgment was found to decrease with age.

Looking at redundancy in temporal sequences of events, furthermore, one might conceptualize the formation of Harlow's learning sets in terms of the reduction of redundant information to a minimum; it is of interest, therefore, that the rapidity of formation of such learning sets is strongly correlated with mental age (cf. Stevenson and Swartz, 1958).

This conception is relevant, incidentally, to Bruner's (1957b) view of perception as an "act of categorization"; as we noted earlier, he has postulated that the amount of redundant information required for veridical identification is inversely proportional to the availability or accessibility of the particular category in the individual's repertoire of perceptual categories. While the intervention of a specific perceptual category cannot be equated to the operation of general symbolic processes, the fact that the action of both can in some sense compensate for lack of redundancy in the stimulus suggests that Bruner, too, is dealing essentially with a dimension running from immediate perception to conceptually mediated judgment.

The Dimension of Selectivity

The ubiquitous interaction between sensory dimensions in virtually every area of perception (psychophysical judgments and illusions, for example) bears ample testimony to the organism's very limited ability to dissociate relevant from irrelevant information at the perceptual level. At the level of thinking, on the other hand, this dissociation represents a *sine qua non* of conceptual functions; the formation of conceptual classes clearly requires the systematic, selective abstraction of relevant (i.e., criterial) from irrelevant information. The same is true in the realm of logical inference, deductive reasoning, mathematical problem solving, and other such manifestations of symbolically mediated behavior.

It is thus noteworthy that one of the major developmental changes that seems to take place in the development of abstract concepts is precisely the differentiation of relevant from irrelevant, but more readily discriminable, attributes. This development is shown in various studies of concept formation (e.g., Vurpillot, 1960); it may also lie at the heart of a

[4] The view proposed here offers a resolution to a rather ticklish question which confronted Attneave (1954) in his attempt to analyze form perception in informational terms. Should a task in which the subject has to predict the "state" of a visual field at a point, on the basis of information obtained at preceding points of a contour, be considered a perceptual or a conceptual one?

problem which Piaget has studied intensively—the development of conservation. Here one aspect of the stimulus, such as number, weight, volume, or quantity, has to be conceived as invariant, in the face of highly visible changes in some other irrelevant attribute with which it is typically correlated. Similar phenomena are involved in the development of the concepts of time, velocity, and movement, as studied by Piaget.

The Dimension of Contiguity

The third dimension is perhaps the most obvious one. Indeed, the major role which spatial and temporal contiguity plays in perception hardly needs detailed discussion. Spatially, we find it illustrated in the Gestalt law of proximity, as well as in the variation of illusions, figural aftereffects, and other perceptual phenomena as a function of the distance between the central stimulus and contextual portions of a field; similarly, figural aftereffects, among other phenomena, demonstrate the relatively limited temporal span over which two stimulus events separated in time interact.

It is characteristic of conceptual processes, on the other hand, that they enable the individual to deal with stimulus information whose components are widely separated in space or time. To give just one example, conceptual groupings can be achieved where the objects to be grouped are not in close spatial relationship and may not even be exposed simultaneously. Here again absolute independence of contiguity represents an ideal which is scarcely, if ever, realized even by adults. Thus, Davidon (1952) has shown that the opportunity for the subject to manipulate the stimulus materials in an object-classification task so as to provide spatial contiguity for the groupings made improves performance significantly; yet the results are perhaps more remarkable for the small size of the effect which manipulation produced.

Davidon's problem would be an ideal one in which to explore developmental changes; it would be hypothesized that with increasing age this factor of spatial proximity in conceptual grouping would steadily decrease in importance. While there is no evidence on this specific point, a variety of related findings can be mentioned. In the realm of perception, first of all, the writer's review of the literature on perceptual development (Wohlwill, 1960) uncovered various examples of developmental changes in the direction of an increasing ability or tendency to relate objects in the stimulus field, independently of their spatial or temporal contiguity. Such a trend appeared, for instance, in studies of size constancy, which for young children deteriorates much more rapidly with increasing distance between the stimuli than for adults, and in the perception of causality, which for children, but not for adults, requires a perceived contact between the objects in order for them to appear as causally related. With increasing age, furthermore, relatively remote visual frameworks exert an increasing influence on perception in diverse situations.[5]

With respect to tasks of reasoning or concept formation, we unfortunately have much less direct evidence of developmental changes indicating a decrease in the role of this factor of contiguity, although what we know of the thinking and problem solving behavior of children is consistent with the assumption of such an age trend. One experimental study that might be mentioned in this connection is that by Kendler and Kendler (1956), who found that the ability of 3- to 4-year-old children to respond inferentially in a Maier-type reasoning task was closely dependent on the temporal sequence in which the steps in the inferential chain were presented. Thus, if children were shown that *A* leads to *B*, *X* leads to *Y* and *B* leads to *G* (the main goal object), and if they were then presented with a choice of *A* or *X*, the frequency of inferential responses (choice of *A*) was considerably higher if the *B-G* experience had immediately followed or preceded the *A-B* experience, than it was where the *X-Y* experience intervened between these two. Inferential choices likewise depended on the *direction* of the temporal sequence, being significantly more frequent when *B-G* followed *A-B* than when it preceded it—pointing to a rather obvious fact—that the temporal order between two events is of importance quite independently of the interval separating them. It would be of interest to determine whether this ordinal factor also decreases in importance with age.

The Resultant of the Three Dimensions: Specificity

Taken together, these three dimensions yield responses of varying specificity, ranging from those of perceptual judgment, in which accuracy is always relative and error is the rule, to the absolute precision and accuracy of the products of conceptual processes. To this extent they do not represent a departure from Brunswik's and Piaget's conceptions of the problem but rather an extension in the direction of continuity of process from perception to conception.

To illustrate the relevance of our dimensions to this specificity criterion, let us compare the assessment of the relative size of two objects through direct perceptual judgment with that achieved by the conceptual process of measurement, i.e., by the intermediary of a ruler. In the former case, the results will be affected by spatial or temporal separation, by variation in irrelevant aspects, and by lack of redundancy, i.e., one-dimensional

[5] Certain perceptual phenomena appear, however, to be exceptions to this developmental trend, notably the role played by the factor of proximity in grouping and the role of spatial and temporal separation between stimuli in apparent movement. Thus, there appear to be definite limitations to the applicability of the principles outlined here.

stimuli yield larger thresholds than two- or three-dimensional ones (Werner, 1957, p. 118)—all of these factors interfering with accuracy and precision of judgment. None of these aspects, on the other hand, influences the results obtained through measurement, the precision of which is limited only by the accuracy of the instrument and the observer's visual acuity in reading it. A very similar kind of comparison could have been made between the assessment of quantity by estimation and by counting. It is thus of no little significance that, in the case of length and in the case of number, perceptual discrimination and conceptual measurement appear to develop in close interdependence (Braine, 1959; Long and Welch, 1941).

Conclusions

Granting for the moment the validity of the dimensions suggested earlier for the representation of important components of developmental change in cognitive functioning,[6] the question of their conceptual fruitfulness arises. Insofar as they do appear to encompass a wide array of phenomena, their status presumably transcends the level of pure description. It is suggested, moreover, that they provide the basis for the construction of a higher-order theoretical framework within which a more systematic and a more generalized approach to problems in this area may be realized.

The major argument in support of this seemingly pious hope is the built-in potential of these dimensions for leading to a set of constructs which can be securely anchored on the stimulus and response sides and which will also facilitate the integration of developmental changes with principles derived from the experimental study of perception and thinking. To give just one example, the conception should prove of heuristic value in the analysis of the perceptual constancies in terms of the role of stimulus variables such as amount of surplus cues, or redundant information, in interaction with organismic variables such as age.

Admittedly, the actual mechanisms mediating the effects of our dimensions of cognitive activity remain quite obscure as yet. Possibly, neurophysiological or cybernetic models of cognitive activity related to the internal activity of the organism in transforming the stimulus input so as to allow for varying degrees of stimulus determination of behavior may provide us with fruitful leads in our quest for such mechanisms. Thus, one might suggest the operation of scanning mechanisms as characteristic of perception, as against digital mechanisms intervening in reasoning.

[6] For an empirical demonstration of the heuristic value of these dimensions in the study of abstraction see the Appendix to this paper. [The Appendix is not included in this volume—Ed.]

The process of developmental change could then be conceptualized in terms of varying forms of interaction between these two.

It would undoubtedly be sheer pretentiousness to elaborate further upon these highly speculative questions at this point, but one may point to certain empirical hypotheses that appear to be implicit in the postulated dimensions of developmental change themselves. For instance, in the previous discussion of selectivity, it was suggested that the problem of the development of conservation might be handled in terms of the dissociation of a particular concept (e.g., number) from irrelevant, though typically correlated and highly visible, perceptual cues (e.g., length over which a row of elements extends). If this dissociation does in fact represent a factor relevant to the psychological process involved in the development of conservation (as opposed to a mere description of this process), it seems to follow that systematically arranged experiences aimed at untying these two variables for the child should at least facilitate the appearance of conservation.[7]

This brings us, lastly, to the more general question of the bearing of the formulation outlined in this paper on Piaget's work in this area. At first sight, it may seem that the two are at variance in several respects, notably in our postulation of essentially continuous dimensions, as opposed to Piaget's discontinuous stages of development, and in the emphasis here on modes of utilizing stimulus information as against Piaget's system of internal structures, operations, and mental actions. Much of this apparent contradiction disappears, however, if one recognizes that Piaget's concern is with changes in the structural characteristics of the products of intellectual activity, whereas the interest here is in the specification of the dimensions and processes of developmental change.

The distinction between these two essentially complementary approaches may be clarified by reference to their respective handling of the role of external environmental factors. Piaget, as is well known, tends to ignore the effects of antecedent conditions and environmental variables in development, relegating them to a place definitely subsidiary in importance to the unfolding of internal structures. This does not mean that he advocates a strict nativist position, for he has frequently emphasized

[7] With respect to the development of number conservation, the writer is currently investigating the role of such experience experimentally, alongside other experiential effects pertinent to the alternative theoretical formulations of Piaget (involving the activation of relevant mental operations, e.g., those of addition and subtraction) and of learning theory (focusing on the role of reinforcement, through direct confrontation with the *fact* of number invariance).

A very similar research project, dealing with experiential effects in the realm of the conservation of weight (under changes in shape) is being conducted in Norway by Smedslund (1959). As regards the effects of dissociating irrelevant perceptual cues, the preliminary report of this author indicates mainly negative results thus far.

the continual interaction between external and internal forces. Nevertheless, his biological orientation and interest in structure leads him to take external factors for granted and to regard the form which this interaction takes as largely predetermined from the start. The only problem, then, is that of specifying the successive stages through which the organism passes; little leeway is left for differential manifestations of external conditions. It is not surprising therefore that his treatment of learning effects in the development of logical processes (1959a) is limited almost exclusively to the activation of previously formed structures bearing a logical relationship to the particular structure under investigation. In comparison, the approach advocated here probably is less adequate to the task of analyzing the structural complexities of intellectual activity and its development; in compensation, however, it allows for a more thorough exploration of functional relationships between antecedent condition and developmental change and should contribute thereby to a more explicit understanding of the processes at work in the interaction between environmental and organismic forces.

PERCEPTUAL CAPACITIES IN EARLY INFANCY*

Eric Aronson and Edward Tronick

Current research in psychology reflects a strong interest in the problem of infant development. A fair amount of this research has been directed at the question of the origins of perception. The question, of course, stems from one of the oldest issues in psychology, that of nativism versus empiricism. As applied to the problem of perception, this issue concerns whether the structural basis of perception is innate or learned. Students of development must consider this question carefully. An understanding of the starting point of development seems essential to any attempt at understanding growth. This does not necessarily imply that the innate-learned controversy must be dealt with directly, but if we presume that development proceeds cumulatively we cannot ignore early infant capacities. They are the determinants of the perceptual and cognitive growth process.

One of the first psychologists to give serious consideration to the ques-

* This selection was written expressly for this volume.

tion of the newborn infant's capacities was Koffka (1959). His position was decidedly nativistic. Noting scattered observational data which suggested that young infants perceive along highly structured stimulus dimensions, Koffka argued:

> Is it not possible that phenomena, such as "friendliness," and "unfriendliness," are very primitive—even more so than the visual impression of a "blue spot?" However absurd this possibility may seem to a psychologist who regards all consciousness as being ultimately made up of elements it ceases to be absurd if we bear in mind that all psychological phenomena stand in the closest relation to objective behavior. "Friendliness" and "unfriendliness" certainly influence behavior, whereas it is not easy to understand how the behaviour of so primitive an organism as the human infant could be motivated by a "blue spot." If phenomena are to be construed from behavior, must we not attribute to them first of all, such properties as might occasion activity? Certainly the fact that something is being done furnishes the basis for an inference that certain phenomena accompany the behavior in question. But this means that we must assume that features like "threatening" or "tempting" are more primitive and more elementary contents of perception than those we learn of as "elements" in the textbooks of psychology. (1959, pp. 149–150)

Koffka's argument did not have much effect on the course that psychology was to follow. More recent data from infant studies, however, do tend to support his thesis: early infant perceptions do seem to be highly structured. It has been observed, for example, that in the first few months of life, the infant displays characteristically different reactions to human and nonhuman forms (Trevarthen and Richards, 1968). It also appears that infants are selectively attentive to the frequency range of human vocalizations (Eisenberg, 1967). As early as the second month of life, the infant appears to be able to distinguish his mother from other forms which are similar in configuration (Carpenter and Stechler, 1967).

Yet while these observations are consistent with a nativist theory, they do not do much to specify the ingredients of such a theory. We would like to know specifically what perceptual structures the infant possesses, either innately or through rapid learning, which enable him to display such precocious perceptual behaviors. It is hardly useful to maintain that the infant is born with the ability to do whatever it is we observe him doing. Our theory then becomes nothing more than a catalogue of infant behaviors. Explanations which deal with capacities, however, and not simply behaviors, are potentially of much greater utility. Our primary concern must be with those structures which provide organization to his experience.

The only major developmental theory which deals directly with the question of perceptual structure is Piaget's. The basic capacities with which his theory is concerned are space, time, and causality. His position, if it can be characterized briefly, is that the infant progresses in a sequence of stages toward a more veridical perception and comprehension of the world. It is by means of developing spatial, temporal, and causal structures that this progress occurs. By the end of the first two years of life (sensorimotor period), Piaget claims, the basic perceptual structures have developed to the extent that the child apprehends a world of objects that have permanent form and substance. The theory is based upon Piaget's intensive and elegant observations of his own three children (1954).

As Bower (1967) has noted, a major flaw in Piaget's observations is the lack of psychophysical control in his test situations. Simply stated, it is difficult to ascertain what aspect of the test situations Piaget's infants were responding to. We would add that in considering only those behaviors which could be recorded under natural observation conditions (for example, head turning, eye tracking, reaching) Piaget introduced a conservative bias into his observations. The infant of less than three months is apparently incapable of much overt behavior. His inability to act appropriately to sensory input does not imply, however, that the input is not perceptually structured.

Consider, for example, the problem of space perception. To what extent does the infant perceive a stable, three-dimensional world? A natural environmental event is typically complex, conveying much information which could conceivably be perceived by the infant. Our question is whether the infant is sensitive to the spatial information contained in the event—whether he registers information about the position of the event relative to him. The evidence suggests that he does. At least as far as two-dimensional space is concerned, this perceptual capacity appears to be either innate or rapidly acquired.

The newborn infant will reliably turn his head to a tactile stimulus that has been presented to a corner of his mouth (Prechtl, 1958). This behavior, sometimes termed the "rooting reflex," suggests the capacity to discriminate right-left spatial information. That is, the infant can perform a head turn that is directionally appropriate to the locus of stimulation. Similarly, newborns will make directionally appropriate head and eye movements in response to a visual stimulus (Peiper, 1963). Again, the behavior is spatially accurate with respect to direction.

A capacity for sound localization is more difficult to recognize, since it is unlikely that an infant in the first three months of life will turn his head to a sound source. We will have reason to discuss this matter at a later point. It does appear, however, that newborns do possess a capacity for sound localization. Wertheimer (1967) observed appropriate eye movements to a toy cricket sound by one newborn. Employing much larger samples, both Bronshtein and Petrova (1967) and Semb and Lipsett (1969) have noted that changing the position of a sound will reelicit a previously diminished sucking orientation in newborns.

A more intriguing question about the infant's space perception is whether he registers distance information. It is perhaps not surprising that the infant discriminates right from left. The retinal image is faithful to the object in two dimensions. Right-left information is thus preserved in the image. Since it is a two-dimensional projection of the object, however, it is not at all clear how it is that we perceive the object as being three dimensional. What can be said of the infant's ability to perceive the object in the third dimension—can he perceive distance? The evidence concerning this question is not entirely consistent.

Haynes, White, and Held (1964) report that the infant does not display lens accommodation to changes in the distance of a visual object until the second month. White and Clark (in press) report failure to observe an eye-blink response to a looming object in young infants. In both of these studies, the fact that the infant does not show specific oculomotor responses to variations in distance implies an inability to perceive distance. However, there is an impressive array of evidence that is not dependent on oculomotor responses that does suggest early sensitivity to distance. This evidence derives from a series of experiments conducted by Bower.

The first set of experiments (1964, 1965c) indicated that infants of less than three months were capable of perceiving the size of an object independent of its distance. In tending to respond to actual rather than projected size, the infants thus displayed some degree of size constancy. Monocular vision alone was sufficient for this discrimination, and the necessary information for distance appeared to be the motion parallax produced by head movements. In subsequent work, Bower (1966a) also demonstrated that three-month-old infants were capable of perceiving the slant of a visual form. They could distinguish between a rectangle viewed from a given angle, and a trapezoid equivalent to the projected shape of the rectangle at that angle. In fact, the infants did not seem to discriminate between the same form at different angles of orientation, indicating very convincingly some degree of shape constancy. In all of these experiments, the response measure was an operantly conditioned head turn.

The fact that three-month-old infants displayed some degree of perceptual constancy suggests that they were able to do more than simply register a third spatial dimension. This distance information was being registered appropriately—specifically, perception of size and shape were reasonably unaffected by distance. We cannot account for such perceptual structuring simply by assuming that distance perception develops within two months. In addition to learning to perceive distance, the infants would have had to learn to correlate distance information with the changes in the retinal image that are created by the same object at different distances. Moreover, the reaching behavior that has long been the presumed learning mechanism for this accomplishment is not typically displayed at three months.

At the very least we would have to conclude that infants perceive distance very early in life, earlier than two months of age. The young infant's perceptual system appears to be registering spatial information in three dimensions. In Piaget's terminology "auditory space," "visual space," and "tactile space" appear to have more structure at birth, or shortly thereafter, than he himself had assumed. Yet for Piaget, the major achievement of the sensorimotor period of development is the gradual integration of sensory information registered by each of the separate modes. The coordination of this spatial information is critical to the eventual perception of a three-dimensional world of objects. To be specific, if an auditory event and a visual event do not coincide spatially, we do not perceive a common object source of these events. A stable world of objects depends, at the very least, upon intersensory congruence of spatial information. Piaget (1954) argues that such integration is not complete until the second year of life. We now have some evidence that this is not at all the case.

Consider, for example, an experiment recently conducted by Aronson and Rosenbloom (1969). In this experiment the question posed was whether young infants would notice a spatial discrepancy between the mother's visual position and the position of her voice. The infants ranged in age from 20 to 58 days. Each infant watched his mother, directly in front of him, while she spoke to him, casually and unrehearsed. Her voice was audible to the infant only by means of a stereo speaker system. By appropriately balancing this system, the locus of the voice could be shifted from straight ahead (where the mother was located) to 90° left or right. As long as the voice emanated from the center, the infants appeared normally alert and relaxed. A shift of the voice to left or right, however, produced dramatic reactions from the infants. They became intensely agitated and visibly upset. In four cases, it was possible to restore calm in the infant, and return him to the experimental situation, where the locus of the voice again was the infant's phenomenal center. However, a shift of the voices location to either side was quickly responded to with upset. Eight of nine infants filmed in this situation cried within four minutes of observation.

These results rather convincingly demonstrate that as early as the first month of life, a coordination of auditory and visual spatial information is part of the infant's perceptual capacity. Piaget himself (1954) has noted that vision and audition are among the first of the senses to become spatially organized, beginning as early as the third month of life. Aronson and Rosenbloom's results suggest a more precocious achievement, and also raise doubts as to whether the capacity is the result of development at all. Bower (1969) has argued that this coordination is innate, indeed that the human organism may be born with an amodal spatial structure requiring no further coordination between auditory, visual, and tactile stimuli. Even if we were to grant the possibility that the infant could learn to coordinate visual and auditory spatial information in so short a time, he would surely need some perceptual structure to allow for such rapid learning. This point requires some elaboration.

The mother-voice experiment indicates that the infant perceives a violation of spatial unity. But it is not immediately obvious why he should expect such unity. Clearly he must be registering some information, independent of spatial information, which specifies a single environmental unit (for example, the mother's voice and her visual image "belong together"). It is possible, for example, that the infant has learned to associate his mother's voice with her image. Assuming he can recognize his mother and her voice, he would have a basis for expecting them to be spatially coordinated. However, subsequent observations have demonstrated that the person talking need not be the mother—a total stranger whose voice is spatially displaced will also upset the baby. Perhaps, then, the infant has developed a less specific association, that human voices are associated with human visual images. This would not be a particularly easy association to form, since the two do not regularly occur together—the infant often hears voices while seeing no one, and sees people who are not talking. Furthermore, the infants did not behave in the mother-voice experiment as if they had such an associative rule. The infants were not at all bothered by a stranger's voice coming from the side, while mother stood, without speaking, in the center. They appeared to attend either to the mother, or the voice, but gave no evidence of disorientation, confusion, or upset. There simply was no evidence that a recognition process, even a very nonspecific one, served to inform the infant of a perceptual unit.

The most likely explanation is that the infant perceives a temporal unity: synchronous lip movements and vocal utterances. The spatial disjunction would thus be perceived as incongruent with this temporal unity. It ought to be demonstrated, of course, that the infant is sensitive to this temporal unity. We are currently testing this hypothesis by electronically delaying the broadcast of the mother's voice while not interfering with the spatial unity. If our reasoning is correct, the infant should again show signs of confusion and upset, since the spatial unity is now violated by the temporal asynchrony.

The point we wish to emphasize is this: some perceptual structure such as we have been discussing is necessary in order to perceive a spatial disjunction. The same structure would be necessary in order to learn to coordinate spatial information from separate modalities. To assert that the four-week-old infant could have learned to coordinate spatial information requires that we specify a mechanism, some other perceptual structure, which would make such learning possible. We then must account for how *that* structure develops, or at least determine whether it develops or is innate. But it is obviously not sufficient to refer to a simple learning process. The behavior of the infant involved here is much too complex to be dismissed as a simple learning achievement.

Of even greater significance than auditory-visual spatial coordination in Piaget's developmental theory is the integration of vision and prehension. This achievement is of considerable theoretical importance because of its implication for the development of the perception of a world of permanent three-dimensional objects. Piaget (1954) argues that visual guidance plays an essential role in early reaching attempts by the infant. He notes that initial reaches only occur when both the hand and the object to be reached are in the visual field, and that, for the reach to be successful, the infant must glance back and forth between hand and object during the reach. These observations suggest the absence of spatial coordination between visual and prehension systems.

However, a recent experiment by Alt (1968) seems to indicate that Piaget's observations are misleading. Alt notes that Piaget underestimated the extent of the infant's visual field. When the infant is oriented to an object at the midline, his hands are almost always within the peripheral visual field. Failures to initiate reaches are thus not explained by a failure to see the hand and object simultaneously. Alt observed a group of infants longitudinally as they progressed in their reaching behavior. He experimentally reduced the infants' visual field by blindfolding one eye, restricting vision to about 40° from the midline. He observed from film records that the first reaches of each infant were quite accurate, and that the infant rarely looked back and forth between hand and object during reaching. Most important, the blindfold experiment showed that the infants did not need to see their hand in order to initiate a reach, nor was their accuracy affected by this visual restriction. The coordination did not require visual guidance at all.

Again it appears that spatial coordination is not primarily dependent upon learning. At least it does not seem that practice in reaching itself is essential to the spatially coordinated reach. This is not to deny that reaching accuracy improves with practice. However, the initial accuracy of the reach, independent of visual guidance, indicates an innate capacity.

Our tentative conclusion is that, contrary to Piaget, the infant has the ability to register spatial information and to coordinate this information as registered by the different modalities. Certainly, the evidence indicates that the emergence of these abilities does not depend on overt action as Piaget hypothesized. The infant displays spatial ability long before he is engaging in much purposeful motor behavior, particularly reaching.

However, if this tentative conclusion is to be accepted, we must deal with a very important question: What is developing perceptually during the first two years of life? If spatial ability is part of the infant's genetic endowment, what accounts for the stepwise progression of behavior observed by Piaget, as in the following descriptive account of auditory-visual integration (Piaget, 1952, pp. 80–83):

1. At one month, a sound disrupts ongoing behavior (such as sucking), apparently commanding the attention of the infant, but no attempt is made to localize the sound.
2. At two months, a sound may elicit directed head turning but this behavior appears to be restricted to the case of a moving object (for example, a rattle). Orientations to points outside the visual field seem restricted to sound sources that were visible immediately prior to the sound's occurrence. [Piaget also observes that if the sound source is immobile when the infant visually apprehends it, he appears unsatisfied that it produced the sound.]
3. At three months, the infant orients to immobile sound sources and appears to be satisfied that he has found the source. Orientations to points outside the visual field are no longer restricted to those objects seen immediately beforehand.

Piaget interprets these observations to mean that spatial coordination is emergent as the child grows. The qualitative behavioral changes indicate the genuine development of a mental structure. We, on the other hand, have been arguing that the perceptual capacity to register spatial information is largely innate. If a spatial structure is not emerging, then what is?

We can quickly rule out an explanation in terms of motor capacity. Piaget dismisses such an explanation himself by demonstrating that whether the response (for example, a head turn) occurs or not is determined by the nature of the test situation. The newborn infant certainly is capable of turning his head (Prechtl, 1958). In fact, we have observed that an infant of four or five weeks can make very rapid lateral head movements to and fro in order to recover a pacifier that has fallen from his mouth.

A more likely explanation is that the infant is gradually increasing in his capacity to integrate information. Here again, we may credit Koffka (1959) with introducing the idea that infant perception is limited by a capacity to process information. Koffka noted that a newborn dog will withdraw his foot in response to a pinprick, and then extend it. The reflex occurs regularly if stimulation is repeated. As stimulation becomes more rapid (by shorter interstimulus intervals) a threshold is approached, beyond which the foot will remain withdrawn. This interval threshold becomes shorter as the pup grows. Koffka likened this phenomenon to the flicker-fusion effect in visual perception (for example, in motion pictures) and assumed that both phenomena were the result of general processing characteristics of the central nervous system. He further assumed that the newborn human must also be subject to a processing limitation. He thus cleverly, if indirectly, arrived at the conclusion that the human infant is relatively limited in its ability to perceive motion. Fast-moving objects will be perceived by the infant as stationary streaks of light.

Koffka's assumption of a processing limitation in infants is supported by a number of observations. When an object moves across the visual field rapidly, the young infant does not display the head and eye tracking that he does with slower velocities. Visual pursuit of faster moving objects develops during the first few months of life (see Preyer, 1893; Shinn, 1893). Optokinetic nystagmus has been observed in newborns, and also appears to be contingent upon velocity; rapid movements across the visual field do not evoke the reflex (McGinnies, 1930). Bower (1967) found that the rate at which an object disappears is an important determinant of whether the infant perceives its continued existence. When an object disappears slowly the infant will search for it, while faster disappearances do not lead to search behavior. All of the fast rates employed in these studies are well within adult processing capacities. The infant, it seems, gradually develops in his capacity to perceive high velocities of motion. The infant will not see a high-velocity motion as a displacement of a stimulus in space, but as a stationary streak, or blur in the visual field.

The effect of a limited processing capacity probably is not restricted to motion perception. Perception theorists, such as Neisser (1967) and Haber (1966), appear to agree that perception is an integrative process. We do not perceive only what we attend to at a given moment in time. For example, we integrate our peripheral field of view into our central, focal field. We see things in a visual context. To what extent does the infant integrate perceptual information? Given a limitation on processing capacity, we would expect perceptual integration to be restricted as well. Peiper (1963) notes that peripheral visual processing does appear to be restricted in the young infant. The eye-blink response appears first to an object approaching along the line of sight, and only later occurs to an object approaching from the periphery. Similarly, as an object moves slowly across the infant's visual field from the periphery, the point at which he becomes distracted and visually orients to the object increases with age; that is, he becomes increasingly sensitive to events in the more extreme peripheral field of vision.

The point of these observations is that an apparent development of spatial structure, as suggested by Piaget, may well be accounted for by the development of more efficient perceptual processing. Behavioral indications of spatial perception, such as directed head turning, need not imply a developmental change in perceptual structure. Instead, we may conceive of an innate spatial structure which spatially organizes whatever information is being processed by the infant. As this processing develops in efficiency, the infant will begin to behave in a manner which suggests greater organization in his perception of space.

We wish to emphasize, however, that this account of early space perception development does not imply that the development is trivial. In terms of the infant's conscious experience, a processing limitation would have as profound an effect upon space perception as if the spatial structure itself were not fully developed. However, the development does not depend upon *specific* kinds of input which would be required if the dimensions of space were learned in the course of development. If experience plays an important role, it must do so by nurturing more efficient processing of perceptual input. Or it may be that an increase in processing capacity is due primarily to internal maturational changes. We simply have no sound basis for answering this question at present.

However the questions raised by this essay may be settled; we must recognize that some very important aspects of the development of spatial abilities remain to be explained. One such problem which has been very carefully studied by Piaget (1954) concerns the child's understanding of the properties of objects. The infant, as evidenced by his search behavior for hidden objects, has difficulty in dealing with spatial displacements. The infant of ten months, for example, looks to where he first found the object, rather than where he has last seen it hidden. Still older infants display the same difficulty when the displacements must be inferred, such as when the object is transported in a container. Neither a perceptual spatial structure nor an increasing capacity for processing perceptual input is sufficient to account for these phenomena.

DEVELOPMENTAL STUDIES OF PERCEPTION*

William Epstein

The potential contribution of developmental studies has been aptly expressed by Hazlitt: "Nothing gives a better insight into the working of the mind than the study of the development of behavior . . . the man who has heard the beginning of a story is ipso facto a more reliable judge of the credibility of the ending than the man who has come in at the middle" (quoted in Munn, 1955, p. VI). In the context of this book, the emphasis will be on perceptual development. More specifically, the question of perceptual development will be explored to determine how this field of investigation can contribute to our knowledge of perceptual learning.

General Observations

The ontogenetic development of perception has been a matter of long-standing interest among European psychologists. The emergence of Gestalt psychology, with its emphasis on innate organizational factors, stimulated efforts to determine whether perception followed a developmental course. Piaget's theoretical writings served as another very important stimulus for investigation. Piaget and his associates have contributed and continue to contribute extensively to the experimental literature. In contrast to the active interest among European psychologists, American psychologists during the first half of this century devoted little attention to the question of the development of perception. However, the post-World War II period has been marked by a notable intensification of interest in the psychology of perception, and with it a new interest in the developmental question. As a result, there exists at present a fairly extensive body of research literature. Summaries of this literature have been presented by Baldwin (1955, Ch. 3), Munn (1955, Ch. 8), E. J. Gibson and Olum (1960), Wohlwill (1960), and E. J. Gibson (1963a).

In this chapter, no effort will be made to duplicate the coverage of Baldwin, Munn, E. J. Gibson, and Wohlwill. There is little to be gained by exhibiting, once again, the general state of disarray in the field. In addition, a good deal of research, summarized by the above reviewers, is characterized by inadequate experimental design and data analysis. As Wohlwill (1960) has observed, the studies place a ". . . heavy strain . . . on the reader's willingness to assume an attitude towards the mere possible" (p. 251). For these reasons, an extensive review of the literature will be eschewed in favor of a more detailed consideration of a number of illustrative studies.

The studies which have been selected are exceptions (certainly not the *only* ones) to Wohlwill's unfavorable characterization. They were selected in accordance with two minimal requirements:

1. Age was one of the independent variables, and an adequate number of values of the age variable were sampled. This can be accomplished either by a cross-sectional technique or a longitudinal technique. The selected studies have employed the former technique, which involves the use of independent age groups.
2. The experimental procedures and especially the constant factors, e.g., instructions, and dependent measures were carefully controlled.

Finally, experiments were favored which articulated or attempted to test an explicit hypothesis about the mechanisms or processes underlying the expected developmental change.

Additional comment is in order regarding two kinds of studies which were excluded for reasons other than those given previously. Investigations of sensory processes, which are considered to be related directly to specifiable properties of isolable receptor mechanisms, were excluded. The selections have been restricted to investigations of the perceptual phenomena which have been emphasized in the earlier chapters.

The second omission requires more extended explanation. Omitted are the growing number of investigations of the perceptual functioning of "naïve" organisms. The organisms in question are either deprived of visual stimulation from birth until the time of testing or are tested at the first practical moment following birth. The best representatives of this approach are the experiments of Walk and Gibson (1961), Fantz (1961b), and Hershenson (1964).

These studies have used the stimulus-preference method, apparently originated by Staples (1932). The S is exposed to a number of discriminable (by an adult S) stimuli, and *E* observes whether S exhibits a preference for one of the stimuli. The preferential response is one already available in S's repertory, prior to the test. Two responses which have been used are fixation of a target and locomotion. If S exhibits a preference, then it is concluded that he has the capacity to discriminate between the stimuli. By suitable variations of the stimuli, it is possible to determine the specific property which is the discriminative cue. Failure to make a preferential response is often taken as evidence that S is unable to discriminate. If a significant preferential response is made by naïve Ss, then it is concluded that learning or previous exposure is not necessary for the perceptual function in question. The absence of a preferential response is taken to imply that learning is necessary for the emergence of the discriminative capacity.

Parenthetically, we may note that this rationale makes a positive conclusion about learning, contingent on demonstration of the null hypothesis. This is never a desirable requirement for support of a thesis. In this case, it is especially prejudicial, because the "nativistic" hypothesis does not make a one-directional prediction. Only significant preference is required, not preference for a given stimulus, designated in advance.

These investigations seem better suited for the exploration of the nativism-empiricism issue than for an examination of the developmental question [see Epstein (1964) for a discussion of the nativism-empiricism question and a summary of the experimental literature]. The finding that prior exposure or learning is not necessary for the original occurrence of discrimination in no way precludes the possibility of developmental changes as the organism ages. These changes may take the form of a capability for increasingly finer discriminations. It is noteworthy that the discrimination of gross differences has been required in most studies of naïve organisms. Perceptual development could consist of increasing differentiation, starting from an innate minimum discriminative capacity. It is also possible that increasing age may be accompanied by the development of error. Thus, the infant might exhibit perfect constancy, but overconstancy may develop as he grows older.

In summary, the findings of significant preference in naïve Ss is compatible with subsequent developmental change. The principal contribution of the naïve organism study, insofar as the developmental question is concerned, is that it allows *E* to extend the range of values of the age variable. But this contribution is not easily realized, since the tasks set for the infants are too easy for older Ss and those suitable for older Ss are too difficult for infants. Thus, there may be no single set of test conditions which can be used over the entire age range.

Nor are the results of a study of naïve organisms more useful in the case of the failure to observe a preferential response. Strictly speaking, under these circumstances, nothing can be concluded about perceptual learning. However, even if the conclusion is accepted that learning is necessary, nothing is known about the course of learning and its underlying mechanisms.

The investigations of naïve organisms are more informative when their procedures are expanded to include systematic variations of the relevant properties of visual stimulation. An example of this type of study is Hess's investigation (1950) of the development of the chick's responses to light-and-shade cues of depth. Chicks were reared under two different conditions of illumination. The control chicks were reared in cages illuminated from above. The experimental chicks were reared in cages that were illuminated from below. Thus, the two groups experienced different distributions of light and shade. A test was designed to explore the chicks' utilization of light-shade distribution as a cue for depth. The test was administered at 1-week intervals, beginning at 1 week following hatching and continuing to 7 weeks of age.

This type of study has several objectives:

1. Determination of the original status of a response
2. Determination of developmental changes in the response
3. An evaluation of the conditions that are responsible for the character of the observations made in exploring the first two questions

Unfortunately, studies of the type performed by Hess are still rare [see Walk & Gibson (1961), Epstein (1964), and Fantz (1965) for reviews of the studies of naïve organisms].

Methodological Considerations

All methodological considerations which apply to the design and interpretation of experimental studies apply to the developmental studies. However, two specific requirements deserve to be reemphasized here. Excellent discussions of these requirements have been provided by E. J. Gibson and Olum (1960) and Kessen (1960).

Control of the Independent Variable. Two preliminary decisions are required of the investigator who wishes to make cross-sectional comparisons among different age groups. First, he must decide how much variation he can tolerate on dimensions other than age. This is almost a meaningless question unless the investigator has some clues concerning the identity of the age-correlated variations which are potentially confounding. He must be able to identify the variables which may be safely disregarded in the selection of Ss as well as the variables which must be taken into account. With this knowledge, he may select his samples at the various age levels so that the values of the relevant variables are randomly distributed within each sample. But this sampling procedure may not always be feasible and sometimes may be virtually impossible. Often *E* must use the restricted number of children who are readily available, regardless of considerations of their representativeness in relation to a larger population. In other cases, the ideal of random distribution is difficult to achieve because the variable in question is tied too closely to age differences. As an example, in a cross-sectional study of distance perception using groups ranging from 6 to 16 years it seems most unlikely that the range of heights could be distributed randomly among the age groups. There is reason to suspect that the angle of observation, which will vary as a function of height, may affect distance judgments. Therefore, it is incumbent upon the investigator who wishes to make statements of the form $R = f(\text{Age})$ to control the variations in height. Harway's experiment (1963), which will be described later, will illustrate how this control can be achieved.

A second decision regarding the age variable concerns the choice of the particular age groups to be included in the study. It is risky to draw conclusions about the course of development from data drawn from very few age groups. The risk is compounded when these few groups are either bunched together on the low end of the age scale or very widely separated. In both cases, the experiment will not be sensitive to the distinctive features which may characterize a developmental trend, e.g., nonlinear characteristics. The cross-sectional study should include a sufficient number of age groups, and the ages that are selected should tap the developmental period at successive stages throughout the full course of development.

Comparability of Conditions. Investigators who have used the cross-sectional approach have frequently introduced variations in the procedures for the different groups. For example, a standard psychophysical method will be used for the older Ss, but a modified version will be used for the young children. Usually, the modification is justified on the ground that the children cannot understand or perform the required operations. Needless to say, modifications, whatever their exigency, can create problems in determining the comparability of the experimental conditions for the different groups. Therefore, two general rules have been proposed by methodologists [e.g., E. J. Gibson & Odum (1960); Wohlwill (1960)] in this field:

1. Differences between the procedures, applied to the various age groups, should be kept to the minimum necessary for the experimental purposes.
2. The modifications should not violate the methodological requirements of acceptable response measurement.

On the other side of the question, the *E* who uses identical procedures for all age groups should ascertain that the instructions and operations are interpreted similarly by all Ss.

Illustrative Investigations of the Development of Perception

This section will be devoted chiefly to a discussion of a few selected investigations of the development of perception. These studies have been concerned with the perceptual phenomena which were treated in the previous chapters: form perception, size-distance perception, and motion perception. All veteran experimenters have ruefully learned that no single experiment can successfully settle a question to everyone's satisfaction. The experiments which will be discussed are no exception to this discouraging rule. Therefore, these studies are not being presented in the fond belief that they offer final, definitive resolutions of their respective problems. The main objective of this presentation is to assess the potential of developmental studies whose primary claim for our attention is not simply a welcoming attitude toward the "mere possible." Consideration of these studies may provide the reader with guidelines for a more systematic developmental study of perception.

Perception of Distance

A thorough review of the developmental studies of the perception of size and distance has been presented by Wohlwill (1963). Our discussion will begin with the investigations of distance perception reported by Harway (1963) and Wohlwill (1963, 1965). Harway's study is especially interesting because, unlike many of the previous investigators, Harway attempted to assess the importance of two rationally selected age-correlated variables. The two variables are height and interocular distance. Both these factors vary systematically with age, and both may affect distance perception. Increasing height may affect apparent distance by decreasing the differences subtended by a specific horizontal stretch at different distances from S. Increases in interocular distance may enhance the effectiveness of binocular depth cues.

There were sixty Ss, grouped at five age levels, with median ages of 5 years 6 months, 7 years 2 months, 9 years 11 months, 11 years 9 months, and 23 years. The judgments followed a procedure introduced by Gilinsky (1951, 1960). The Ss stood at the end of a grass field and directed *E* to mark off successive 1-foot intervals. A 1-foot ruler, placed on the ground at S's feet, served as a standard. S was instructed to tell *E* when a marker in the field had been moved ". . . as far . . . as the ruler is long." The nearest 1-foot stretch began 1 foot from S, and nineteen successive 1-foot estimates were obtained for most Ss. (Some Ss provided such gross overestimates that the entire field was traversed before nineteen units were judged.) Each S provided judgments under two conditions: one at normal height, the other at an adjusted height. For the latter condition, the children stood on an adjustable platform, so that the total height was 5 feet 6½ inches for the children in all four groups. The adults' adjusted height was accomplished by requiring them to kneel.

Harway's task is analogous to that posed in a size-constancy experiment. The principal difference is that a horizontal extent is judged at varying distances, rather than a frontoparallel extent. In Harway's experiment, perfect distance constancy would be evidenced by invariantly accurate estimates of the successive 1-foot extents. Progressive underestimation of successive 1-foot extents is equivalent to progressive underconstancy. Figure 1 shows the average constant error for successive distance judgments obtained under the condition of normal height. The curves on the left of the graph represent the visual-angle functions for a 1-foot ruler at S's feet and nodal points at four representative heights.

Figure 1 shows that the 1-foot extents were progressively underestimated with increasing distance. Unfortunately, the procedure of obtaining judgments of increasingly distant extents, in ordered succession, confounds the distance variable with a practice variable. Figure 1 also suggests that the rate of change (slope) with increasing distance was greater for the three younger groups as contrasted with the two older groups. An analysis of variance confirmed these observations and also showed that the height at which the judgments were made did not affect the judgments. The pattern of constant errors was the same for the normal-height and adjusted-height conditions. Further evidence on this point and on the influence of interpupillary distance was obtained by

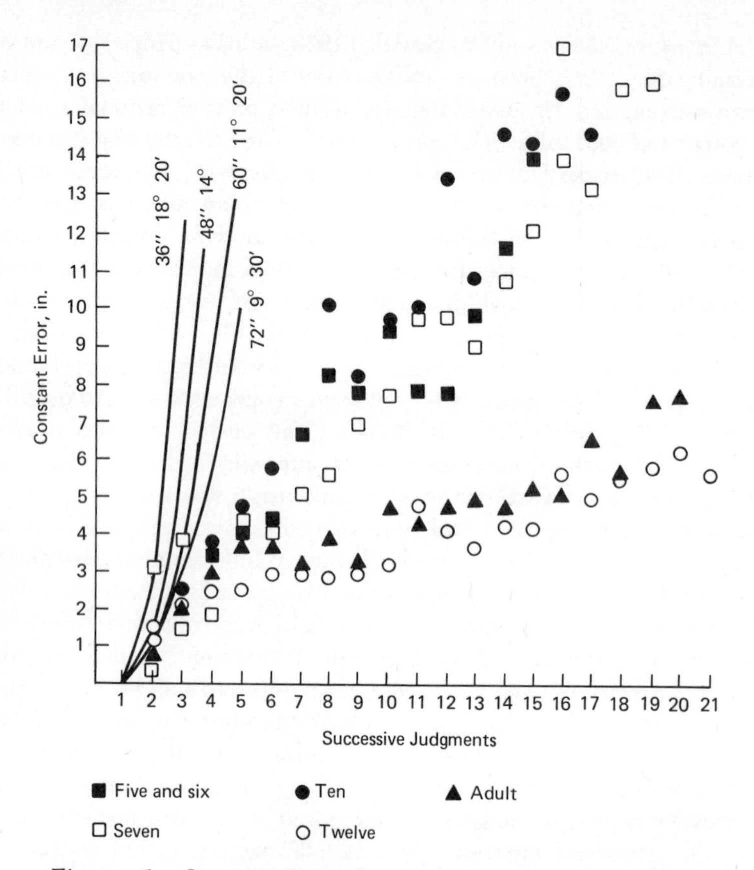

Figure 1 Constant Error for Successive Distance Judgments. The Line Curves at the Left Represent Visual-Angle Functions for a Hypothetical S Whose Eyes Are 36, 48, 60, and 72 Inches above the Ground. (From N. I. Harway, "Judgment of Distance in Children and Adults," *Journal of Experimental Psychology*, vol. 65 (1963), p. 387. Copyright 1963 by the American Psychological Association, and Reproduced by Permission.)

computing two correlations for each group: (1) between the distribution of normal heights and distance judgments and (2) between the distribution of interocular distances and distance judgments. No significant correlations were found. In other words, when age is held constant, distance judgments are not associated with height or interocular separation.

Harway's results reveal a developmental trend in the perception of distance. Underestimation of distance extents is greater and more closely correlated with visual-angle requirements for Ss between 5 and 10 years old than for Ss 12 years and older. Figure 1 indicates that a marked change in the magnitude of underestimation occurs between the ages of

10 and 12 years. Harway offers no hypothesis to account for the developmental results.

How general are Harway's findings? Will similar results be obtained for a variety of task variables and over a broader range of distances? A cursory review of the literature on space perception (Denis-Prinzhorn, 1960; Epstein, 1963) provides ample reason for believing that distance judgments will vary as a function of a number of environmental and task variables. An experiment by Wohlwill (1963) conducted under conditions which differed greatly from Harway's study will serve to demonstrate this fact.

The experimental setting and procedure were those used in Wohlwill's investigation (1964) of the effects of practice on distance judgments. The experimental environment consisted of a large box, into which S looked monocularly. By inserting appropriately designed panels, the optical texture of the floor of the box could be varied from zero texture through varying degrees of density and regularity. The task required S to bisect a 90-centimeter extent of the floor. The extent was demarcated by two red arrows that touched the floor, the nearer one located 63 centimeters from the eyepiece. S was instructed to direct *E* to stop a moving pointer when it reached the objective midpoint between the two arrows. Ascending and descending judgments were obtained for each of the six floor panels. A practice session preceded the test judgments. This was done in order to guarantee that the children understood the task. The Ss were five groups of 24 Ss each, taken from grades 1 (mean ages 6 years 10 months), 2 (8 years 10 months), 5 (11 years), 8 (13 years 11 months), and 11 (16 years 10 months).

The results showed significant main effects of age and texture, as well as a significant interaction. The interaction effect was due mainly to a difference between the older and younger groups for only one texture (maximal density and regularity). Therefore, the texture variable will be disregarded, in order to exhibit the age function more clearly. Figure 2 presents the mean errors for the bisection judgments for each age group, for all textures combined. The kindergarten (K) data are from a supplementary study; the adult data are from another study (Wohlwill, 1964). Positive errors indicate overconstancy: S judged the midpoint to be back of its objective midpoint. Negative errors indicate underconstancy: S judged the midpoint to be to the front of the objective midpoint.

Figure 2 shows an overall trend from underconstancy for the two younger groups to overconstancy for the older groups. In contrast to Harway's findings of increasing accuracy with age, Wohlwill found decreasing accuracy with age. An additional developmental trend was noted in the differences between the intervals of uncertainty for the groups. The interval decreased steadily with age. Older Ss displayed greater precision in their judgments, as well as greater error.

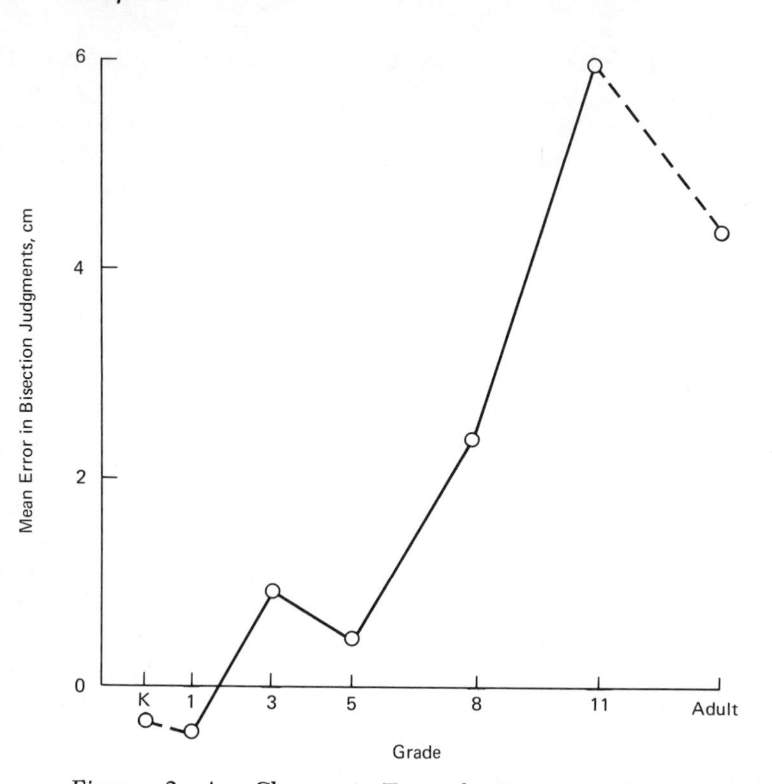

Figure 2 Age Changes in Errors for Bisections of a 90-Centimeter Distance. (From J. F. Wohlwill, "The Development of 'Overconstancy' in Space-Perception," in L. P. Lipsit and C. C. Spiker, eds., *Advances in Child Development and Behavior*. New York, Academic Press, 1963, p. 298. Reprinted by Permission of the Publisher and Author.)

A similar study, involving the perception of relative distance in photographic slides, has been reported by Wohlwill (1965). The experiment was conducted to confirm the findings of the earlier (Wohlwill, 1963) study, and in particular to reexplore the possibility of an age-texture interaction. The hypothesis was ". . . that age and texture would interact, with age differences being maximal under low-texture conditions and minimal under high-texture conditions" (Wohlwill, 1965, p. 164). The chief differences between this study and the one described in the preceding paragraph are the following:

1. The viewing-box scene was not exposed directly. Instead, S was shown a projected photographic slide of the scene. The textured panels of varying regularity and density were photographed under conditions that were designed to duplicate the optic array available to S's eye at the observation aperture of the viewing box. It was hoped, that the use of two-dimensional

projections would force S to rely on the texture gradient and that this condition would be favorable for exhibiting an interaction between age and texture. Instead of the pointers to demarcate the standard extent, the figures of a cow and a horse marked the near and far boundary of the extent.

2. The second main difference was the method of obtaining bisection judgments. The procedure was a modified version of the method of constant stimuli. This modified method, which was invented by Piaget, is called the "clinical concentric" method. For each texture, two series of slides were prepared. Each slide showed the "cow" and "horse" against one of the backgrounds and a fence located between them. The position of the fence varied from slide to slide. The slides were presented in each of two orders: in one order the positions of the fence converged concentrically on the true midpoint, and in the other order the fence diverged concentrically from the midpoint. The instructions were as follows: "Do you see the cow in front and the horse way in back, and the fence in between? I want you to make believe that you were standing right in the field in which this picture was taken. Now tell me, if you were looking at the cow and the horse in that field, which do you think would be closer to the fence" (Wohlwill, 1965, p. 168). The slides were exposed in "rapid-fire sequence." Each S made judgments for four texture conditions.

There were 120 Ss, evenly distributed among five age groups; grade 1 (6 years 11 months), grade 4 (9 years 8 months), grade 8 (13 years 8 months), grade 11 (17 years), and adults. The results did not agree with the findings of the earlier study by Wohlwill (1963). All Ss exhibited underconstancy, i.e., displacement of the apparent midpoint toward the front. The constant error diminished as the texture became more determinate, e.g., high density and high regularity. The error also decreased as age increased, the greatest diminution occurring between grade 8 and grade 11. However, the changes correlated with age were not as pronounced as in the earlier study. These results seem to be in greater accord with Harway's findings than with Wohlwill's previous experiment.

These studies, and those (e.g., Denis-Prinzhorn, 1960) which have not been described, permit only a limited conclusion. The evidence suggests that changes in the accuracy of distance perception accompany changes in age. However, the direction and rate of change cannot be presently stated. It seems likely that the exact characteristics of the developmental trend which is observed will depend on features of the experimental situation. This remark certainly must rank as a scientific truism. Nevertheless, to the degree that these contingencies are not simply artifactual, they present problems for the generalized developmental hypotheses. To be useful, these hypotheses must specify precisely those conditions under which a predicted shift will be observed. Stated somewhat differently, the developmental theory of distance perception must explain the interaction between age and the task and environmental variables.

Perception of Size with Distance Variable

If distance judgments vary as a function of age, so should size judgments (Epstein, Park, & Casey, 1961). Most of the studies of this question have only limited value. Their limitations stem from the restricted conditions that have been studied. For example, a great number of experiments purporting to study size constancy have not varied distance. Instead, the standard and variable are placed at two different distances, which remain the same throughout the experiment [see Wohlwill's (1963) review.] This procedure provides only minimal information about the pivotal relationship involved in size perception.

Two developmental studies which have investigated size perception over a range of distance have been reported by Beyrl (1926) and Zeigler and Leibowitz (1957). Beyrl used the method of constant stimuli to obtain size judgments from seventy-five children between the ages of 2 and 10 years and from five adults. The experiment was conducted in a well-illuminated room. The standard was a disk or cube located at a distance of 1 meter from S. The comparison stimuli were presented at each of seven distances, ranging from 2 to 11 meters. Although the procedure involved a great number of comparisons, Beyrl found that the children remained motivated and attentive throughout. The results showed an overall tendency for size underestimation which decreased with age. More interesting is an apparent interaction between distance and age. The effects of increasing distance diminished as age increased. For example, the matches for a 10-centimeter standard disk ranged from 12.25 to 18 centimeters, 10.5 to 15 centimeters, 10.25 to 12 centimeters, and 10 to 10.5 centimeters for the 2-year, 5-year, 9-year, and adult groups, respectively.

Zeigler and Leibowitz (1957) obtained a similar pattern of results. The procedure was modeled after that used in the classic experiment by Holway and Boring (1941) on the determinants of perceived size. The S sat at the junction of an L-shaped arrangement of two alleys. The comparison alley contained a dowel located at a distance of 5 feet from S's eye. The length of this dowel could be varied continuously. In the standard alley, the standard dowels were presented at distances of 10, 30, 60, 80, and 100 feet. Following the procedure of the Holway-Boring study, the lengths of the standards were selected so that at the respective distances the standards subtended the same visual angle, 0.96°, at S's eye. Normal illumination prevailed, and S viewed the scene with unimpeded binocular vision. The instructions for both children and adults were as follows: "I am going to move this stick (comparison) up and down. I want you to tell me when it looks as high as the one out there (standard)" (Zeigler and Leibowitz, 1957, p. 106).

There were only thirteen Ss, one 7 years old, five 8 years, two 9 years, and five men ranging from 18 to 24 years. Needless to say, this is not an adequate sampling of Ss for the purpose of demonstrating a developmental function. However, there was evidence of a marked difference between the children and the adults. The adults showed a high degree of constancy, although there was a small negative constant error at the three farthest distances. The children also exhibited constancy at the shortest distance. However, for the remaining distances, their matches were underestimations. The magnitude of these underestimations increased with distance. Thus, the same interaction between age and distance which was observed by Beyrl was also obtained in the Zeigler and Leibowitz study.

These two experiments have produced comparable results despite the procedural differences between them. One difference which should be noted concerns the relative distance of the standard and variable. In Beyrl's study, the standard was near and the variable was distant; in the study by Zeigler and Leibowitz, the positions of the standard and variable were the reverse. Therefore, the results are obtained even when the "error of the standard" (see Vurpillot, 1959) should produce overestimation of the standard.

We have considered experiments that examined perceived size and perceived distance separately. By comparing the results of the separate experiments, it may be possible to derive certain inferences about the development of the relationship between perceived size and perceived distance. However, the inconsistent results of the experiments which have examined the dependence of perceived distance on age prohibit ready inferences about the size-distance relationship. The only experiment to obtain both size and distance judgments is an investigation by Wohlwill (1962a) of the responses of children and adults to an illusion of perspective. However, since no significant developmental trends were noted, this experiment does not contribute to our question.

One type of experiment which could contribute significant information is an expanded Holway-Boring experiment. Like the original Holway-Boring (1941) study, the distance of the standard and the availability of distance cues should be varied. However, the proposed experiment would vary the age of the Ss and the attitudes of judgments, and obtain distance judgments. In addition to providing data for the assessment of the size-distance relationship, this study would enable us to examine the possibility that there are different developmental trends for the various distance cues, as well as differential trends for the various attitudes of judgments.

The Perception of Form

There have been a variety of emphases in the developmental investigations of form perception. One type of study has focused on developmental differences in part-whole or figure-ground perception (e.g., Elkind, Koegler, & Go, 1964; Ghent, 1956). Unfortunately, many of the studies have used meaningful, representative materials as stimuli. For example, overlapping outline drawings of a violin, drum, and clarinet were used in one study. The use of meaningful forms introduces an obvious confounding of differential perceptual organization with differential availability of verbal designations of the various alternatives. Using nonrepresentative forms enables *E* to minimize this problem. A number of studies that have used nonrepresentative forms (e.g., Ghent, 1956, Exp. II) have followed a procedure that is poorly suited to the study of form perception. The procedure was that of Gottschalk's embedded-figures test. The Ss were told to trace the part of the complex figure which looked exactly like the simple figure. This task is too complex for the intended purpose of determining differences in the capacity for form discrimination. It probably would be useful to begin by studying this question at a more primitive level. Thus, the rules of figure-ground articulation (Koffka, 1935, pp. 190–196) might be individually investigated to determine the presence of developmental trends. The studies (e.g., Rush, 1937) of age-correlated changes in the effectiveness of various grouping principles should be reexamined for their relevance to this question.

A second group of experiments has been concerned with the role of spatial orientation in the discrimination of forms. Several different questions have been examined:

1. Are unfamiliar, nonrepresentative geometric forms seen as having a "right side up" by Ss of all ages?
2. Do changes in the orientation of a form affect its recognizability in the same way for children and adults?
3. Are children able to discriminate between forms which differ only in orientation?

Although these questions have a long history, no conclusive answers are available. A number of studies by Ghent (Ghent, 1960; Ghent, 1961;

Ghent, 1964; Ghent and Bernstein, 1961) illustrate the experimental approaches to these questions.

The experiment which has been selected for detailed examination is a study of the development of the ability to discriminate letterlike forms. This study by E. J. Gibson, J. J. Gibson, Pick, and Osser (1962) is notable for its careful attention to the properties of the stimulus forms, as well as a general thoroughness of analysis not often matched in the developmental literature. Another point of special interest is the writers' interpretation of their results in terms of the differentiation theory of perceptual learning [J. J. Gibson and E. J. Gibson (1955); . . .] .

The experiment was designed to obtain form matches on a recognition test in which the alternatives were four main types of transformations. The critical data were the "confusion errors" and their distribution among the transformation types. A confusion error is an incorrect selection, i.e., identification of a variable as a standard. Developmental trends were determined for the various transformations.

The standard forms are shown in the left-hand column of Figure 3. The standards were constructed according to rules of construction derived from an analysis of the structural features of English letters. They may be considered to be artificial graphic forms. Each row in Figure 3 pictures the twelve alternatives for the standard in the first cell of the row. The alternatives represent four types of transformation:

1. Columns 1 to 3: transformation of line to curve or curve to line
2. Columns 4 to 8: transformations of rotation or reversal
3. Columns 9 and 10: perspective transformations, a 45° slant left and a 45° backward tilt
4. Columns 11 and 12: topological transformations, a break and a close

As Figure 3 shows, each array of alternatives had from one to three copies of the standard.

Figure 4 shows the display apparatus for the discrimination test. *E* put a standard in the center of the upper row of the display. Then *E* indicated the appropriate row and S scanned the row, searching for any form which was "exactly like the standard." No correction was provided. In order to ensure that the task was understood by all Ss, several demonstration tests were administered. The Ss were 167 children, distributed among five groups, ages 4 to 8 years (see Table 1 for the number in each group.)

Two kinds of errors were recorded: (1) omission errors—failure to select the true match—and (2) confusion errors—selection of a variable

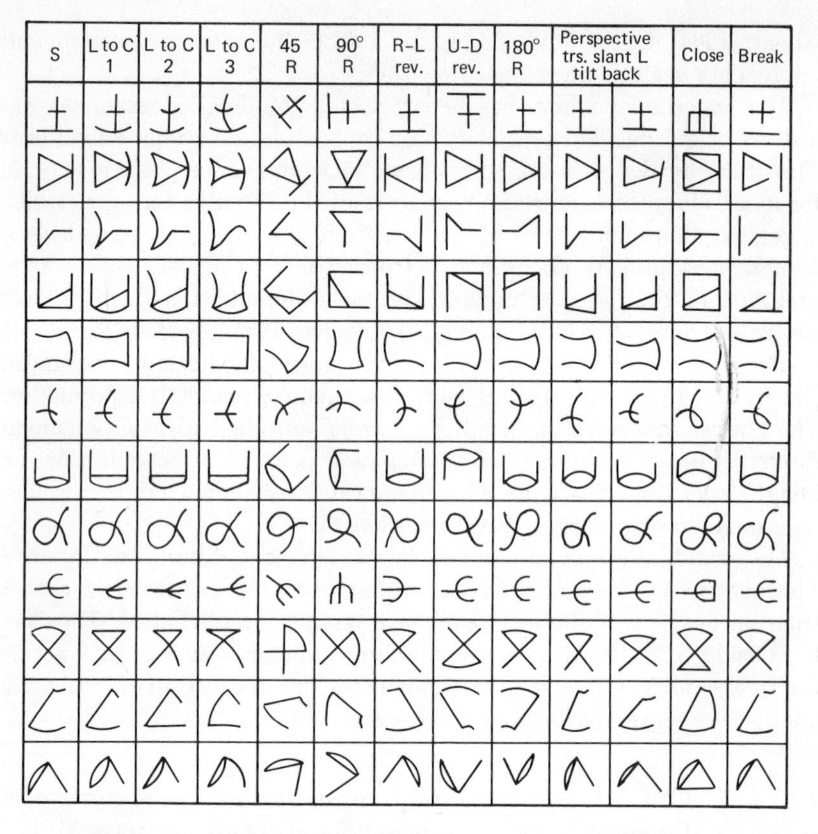

Figure 3 The Standard Forms and Their Transformations in the Study by E. J. Gibson *et al.* of the Development of the Ability to Discriminate Letter-Like Forms. (From E. J. Gibson, J. J. Gibson, A. D. Pick, and H. Osser, "A Developmental Study of the Discrimination of Letter-Like Forms," *Journal of Comparative Physiological Psychology*, vol. 55 (1962), p. 898. Copyright 1962 by the American Psychological Association, and Reproduced by Permission.)

as a match for a standard. The S was free to make several selections for each standard on a single test trial. Therefore, on a given test trial, S's responses could include a number of confusion errors, in addition to the correct match. In fact, the mean total number of omission errors for the 4-year group was only 1.31, while the mean number of confusion errors was 58.12 for the same group. Since omission errors were very infrequent, they were disregarded in favor of an analysis of the confusion errors.

The mean number of confusion errors decreased steadily with increasing age. The greatest decrease occurred between 4 and 5 years, and the smallest decrease between 5 and 6 years. Thus, a developmental trend is evident. However, this undifferentiated analysis is not very instructive. More informative is an analysis of errors by transformation, over the age

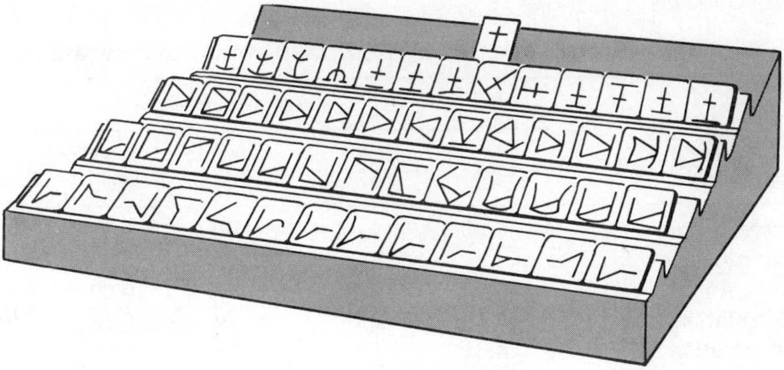

Figure 4 Apparatus For Displaying Forms in the Matching Task. (From E. J. Gibson *et al.*, 1962, p. 899).

scale. The results of this analysis are presented in two ways: in Table 1 and in Figure 5. Table 1 shows the age trends for each of the twelve transformations separately. Figure 5 shows the age trends for the twelve transformations consolidated into four curves, each representing one of the four main transformation types.[1]

It is obvious, on inspection, that the confusion errors were not distributed equally among the transformations. Furthermore, the rates of decrease were different for the various transformations. In other words, the different transformations were not equally discriminable from the standard. Nor did increasing age result in equivalent improvements for all transformations. For example, perspective transformations were most difficult to distinguish from the standard. Confusion errors of this type were very frequent, even for 8-year-old children. In contrast, errors involving line-to-curve transformations were never more than half as frequent as the errors involving perspective, and they occurred only infrequently for the oldest Ss. A similar ranking of the relative frequency of errors for the four transformations was obtained in a supplementary study, which used real letters as standards.

E. J. Gibson et al. (1962) attribute the developmental differences to a learning process of the sort proposed in an earlier analysis of perceptual learning (J. J. Gibson & E. J. Gibson, 1955). "It is our hypothesis that it is the distinctive features of grapheme patterns which are responded to in discrimination of letter-like forms. The improvements in such discrimination from four to eight is the result of learning to detect these invariants and of becoming more sensitive to them" (E. J. Gibson et al., 1962, p. 904).

[1] For justifications of this procedure, the reader should see the original source. This matter, as well as other details of statistical nature, have been omitted from our description.

Table 1

MEAN ERRORS MADE FOR EACH TRANSFORMATION BY AGE GOUPS[a]

	Age Groups				
Transformation	4 ($N=$ 25)	5 ($N=$ 35)	6 ($N=$ 29)	7 ($N=$ 30)	8 ($N=$ 32)
Curve to line (1)	5.85	4.06	4.00	2.53	1.28
Curve to line (2)	4.42	2.60	2.69	1.33	0.53
Curve to line (3)	3.04	1.46	1.76	0.60	0.31
45° rotation	5.19	2.14	1.79	0.53	0.78
90° rotation	4.31	1.48	1.28	0.03	0.34
Right-left reversal[b]	6.56	3.96	2.07	0.97	0.59
Up-down reversal[b]	6.47	3.55	2.44	1.56	1.08
180° rotation[b]	5.24	2.74	1.10	0.14	0.38
Perspective, hor.	9.88	9.20	9.69	9.27	7.34
Perspective, vert.	9.23	8.97	9.31	8.20	6.81
Close	1.19	0.69	0.83	0.43	0.31
Break	2.62	1.86	1.86	1.07	0.59

[a] From E. J. Gibson et al., 1962, p. 901.
[b] These figures have been corrected to allow for the fact that opportunities for error were less than for the other transformations.
NOTE: $N=$ number of children in each group.

This is a plausible d scription of the process which is responsible for the decreasing number of confusion errors as age increases, but an additional assumption is required to explain the distribution of errors among the transformation types and their different rates of decrease. The assumption is made that in addition to learning to abstract and respond to the invariants, S also learns that certain variations can be tolerated, with no consequences for shape identity. Examples of tolerable variations are the perspective transformation and the right-left reversal. On the other hand, a topological transformation is significant and will often alter the identity of the object, e.g., C and O.

To account for their findings, E. J. Gibson et al. suggest that confusion errors are frequent for those transformations which are usually tolerated in the identification of nongraphic, solid forms. Nor will these errors drop out with increasing age. The data for the perspective transformation fit this interpretation. Confusion errors will be infrequent for transformations that involve distinctive features and will decrease with increasing age. The results for the topological (break and close) transformations illustrate this trend.

A general question may be raised concerning the conditions that are necessary for the type of perceptual learning which is hypothesized.

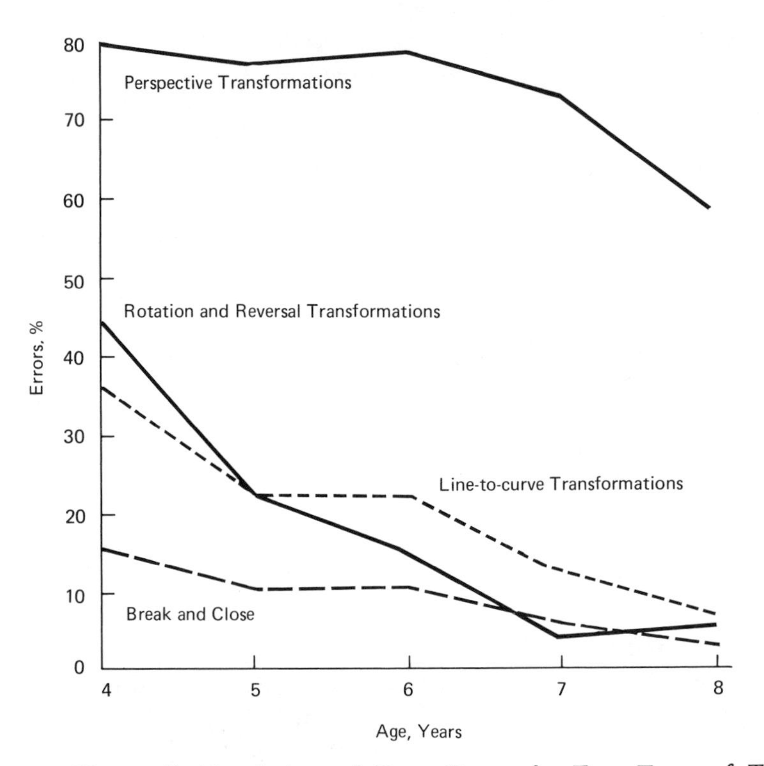

Figure 5 Developmental Error Curves for Four Types of Transformation. (From E. J. Gibson *et al.*, 1962, p. 901.)

E. J. Gibson et al. (1962) deal with this question in a manner consistent with their general view: "It may be that the child learns which varying dimensions . . . are significant and which are not by simply looking repeatedly at many samples containing both varying and invariant features. The distinctive features . . . are not taught but they are nevertheless learned" (p. 905). The results of the investigations of the hypothesis of acquired distinctiveness do indeed suggest that sheer repeated observation may facilitate discrimination, but one can legitimately wonder whether unreinforced observation is a sufficient condition for the improvement with age. The fact that the normally extended developmental period can be compressed into a relative brief period by the application of operant conditioning procedures suggests that differential reinforcement may be a factor.

A good example is the success of Bijou and Baer (1963) in training young children to discriminate mirror images. Cross-sectional comparisons show a developmental trend. Vernon (1957, p. 25) reports that 5-year-old children are simply unable to distinguish between mirror reversals. Yet, using operant conditioning procedures, Bijou was able to train 5-year-olds to the level of flawless performance. Thus, two interpretations of the development of form discrimination can be contrasted:

1. The extended developmental period is necessary in order to expose S to a sufficient number and variety of forms.
2. The period is extended because fortuitous reinforcement schedules and uncontrolled stimulus settings are inefficient for the purpose of establishing discrimination.

Apparent Movement

Brenner's investigation (1954) of the developmental changes in the threshold for apparent simultaneity and apparent movement will be our final illustration. The stimulus was an inverted V. Under optimal exposure frequency the V appears to perform a pendulum-like movement. With more rapid frequencies a report of simultaneity is obtained, while slower frequencies result in reports of succession. Brenner's procedure was to start with a very high frequency (descending trials) or a very low frequency (ascending trials) and to continue altering the frequency until S reported the two changes in the appearance of the display, e.g., descending: simultaneity → pendulum movement → succession. The data were used to determine the lower and upper thresholds for apparent movement. There were 111 Ss, distributed among fourteen groups, ranging in age from 3 to 19 years. With the exception of the youngest group ($N = 3$), each group had from seven to ten Ss.

Figure 6 shows the mean lower and upper thresholds as a function of age. The thresholds are expressed in terms of frequencies per second. It is obvious that the thresholds change with age. With increasing age, there is a gradual increase in the optimal frequency for the perception of apparent movement. The young children report apparent movement at frequencies which yield succession reports for older Ss, and the young children tend to report simultaneity at frequencies which yield apparent motion in older Ss.

Brenner suggests two hypotheses to account for her findings. Neither of these hypotheses is convincing. The first hypothesis, originally proposed by Meili and Tobler (1931), states that ". . . both apparent simultaneity and apparent movement are gestalt processes, and that children had greater facility in forming a gestalt. . . . Strength and speed of closure might be related to the ability to see apparent movement" (Brenner, 1957, p. 172). Why this relationship is to be expected is not intuitively clear. In any event, Brenner (Exp. II) subsequently found that, contrary to expectation, the time required to achieve closure on a test similar to the Street completion test decreased steadily with age. Therefore, Brenner concluded that the hypothesis of Meili and Tobler is not tenable.

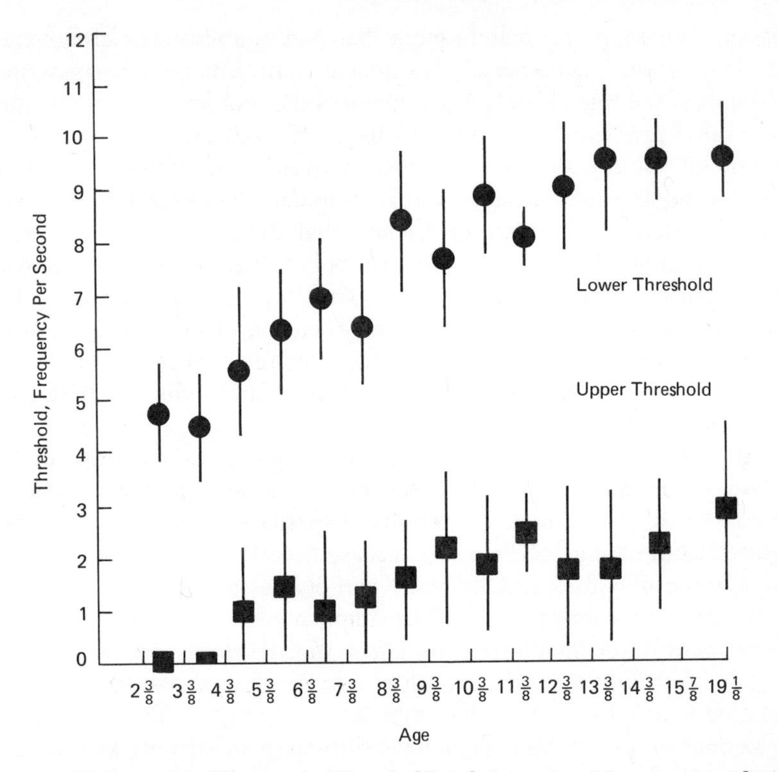

Figure 6 Changes in Thresholds of Apparent Movement with Age. Filled Squares Are the Mean Upper Thresholds for Each Age Group; Filled Circles Are the Mean Lower Thresholds for Each Age Group. The Bars Represent 1 SD on Either Side of the Mean. (From M. W. Brenner, "The Developmental Study of Apparent Movement," *Quarterly Journal of Experimental Psychology,* vol. 9 (1954), p. 171.)

The second hypothesis proposed by Brenner is that "the discriminative ability" increases with age and experience and for this reason higher frequencies are required for the illusory appearances. Brenner does not elaborate this hypothesis. One possible construction is that the illusions of movement and simultaneity may be construed as failures to discriminate temporal succession—a discriminative ability which grows with age and practice. Brenner's hypothesis, however it is interpreted, has not been tested.

Evaluation and Proposals

In the introduction to this chapter, we observed that developmental studies can make important contributions to an understanding of perceptual learning. That this promise has not been realized does not alter our assessment. But to fulfill their potential these studies must broaden their scope to include more than the establishment of age functions for various phenomena. The demonstration that one may write a statement of the form $R = f(\text{Age})$ only sets the problem for investigation. The crucial problem is one of identifying the processes responsible for the change. Of course, these processes require time, and so the process variable and the age variable are confounded. Nevertheless, it is generally accepted, in the logic of science, that the passage of time per se is never a sufficient condition. Time is only the necessary condition for the occurrence of the critical process. Therefore, the significant analysis of any time-correlated phenomenon will focus on these critical processes. Until the developmental studies complement their current emphasis by a focus on process, their contribution to an understanding of perception will not be significant.

Stated most simply, the argument of the previous paragraph is that once a developmental trend is discovered, the next step should be the formulation of experimentally testable hypotheses to account for these age-correlated changes. There are at least two acceptable procedures for testing these hypotheses: One is to manipulate the variables (task, environmental, or subject) which are implicated by the hypothesis and observe how these manipulations affect the developmental trends. For example, *E* might hypothesize that the developmental trend in size judgment is due to the fact that with increasing age increasingly greater proportions of the Ss at each age level shift from an analytic to a perspective attitude (Epstein, 1963). This hypothesis can be tested by varying the instructions so as to emphasize various attitudes. Needless to say, the postulated difference in attitude preference will also require explanation. This example will alert the reader to a potential limitation of this general approach. The underlying difference, which is presumed to be responsible for the age-correlated changes, *itself* may be the result of time-related changes. Thus, in our example, the perspective attitude may be a by-product of a mature level of intellectual conceptualization and therefore may be beyond the grasp of young children. A judgment about the severity of generality of this limitation should be reserved until the approach has been given a more extensive trial.

Another approach to the testing of developmental hypotheses was alluded to in the earlier reference to Bijou's experiment on the discrimination of mirror images. Bijou and Baer (1963) have introduced the term "constructed history" to refer to this approach. The following is a general summary statement of the constructed-history approach:

1. Formulate a hypothesis which refers to process variables which can be controlled by *E*.
2. Select Ss who cannot perform the perceptual task in question.
3. On the basis of the hypothesis, prepare a program of training aimed at enabling the Ss to perform the task.

4. When the program is successful, *E* will have enhanced the plausibility of his hypothesis. His hypothesis describes one of the ways in which the differences between age groups could have developed.

The current interest in this approach is tied to the entry of the operant conditioning specialists into the field of developmental psychology. However, the approach is not a contemporary innovation. The widely known study by Watson and Rayner (1920) on the development of fear reactions is an early illustration of this approach. For a variety of reasons, the constructed-history approach has not been prominent in developmental analyses. One of the principal reasons has been a recognition that experiments like Bijou's cannot establish that normally extended extralaboratory development actually does proceed on the basis of the processes implied by the experimental program. However, it has already been noted that the advocates of the constructed-history approach do not claim this accomplishment. "Such an experimental history would not be expected to coincide with the way children of comparable age learn the same task in their everyday experiences. . . . It would, however, provide *one* account of the variables of which such behavior is a function" (Bijou & Baer, 1963, p. 203). Any conclusions about naturally occurring development will always involve extrapolations from a narrower or more restricted set of observations. Therefore, the risks involved in extrapolating from the data of a constructed-history experiment are not peculiar to this procedure.

Two procedures for testing developmental hypotheses have been described. Other approaches could be described, but, instead, let us consider the hypotheses themselves. The most appropriate source of hypotheses is the viable body of theory concerned with perception. Advocacy of a given theoretical formulation will usually require a specific view of the nature of perceptual development. This view is likely to have testable hypotheses. To illustrate these remarks about the origins of hypotheses, we will consider briefly three of the contemporary theoretical alternatives: (1) J. J. Gibson's psychophysical theory (1959), (2) Brunswik's probabilistic functionalism (1956), and (3) the operant conditioning model (Bijou & Baer, 1965).[2]

[2] Obviously this selection is not exhaustive. Other alternatives could be considered. Piaget's formulations immediately come to mind. Piaget has been principally interested in the processes which underly the qualitative development of intellectual structures, but he has maintained an active interest in perceptual theory and experimentation. An excellent summary of Piaget's thinking has been presented by Flavell (1963, pp. 226–236, 350–356). Additional discussion of several details of Piaget's views on perception is available in articles by Vurpillot (1959) and Wohlwill (1962b). There are other theoretical approaches which deserve attention, notably those advocated by Wapner and Werner (1957) and by Witkin, Dyk, Faterson, Goodenough, and Karp (1962).

Psychophysical Theory

The hallmark of J. J. Gibson's theory is his generalized psychophysical hypothesis. Simply stated, this hypothesis asserts that visual perception is a function of optical stimulation. To maintain the integrity of this proposition, in the face of evidence of changes in perception without physical changes of stimulation, the differentiation hypothesis of perceptual learning was proposed [J. J. Gibson & E. J. Gibson (1955); also see Chapter 6]. As was noted in our description of the experiment by E. J. Gibson, J. J. Gibson, Pick, and Osser (1962), the same hypothesis can be applied to age-correlated changes. Perceptual development is a process of learning to respond to aspects of stimulation which were not responded to previously. Perceptual development is the learning of new stimuli, not the learning of new responses.

Probabilistic Functionalism

For Brunswik (1956), perception is a ". . . ratiomorphic subsystem of cognition" (p. 146) involving three relationships, all of which are probabilistic. The relationship are ones of ". . . probable partial causes and probably partial effects." The three relationships are the correlations between the distal source and the proximal stimulation, between the proximal stimulation and perception, and between perception and the distal source. These three types of covariation are called "ecological validity," "criterial validity," and "functional validity," respectively. Information about these relationships is acquired by a form of generalized probability learning. Brunswik finds the existence of developmental changes in perception entirely compatible with his theoretical views. In fact, Brunswik (1928) performed an early developmental study of brightness constancy, and Beyrl's investigation (1926) of size constancy was performed under Brunswik's direction.

Brunswik's account of Beyrl's results stresses acquired differences in attitude of observation. But alternative hypotheses can be formulated, stressing changes in the amount and accuracy of information about ecological validities, as well as acquired, age-correlated changes in criterial validity. As we grow from infancy to adulthood, we learn more about the objective relationships between variations of proximal (optical) stimulation and variations of the distal (external) source. In addition, we develop tendencies to repose different degrees of confidence in the various proximal cues. These confidences, which are referred to as criterial validities, are not simply reflections of ecological validities. Other variables affect criterial validity, so that criterial validity will, most often, be correlated with ecological validity only partially. To be of significant value, these speculations need to be transformed into explicit and testable hypotheses. The study of thinking by Bruner, Goodnow, and Austin (1956), which is based on Brunswik's conceptual framework, may provide useful leads.

Operant Conditioning

Perception may be analyzed within the system of principles advocated by the proponents of the operant conditioning model of behavior change. The most important principle is that differentiated responses which are not part of the genetic endowment are established by the deliberate or adventitious administration of selective reinforcement. The developmental changes of perception are simply instances of the general class of behavior modification. A clear exposition of this theoretical view is found in the discussion by Bijou and Baer (1965) of ". . . perceiving behavior to physical stimuli" (Ch. 9). The following excerpt might serve as a hypothesis to account for the developmental changes in form discrimination: "Through selective reinforcement . . . reactions come under the control of one or a combination of several dimensions or components of a complex stimulus. For this kind of discriminative behavior which is often called abstracting behavior to come into existence, it is necessary for the mother, father, and other members of the family to arrange antecedent and consequent stimuli so that responses to the critical dimension of a stimulus (e.g., its triangularity feature) are systematically reinforced and responses to all other dimensions (e.g., its size, color, location, and spatial orientation) are regularly followed by neutral or aversive consequences" (Bijou & Baer, 1965, pp. 154–155). Bijou and Baer (1963, p. 210) contend that the general form of this hypothesis could be similarly applied to the developmental changes in other perceptual tasks, e.g., depth perception and motion perception.

Three hypotheses concerning developmental change have been presented. No effort will be made to assess their relative merits. In a field of investigation which has provided only infrequent instances of hypothesis testing, every plausible hypothesis should be treated kindly. An occasion for evaluation will be on hand after appropriate experimentation has been completed. Again, a distinction must be maintained between demonstrating an age-correlated change and explaining the change. All too often the argument has taken the form "My theoretical formulation leads to the prediction of an age-correlated change in perceptual function X. The change is observed, so the formulation is confirmed." This argument is unconvincing, because the gross prediction of change can be derived, with equal plausibility, from other theoretical formulations. The hypotheses, and the studies which are designed to test them, need to be more analytic.

part three

LANGUAGE

The study of language has three foci: the sounds which we use to communicate with one another, the logic or grammars which govern our combinations of words, and the relation of words to meaning. The language of a child, to paraphrase Anderson, first appears between the eighteenth and twenty-eighth month, and is to be distinguished from the signals which many animals make. The child comes to manipulate words deliberately as a means of controlling others and himself, and thus frees words from their immediate references and uses them in and for themselves. Our study of a child's language must incorporate all three foci of study if we are to account for its origin, its emergence, and its influence upon the child's behavior.

The first essay of Part Three, Susan Houston's "The Study of Language: Trends and Positions," identifies four theoretical approaches to language acquisition and then sketches their emergence from a dual historical tradition of speculation and structural description. Like Ausubel, she classifies these theoretical positions according to their views on the relationship between heredity and environment.

Hans Furth, in his "Piaget's Theory of Knowledge: The Nature of Representation and Interiorization," considers ways in which theorists have regarded representation and then contrasts the more traditional associationist notion of meaning with Piaget's theory. In particular, he addresses himself to the problem of meaning: the nature of relationships between knower and concept; knower and symbol; concept and symbol; the real event and concept; the real event and symbol.

The third essay in Part Three, Lenneberg's "The Natural History of Language," is concerned principally with spoken language, not with language systems as they appear in other animals. Lenneberg argues that man has a biological and evolutionary predisposition to language, and that the onset of language may reasonably be attributed to a maturational process.

Bruner takes the view that language stems from a primitive and innate symbolic activity which, through acculturation, gradually becomes specialized. In his chapter from *Studies of Cognitive Growth* he goes beyond Lenneberg in discussing some universal properties of protosymbolic activity which may underlie language. Bruner draws our attention to the often confusing relationships between human development, the structure of language, and the organizing influences of cultures.

The fifth essay in this part reviews research literature pertaining to children's linguistic behavior before the age of 18 months. "The Prelinguistic Child," by the Kaplans, discusses alternate approaches to the question of when language acquisition begins and then refers to studies which relate language production to language comprehension.

The final essay, "The Status of Developmental Studies in Language" by Jacqueline Sachs, presents a case for regarding language as a system of rules about word use. She reviews studies of rule learning and rule change which are of interest both in themselves and in light of Kessen's criteria for a cognitive theory.

Some Topical Readings

SURVEY OF THEORETICAL POSITIONS

McNeill, David, "On Theories of Language Acquisition," in Dixon and Horton eds., *Verbal Behavior and General Behavior Theory*. Englewood Cliffs, N.J.: Prentice-Hall, 1968.

COMPLEXITY OF PROCESSES

Furth, H. G., *Thinking Without Language*. New York: Free Press, 1966.
Katz, Jerrold, *The Philosophy of Language*. New York: Harper & Row, 1966.

COMPARATIVE STATUS

Lancaster, J. B., *Emergence of Language in Primates*. New York: Holt, Rinehart and Winston, Inc., 1969.

INTERRELATIONSHIPS

Bronowski, J. and U. Bellugi, "Language, Name, and Concept," *Science*, 168 (1970), pp. 669–673.

Goodnow, Jacqueline, "Problems in Research on Culture and Thought," in *Studies in Cognitive Development*. New York: Oxford, 1969.

Greenfield, T. P., and J. Bruner, "Culture and Cognitive Growth," in Goslin ed., *Handbook of Socialization Theory and Research*. Chicago: Rand McNally, 1969.

INFANT RESEARCH

Rebelski, F., Starr, and Luria, "Language Development, the First Four Years," in Y. Brackbill ed., *Infancy and Early Childhood*. New York: Free Press, 1967.

LATER RESEARCH

Slobin, Dan, *Ontogenesis of Grammar*. Berkeley, Calif.: University of California Press, 1970.

Additional Readings in Language and Language Development

Alberts, E., *Semiotics and Semantics: A Partial Interdisciplinary Bibliography to 1967*. Northwestern University, 1968.

Berko, J., "The Child's Learning of English Morphology," *Word*, 14 (1958).

Blumenthal, A., *Language and Psychology*. New York: Wiley, 1970.

Braine, M., "The Ontogeny of English Phrase Structure," *Child Development Monograph* (1963).

Brown, R., *Social Psychology*. New York: Free Press, 1965.

Brown, R., *Words and Things: An Introduction to Language*. New York: Free Press, 1968.

Brown, R., *Psycholinguistics*. New York: Free Press, 1970.

Brown, R., and U. Bellugi, "Three Processes in the Child's Acquisition of Syntax," *Harvard Educational Review*, 34 (1964), pp. 133–151.

Brown, R., and J. Berko, "Word Association and the Acquisition of Grammar," in D. Palermo and L. Lipsitt, eds., *Research Readings in Child Psychology*. New York: Holt, Rinehart and Winston, Inc., 1963.

Carroll, J. B., *The Study of Language*. Cambridge, Mass.: Harvard University Press, 1953.

Carroll, J. B., "Language Development," in AERA's *Encyclopedia of Educational Research*. New York: Macmillan, 1960.

Carterette, E. C., ed., *Brain Functions: Speech, Language and Communication*, Volume III. Berkeley, Calif.: University of California Press, 1961.

Cassirer, E., *Philosophy of Symbolic Forms*, Vol. 1, *Language*. New Haven, Conn.: Yale University Press, 1923.

Cassirer, E., *Philosophy of Symbolic Forms.* New Haven, Conn.: Yale University Press, 1957.
Cazden, C., "Environmental Assistance to Child's Acquisition of Grammar." Unpublished doctoral dissertation, Harvard University, 1965.
Cherry, C., *On Human Communication.* Cambridge, Mass.: M.I.T. Press, 1957.
Chomsky, C., The *Acquisition of Syntax in Children from 5–10.* Cambridge, Mass.: M.I.T. Press, 1970.
Chomsky, N., *Syntactic Structure.* The Hague: Mouton, 1957.
Chomsky, N., "Review of Verbal Behavior," *Language,* 35 (1959), pp. 26–58.
Chomsky, N., *Acquisition of Syntax in Children from 5–10.* Cambridge, Mass.: M.I.T. Press, 1970.
Church, J., *Language and the Discovery of Reality.* New York: Random House, Inc., 1961.
Cofer, C., and B. S. Musgrave, *Verbal Behavior and Learning.* New York: McGraw-Hill, 1963.
DeCecco, J., *The Psychology of Language, Thought, and Instruction.* New York: Holt, Rinehart and Winston, Inc., 1967.
Deese, J., *Structure of Associations in Language and Thought.* Baltimore: The Johns Hopkins Press, 1965.
Deese, J., *Psycholinguistics.* Boston: Allyn and Bacon, 1968.
Delaguna, G. A., *Speech: Its Function and Development.* New Haven, Conn.: Yale University Press, 1927.
Dixon, T. R., and D. L. Horton, *Verbal Behavior and General Behavior Theory.* Englewood Cliffs, N.J.: Prentice-Hall, 1968.
Ervin, S., and G. Miller, "Language Development," *NSSE Yearbook, Child Psychology,* 1963.
Ervin-Tripp, S., and D. Slobin, "Language Development," *Annual Review of Psychology,* Palo Alto (1966).
Feigenbaum, E., and J. Feldman, eds., *Computers and Thought.* New York: McGraw-Hill, 1963.
Fodor, J., and J. Katz, eds., *Readings in the Philosophy of Language.* Englewood Cliffs, N.J.: Prentice-Hall, 1964.
Furth, H., *Thinking Without Language.* New York: Free Press, 1966.
Greenberg, J. H., ed., *Universals in Language.* Cambridge, Mass.: M.I.T. Press, 1963.
Guilford, J. P., "Morphological Model for Human Intelligence," *Science,* 131 (1960), p. 1318.
Jakobson, R., and M. Hakke, *Fundamentals of Language.* The Hague: Mouton, 1956.
Head, H., *Aphasia and Kindred Disorders of Speech.* New York: Macmillan, 1926.
Henle, P., *Language, Thought and Culture.* Ann Arbor, Mich.: University of Michigan Press, 1966.
Hildum, D., *Language and Thought.* Princeton, N.J.: Van Nostrand, 1967.
Hockett, C., "Origins of Speech," *Scientific American,* 203 (1960), pp. 89–96.
Hunt, J. McV., *Intelligence and Experience.* New York: Ronald, 1961.
Lee, L., "Developmental Sentence Types," *Journal Speech and Hearing Disorders,* 31, no. 4 (1966).

Lewis, M. M., *Infant Speech: A Study of the Beginnings of Language.* New York: Harcourt, 1936.
Luria, A. R., *Role of Speech in Regulation of Normal and Abnormal Behavior.* New York: Pergamon, 1961.
Luria, A. R., and F. Iayudovich, *Speech and the Development of Mental Processes in Children.* London: Staples Press, 1959.
Lyons, J. L., and R. J. Wales, *Psycholinguistic Papers.* Edinburgh 8: Edinburgh University Press, 1966.
McCarthy, D., "Language Development in Children," in L. Carmichael, ed., *The Manual of Child Psychology.* New York: Wiley, 1954.
McCarthy, D., "Research in Language Development: Retrospect and Prospect," *Child Development Monograph,* 24, no. 5 (1959).
McCarthy, D., "Affective Aspects of Language Learning," in A. H. Kidd and J. L. Rivoire, eds., *Perceptual Development in Children.* International Universities, 1966.
McNeill, D., *Creation of Thought.* Edinburgh Conference, 1968.
McNeill, D., *The Acquisition of Language.* New York: Harper & Row, 1970.
Menyuk, P., *Sentences Children Use.* Cambridge, Mass.: M.I.T. Press, 1969.
Merleau-Ponty, M., "Conscience and Language Acquisition," *Bulletin of Psychology,* 18 (1964), pp. 226–259.
Miller, G., *Language and Communication.* New York: McGraw-Hill, 1951.
Miller, G., "Speech and Language," in S. S. Stevens, ed., *Handbook of Experimental Psychology.* New York: Wiley, 1951.
Miller, G., "Some Psychological Studies of Grammar," *American Psychology,* 17 (1962), pp. 748–762.
Morris, C. R., *Signs, Language, and Behavior.* Englewood Cliffs, N.J.: Prentice-Hall, 1946.
Mower, O. H., *Learning Theory and the Symbolic Processes.* New York: Wiley, 1960.
Oldfield, R. C., *Language.* Baltimore: Penguin, 1966.
Osgood, C., and T. Sebeok, *Psycholinguistics, A Survey of Theory and Research Problems.* Bloomington, Ind.: Indiana University Press, 1965.
Piaget, J., *Language and Thought of the Child.* New York: Harcourt, 1926.
Rosenberg, S., and S. H. Koplan, *Developments in Psycholinguistics Research.* New York: Macmillan, 1968.
Skinner, B. F., *Verbal Behavior.* New York: Appleton, 1959.
Smith, F., and G. Miller, eds., *Genesis of Language.* Cambridge, Mass.: M.I.T. Press, 1966.
Staats, A., *Learning, Language, and Cognition.* New York: Holt, Rinehart and Winston, Inc., 1968.
Templin, M., *Certain Language Skills in Children.* Minneapolis: University of Minnesota Press, 1950.
Terwilliger, R., *Meaning and Mind.* New York: Oxford University Press, 1968.
Vygotsky, L. S., *Thought and Language.* Cambridge, Mass.: M.I.T. Press, 1962.
Weir, R. H., *Language in the Crib.* The Hague: Mouton, 1962.

THE STUDY OF LANGUAGE: TRENDS AND POSITIONS*

Susan H. Houston

Since the third century B.C., a dual tradition of language study has existed. On the one hand, there has been speculation about such topics as the origins of language, the differences between child and adult language, and the relationship of speech to thought; on the other hand, there has been structural description of contemporary spoken languages. Throughout history, one of these traditions has tended to dominate the study of language; rarely have the two been pursued with equal vigor. From the time of the ancient Greeks until the nineteenth century, grammar was regarded as a proper focus of inquiry for the logician and philosopher. The study of language was preoccupied either with isolating the logical principles upon which grammar was purportedly based (see Robins, 1951—for example, p. 14 ff.), or with solving, through philosophical argument, such problems as the origins of grammatical gender or the connection of names with objects. As Robins notes, during medieval times "grammar could be and was studied as a means of reading Latin and as a branch of speculative philosophy, a key to understanding the nature and working of the human mind" (p. 75). The seventeenth- and eighteenth-century philosophers who studied language considered it their goal to produce a general theory of language, a universal grammar which would serve both to explain the specific workings of lexical and syntactic problems which interested them and to provide insight into the functioning of the mind. Descartes and his followers, especially, believed it possible to discover a universal grammar (see Chomsky, 1966).

During the nineteenth century, the focus of linguistics began to move from grammar to comparative Indo-European studies and the development of a science of phonetics (see Pedersen, 1931, for an account of this period in linguistics). Beginning with the publication of Bloomfield's

* This selection was written expressly for this volume.

Language in 1933, rigorous structural linguistic study came into prominence. Bloomfield's work, one of the cornerstones of structural linguistics, emphasized the importance of behavioristic concepts—concepts devoid of introspection. Philosophical grammar, a universal theory of language, and psycholinguistics in general were given very low priority by linguists, although such topics were being studied by specialists in other fields. The description of current spoken languages became the central purpose of language study during the structuralist era, and an underlying theory of language was not regarded as a necessary prerequisite.

About the mid-fifties the spotlight shifted away from the structural linguists. Harris (1951), Chomsky (1957), and their followers initiated a return to the speculative tradition and an interest in language universals. At the same time, linguists began to recognize the contributions that other disciplines were offering toward the development of a general theory of language. Questions of interest to seventeenth- and eighteenth-century thinkers, which had been largely set aside by the structuralists, were revived as the central issues of language study.

This brief outline of phases in the history of linguistics, and the kinds of evidence accumulated in each phase, is important when we consider the alignment of language theorists on major issues in the field. It is not surprising that certain theoretical stances tended to dominate the field during each phase. For example, during the structuralist era there was an emphasis on description of language as observable behavior and a corresponding unwillingness to speculate on the psychological entities corresponding to units of language. Those psychologists who studied language acquisition during this era tended to reject such notions as internalized linguistic rules or language-specific learning mechanisms. This similarity in the approaches of linguists and psychologists was not due to a high degree of communication between the two fields; perhaps it reflected a general trend in behavioral science. The periods of prominence in linguistics and other fields do not always coincide, of course, but the theoretical and methodological revolution in linguistics which brought about the generative grammars of the 1950s occurred at a time when similar upheavals and shifting of positions were taking place in psychological learning theory.

It is possible to classify the major theoretical positions within linguistics by means of a number of schemata. One such schema, developed by Ausubel (1957), classifies positions according to their views on the relationship between heredity and environment in human development. It is reasonable to apply this schema to theoretical positions on language behavior as well, although there are other issues which differentiate schools of thought in this field. Many of the most important conflicts in the study of language center around the question of whether language is a learned, inherited, or interactive phenomenon.

In Ausubel's schema, there are four main approaches to the study of language, each of which contains a continuum of theoretical positions. In historical order these approaches have been what Ausubel calls behaviorism, cultural relativism-determinism, interactionism, and preformationism-predeterminism.

The various positions included in the last of these approaches have often been described by linguists as *developmental psycholinguistics.* This is because preformationists and predeterminists think language changes qualitatively rather than quantitatively with the age of the speaker. It should be recognized that the term *developmental* may be used here somewhat differently from the way it is used in other branches of sociobehavioral science.

The four approaches to language theory range from environmentalist to hereditarian positions, although on the whole there are some complicating factors. Cultural relativism-determinism, for example, is based on somewhat different concepts from the others, as will be seen. However, each of the four approaches is characterized by certain hypotheses concerning the beginnings of language use and, further, by the answers which proponents give to a number of basic questions about the theory of language.

In particular, six questions will serve to accent critical issues in language theory. The first is, "How and why is language acquired?" This question is very revealing since answers to it indicate the relative emphasis each approach places on the heredity-environment issue. Second, there is the question, "How is the concept *language* best defined?" Responses to this question indicate the views within each position as to whether language should be studied as a psychological entity, a cultural phenomenon, a learned series of stimulus-response bonds resulting in verbal behavior, and so forth. Clearly, the definition of language advanced by each position will provide a key to the subsequent analysis of linguistic events. A third question is, "What is the best device or technique for describing language?" This question, closely related to the problem of definition, focuses upon the modes of inquiry employed by the linguist or other language scientist. Fourth, "To what extent is language peculiar to man?" Answers to this question reflect different views of the nature of animal communication and its relation to human language. A fifth question, "Are there significant universals of language?" is designed to elicit views concerning the importance of innate components of language. Finally, a sixth question, "What form does the speaker's internalized competence take?" reveals differences between theories about the psychological nature of language.

Following a brief introductory comment on each of the four approaches, answers to these six questions will be examined; in this way a concise summary of each position, as well as a convenient means of comparisons among them, may be achieved.

Behaviorist Approaches

The behaviorist approaches to language include a variety of environment-oriented positions. Behaviorist thinking in both psychology and linguistics during the early part of this century sought to bring scientific rigor to investigations of language and other human activities. In psychology, behaviorism expressed the assumption that all learning is based upon a few relatively simple principles, and no special human learning mechanisms are necessary. Indeed, the behaviorist laws of learning represent a theory of interspecies universals. All items of observable behavior are presumably related to learning mechanisms. However, the key word here is *observable*. The strict behaviorist is highly reluctant to allow the existence of internalized perceptual, linguistic, or cognitive systems. (For a further explanation of the learning principles upon which behaviorist psychology is based, see Hilgard, 1956, pp. 82 ff.; W. F. Hill, 1963, pp. 57 ff.; and similar texts.)

With regard to language acquisition, most behavioral theories presume that the child possesses only those inborn mechanisms which underlie the capacity to form stimulus-response bonds and to articulate speech sounds. For example, Skinner contends that verbal behavior differs from other sorts of behavior chiefly in being reinforced through the mediation of other persons (see Skinner, 1957, pp. 21–28). Language is regarded as a collection of verbal operants, or responses, acquired through operant conditioning and emitted as a function of various sorts of stimuli (Skinner, 1957, p. 14).

The behaviors constituting language were viewed by early behaviorist theorists as relatively simple. They must be (and presumably universally are) taught to the child by specific reinforcement, at least in the early stages of learning; a child not so reinforced would theoretically not learn to speak (Mowrer, 1960, pp. 82–83). The parent is usually the conditioning agent who reinforces the infant's natural babbling, especially sounds which resemble words of adult language, and gradually becomes willing to accept only progressively better approximations to adult language. Mowrer argues that as a child grows older he will become habituated to the use of language, perhaps recognizing its utility as a means of gaining control over his environment (Mowrer, 1960, pp. 85–86). Habituation may occur in order for the child to meet his growing communicative needs (Chase, 1966, pp. 260–262), or simply because of continuing reinforcement by parents and teachers (Staats and Staats, 1964, pp. 173–177, for example).

It should be noted that most applied linguistic or language-instruction theory continues to be based upon this reinforcement model of language acquisition (see Rivers, 1964, pp. 19–20, and 1968, pp. 38–39; Lado, 1964, p. 39). Generally speaking, behaviorist theorists do not differ greatly about language learning; they tend to diverge mostly in their accounts of the reasons behind the process, the motivation of the conditioning agent, the form of reinforcement experienced by the child, and their concepts of what sort of system the child actually acquires. Answers to the six questions on language mentioned earlier may help to clarify some of these differences.

Summary of Behaviorist Approaches

How and Why Is Language Acquired? As previously stated, the predominant, behavioristic reply is that language is acquired through the selective reinforcement of natural infant vocalizations and the gradual shaping of vocal behavior through operant conditioning techniques employed by parents and teachers. This response is of practical significance in the treatment of childhood language disorders, for example, for it implies that where language development is retarded or otherwise aberrant for nonphysiological reasons, the clinical psychologist should examine the kind of reinforcement given the child for verbal responses, the motives of the parent in carrying out the conditioning in a faulty manner, and similar factors.

There are differences among behaviorists concerning later stages of the language-acquisition process. Skinner (1957) feels that different types or classes of verbal operants (roughly, parts of speech) are acquired for different reasons (specifically, they are under the control of different categories of stimulus); whereas Mowrer (1960, pp. 70–162) tends to ascribe all language learning to the same processes.

A further divergence is a two-stage proposal by Osgood (1957, pp. 5–14). Osgood derives a mediated stimulus-response paradigm for the learning of words, whereas most other behaviorists tend to base their descriptions on a one-stage model. Osgood's proposal fostered development of a device for the measurement of connotative meaning, the Semantic Differential, which he claims provides an explanation for most of the common learning phenomena observed in verbal learning experiments (see Staats and Staats, 1964).

How Is the Concept *Language* Best Defined? Both Skinner's one-stage stimulus-response paradigm and Osgood's two-stage model of language acquisition (1957, especially pp. 161 ff.) define language as a network of interclass associations in which classes are defined by the possible context of their occurrence (for discussion of this viewpoint, see McNeill, 1968, pp. 408–409). The child acquires response chains directly from the language of those around him without reorganizing the utterances he hears or subjecting them to further analysis. For example, if a child hears such phrases as *the table, a table, my table,* and *that table,* he will gradually learn to classify the words *the, a, my,* and *that,* together as a category which can elicit the word *table* (and other words of the same class as *table*) as a response (Staats and Staats, 1964, pp. 171–172). Thus what the child learns is a linear series of associations. Sentences such as "John is easy to please" and "John is eager to please" are presumed to possess the same structure, and the words *easy* and *eager* are assigned to a single category since both occur in the same context.

A somewhat more complicated behavioral description of language has been put forth by M. D. S. Braine (1963a, 1963b, and 1965). Braine argues that language possesses a hierarchical structure determined by the location of units within phrases and of phrases within sentences. He holds that the acquisition of language is not an accumulation of simple linear associations but is based upon learning the relative positions of constituents of larger constructs. Although this description has certain parallels in arguments about the nature of intelligence, most behaviorists tend to subscribe to the one- or two-stage stimulus-response model already described.

What Is the Best Device or Technique for Describing Language? The grammar which accompanies the behaviorist approach to language is, in its most traditional form, an Item-Arrangement (as contrasted to Item-Process) grammar, the purpose of which, as the name implies, is to identify the units or items comprising the language and their arrangement into grammatical structures or utterances. Structural linguists share with behaviorist psychologists such as Skinner the use of taxonomic or classificatory presentation of data, and the concomitant belief in unambiguous, logical decision procedures for linguistic or psycholinguistic analysis. The item-arrangement grammar, in short, does not constitute a theory about how speakers produce language in any literal sense.[1] Nor is it particularly suitable for demonstrating complex relationships among sentences or sentence types, since it is designed for analysis of spoken verbal behavior only.

To What Extent Is Language Peculiar to Man? Although many behaviorists of the 1930s and 1940s described language acquisition in terms of universal laws of learning, later theorists were impressed by the fact that language is used among humans alone, and have since attempted to describe patterns of learning peculiar to man. For example, although Osgood's two-stage stimulus-response paradigm is not theoretically limited to human learning, he implies that humans acquire this sort of learn-

[1] There is, however, an offshoot of structural linguistics called Tagmemics developed by K. Pike (1967) in which the form of grammatical descriptions attempts to mirror the response-class structuring presumed to underlie language acquisition and production.

ing more readily than do other species. Braine's contextual generalization is not only clearly limited to human learning, but appears relevant to few situations other than language acquisition. It is more common, however, for behaviorists to note that other species lack either some of the necessary neural apparatus for language or the motor connections needed for proper articulation of speech sounds (Mowrer, 1960, p. 74). Since the behaviorist stresses the verbal behavior of language rather than some internal capacity, inability to produce verbal behavior is equivalent to inability to form language. Thus, there is no need for a more complete explanation of why only humans use language.

It is characteristic of behaviorists to postulate the existence of systems of communication among other species, typically primates, as differing from human language only quantitatively (Bloomfield, 1933, p. 27; Skinner, 1957, p. 464). The obvious implication is that behaviorist-oriented psychologists and ethnologists do not often pursue detailed analysis of communication among different species. The major studies of animal communications have come from theoretical approaches other than behaviorism.

Are There Significant Universals of Language? Although they operate from essentially the same theory of language and language learning, behaviorist psychologists and structural-descriptive linguists differ somewhat on the issue of linguistic universals, largely because of their interests in different aspects of the problem. As we have seen, behaviorists postulate universal laws of learning as the foundation for language acquisition. These laws are universal in the strictest sense, common not only to all languages but to acquisition of all learned behavior among—depending on the individual theorist—primates, vertebrates, or even all organisms whose behavior is subject to modification. In human language the learning of response classes and the associations among them is the process by which all languages are structured, although the particular taxonomy of each language is often considered peculiar to that language.

In contrast, structural linguists have focused on universals of linguistic structure rather than on language acquisition and production. They are inclined to believe that the diversity among languages is far more significant than the similarity, and that the search for linguistic universals is by nature unscientific (see Bloomfield, 1933, pp. 5, 8; see also Chomsky, 1966, p. 101, Note 83). The main reason for this belief is that structuralism, in its attempts at rigorous and objective linguistic analysis, describes observable verbal behavior, not its underlying structure. Indeed, the most notable fact about language, structuralists feel, is its seeming diversity. A basic tenet of structural linguistics is that each language must be analyzed in terms of its own features, without reference either to past stages of that language or to patterns found in other languages.

What Form Does the Speaker's Internalized Competence Take? Although structural-descriptive linguists tend to eschew speculation on this question, in general they accept the behaviorist description of linguistic competence as consisting of a very large number of stimulus-response associations among grammatical classes or response types. Since these response types are unique to each language, they are presumably acquired entirely through learning. The speaker's competence need not be in any way prestructured or innately formed.

Cultural Relativist-Determinist Approaches

The cultural-relativist and cultural-determinist approaches toward language were influenced by the same general traditions as behaviorist thought, modified by anthropological study. They take a slightly less strict environmental approach than behaviorism, including the concept of an innate and species-wide consciousness or "mind." Cultural relativism-determinism is by far the most prevalent position among even contemporary anthropologists, representing a countertrend to early ethnocentric descriptions of cultural patterns. In its general form, without specific reference to language, the cultural-determinist approach asserts the predominating influence of social and cultural patterns upon human behavior, minimizing the importance of individual differences and heredity. The cultural-relativist approach further asserts that each person is the product of his specific sociocultural history, and therefore is unique.

The linguistic parallels to these approaches found expression in the works of Boas and Sapir in the first three decades of this century. Boas emphasized the uniqueness of individual linguistic systems and stressed the difficulty of expressing in one language the dominant cultural themes set forth in any other language. Sapir regarded language as infinitely variable among different social groups. He explicitly denied the possibility that man possessed a biological predisposition for language acquisition (1921, pp. 3–23). It was his belief that perceptions of the real world were largely shaped by language and consequently that "the worlds in which different societies live are distinct worlds" (1949, p. 69).

The relationship of these positions to that of the structural linguist can readily be seen, since both stress the uniqueness of each linguistic system and doubt that techniques for describing language universals can be identified. Also implicit in Sapir's position is the argument that the individual contributes little to his own linguistic and perceptual development; it was his belief that individual differences are diminished by the linguistic patterns common to the culture.

A linguist more frequently identified with the cultural relativist-determinist position is B. L. Whorf. Whorf, with Sapir, was responsible for propounding the hypothesis of linguistic relativity, also known as the Whorfian Hypothesis: That language is the single most important influence upon the development of our thought patterns and perceptions of the world, and that different linguistic systems lead inevitably to different ways of thinking. He contended that Western scientific and logical concepts reflect the language patterns of their originators. Although he did not deny the existence of a universal psychological foundation for language and cognition, Whorf conceived of such a universal only as an abstraction at the level of "consciousness" or "mind." He did not believe that the innate potential for thought includes the development of perceptions such as those described by Gestalt psychologists (Wertheimer, 1923, for example), since he believed that without language the universe would be perceived as an undifferentiated continuum of sensory impression (Whorf, 1956, p. 239).

Although cultural relativists and determinists parallel the behaviorists in emphasizing the differences rather than the similarities among languages, they differ in the ways they emphasize these differences. First, cultural relativists commonly are less environmentalistic than behaviorists. The former are willing to concede the existence of at least a core of basic hereditary structures for language (these may take the form of predispositions to learn language or innate capacity to develop thought based upon language). Second, while the cultural relativist or determinist holds that the kinds of verbal response which compose language differ for each linguistic system, the behaviorist contends that the same kind of associative processes obtain throughout human language, with major differences occurring only in the linguistic items composing response classes in each language.

It is clear that the cultural relativist-determinist and the behaviorist regard the topic of language from different perspectives: the behaviorist is interested in language as observable vocal behavior; the cultural relativist or determinist is more concerned with the relationships between language and cognition. However, the two approaches share many common theoretical roots, and their respective conclusions about language complement more often than contradict each other.

Summary of Cultural Relativist-Determinist Approaches

How and Why Is Language Acquired? For the cultural relativist-determinist, language acquisition is a social necessity and a fundamental factor in the growth of individuality (Sapir, 1949, p. 19). Particular mechanisms of language acquisition are not a critical concern, since children do not have to be taught language in order to acquire it—a logical viewpoint, considering the anthropological origins of this approach and the observation by field anthropologists of cultures in which children are not specifically taught to use language (as they are in ours). Most cultural relativists-determinists argue that children learn language naturally, and that language acquisition parallels and greatly influences the development of rational thought.

How Is the Concept *Language* Best Defined? Unlike the behaviorist, the cultural relativist-determinist does not view language as a collection of learned responses. More significant, he would not agree with the behaviorist that language is equivalent to speech, specifically, that it is entirely external. To the cultural relativist-determinist, language is primarily a conceptual system, intertwined with and perhaps indistinguishable from, thought. Words are not neat and consistent packages of symbol-referent association; the cultural relativist or determinist notes that speakers often confuse symbol and referent, and treat symbols that have no concrete referents as though they were connected with actual objects (Whorf, 1956, pp. 135 ff.). In fact, the field of General Semantics, an outgrowth of this position on language, is based upon the study of the connections between word and object, and the problems resulting from confusion of the two (see, for example, Hayakawa, 1941; and Lee, 1941). But cultural relativists and determinists do agree with behaviorists on the arbitrary nature of language. As proof the cultural relativist often cites the extreme diversity of languages around the world and their apparent lack of structural universals.

What Is the Best Device or Technique for Describing Language? Since cultural relativism-determinism is based on anthropological studies of various cultures, it relies upon field techniques of language description. Although there is no specific theory of language underlying field linguistics, its procedures reflect the principles of structural linguistics and therefore parallel structuralist-behaviorist theory. Anthropological linguists such as Sapir and Whorf applied the standard taxonomic methods of description to their data. They both strongly emphasized the importance of understanding the world-view peculiar to each language, and thus placed linguistic description in its sociocultural framework. Whorf, in particular, utilized psychological criteria in setting up his linguistic categories, and considered as a main goal of linguistic description the determination of the thought patterns and perceptions of the world which correspond to the structures of a language.

To What Extent Is Language a Species-Specific Behavior? Unlike the behaviorists, who consider language acquisition an elaboration of universal principles of learning, cultural relativists-determinists hold that language is exclusively a human phenomenon, the major distinction between human and animal behavior. This position is based not upon study of the language-learning process but upon examination of the uses and functioning of language. Specifically, it is the capacity for symbolic behavior that animals are said to lack, not any learning mechanism peculiar to linguistic systems.

Are There Significant Universals of Language? In general the cultural relativist-determinist's answer to this question would simply be "no," given the incredible diversity of existing languages. Perhaps the only significant linguistic universal he would recognize is that language functions everywhere as the means by which concepts and perceptions are formed. Structurally speaking, the cultural relativist does not even accept the existence of universal categories such as the noun phrase or predicate phrase; in recent terminology one would say that he studies only surface structure without searching for underlying deep-structure regularities.

What Form Does the Speaker's Internalized Competence Take? This is another topic with which the cultural relativist-determinist has not been greatly concerned. He would probably maintain that the speaker masters not only a set of arbitrary patterns, but an entire symbolic and conceptual system. Internalized language is seen as the substratum of all thought rather than being solely linguistic in nature. Whether the internalized competence is in the form of response chains or rules is thus irrelevant, since the cultural relativist-determinist tends to see language as integrated with the rest of human behavior and thought.

Interactionist Approaches

Interactionist positions on the acquisition and use of language are far more developmentally oriented than either of the former two approaches, and may be said to be closer in spirit to current trends. Although much of the work in this area first appeared in the 1930s during the era of structuralism and behaviorism, its implications were not realized until much later.

Piaget is now perhaps the best-known proponent of interactionism with respect to language. He argues that growth and development occur through biological interaction between man and his environment—through the complementary processes of the organism's adaptation to the external environment and to internal organizations by means of cognitive schemata. Involved in external adaptation are two subprocesses, assimilation and accommodation, by which new elements from the environment are absorbed into the organism's total structure and integrated into its functioning (Piaget, 1952, pp. 3–20). A child develops the necessary schemata for language as a function of his neurological maturation and the environment to which he is exposed.

Interactionists such as Piaget typically study language not as a unique phenomenon, but as part of the entire scheme of human development. According to Piaget the child passes through a number of fixed developmental stages—each characterized by certain types of cognitive organization and behavior, both linguistic and nonlinguistic; language develops along with the child's capacities for logical thought, judgment and reasoning, and reflects these capacities at each stage. Developmental progress is thus caused by a combination of heredity, growth, and experience, or what Hebb (1966, p. 157) has termed psychological maturation.

While the cultural relativists and determinists regard language as the critical shaping factor for thought, the interactionists see language merely as a medium of representation which is influenced throughout the child's growth by his overall capacities for thought and logic, his perception of the environment and of other people in it and, significantly, his perception of himself vis-à-vis others. A central factor in the child's environment is the existence of other people. Language helps a child to see others as separate from himself—a perception which is achieved only gradually. Piaget, though he does not equate language and thought, considers that they influence and reflect each other. Although he states that the child's illogicalities of perception and of language usage may not always coincide in time (1955a, p. 141), he argues that attainment of rational and logical thought and of adult language ability are parts of the same developmental process. Other interactionists, such as Vygotsky, regard language and thought as being separate in early childhood, developing independently and becoming interrelated as the child grows older (see Vygotsky, 1962, pp. 41–44). More recent interpretations of the interactionist approach, however, incline toward Piaget's view of development of thought and language as parallel and interconnected (for example, Bruner, 1966b, pp. 14–17), and of the development of language as the key to the child's socialization.

Summary of Interactionist Approaches

How and Why Is Language Acquired? Generally speaking, interactionists believe that children have a biological predisposition for language, and view language acquisition as both a gradual internalization of linguistic structures and a growing awareness of the social and communicative functions of language (Bruner, 1960, p. 8; Piaget, 1967, pp. 18–22). Linguists who hold interactionist positions regard language acquisition not as dependent upon conditioning or training by the mother, but rather as a process by which verbal information received from the mother and other sources is adapted by the child and gradually integrated into a linguistic system (Hockett, 1958, pp. 353 ff.). In other words, the basic ability to acquire a language, as well as certain facets of linguistic structure itself, are assumed to be innate, and the process of acquiring a language is considered to be a function of both maturation and learning. The interactionists argue that language is learned to serve the communication needs of the growing child—a kind of learning dependent both upon neurological maturation and upon a multitude of learning encounters. In no sense is language considered a matter of obvious response categories.

How Is the Concept *Language* Best Defined? In some respects, the interactionists define language in ways akin to the definition of cultural relativists such as Whorf. Language is the means for communicating perceptual and symbolic information, rather than a collection of behaviors or linguistic patterns. To the interactionist, language is both a form of behavior and an internalized system: as a form of behavior, it provides an observable key to the child's concepts of reality at each stage and serves as the chief instrument by which the child relates to other people; as an internalized system, language provides the means by which the child can manipulate objects temporally and spatially absent (Lewis, 1963, p. 83), form concepts of causality and logical relations, and think about his world.

What is the Best Device or Technique for Describing Language? Since most proponents of interactionism are psychologists, it is predictable that linguistic description per se plays little part in interactionism. A description of language within this system would concentrate on linguistic ontogeny, noting the kind of utterance characteristic of each developmental stage and the type of cognition and reasoning it implies. Most interactionist studies are designed to show children's concepts of space, time, causality, and other similar concepts at different ages rather than focusing on language. Language is elicited as the means by which these concepts are expressed, and is examined for evidence of specific cognitive processes. In general, language is not itself the chief object of study, and interactionists have rarely concerned themselves with problems of adult language.

To What Extent is Language a Species-Specific Behavior? There are several interactionist viewpoints concerning the species-specificity of language. Piaget considers language a human phenomenon not approximated by animal communication. He bases his concept of the mind upon various hereditary elements specific to man; language is species specific because reason is species specific. In contrast, Vygotsky is more ready to find approximations to human language among higher primates (1962, p. 41), although he states that animal communication takes place without intellectual activity and lacks the capacity for symbol manipulation. The position of Bruner and Hockett is more complex: they posit the existence of prototypical symbolic universals derived from a comparative analysis of the component factors in both human language and animal communication (Hockett, 1960 and 1963; Bruner, 1966b, pp. 76–78). The conclusion of their analysis is that although human language and animal communication share certain features; they are unlike enough for language to be considered species specific.

Are There Significant Universals of Language? A biological predisposition for language is presumed by interaction theorists to underlie the acquisition and use of all human language and thought. Indeed, the developmental stages through which, interactionists believe, children pass are comparable to psychophysical universals, inasmuch as they are unvarying. For example, Piaget argues that egocentric speech is a general characteristic of children until they are about 7½ years old regardless of the particular language they speak. In general, the larger the hereditary component proposed in a theory of human development, the greater the emphasis on behavioral universals.

What Form Does the Speaker's Internalized Competence Take? There is no single theory of internalized competence among interactionists. Piaget's view resembles the view of the cultural relativists in that he regards the speaker as possessing a total conceptual system, developed with and inseparable from the rest of his behavior. Thus it is not especially relevant to inquire whether linguistic competence is stored as rules, response chains, or categories, since it is presumably integrated within a larger cognitive system. Bruner, on the other hand, accepts the linguistic notion of a separate linguistic competence involving the capacity for forming a complex hierarchical system which functions through generative operations (1960, p. 8). However, the views of Piaget are perhaps more typical of the interactionist approach.

Preformationist-Predeterminist Approaches

The theoretical positions characterized as preformationism-predeterminism are the current views among linguists. In general, these positions are strongly oriented toward the role of heredity in the language-acquisition process, although there is a rather wide variety of theories within this approach.

The preformationist-predeterminist approach toward language is interesting because of its long history and its forms of expression. Many of the major postulates of this approach were brought together in the mid-seventeenth century by Descartes and his followers (see Chomsky, 1966, for discussion), and their most articulate expression appears in the Port-Royal *Grammar* of 1660. Descartes' theory of language remained popular among philosophers through the eighteenth century. By the nineteenth century, however, the rise of comparative linguistics in Europe brought about change in the methods of discussing language, change which foreshadowed the structural-behaviorist views of the early twentieth century.

Cartesian-influenced philosophers such as Cordemoy in the seventeenth century and Beauzée, in the nineteenth (see Cordemoy, 1666; Beauzée, 1767; see also Chomsky, 1966, for discussion of these works) stressed the species-specificity and innate basis of language. Man was thought to have a unique capacity for language acquisition, common to all members of the species and uncorrelated with intelligence or environmental circumstances. Man's capacity for language was different from the capacity to articulate speech, since even parrots can "talk." Cordemoy and others found in the creative and innovative aspects of language one of the chief distinctions between human language and animal communication.

Seventeenth- and eighteenth-century language philosophers were concerned with the complexity of language. The Port-Royal *Grammar* approximates some points of recent generative linguistic theory in its discussion of the ways in which sentences are formed and related to one another. Indeed, the notion of a distinction between deep and surface structure, a key concept of generative linguistics (Chomsky, 1965, pp. 15 ff.; or 1967, pp. 397–404), was first set forth in this work, as was the general concept of linguistic transformation.

As mentioned earlier, interest in the psychological aspects of language and in the discovery of universal principles of language acquisition and operation declined greatly in popularity during the first half of the twentieth century. During the late 1950s, however, work by Harris, Chomsky, and others helped to shift interest from structural description to a more speculative interest in language. This shift brought about a resurgence of interest in the kind of questions that today would be called psycholinguistic.

One of the main differences between structural and the more recent generative linguistics is that the latter emphasize the similarities rather than the differences among languages—an emphasis made possible by the concept of deep structure, which many linguists presume is innate and contains features of linguistic universality (Chomsky and Halle, 1968, pp. 4–5; McNeill, 1968, pp. 412–414). There are a large number of universal features at all levels, ranging from universals of language acquisition to structural universals of linguistic systems. Further, it has been pointed out that children could not acquire all these universals directly from the primary linguistic environment (Chomsky and Halle, 1968). A major portion of linguistic structuring, therefore, as well as the capacity for acquiring language in the first place, is believed to be innate. This basic assumption is made by most linguists and psycholinguists. Among the preformationists-predeterminists differences of opinion center chiefly around the amount of language thought to be innate, and the nature of evidence concerning this question. The following summary may clarify the salient points of the various positions within this approach.

Summary of Preformationist-Predeterminist Approaches

How and Why Is Language Acquired? Language is presumably acquired as a function of an innate biological propensity in man (see especially Lenneberg, 1967). As in the cultural relativist-determinist position, language learning is considered a natural behavior which children engage in without instruction, much as they learn to walk. Evidence includes the facts that (1) all children learn language, given normal neurological equipment, regardless of the complexity of the language or the environmental circumstances, and (2) that language learning seems to take about the same length of time for all children. All children apparently master the basics of their native language between the ages of 4 to 6.

The notion of an innate predisposition to learn language does not imply that language is inherited in its entirety, however. The term *preformationism* has often been exaggerated, and it has been juxtaposed with *predeterminism* in this paper to indicate that there is a continuum of positions concerning the influence of heredity upon language acquisition and use. McNeill and others, for example, contend that the child is born with a set of unified neurological structures or processes specifically intended for language learning. These structures have been termed collectively the Language Acquisition Device or LAD (McNeill, 1966b, pp. 38 ff.). Within the LAD model are generalizations about linguistic structuring which have the characteristics of universals. Presumably, the LAD is able to scan the linguistic input data—that is, the language heard by the young child—for material corresponding to these universal features, and to organize this input in such a way as to compose a system. This system does not take the form of stimulus-response bonds or other associative chains (see Braine, 1963a and 1965, for a slightly different viewpoint), but is expressed at each point in time as a grammar. The production of utterances or speech thus occurs in a systematic and generally regular fashion allowing, however, for the influx nature of the whole acquisition process (Brown and Bellugi, 1964b; Miller and Ervin, 1964; Brown and Fraser, 1964).

Although preformationist-predeterminist theories incline toward the hereditarian, they do not go so far as to suggest specific propensities for the learning of a particular language. They consider that the biological basis for language acquisition is general, containing only those features common to language as a whole. The environment plays a fairly obvious role in the language-acquisition process; according to this view, the child cannot learn a language without hearing one and will learn any language which he does hear (possibly any number of languages). However, the linguistic system formed by the child will eventually correspond in certain specific ways to the language which predominantly surrounds him.

That the child eventually speaks the language he hears is not considered altogether self-evident. Indeed, only a minimal amount of child language acquisition is attributed to direct imitation of adult speech, or what Skinner termed echoic behavior (Skinner, 1957, pp. 55–65). Various studies of child language (for example, Brown and Bellugi, 1964b) indicate that much of what children say reflects what is presumably their internal grammar at that stage, not the speech of others. In fact, Brown and Bellugi's study found that parents imitate their children's speech more than the other way around. Children have been found to be notoriously poor at imitating even when directed to do so. Studies have shown that when children do imitate the sentences of their parents, they typically repeat only a selection of the words; they tend to omit articles, prepositions, conjunctions, and the like. This aspect of child language cannot be adequately explained by structuralist-behaviorist theories based on stimulus-response conditioning (but see Staats and Staats, 1964, pp. 173–177).

How Is the Concept *Language* Best Defined? Language is defined as a system which associates sounds and meanings in particular ways. These associations entail more than verbal behavior. Central to the preformationist-predeterminist approach is the distinction between linguistic competence, or the underlying internalized knowledge of a language, and linguistic performance, or actual speaking and understanding behavior. Although competence must exist in order for performance to take place, competence is far more extensive and complicated than performance at all stages of development.

Implicit in this definition of language is the concept of infinite variety or plasticity. Although the internalized component of language is finite, linguistic performance is potentially infinite: We can produce and understand novel utterances or sentences we have never spoken or heard before. Similarly, there is no "longest sentence" within a language, because a longer one can always be produced. Preformationists-predeterminists maintain that, were language formed from stimulus-response bonds in the formal sense, this sort of unlimited production would be impossible: it is clearly not possible for a child to be conditioned to each of an infinite series of stimulus-response bonds. Actually, adherents of this approach note that it would be impossible to learn even the amount of language known by a 4-year-old child by stimulus-response conditioning (see Miller, Galanter and Pribram, 1960, pp. 139–148).

What Is the Best Device or Technique for Describing Language? A generative grammar is considered the most appropriate device for describing language. A transformational generative grammar has three main components: a syntactic component producing deep structures and surface structures; a semantic component which operates on deep structures; and a phonological component associated with surface structures. It is presumed that this grammar takes the form of a more or less ordered set of rules. (There is disagreement on the necessity for, or the extent of, ordering rules). This set of rules is very large for a given language, but it is finite. The generative grammar is thus essentially a competence model, not a performance model. In order to make it into a performance model, more information in the form of rules or perhaps of further components would have to be integrated into it.

A generative grammar functions not by stringing words together but by analyzing utterances in terms of abstract grammatical categories.

Among the advantages of this descriptive technique are its representation of language as hierarchical and multilevel rather than linear, its surface structure–deep structure distinction, and its general conformation to current notions of how a language is actually organized within the mind of a speaker.

Within the last twenty years, a number of models of generative grammar have been proposed, although only one, the transformational or Chomskian model, has gained currency for linguistic description or psycholinguistic analysis. In recent times there have been a number of psychological experiments using the transformational generative model, but as yet no conclusions about the adequacy of the model are possible.

To What Extent Is Language a Species-Specific Behavior? Language is regarded as a phenomenon specific to man, not approximated by any other species. Although this view largely conforms to the interactionist's, the grounds for it among modern linguists differ considerably from those of the older view. Preformationist-Predeterminist language is species specific because other species lack the biological mechanisms necessary for the formation of language. While the behaviorists argue that language, learned through universal conditioning techniques, can be conceivably approximated by other species capable of undergoing such conditioning, the preformationists-predeterminists hold that language acquired through the operation of a specific mechanism such as LAD, designed exclusively for linguistic processes, cannot.

Are There Significant Universals of Language? Adherents of the preformationist-predeterminist approach contend that the study of linguistic universals will divulge essential information about the psycholinguistic and psychophysical makeup of the species. There are several ways of viewing the question of linguistic universals, of course. One proposition is that there are two categories of linguistic formal and substantive universals: one relevant to the structuring of language, and one relevant to its acquisition. Substantive universals are those items at a particular level of language which must be drawn from a fixed class or pool of such items (Chomsky, 1965, pp. 28–29). For example, the theory of distinctive features of phonology (roughly, that the sound systems of all languages are made up of the same kinds of units), or the theory that all languages have certain grammatical categories such as noun phrases and predicate phrases, are statements of substantive universals. Formal universals state that the grammars of all languages must meet certain formal conditions, such as having a transformational component, cyclical phonological rules, and certain other technical properties.

The difficulty in this proposition lies in determining precisely which of the known or postulated linguistic universals are innate to the LAD of the child, and which are exogenously acquired. The strongest or most preformationist position is that all formal and substantive universals are built into the child's LAD—including such distinctions as noun/verb, animate/inanimate, present/past and so forth—and that all the child has to do is scan the linguistic environment for examples of universal categories and integrate them into his system. Proponents of this position include Chomsky (1968a, pp. 66–68) and McNeill (1966b, pp. 49–53). An alternative hypothesis is that the child's preprogramming is not for specific categories such as noun/verb and the like, but rather for the general ability to acquire this kind of category (Slobin, 1966, p. 89).

Again, the innate component of language may not be a single linguistic package, but a series of species-specific competences active in the acquisition of language as well as of other characteristically hierarchical forms of human intellect.

In regard to universals of language acquisition, preformationists-predeterminists accept as invariant among normal members of the species: the occurrence of the language acquisition process itself, under a limitless variety of conditions both of environmental circumstance and of directed teaching or training; the relatively constant length of time this process takes, irrespective of the structures of the language being learned; and perhaps also stages within the language acquisition process. There is, for example, some evidence to suggest that all young children begin using fixed word order in their language, whether or not the language they are learning actually has constant word order as a grammatical device. Moreover, there is evidence that all young children have an initial grammar consisting of two main classes of grammatical unit, formed possibly on the grounds of ordering (Braine, 1963a), or according to more complex approximations found in adult language (McNeill, 1966b, pp. 20–26). The existence of these and similar universals of language acquisition are cited as conclusive evidence of an important hereditary component in language.

What Form Does the Speaker's Internalized Competence Take? The problem of internalized competence is closely related to that of language universals and the proper form of linguistic description. However, the degree to which statements such as Chomsky's that "obviously, every speaker of a language has mastered and internalized a generative grammar that expresses his knowledge of his language" (1965, p. 8) are to be taken literally is rarely specified by the preformationist-predeterminist. In fact, this topic poses some problems for the theoretical psycholinguist seeking to relate generative grammars to models of internalized competence, because the application of the descriptive technique to the latter use is not obvious. (See tables 1 to 3.)

Conclusion

In this brief review, four approaches to the study of language have been described in roughly historical order. Within each approach, several positions have been indicated. Each of the four approaches have been considered in terms of six questions or issues which not only divide theorists but also provide boundaries to their scope of observation.

Table 1

OVERVIEW OF THEORETICAL POSITIONS ON LANGUAGE DEVELOPMENT
GENERAL CHARACTERISTICS

Behaviorist	Cultural Relativist-Determinist	Interactionist	Preformationist-Predeterminist
1. Environmental orientation 2. All learning is based on a few basic principles 3. Language development is part of a universal learning system Anything that learns, learns the same way Anything that anyone learns, is learned the same way	1. Environmental orientation Less emphatic than behaviorists 2. Theory and methods borrowed from anthropology 3. Influence of culture on language and influence of language on thought 4. Different cultures, languages and societies are unique	1. Developmental orientation 2. Rooted in European psychological and developmental thought 3. Concern with development of perceptions and relation between language and cognition Growth and development take place through organism's adaptation to	1. Hereditary orientation 2. Concern with generative linguistics Distinction between deep and surface structure Determination of ways sentences are formed and related to one another Analysis of linguistic transformation

4. Language mechanisms are simple Stimulus-response Operant behavior 5. Language is an acquired function Only innate aspect is ability to deal with stimulus-response Shared with all other creatures 6. Interest in language as an observable behavior	Kinds of verbal responses that compose language differ for each linguistic system Perceptions of real world are shaped by language 5. Individual contributes little to his own linguistic and perceptual development 6. No biological predisposition for language acquisition	environment and organization of conceptual schemes Language part of general scheme of human development 4. Continuous interaction between hereditary structure of organism and input from environment 5. Concern with linguistic ontogeny Language develops as child passes through endogenously motivated stages Language develops along with capacity for logical thought 6. Belief that language and thought influence and reflect each other	3. Emphasis on similarity among languages 4. Goal of language study is development of language theory 5. Linguistic structuring and acquisition is innate 6. Concern with question of internalized competence Syntactic structures Biological foundations

Table 2

OVERVIEW OF THEORETICAL POSITIONS ON LANGUAGE DEVELOPMENT REPRESENTATIVE THEORISTS

Behaviorist	Cultural Relativist-Determinist	Interactionist	Preformationist-Predeterminist
L. Bloomfield 1. Structural linguist 2. Dealt with language as a science Stressed importance of eliminating introspective, mentalistic concepts from linguistics 3. Viewed mechanics of language as fairly simple Language a form of behavior that had to be taught by specific reinforcement Parent is conditioning agent Reinforce accidentally occuring sounds that resemble adult language Language becomes a habit Used for communication and for controlling environment 4. Item-arrangement grammar Enumerate items comprising language	*Benjamin Whorf* 1. Set forth hypothesis of linguistic relativity 2. Language most important influence on development of thought patterns and perceptions Different linguistic systems lead to different ways of thinking Language patterns responsible for Western scientific and logical concepts 3. Without language, world would be perceived as undifferentiated continuum of sensory impression 4. Words not just results of symbol-referent association Speakers confuse symbols and referents Treat symbols as though connected with actual objects 5. Language arbitrary Lacks biological basis	*Jean Piaget* 1. Developmental orientation 2. Growth and development take place through organism's adaptation to environment and organization of conceptual schemes Involves assimilation and accommodation New elements from environment absorbed into organism's total structure Integrated into functioning Represents continuous interaction between hereditary structure of organism and environment 3. Language part of general scheme of human development Concerned with linguistic ontogeny Child passes through number of endogenously motivated developmental stages	*Noam Chomsky* 1. Deals with generative linguistics Ways sentences are formed and are related to one another Deep and surface structure Linguistic transformation Concern for how language is organized within the mind of the speaker 2. Interested in psycholinguistics 3. Goal of language study should be development of language theory 4. Emphasis on similarities among languages Innate acquisition Linguistic universals 5. Two categories of linguistic universals relevant to structuring of language *Substantive*

Arrange into grammatical utterance
Present data in taxonomic fashion
Assign linguistic elements to specific level of phonology, morphology, or syntax and to specific units within each level
No level-mixing
No ambiguity of classification

5. Concern for interlanguage diversity
    Search for universals unscientific
    Concentrate on description of observable linguistic behavior
    Each language should be analyzed only in terms of its own features
    Items composing each language and their arrangement into grammar patterns are unique to the language

Seen in diversity of languages around world
Lack of conceptual framework

6. Important to understand world view peculiar to each language
    Place linguistic descriptions in sociocultural framework
    Use psychological criteria to set up linguistic categories
7. Goal of linguistic description to determine thought patterns and perceptions which correspond to structures of language in question

Invariant throughout species
Uniform in ordering and age of onset
Each characterized by certain kind of cognitive organization and behavior
Language develops along with child's capacity for logical thought, judgment, reasoning
Reflects these capacities at each stage

4. Language and thought influence and reflect each other

Items of particular kind are drawn from fixed pool of such items
Theory of distinctive features of phonology
All language must have certain grammatical categories
Noun phrase, predicate phrase
Formal
All grammars of all languages meet formal conditions
Transformational component
Phonological rules

6. Formal and substantive universals built into biological language-learning mechanism
    Noun/verb; Present/past
    Child scans linguistic environment for examples of universal categories
    Integrates into system
7. Every speaker has mastered and internalized generative grammar that expresses language knowledge

Table 3

OVERVIEW OF THEORETICAL POSITIONS ON LANGUAGE DEVELOPMENT
POSITIONS ON SIX QUESTIONS CONCERNING LANGUAGE THEORY

Question	Behaviorist	Cultural Relativist-Determinist	Interactionist	Preformationist-Predeterminist
1. How and why is language acquired? 2. How is the concept *language* best defined? 3. What is the best device or technique for describing language 4. To what extent is language peculiar to man? 5. Are there significant universals of language? 6. What form does the speaker's	1. Language acquired through selected reinforcement of natural babbling and shaping of such vocal behavior through operant conditioning Child learns what he is taught 2. Language a series of responses to stimuli Is equivalent to speech 3. Language described by taxonomy or classification List parts of language 4. Language a species-specific behavior Other species lack neurological apparatus for language	1. Language acquired as social necessity Acquisition parallels development of thinking Does not have to be taught 2. Language a conceptual system Closely related to thought Words associated with objects Not just results of symbol-referent associations 3. Language described by taxonomy with addition of psychological criteria Psychological criteria used to set up linguistic categories	1. Language acquired because of biological predisposition to language Acquisition not dependent on training Information received from others adapted by built-in genetic language-learning mechanism and integrated into system 2. Language defined in terms of conceptual and perceptual framework Form of behavior Internalized system 3. Language described in terms of linguistic ontogeny Note utterances	1. Language acquired because of biological propensity of man to do so Has neurological structures specifically intended for language learning Does not have to be taught language or reinforced for acquiring 2. Language a system that associates sounds and meanings in a particular way Not equivalent to verbal behavior Characterized by variety 3. Language described in terms of generative grammar

internalized competence take?		Important to understand world view peculiar to each language	characteristic of each developmental stage and type of cognition implied Language the means by which concepts of space, time, causality are expressed	Hierarchical and multileveled Three main components Syntactic Semantic Phonological Concern for organization of language in speaker's mind
		4. Language a species-specific behavior Major factor dividing humans from animals	4. Language a species-specific behavior Come from intellectual character of humans	4. Language a species-specific behavior Other species lack biological mechanisms necessary for language
	5. No universal grammar Search for universals unscientific Interlanguage diversity more prevalent than interlanguage similarity Language best considered in terms of its own structure	5. No universal grammar Individuals speaking one language live in different universe from those speaking another	5. Are universals in language development All children use language about same way Children learn language at about same ages	5. All significant features of language universal Substantive universals All languages have grammati-categories Built into biological language-learning mechanism Child scans linguistic environment and integrates Formal universals All grammars meet
	6. One talks by making stimulus-response associations among grammatical classes or response types Classes unique Classes acquired through learning	6. Speaker has mastered entire symbolic and conceptual system Innate structure for language acquisition Internalized language substratum for thought	6. Several views: *Piaget*: Speaker has total conceptual system Developed with, inseparable from rest of behavior Linguistic compe-	

Table 3—(Continued)

		tence integrated with cognitive process	formal conditions
			Transformational components
			Phonological rules
		Bruner: Separate linguistic competence in complex hierarchical system	Child has general ability to acquire these categories
			6. Every speaker of language has mastered and internalized generative grammar that expresses language knowledge

PIAGET'S THEORY OF KNOWLEDGE: THE NATURE OF REPRESENTATION AND INTERIORIZATION*

Hans G. Furth

For well over 300 years the notion of representation has taken on increasing significance for a theory of knowledge. The emphasis on representation started as a reaction to a theory of intentionality which had derived from Aristotle and the scholastics. During the late middle ages this theory had declined into a veritable play on words in which theological preoccupations and an unchecked tendency for formulating endless distinctions and reifications was rampant. With the advent of Occam's position that concepts which previously were held to be universal and eternal were nothing but a name, a "flatus vocis," the time was ripe for a reappraisal. Descartes is commonly recognized as being the person who gave philosophy the new direction it needed.

This paper first focuses on different meanings attributed to the word "representation" by examples from Descartes and other philosophers of the idealistic type. Subsequently the interpretation given this term in the English empiricist tradition is discussed. This tradition was taken over into the mechanistic-positivist atmosphere that surrounded the birth of empirical sciences concerned with human knowledge, as linguistics, psychology, anthropology, semiotics. With this as background, the place and meaning of the two words "representation" and "interiorization" in Piaget's theory will be reexamined. Two main conclusions are suggested: (*a*) Piaget's system is able to incorporate competing and frequently confused meanings of the words by assigning them to clearly different functions and different genetic derivations; (*b*) in view of the inherent ambiguities of the words and the misunderstandings which ensue, it would be advisable to recognize explicitly the sense in which the words are taken or even limit their use to one specific meaning.

The discussion starts with the word "representation" and brings in "interiorization" later on in connection with specific interpretations of

* From Hans G. Furth, "Piaget's Theory of Knowledge: The Nature of Representation and Interiorization," *Psychological Review,* vol. 75, no. 2 (1968), pp. 143–154. Copyright 1968 by the American Psychological Association, and reproduced by permission. This article is the substance of a paper presented at the International Center for Genetic Epistemology, Geneva, Switzerland and was written during the author's sabbatical stay at the Center. The author thanks Piaget, H. Sinclair, and M. Chandler for critical discussion and reading of the manuscript.

representation. The term "representation" can be understood in either an active or a passive sense; the passive sense can be further divided into a narrow configurative and a wider or general significative sense. The primary meaning of the term is active: "to make something present by means of . . ." ("rem praesentem facere"). In this case the person is the subject of the activity and a mediating instrument is implied. "The deaf person reveals his knowledge about the alleged event by means of natural gestures" is an example of the active sense of representation: a person who knows something about an event communicates this knowledge to others by means of gestural representation. In its passive sense the mediating instrument becomes the subject of the sentence, that is, "something stands in the place of something else." In the narrow configurative sense we have this example: "A map represents the outlay of the city;" and in the wide significative sense there is: "The letter X represents those children of the city who are between 6 and 10 years of age." Note that in the configurative sense there is an inherent correspondence between the drawing of the map and the real thing or that knowledge of the representation by itself gives or implies a corresponding information about the real thing. In the general significative sense, the "X" as such has no intrinsic relation to the real thing and knowledge of "X" by itself provides no information.

One can notice already that in the passive sense of "A thing that represents something else" the words "to signify" can readily be substituted for "to represent." Thus representation seems to take on the added property of signifying and comes to have the same meaning as signification. In this view representation in the wide sense refers to signification generally while representation in the narrow sense means a special kind of signification, namely a signification that is mediated by some configurative correspondence between the representation and the thing. This ambiguity in meanings could only arise once the basic active sense of the notion of representation had been lost sight of.

In the French Cartesian tradition the expression "l'idée représentative" became part of the philosophical vocabulary. For Descartes the idea is *that which* we know, the direct object of knowing. It has its efficient factual cause in the real thing that enters our senses. Its formal cause, that is, that which explains the specific nature of the idea, is found in the general idea of the self-as-knowing within which are implicit all possible ideas.

For Descartes the thing-concept relation is an efficient causality relation which does not imply any intrinsic connection between the two terms. An efficient cause is frequently only a signal of its effect with no inherent relation between the nature of this signal and its effect. A broken window can be the result of a variety of efficient causes and tells us very little about the specific nature of the causal event. Descartes, whose philosophy did not envisage knowledge as based on efficient causality alone, realized that his theory provided no intrinsic assurance that the ideas which we knew represented the real thing in any relevant or nontrivial way. Therefore he has recourse to God, to his veracity and his goodness, suggesting that the will of God is the basic ground for our belief that knowledge is trustworthy.

Descartes says in his *Meditations* "There is in me some faculty . . . able to produce these ideas without the help of any exterior thing, as . . . when I sleep they produce themselves in me without the help of the objects which they represent, while I remain convinced that they are caused by these objects . . . [111, 9]." Here we see various meanings of the word representation intermingled, there is the active faculty to produce, the passive images of dreams that represent external objects and it is intimated that the ideas are similar to images in their ability to represent. Note also Descartes' reference to the efficient causality of things vis-à-vis images, and by analogy, vis-à-vis ideas.

Leibniz continued to use the word representation in a passive sense of correspondence. He explains that the monad constitutes "a representation of many things in one only." Wolff, who popularized Leibniz in German, translated representation as "Vorstellung" and there was then a mutual influence on the meaning of the word "representation" between German and French philosophers. In both these languages "representation" has come to have two meanings, a wide and a narrow one. (1) In the wide sense, it is any kind of knowing, of "putting a thing before one's mind," without the thing being present to the senses; this can involve a sensory representation or take place in a nonsensory manner. (2) In the narrow sense it is limited to the sensory manner of representation, or to the making present of sensory content. Notice that because of the above mentioned confusion attached to the passive sense of the term representation, Meaning 2 is now a subpart of Meaning 1. For the notion 1 minus 2 there is no special word in French or in English but the Germans can speak of "Unanschauliches Denken." Hence the following semantic muddle: "Vorstellendes Denken" (representational thinking in the wide sense) is divided into "Vorstellendes Denken" (in the narrow sense) and "Unanschauliches = Unvorstellendes Denken!"

While this development was taking place on the continent, English empiricism followed Descartes' lead and began treating concepts like images. Locke still used "idea" in a sense similar to Descartes as referring to any object of knowing, imaginal or nonimaginal; but the empiricist tradition which he founded was soon satisfied with an efficient causality of knowledge. Therefore Locke's followers, such as Hume, considered ideas like superfaint copies of the real thing, with images being faint copies. The difference between an idea or an image became merely a difference in degree of acuity. For 19th-century associationists whose views deeply influenced the founding fathers of scientific psychology, images and concepts were internal representations in the copy sense, caused by internalized perception. In the English language today representational thinking means distinctly thinking in images, of whatever type. Translators beware! Note that in this tradition the external thing is regarded both as efficient and as formal cause of knowledge and such a synthesis of efficient and formal causality constitutes what Leibniz had called the principle of sufficient reason.

To summarize so far, it is suggested that from Descartes onward philosophical theories considered ideas as objects of knowledge. Ideas as distinct from percepts were called representations in the sense of referring to sensorially not present reality states. The French-German idealistic tradition continued to use the term "representational thinking" in a wide sense that did not by itself involve a sensory representation. Since it did not concede that nonperceptual knowledge had its sufficient determinant in externally caused information, it had recourse to a sphere of an "idealistic" reality of categories and essences. In this respect it is not amenable to empirical observation and retains in all its forms, past or present, a flavor of preformation, of a priori and of "supra"-scientific.

The English empiricist tradition, however, came to use the term representation in the narrow configurative sense with the implication that ideas or concepts were nothing but imaginal representations. In this view knowledge is directly caused by external events to which the organism responds and no other "formal" explanation is required.

Psychology as an empirical science developed within the causal-mechanistic framework of English empiricism. When psychologists did not relegate the processes of intelligence to a *Geisteswissenschaft,* they almost invariably employed a representational theory of knowledge that assigned a crucial role to the mediating representation, taken in the passive sense of the term. The internal sign, shown in Figure 1, as that which represents reality, takes the place of things that are outside. Knowledge is likened to a perception of and reaction to interior signs or perhaps to a manipulation of signs that mirror reality. Only clear causal relations are at work: The thing or the exterior sign causes the interior sign by interiorization. The person perceives the sign which in this manner becomes the functional object of knowledge. The sign's power to have the person react to it as to the real thing is the peculiar characteristic of representation which, according to a mediating representation theory, explains knowledge.

Consistent with the mechanistic-causal model, most of those who reason this way will reject a signal-symbol distinction or at best call a symbol the signal of a signal. In any case, the functioning of a signal, for example, a learned association, is conceived in terms that are quite similar to the functioning of the interior mediating sign in knowledge. Consequently the representation theory of knowledge is confined to a signal knowledge regardless of whether it labels the mediators images, words, or fractional stimuli. Characteristically it looks for external causes

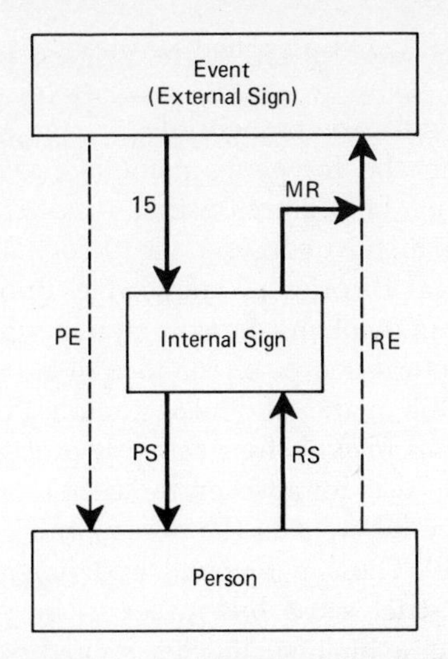

Figure 1 Diagram of Mediating-Representational Knowing. (PE = Perception of Event; RE = Reaction to Event; IS = Internalization (Real) of Sign; PS = Perception of Interior Sign; RS = Reaction to Interior Sign; MR = Mediating Representation.)

that connect the internal signals with the real things. The combination of external factors that connects signs and things and of internalization which makes an overt sign covert, provides a sufficient explanation of knowledge. The covert sign, as an internal representation, is the key to knowledge and internalization can now be seen as an important explanatory notion for knowledge, second only to representation.

Internalization in this connection means that something that has been external and observable is gradually withdrawn and takes place in a covert manner. What has been internalized can perhaps be introspectively experienced or registered by special instruments. Essentially, internalization as such does not change the nature of the external act, except perhaps by speeding it up or by enlarging its spatial extension. An internally recalled sequence of events can proceed at a rate that far exceeds normal limits, for example, the imagined mountain trip can encompass in one sweep much more than is accessible to an observer standing in one spot.

Two points can be made concerning the representation-internalization model of knowledge as sketched above. First, this way of thinking strikes the ordinary person in our society as entirely reasonable. It has behind it a history of thinking that goes back for centuries, and it seems as much a part of our cultural heritage as other customs deeply ingrained in our society. As a model it does not seem to call forth serious questions. Open questions relate only to the nature of the internal signs, whether they be images, words, or just neural connections, and to the nature of motivating factors that connect a sign with the corresponding reality. The second point is simply that the explicit or implicit acceptance of such a model makes a more complex model, such as the operative model of Piaget, unnecessary if not incomprehensible.

It remains to relate the foregoing model to some current types of psychological theorizing before contrasting it with Piaget's theory. For this purpose we shall in turn consider the theory of Gestalt as well as later forms of structural theories of knowledge. Subsequently, attention will be given to recent neobehavioristic trends which are particularly concerned with the motoric source of complex thinking processes. Finally, special mention will be made of recent theories that rely heavily on verbal mediating support in explaining thinking processes.

The gestalt tradition was founded on the notion of cognitive structure. The gestalt is a totality inherent in the perceptually given data to which the organism responds. These perceptual gestalten were not considered to be learned in the strict sense but rather to be preformed in such a manner that there was a biologically determined isomorphism between external reality, neurological structure, and perceptual response. When the notion of structure was transferred to the field of learning, the preformation perspective was dropped and the acquisition and internalization of cognitive structures through reinforcing conditions was emphasized. The internal sign of Figure 1 would aptly illustrate the mediating function of Tolman's "cognitive map" insofar as it controls outward motoric behavior.

Numerous theorists of the so-called cognitive school of psychology have amplified the notion of internal structure in many important respects. For example, Miller, Galanter, and Pribram (1960) introduced a more tightly reasoned connection between the internal image and outward behavior with their hypothesis of a Test-Operate Test-Exit plan. Bruner (1966a) related cognitive growth to the development of various techniques by which humans internally represent their experience of the world and organize these representations for future use. For the present discussion it is important to realize that in all these theories, including also the recent cybernetic and computer-based models of intelligence, the internal representation of outside reality is not only crucial but constitutes the chief explanatory factor for intelligent behavior. The internalized sign becomes here a kind of representational sediment which is available for the control of complex behavior. Knowledge is conceived as coextensive with the internalized representations. It does not seem farfetched to relate the ideas and images of the empiricist philosophical tradition—the ideas being derived from experience and constituting the objects of knowledge—with the internal mediators of the representational theories of knowledge as outlined in Figure 1.

Another tradition that has strong contemporary support derives from the strict stimulus-response theory. Enlarging on the theoretical variables of covert and cue-producing stimuli and responses, psychologists like

Osgood (1953) and Berlyne (1965) emphasize the symbolic character of cognitive behavior. Osgood postulates the presence of fractional anticipatory goal responses; Berlyne suggests chains of symbolic covert responses made up of situational and transformational representations. The internal stimulus-response processes are conceived as determinants of outward behavior or as variables that intervene in a complex network of input and output relations. These theorists would hold that the covert connections which determine thinking behavior are basically not different from the overt connections of outward behavior that can be observed and experimentally controlled. Consequently they would be less inclined to speak of an internalized structure, or rather they would analyze the internal sign of Figure 1 into an intricate sequence of covert stimulus and response connections.

Together with the cognitive and neobehavioristic traditions emerged a third position which considered verbal language as the decisive factor in human intelligent behavior. This trend can be seen as an outgrowth of logical positivism, a philosophical view which holds that the truth of logic and science is essentially a matter of correct use of language. This theory is presented in different forms, such as verbal learning or as second signal system. The decisive point for a general psychological theory of knowledge seems to be its insistence that internalized language is at the base of intelligent behavior. The internal sign of Figure 1 would then primarily consist of linguistic elements which mediate in and determine outward behavior.

Piaget's theory of operative knowledge is unique in dispensing with a mediational representation as far as the essential aspect of critical, objective knowing is concerned. He can describe the structures of his three developmental stages without mentioning the word representation or internalization.

Consider Piaget's diagram of knowing as sketched in Figure 2. The essential point in the "knowing circle" is the internal structure. The circle assimilates or incorporates the real event into the structure and at the same time accommodates the structure to the particular features of the real event. Only through the closing of the circle is the real event turned into an object of knowing, that is, an event-that-is-known, and the structure into an active knowing structure. Moreover, the growth and development of the internal structure is primarily due to the coordinating abstraction which feeds back from the knowing activity itself to the enrichment of the structure. Note also that the internal structure is not something that at any time was external and gradually became internalized, rather it is developmentally and phylogenetically related to the living organization itself which at no level can be considered as being outside the living organism.

The solid broad lines of the diagram schematize sensory-motor knowing. At that stage the knowing circle is only closed when it is part of an

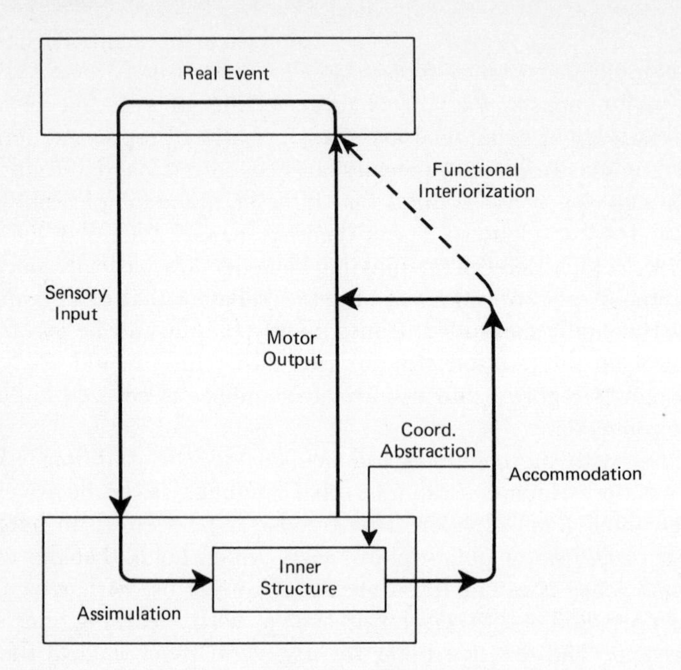

Figure 2 Diagram of Piaget's Theory of Knowing.

external motor reaction; in other words, knowing is on a practical level only. In fact, sensory-motor knowing is identical with the available coordinations that become manifest in adapted exterior actions.

However, beginning with the transition to the operational period, one can observe the first knowing activities that are no longer inherently tied to the child's own external actions. The most basic of these activities is the formation of the object-as-such or what Piaget also calls the scheme of the permanent object. It is the beginning of objectivation, the beginning of a different kind of knowledge in comparison with sensory-motor, practical knowledge. This "objective" knowledge will lead, according to Piaget, to the concrete operational and finally the formal operational stage.

This new mode of knowing, indicated by the broken line of Figure 2, is characterized by a growing dissociation between the knowing act and its external particular manifestation. It is apparent that by means of the broken line the knowing circle can be closed without an external action. What I like to call *functional interiorization* is thus the specific condition of a knowing in which the external motor reaction is no longer an essential prerequisite, although it may well remain an accompanying phenomenon. It is this condition which Piaget has in mind when he speaks of interiorization in connection with an operational act. As compared to a sensory-motor act, what is interiorized is not the structure that always is internal, but the result of this inner structure. In the case of an active sensory-motor scheme there issues an external act, in the case of an operational scheme an internal knowing.

Piaget refers to the internal knowing by different terms: *operations* when he emphasizes their being part of a reversible structure; *judgment* when he considers the assimilatory activity that assigns an event as belonging to a structure; *concept* when he focuses on the operational scheme as the common source of assimilations. Psychologically all these terms partake of an identical reality status. An active structure, an operation, a concept, or a judgment are for Piaget one and the same reality and not different reified entities. Moreover, these notions enjoy no reality status of their own, they are merely ways of expressing the only real event which exists, namely the fact that "a person knows something." They have, if you like, a "knowing" existence, as long as it is understood that this knowing is one of the modes of existence belonging to the living organism as a whole.

Is there no room for representation in Piaget's model? Certainly, by means of the representational function the person can make present to himself events that are not present to the senses. Piaget employs representation in the active sense and relates it to the symbolic or semiotic function of intelligence. It is a specialized capacity that lies midway between operational activities and motoric output. Piaget refers to the product of this activity as a symbol (or a linguistic sign). While Piaget distinguished an imaginal symbol from a linguistic sign it must not be forgotten that for him both are representational in the sense of being differentiated signs and therefore they are not essential elements of operative knowing. As will be pointed out later on, knowing in Piaget's theory is never a mere matter of representation. For a fuller elucidation of Piaget's view on the nature of symbols and on the distinction between the plane of operativity and the plane of representation inherent in the figurative aspect of a symbol, see Furth (1967, 1968).

Every symbol has two differentiated aspects, a figurative aspect which refers to some sensory or motoric event in itself and an operative aspect which refers to meaning, that is, to its significate. In its figurative aspect a symbol has a different reality status than the operative knowing. A symbol is, or is experienced as, some thing and Piaget suggests that it derives developmentally by internalization from external motoric imitation. In its operative or meaning aspect it joins the operational circle of knowing. Thus for Piaget the direct significate of a symbol is a knowing or a concept and it is only through this concept that it can be said to represent an external thing.

If these fundamental notions of Piaget's theory are grasped, one can proceed to specify the nature of the various relations: (*a*) knower to concept and (*b*) to symbol; (*c*) symbol to concept; and finally the relations of (*d*) the real event to concept and (*e*) to symbol. The word concept is here used for the active knowing operation as mentioned above and the real event is taken to include both physical and symbolic external events as well as internal events; in short, anything that can become an object of knowledge.

(*a*) The relation of the knower to concept does not imply two factually different entities. To understand this statement better, the reader is referred to the above-mentioned elaboration on *operations, judgments,* and *concepts.* The concept has as much reality as the person who is conceiving it and the concept stays fully within the plane of operative knowing. (*b*) The relation of knower to an active symbol is partially different insofar as the symbol in its figurative aspect is on a different plane of reality which we can call the plane of representation proper. The genetic derivation of the figurative aspect from motoric imitation was already mentioned. Here we can add that in symbol formation the knower "makes a known thing present" by means of a figurative thing which thereby becomes a representational symbol. From this it follows that (*c*) the relation symbol-concept is the essential meaning relation which expresses the dependence of a symbol on operativity.

Leaving the most difficult to the last, (*e*) the relation symbol-real event can be called a relation of representation as long as it is understood that representation is not passively inherent in the symbol but only functions via the operation which has produced or comprehended the figurative thing as a symbol in the first place. Consider now the relation (*d*) concept-real event. This is not a relation of representation except when the word "representation" is taken in the widest and most general sense of signification. But actually this use of the term is here doubly inappropriate since according to Piaget we would have to call concept an undifferentiated signal. It is undifferentiated because it is not first known in itself and subsequently leads to the knowledge of the event. Second, while all significates are in the last analysis concepts, in this case one would have to call a concept a signal or symbol and the real event the significate. For these reasons one should not describe the relation concept-event as a relation of representation. I submit that this is a relation of a special kind, different from causal and representational relations. Indeed, this is the essential knowing relation which Aristotelian philosophers referred to by the term intentionality. Since this is just another word for this relation of knowing, there is no particular advantage in substituting an unfamiliar and possibly misleading term instead of calling it what it is, a relation of knowing.[1]

It would seem that this short summary would suffice to highlight the difference between Piaget's theory and the current representation theory

[1] One recalls that Brentano used the term "intentional inexistence" to differentiate the psychic from the extension of the physical. Such terminology has not helped to make his ideas comprehensible.

of knowledge. Unfortunately, Piaget's writings are at times not as lucid as one could wish and his choice of terms frequently leads to misunderstandings.

Concerning internalization there is no doubt that for Piaget the *real* interiorization of an imitative movement is something different from the *functional* interiorization of a sensory-motor act. Yet he keeps using the same French word "intérioriser" and does not warn the reader to interpret it differently. In English we do have two words that could be employed for the two different meanings, at least in connection with Piaget's theory. We could use "interiorize" for the functional dissociation between general schemes of knowing and external content and the word "internalize" for the real literal diminutions of imitative movements that according to Piaget lead to internal images or internal language.

With regard to the word "representation" Piaget has only once made an explicit statement distinguishing representation in the wide and in the narrow sense:

> In fact, the word "representation" is used in two different senses. In the wide sense, representation is identical with thought, that is, with all intelligence which is not simply based on perceptions or movements (sensory-motor intelligence), but on a system of concepts or mental schemes. In a narrow sense, representation can be limited to the mental image or to the memory-image, that is, to the symbolic evocation of absent realities. Moreover, it is clear that these two kinds of representation, wide and narrow, are related to each other insofar as the concept is an abstract scheme and the image a concrete symbol; even though one no longer reduces thought to a system of images, it is conceivable that all thought is accompanied by images. For if thinking consists in relating significations, the image would be a "signifier" and the concept a "significate" [Piaget, 1951b, p. 68].

This distinction between representation in the wide and narrow sense conforms to the earlier mentioned double sense of the word representation that is common to the French but unfamiliar to the Anglo-Saxon philosophical tradition. In the narrow sense there is for Piaget a sensorial, that is, an imaginal representation; in the wide sense representation stands for any kind of thinking that is not entirely based on perceptual or motoric involvement. Thus representational thinking in the wide sense means simply operative thinking above the sensory-motor stage with no intrinsic connection with representational thinking in the narrow sense. Unfortunately neither here nor in subsequent works does Piaget consistently keep the two meanings separate and his failure to mention explicitly the place of arbitrary linguistic signs can easily lead to misinterpretations. It could be thought that representation in the wide sense means just such conventional signs. This is, however, not the case: Piaget's distinction between representation in the wide and narrow sense has nothing to do with the distinction between what Piaget calls an imaginal symbol and an arbitrary linguistic sign.

In fact, I would not hesitate to call Piaget's distinction between the two meanings of representation ontological, as implying different levels of reality. Representation in the wide sense has to do with the plane of operative knowing, representation in the narrow sense with the plane of representation proper. It is instructive to note that Piaget holds to this distinction and gives it scientific underpinning; a distinction that the philosophical tradition has blurred since Descartes first introduced "l'idée représentative" as the object of knowledge. While in the idealistic tradition the notions of the idea and representation were—uneasily and not unambiguously—differentiated, the empiricist tradition, on the other hand, rejected completely a difference in status between knowing and imaginal representation and considered knowledge as being essentially representational and as entirely determined by mechanistically conceived causal factors.

Piaget, however, has never lost sight of the basic distinction between the relation of knowing which is on the level of action and constructs the objectivity of a reality state and the relation of representation which focuses on the reality state as such. At the same time he likes to emphasize the relation between knowing on the one hand and representation in the narrow sense on the other hand. While the *fact* of this relation is obvious, the *how* is the crucial question on which he differs from others. In the passage quoted Piaget somewhat beclouds the issue by adding that conceivably all thought is accompanied by images and calling images symbolic signifiers and thought or concepts significates. Taken literally, such a sentence would place Piaget on the Wundt-Tichener side of the imageless-thought controversy. On the contrary, Piaget stresses in this passage that the meaning of a symbol is to be found in the operative scheme. His additional remark about the presence of images, here and in other places, seems to refer to the everyday, global situation of thinking behavior in which some representational and imaginal elements in the narrow sense are present as ordinary auxiliary concomitants, as indispensable means of communication or of interior attention, but not as constitutive elements of thinking proper.

The plausibility of such an interpretation can be illustrated by numerous texts of which the following two are recent examples. In their book on the mental image, Piaget and Inhelder (1966b, p. 446) conclude, "the image is not an element of thought, but functionally similar to language . . . it can be in spatial domains a better symbolic instrument to signify the content of operational thinking." The authors continue:

> All representational knowledge (this term being taken in the wide sense of thought, as distinct from sensory-motor and perceptual knowledge)

> supposes the activation of a symbolic or better, semiotic function. . . . Without this semiotic function thought could not be formulated or put into an intelligible form, neither for others nor for the self (inner language, etc.).

When the authors speak here of representational knowledge they refer to thinking that is put into communicable form. While a symbolic articulation is certainly a true and vital characteristic of all thinking, taken in the totality of the behavioral situation, it is not for their theory a constitutive element of thinking. In a contemporaneous work in which the word "representational" is employed throughout in a sense that is not consistently or easily definable, Piaget and Inhelder (1966a) assert after mentioning the "imageless thought" controversy:

> . . . one can have an image of an object, but the judgment itself which affirms or denies its existence is not an image. Judgments and operations have no imaginal component but this does not exclude that images can play a role, not as being elements of thinking, but as being supportive symbols that complement the function of language [p. 55].

These quotations illustrate the type of possible misunderstandings due to a somewhat ambiguous use of words.

For the representation theory of knowing, the representational sign is considered as an object of knowing, as that to which the person responds. The sign is said to have become internal by a real process of internalization and it constitutes an internal copy or representation of an object that is originally external to the organism. This is the crux of the empiricist position: Knowledge has its adequate source in external reality or external actions and resides in internal re-presentations.

For Piaget's theory at all developmental levels knowledge is basically linked to the biological internal organization. Knowledge does not merely derive from the taking in of external data; the organism in interacting with the environment transforms or constructs external reality into an object of knowledge (Furth, 1968). It would be helpful if in Piaget's theory the use of the English word "representation" were limited to the narrow sense. Representation would then always refer to the direct product of the symbolic or semiotic function (e.g., image, language) and would be connected in its figurative aspect with the real, literal internalization of external actions. In contrast, the operative aspect of thinking, which for Piaget is the essential aspect of logical or prelogical constructive knowledge, would then be seen as clearly separate from representation without denying that both aspects are parts of the global behavior of intelligent thinking. The increasing developmental dissociation between the generalizable forms of internal schemes and a particular content is the meaning of the functional interiorization that leads from external sensory-motor acts to internal operations. If Piaget uses expressions like "thinking in symbols or in words," this is to be understood in the sense that thinking makes use of representational instruments, not that any representation is either a constitutive element or the object of thinking. For Piaget operations or concepts as such are not reified objects which we know but rather that through which we interact intelligently with the world and society and constitute them as an objective reality vis-à-vis our own person.

Recall the interminable speculations and verbal arguments concerning a theory of knowledge based on philosophical preconceptions. When one compares these abstractions with the fruitful results of scientific thinking that is open to a search for facts and recognizes the legitimacy of many philosophical questions, Piaget's theory can be appreciated as productive of relevant factual data on human knowing and as incorporating trends that have been expressed in philosophical theories throughout the centuries. Only now we are no longer dealing with a philosophy that is primarily subjectively determined but with a scientific theory that is open to critical objective verification.

THE NATURAL HISTORY OF LANGUAGE*

Eric H. Lenneberg

Characteristics of Maturation of Behavior

Why do children regularly begin to speak between their eighteenth and twenty-eighth month of life? Surely it is not because all mothers on earth initiate language training at that time. There is, in fact, no evidence whatever that any conscious and systematic teaching of language takes place, just as there is no special training for stance or gait. Superficially it is tempting to assume that a child begins to speak as soon as he has a "need" for it. However, there is no way of testing this assumption because of the subjectivity of the notion "need." We have here the same logical difficulties as in testing the universality of the pleasure principle as the prime motivation. To escape the inevitable circularity of the argument, we might ask, "Do the child's needs change at a year-and-a-half because his environment regularly changes at that time or because he himself undergoes important and relevant changes?"

---

* Reprinted from *The Genesis of Language*, by Frank Smith and George A. Miller (eds.), by permission of the M.I.T. Press, Cambridge, Massachusetts. Copyright © 1966 by the Massachusetts Institute of Technology.

Society and parents do behave somewhat differently to an older child, and thus there are some environmental innovations introduced around the time of the onset of speech; yet the changes of the social environment are to a great extent in response to changes in the child's abilities and behavior. Quite clearly the most important differences between the prelanguage and postlanguage phases of development originate in the growing individual and not in the external world or in changes in the availability of stimuli. Therefore, any hypothesis that pivots on an assumption of need may be restated: the needs that arise by eighteen months and cause language to develop are primarily due to maturational processes within the individual. Since needs *per se* can be defined only in a subjective and logically circular manner, it is futile to begin an inquiry into the relevant factors of speech development by the adoption of a need hypothesis. Instead, one must try to understand the nature of the maturational processes. The central and most interesting problem here is whether the emergence of language is due to very general capabilities that mature to a critical minimum at about eighteen months to make language and many other skills possible, or whether there might be some factors specific to speech and language that come to maturation and are somewhat independent from other, more general processes.

Unfortunately, the importance and role of maturation in the development of language readiness cannot be explored systematically by direct experiment, and we are reduced to making inferences from a variety of observations and by extrapolation. The difficulty is that we cannot be certain what kind of experiments or observations to extrapolate from. Behavior is far from the monolithic, clear-cut, self-evident phenomenon postulated by psychologists a generation ago. Different aspects of behavior make their emergence at different periods in the life cycle of an individual and for a variety of causes. Further, the spectrum of causes changes with species.

The hallmarks for maturationally controlled emergence of behavior are four: (1) regularity in the sequence of appearance of given milestones, all correlated with age and other concomitant developmental facts; (2) evidence that the opportunity for environmental stimulation remains relatively constant throughout development but that the infant makes different use of such opportunities as he grows up; (3) emergence of the behavior either in part or entirely, before it has any immediate use to the individual; (4) evidence that the clumsy beginnings of the behavior are not signs of goal-directed practice.

Points (1) and (2) are obvious and need no elaboration. Point (3) is a commonplace in the embryology of behavior. A vast array of motor patterns may be observed to occur spontaneously or upon stimulation in embryos long before the animal is ready to make use of such behavior. The so-called *Leerlaufreaktion*, or vacuum activity, observed by ethologists is another example of emergence of behavior at given developmental stages and in the absence of any use or need fulfillment (for details see Hess, 1962; Lorenz, 1958).

Point (4), the relatively unimportant role of practice for the emergence of certain types of behavior with maturation, has been amply demonstrated in animals by Carmichael (1926, 1927), Grohmann (1938), and by Thomas and Schaller (1954). Similarly, children whose lower extremities have been immobilized by casts (for the correction of congenital hip deformations) at the time that gait normally develops can keep perfect equilibrium and essentially appear "to know" how to walk when released from the mechanical handicap, even though their muscles may be too weak during the first weeks to sustain weight over many steps.

Generally there is evidence that species-specific motor coordination patterns (*Erbkoordination*) emerge according to a maturational schedule in every individual raised in an adequate environment. The emergence of such patterns is independent of training procedures and extrinsic response-shaping. Once the animal has matured to the point at which these patterns are present, the actual occurrence of a specific pattern movement may depend on external stimuli or internal ones (for instance certain hormone levels in the blood) or a combination of the two (Lehrman, 1958*a,b*).

The aim of these comments is to direct attention to *potentialities* of behavior—the underlying matrix for behaving—instead of to a specific *act*. If we find that emergence of a certain behavior may be partially or wholly attributed to changes within the organism rather than causative changes in the environment, we must at once endeavor to discover what organic changes there are. Unless we can demonstrate a somatic basis, all of our speculations are useless.

The four characteristics of maturationally controlled emergence of behavior will now be employed as touchstones, so to speak, in a discussion of whether the onset of language may reasonably be attributed to a maturational process.

Emergence of Speech and Language

The Regularity of Onset

The onset of speech consists of a gradual unfolding of capacities; it is a series of more or less well-circumscribed events that take place between the second and third year of life. Certain important speech milestones are reached in a fixed sequence and at a relatively constant chronological age. Just as impressive as the age constancy is the remarkable synchronization of speech milestones with motor-development milestones, both of which are summarized in Table 1.

Table 1

SIMULTANEOUS DEVELOPMENT OF LANGUAGE AND COORDINATION

Age in Months	Vocalization and Language	Motor Development
4	Coos and chuckles.	Head self-supported; tonic neck reflex subsiding; can sit with pillow props on three sides.
6 to 9	Babbles; produces sounds such as "ma" or "da"; reduplication of sounds common.	Sits alone; pulls himself to standing; prompt unilateral reaching; first thumb opposition of grasp.
12 to 18	A small number of "words"; follows simple commands and responds to "no."	Stands momentarily alone; creeps; walks sideways when holding on to a railing; takes a few steps when held by hands; grasp, prehension, and release fully developed.
18 to 21	From about 20 words at 18 months to about 200 words at 21; points to many more objects; comprehends simple questions; forms two-word phrases.	Stance fully developed; gait stiff, propulsive, and precipitated; seats himself on child's chair with only fair aim; creeps downstairs backward; has difficulty building tower of three cubes; can throw a ball, but clumsily.
24 to 27	Vocabulary of 300 to 400 words; has two- to three-word phrases; uses prepositions and pronouns.	Runs but falls when making a sudden turn; can quickly alternate between stance, kneeling or sitting positions; walks stairs up and down, one foot forward only.
30 to 33	Fastest increase in vocabulary; three- to four-word sentences are common; word order, phrase structure, grammatical agreement approximate language of surroundings, but many utterances are unlike anything an adult would say.	Good hand and finger coordination; can move digits independently; manipulation of objects much improved; builds tower of six cubes.
36 to 39	Vocabulary of 1,000 words or more; well-formed sentences using complex grammatical rules, although certain rules have not yet been fully mastered; grammatical mistakes are much less frequent; about 90 per cent comprehensibility.	Runs smoothly with acceleration and deceleration; negotiates sharp and fast curves without difficulty; walks stairs by alternating feet; jumps 12″; can operate tricycle; stands on one foot for a few seconds.

The temporal interlocking of speech milestones and motor milestones is not a logical necessity. There are reasons to believe that the onset of language is not simply the consequence of motor control. The development of language is quite independent of articulatory skills (Lenneberg, 1962), and the perfection of articulation cannot be predicted entirely on the basis of general motor development. There are certain indications for the existence of a peculiar, language-specific maturational schedule. Many children have a word or two before they toddle and thus must be assumed to possess a sufficient degree of motor skill to articulate, however primitively; yet the expansion of their vocabulary is still an extremely slow process. Why could they not rapidly increase their lexicon with "sloppy" sound symbols, much the way a child with a cleft palate does at age three? Similarly, parents' inability to train their children at this stage to join the words *Daddy* and *bye-bye* into a single utterance cannot be explained on the grounds of motor incompetence because at the same age children babble for periods as long as the duration of eight or ten syllables. In fact, babbled "sentences" may be produced, complete with intonation patterns. The retarding factor for language acquisition here must be a psychological one or perhaps better a cognitive one, and not mechanical skill. Around age three manual skills show improved coordination over earlier periods, but dexterity is still very immature on an absolute scale. Speech, which requires infinitely precise and swift movements of tongue and lips, all well-coordinated with laryngeal and respiratory motor systems, is all but fully developed when most other mechanical skills are far below their levels of future accomplishment. The evolvement of various motor skills and motor coordinations also has specific maturational histories, but the specific history for speech control stands apart dramatically from histories of finger and hand control.

The independence of language development from motor coordination is also underscored by the priority of language comprehension over language production. Ordinarily the former precedes the latter by a matter of a few months, especially between the ages of eighteen and thirty-six months. In certain cases this gap may be magnified by many years (Lenneberg, 1964). Careful and detailed investigations of the development of understanding by itself have only been undertaken in more recent years (Brown and Bellugi, 1964b; Ervin, 1964; Ervin and Miller, 1963). The evidence collected so far leaves little doubt that there is also an orderly and constant progression in this aspect of language development.

The development of children with various abnormalities affords the most convincing demonstration that the onset of language is regulated by a maturational process, much in the way the onset of gait is dependent upon such a process, but that at the same time the language-maturation process is independent of motor-skeletal maturation. In hypotonic children, for instance, the musculature in general is weak, and tendon reflexes are less active than normal. Hypotonia may be an isolated phenomenon that is quickly outgrown or it may be a sign of disease, such as muscular atrophy, carrying a bad omen for the child's future motor development. Whatever the cause, muscular development alone may be lagging behind other developmental processes and thus disarrange the normal intercalation of these various processes. Here, then, speech and language emerge at their usual time, while motor development lags behind.

On the other hand, there are some children with normal intelligence and normal skeletal and motor development whose speech development alone is markedly delayed. We are not referring here to children who never learn to speak adequately because of acquired or congenital abnormalities in the brain, but of those who are simply late speakers, who do not begin to speak in phrases until after age four, who have no neurological or psychiatric symptoms that can explain the delay, and whose environment appears to be adequate. The incidence of such cases is small (less than one in a hundred), but their very existence emphasizes the independence of language-maturation processes from other processes.

There are also conditions that affect all developmental processes simultaneously. These are diseases in which growth and maturation are retarded or stunted through a variety of factors, for instance, of an endocrine nature as in hypothyroidism; or retardation may be due to an intracellular abnormality, such as the chromosomal disorder causing mongolism. In these cases all processes suffer alike, resulting in general "stretching" of the developmental time scale but leaving the intercalation of motor and speech milestones intact (Lenneberg, Nichols, and Rosenberger, 1964). The preservation of synchrony between motor and speech or language milestones in cases of general retardation is, I believe, the most cogent evidence that language acquisition is regulated by maturational phenomena.

The evidence presented here rules out the possibility of a direct causal relationship between motor and speech development. Normally, growth and maturation proceed at characteristic rates for each developmental aspect. In the absence of specific retardations affecting skills or organs differentially, a picture of consistency evolves, such as represented in Table 1 or in the many accounts of normal human development (Gesell and Amatruda, 1947; McGraw, 1963).

The use of the word "skill" brings out another interesting aspect of the emergence of speech. With proper training probably everybody could attain some proficiency in such diverse skills as roller-skating, sketching, or piano-playing. However, there are also vast individual differences in native endowment and considerable variation with respect to the age at which training is most effective. Perfection can rarely be expected before the teens. The establishment of speech and language is quite different; a much larger number of individuals shows equal aptitude, absence of the skill is rare, and onset and fluency occur much earlier, with no particular training required.

Nevertheless, individual differences in time of onset and reaching of various milestones exist and need to be accounted for. The rate of development is not constant during the formative years, and there may be transient slowing in the rate of maturation, with subsequent hastening. This is hardly surprising in view of the complex interrelation of intrinsic and extrinsic factors that affect development. Nevertheless, there is a remarkable degree of regularity in the emergence of language. Figure 1 illustrates the regularity in the attainment of three major language-development levels and Figure 2 the sudden increase in vocabulary size, particularly around the third birthday.

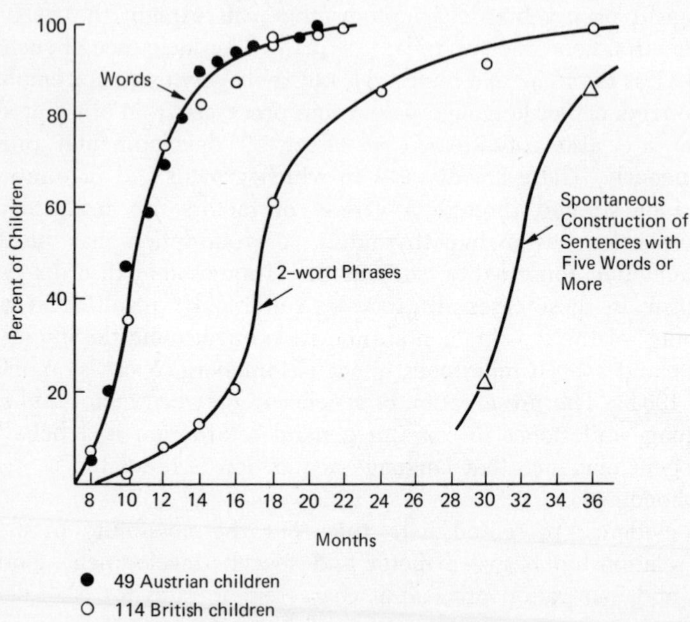

Figure 1 Emergence of Various Developmental Milestones in the Acquisition of Language. [The Information about the Austrian Children is from Bühler, 1931. The data on the British Children Is from Morley, 1957. The Information about the American Chlidren Is Based on Lenneberg's Observation of Children in Boston.]

In a survey of 500 middle- and lower-class children in the Boston area, examined in connection with an epidemiological study, I found that nine out of ten children had acquired all of the following verbal skills by the time they reached their thirty-ninth month: ability to name any object

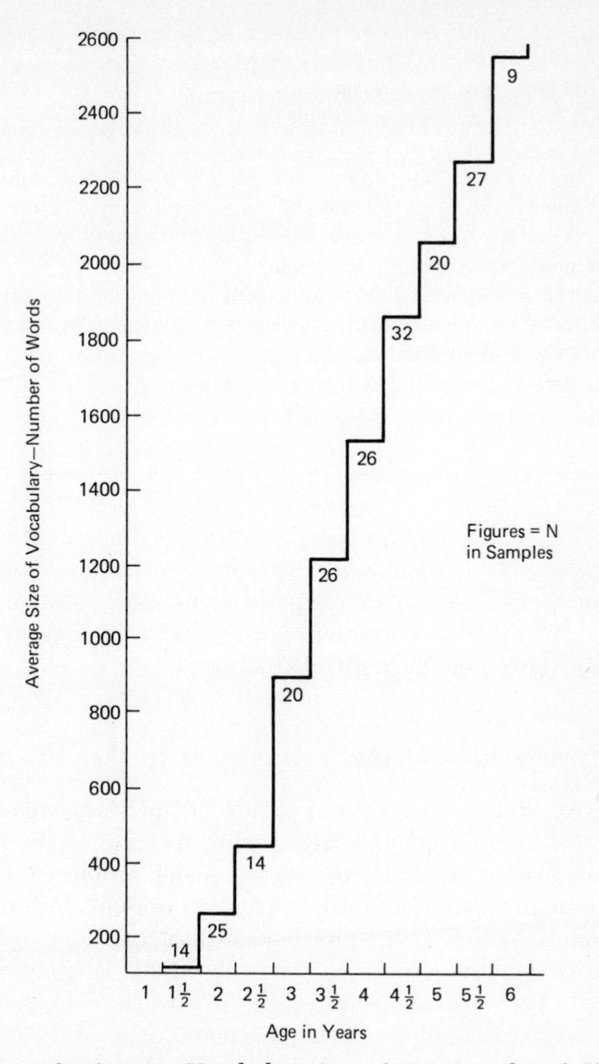

Figure 2 Average Vocabulary Size of Ten Samples of Children at Various Ages.

in the home, fair intelligibility, ability to comprehend spoken instructions, spontaneous utterance of syntactically complex sentences, and spontaneity in oral communication. The field observations were made in the child's home by specially trained social workers who worked with a screening test and a schedule of questions. Any child who was found or suspected to fall short of these standards was referred by the social worker to my office where he was examined by a speech therapist, an audiologist, and by myself. Fifty-four children were thus referred and found to fall into the following classifications:

Table 2

DISTRIBUTION OF CAUSES FOR FAILING LANGUAGE-SCREENING TEST
(given to 500 children at about the third birthday)

Cause	Number
1. Uncooperative child but, upon more intense examination, apparently normal speech development (health good, environment adequate).	7
2. Poor articulation but otherwise normal onset of language milestones (health good, environment adequate).	29
3. Various types of speech defects associated with psychiatric conditions.	9
4. Speech defects associated with other behavioral disorders due to gross environmental abnormalities.	2
5. Speech defects associated with nervous system disease.	3
6. Delayed onset of speech, unexplained (health good, environment adequate).	4

Differences in age of onset become much less dramatic if we scrutinize these statistics. Of 486 children who were free from nervous or mental disease and were raised in an adequate environment—all children in the sample except those of Groups (3), (4), and (5)—only 33 (less than 7 per cent) were below the norm of attainment.

The Relation of the Environment to the Age of Onset

It is obvious that a child cannot acquire language unless he is exposed to it. Apart from this trivial point the role of the environment is not immediately clear. There are two major problems: How are the infant's eventual capabilities for language acquisition affected by environmental variations during his prelanguage life, and what influence does the environment have upon the age at which language capabilities appear? We must emphasize once more that we are concerned here with potentialities, not actually occurring behavior. Many of the earlier studies failed to make this distinction. Subnormal speech habits may not be used as evidence for subnormal capacity. Most language tests assess the quality of existing language development but not whether children are actually not capable of taking advantage of existing stimulation. It is a reasonable assumption that in most instances in initially poor language, environment does not cripple the child's basic potentialities forever. If the social environment is enriched early enough, he will at once improve his language habits. The important point here is that intuitively the notion can be accepted that language potentialities do develop regularly and in spite of certain environmental deprivations. A closer look at the empirical investigations supports precisely this view.

In most countries families consist of many children where the birth of one is quickly succeeded by the birth of another. The social environment of the first child is clearly different from that of a subsequent child. This makes possible empirical research into the relation between age at attainment of milestones and social environment. Morley (1957) contributed statistics showing that the age of emergence of single words, of phrases, and of intelligible speech is no different for first than for subsequent children.

Even for single or first children the environment is not always the same. Mothers vary greatly in their attitudes toward their children. Some use baby talk, others are very silent. Some mothers bring a natural maternal warmth and certainty to the nursery, and others are ill at ease with their first children. Some children are welcome additions, but others are not. Further statistics compiled by Morley indicate that variables such as "mother's ability to cope," loss or temporary absence of either parent, or socioeconomic class are not predictive of the age of emergence of various milestones in speech development.

Morley's findings are not contradicted by studies that report differences in speech habits of children of the upper, middle, and lower class (Bühler, 1931; Irwin, 1948; and many others). These are usually cross-sectional studies in which the nature and quality of speech is compared with a norm, but the age of onset of certain speech phenomena is not determined. Morley found that the language habits that emerged at the common time soon showed signs of impoverishment in the underprivileged, and unintelligibility occurred more commonly in second and subsequent children than in first. Thus the influence of the environment upon speech habit is undeniable, even though the onset of speech habits is relatively unaffected.

The differences observed in the speech habits of upper- and lower-class children are, actually, difficult to evaluate because of the many covarying factors. For example, the influence of malnutrition and of diseases that delay development is higher among poorer children, who may also be emotionally less amenable to testing situations than those from more carefree homes. In attempts to estimate the child's vocabulary by means of flash cards, an assumption is made that the relative frequency of occurrence of words in the vocabularies of upper and lower class is identical. This is not necessarily valid.

The role of environment is documented most drastically by the studies on the development of children in orphanages (Brodbeck and Irwin, 1946; Dennis and Dennis, 1951; Fisichelli, 1950; Goldfarb, 1943; McCarthy, 1954). Leaving aside the old question of the possible differences in biological stock between this and the general population, there is no denying that the institutional life leaves its mark on speech and language habits. On the other hand Goldfarb (1945a) and Dennis and Najarian (1957) have given an illustration of covert processes in the unfolding of potentialities. Children reared in orphanages are frequently below average in speech and motor development when tested at three, but when retested at six or seven they are found to have caught up with the control population. As soon as their environment is enriched, perhaps through greater freedom to move around, they are able to make use of the greater stimulus availability.

Lenneberg, Rebelsky, and Nichols (1965) have studied emergence of vocalization during the first three months of life as it relates to the parents' speech and hearing. Children of congenitally deaf parents (deaf father and mother) were compared with children born to hearing and normally speaking parents. Among the six deaf families studied, five of the babies had normal hearing, but in one case the baby was born deaf. Twenty-four-hour tape recordings were made biweekly in the child's home and compared qualitatively and quantitatively. The environment of the two groups of children differed in two ways: (1) the amount, nature, and occasion of adult vocalization heard by the babies differed significantly, and (2) the baby's own vocalizations could never be responded to by a deaf mother who, we discovered, could not even tell whether her child's facial expressions and gestures were accompanied by silence or noise. While babies born to hearing parents appeared to vocalize on the occasions of adult vocalization, the babies born to deaf parents did not. Nevertheless, they made as much noise and went through the same sequence of vocalization development with identical ages of onset (for cooing noises) as the control group.

I have also been able to observe older hearing children born to deaf couples, though I have not undertaken a statistical study. These observations, on a dozen children in five families, leave no doubt that language onset is never delayed by this dramatically abnormal environment, even though the quality of vocalization of the preschool children tends to be different; children very soon became "bilingual" in the sense that they use normal voice and speech for hearing adults and abnormal voice and "deafisms" for their parents.

How universal in human society is the onset of speech? Do cultural attitudes toward child-rearing influence its emergence, or are there languages so complicated that no one can master oral communication until puberty or so primitive that the entire system is learned by every child before he begins to toddle? There are many studies on child development in primitive cultures, and most authors have described every minute deviation from the norm of western society. Strangely enough the onset of speech has rarely been a subject of a detailed study in the anthropological literature (but see Austerlitz, 1956; Kroeber, 1916). Apparently no field worker has even been struck by any discrepancies between the vocalizations or communicative behavior among the children of "primitive" and "western" man. Lenneberg, Putnam, Whelan, and Crocker (in preparation) have investigated this problem further by direct field observations among the Dani of Dutch New Guinea, the Zuñi of the American Southwest, and the Bororo and some Gê tribes of Central Brazil. In these investigations children were given tests of sensorimotor development, such as coordination for reaching, nature of the grasp, and the ability to walk, stand on one leg, or throw a ball. Tape recordings were made of the vocalizations of the babies before they appeared to be in possession of the common language, as well as of their utterances throughout their physical development to age three. In addition, information was obtained from native informants about the linguistic competence of the various children studied, their fluency, types of mistakes in articulation, syntax, and choice of words, and on the parents' attitude toward their children's speech development. In some instances the chronological age of the child studied was not known, and therefore neither the motor nor the language achievements could be compared directly with the age of emergence in American children. But the chronological age was not as important as the question whether developmental progress, gauged upon the emergence of definite motor skills, marked also the beginning and major milestones of the child's speech development and whether the concordances between speech and motor development observed in western children were also found in children of these cultures. As far as we can judge from the analysis of the material so far, the answer is clearly yes. The first words appear at about the time that walking is accomplished, and by the time a child is able to jump down from a chair (or its equivalent), tiptoe, or walk backward three yards, he is reported by the informants to be communicating fluently, even though certain inaccuracies and childlike usages seem always to persist for a longer period. Anthropologists have pointed out that the label "primitive language" is misleading when applied to any natural living language. The developmental studies here support this view in that they indicate that no natural language is inherently more complicated or simpler to learn by a growing child than any other language.[1] There seems to be no relation between progress in language acquisition and culturally determined aspects of language.

In summary, it cannot be proved that the language environment of the growing child remains constant throughout infancy, but it can be shown that an enormous variety of environmental conditions leaves at least one aspect of language acquisition relatively unaffected: the age of onset of certain speech and language habits. Thus, the emergence of speech and language habits is more easily accounted for by assuming maturational changes within the growing child than by postulating special training procedures in the child's surroundings.

[1] Slobin . . . reported, that several aspects of Russian morphology and syntax are not learned by the Russian child until after entering school. However, he informed me in personal conversation that the forms acquired later are rarely used (or ever used correctly) in colloquial discourse.

The Role of Utility in the Onset of Speech

There is evidence, though only of a circumstantial nature, that language does not emerge as a response to an experienced need, as a result of discovery of its practical utility, or as a product of purposive striving toward facilitated verbal communication. I have made tape recordings of the spontaneous noises made during play by congenitally deaf children. In two instances periodic sample recordings were made of deaf children of deaf parents from the first month of life. Sixteen other deaf children were recorded between their second and fifth years. In most instances follow-up recordings are available throughout an eighteen-month period. All eighteen children vocalized often during concentrated play; the quality of their voices was quite similar to that of hearing children, and in certain respects the development of their vocalizations was parallel to that observed in hearing children though the deaf did not develop words. Nevertheless, cooing appeared at about three months, babbling sounds were heard at six months and later, and laughter and sounds of discontent were virtually identical with those of the hearing population. It was particularly interesting to note that many of the deaf children, during their spontaneous babbling, would produce sounds that were well-articulated speech sounds such as /pakapakapaka/. This is not to say that there was no difference between the deaf and hearing children over six months. The deaf had a tendency to engage in certain types of noise more persistently, while the hearing would tend to go frequently through a wide range of different types of babbling sounds, as if to run through their repertoire for the sheer pleasure of it. No precise quantitative measures of amount of vocalization could be made on the children after the first three months, but subjectively, the hearing children were much more vocal in the presence of others than the deaf.

A healthy deaf child two years of age or older gets along famously despite his total inability to communicate verbally. These children become very clever in their pantomime and develop techniques for communicating their desires, needs, and even opinions, that leave little to be desired. There is no indication that congenital peripheral deafness causes significant adjustment problems within the family during preschool years. This observation has an important bearing on the problem of motivation for language acquisition. Language is extremely complex behavior, the acquisition of which, one might have thought, requires considerable attention and endeavor. Why do hearing children bother to learn this system if it is possible for a child to get along without it? Probably because the acquisition of language is not, in fact, hard labor—it comes naturally—and also because the child does not strive toward a state of perfect verbal intercourse, normally attained only two years after the first beginnings.

The Importance of Practice for the Onset of Speech

Closely related to the question of utility is the problem of practice. Do cooing and babbling represent practice stages for future verbal behavior? We have every indication that this is not so. Occasionally the natural airways above the voice box become narrowed because of swelling in connection with disease, and an opening must be made into the trachea below the larynx for insertion of a tube through which the patient can breathe. This prevents the patient from making sounds because most of the expired air escapes before it can excite the vocal cords. I have examined a fourteen-month-old child who had been tracheotomized for six months. A day after the tube had been removed and the opening closed, the child produced the babbling sounds typical of the age. No practice or experience with hearing his own vocalizations was required.

Comparable observations may be made on children not older than twenty-four months who are admitted to pediatric hospitals because of severe physical neglect by the parents. Characteristically, they are on admission apathetic, unresponsive babies who seem to be grossly retarded in their motor, social, and sound development. After a few weeks of hospital care they blossom out, begin to relate to the nursing staff, and make all the noises that are heard in infants of comparable age. If the neglected child is over three or four years of age, environmental deprivation will have contributed to severe emotional disturbance often more typical of psychotic conditions (Davis, 1947). However, some children with psychoses, regardless of whether parental neglect was a contributing factor, give excellent demonstrations of "subclinical" language development. There are children who fail to communicate with the world around them, including their own parents, and who give an impression of muteness and incomprehension from their second year of life on. Yet in response to treatment, or even spontaneously, some will often snap out of their state of isolation and almost miraculously begin to talk fluently and up to age level (Luchsinger and Arnold, 1959, and my own experience). Practice is, of course, not the same as learning. In these children it is fair to say that they have not practiced speech and language in the same manner as normal children might, but we cannot say that they have not undergone years of learning. They simply had not chosen to respond.

"Wolf Children"

It is difficult to refrain from referring to the stories of children supposedly reared by wolves and other cases of extreme neglect. Yet a careful analysis of this literature has convinced me that even the most fundamental information is usually missing in the descriptions or omitted altogether from the case reports. The children are invariably discovered by well-meaning but untrained observers, and the urgency for getting help to the victims is so overwhelming that the scientifically most important first months are least well documented. The nature of the social and physical environment is never clear, and the possibility of genetic deficiencies or congential abnormalities can never be ruled out. One child reported by Davis (1947) was discovered at age six without speech but was said to have made very rapid progress, going through all the usual baby language stages, and within a period of nine months attained complete mastery of speech and language. In the same article a comparable case is described, also discovered at age six, but this child only began to speak at age nine. At the time of her death at ten-and-one-half, she could name people and communicate her needs by a few sentences. The behavioral descriptions of this child point to severe psychosis and feeblemindedness. Descriptions of children supposedly reared by wolves or growing up in forests by themselves are plentiful, but none is trustworthy (Koehler, 1952). Singh and Zingg (1942) have collected the entire material, and an excellent commentary may be found in Brown (1957). The only safe conclusions to be drawn from the multitude of reports is that life in dark closets, wolves' dens, forests, or sadistic parents' backyards is not conducive to good health and normal development. It is impossible to say why some of these children are capable of overcoming the insults inflicted upon their early health while others succumb to them. The degree and duration of neglect, the initial state of health, the care provided for them after discovery, and many other factors are bound to influence the outcome; in the absence of information on these points, virtually no generalizations may be made with regard to human development.

We started by developing criteria for the distinction of behavioral emergence due to changes of capacity within the growing individual (regularity in onset, differential use of environmental stimulation with growth, independence from use, and superfluousness of practice). Applying these criteria to language we have found strong suggestions that the appearance of language is primarily dependent upon the maturational development of states of readiness within the child, assuming the existence of an adequate environment.

Age Limitation to Language Acquisition

Complementary to the question of how old a child must be before he can use the environment for language acquisition is that of how young he must be before it is too late to acquire speech and language. There is evidence that the primary acquisition of language is dependent upon a certain developmental stage which is quickly outgrown at puberty. I have presented detailed evidence for this elsewhere (Lenneberg, 1966) and shall confine myself to a few summary statements. The evidence is largely based on clinical experience with acquired aphasia.

The Relation of Age to Recovery from Acquired Aphasia

The outlook for recovery from aphasia varies with age. The recovery chance has, so to speak, a natural history that is the same as the natural history of cerebral lateralization of function. Aphasia is the result of direct, structural, and local interference with the neurophysiological processes of language. In childhood such interference cannot be permanent because the two sides are not yet sufficiently specialized for function, even though the left hemisphere may already show signs of speech dominance. Damage to it will interfere with language, but the right hemisphere is still involved to some extent with language and so there is a potential for language function that may be strengthened again. In the absence of pathology, a polarization of function between right and left takes place during childhood, displacing language entirely to the left and certain other functions predominantly to the right (Ajuriaguerra, 1957; Hécaen and Ajuriaguerra, 1963; Teuber, 1962). If, however, a lesion is placed in either hemisphere, this polarization cannot take place, and language function together with other functions persist in the unharmed hemisphere.

Notice that the earlier the lesion is incurred, the less grave is the outlook for language. Hence we infer that language-learning can take place, at least in the right hemisphere, only between the ages of two to about thirteen. That this is probably also true of the left hemisphere follows from observations on language development in the retarded and in the congenitally deaf, discussed later.

A unique pathological study of congenital aphasia was reported by Landau, Goldstein, and Kleffner (1960). This was a child who died of heart disease at age ten. This patient, in contrast to the cases discussed so far, had not begun to develop speech until age six or seven. At that time he was enrolled in a class for congenitally aphasic children at the Central Institute for the Deaf. By age ten, the authors report, "he had acquired considerable useful language." A postmortem examination of the brain revealed bilateral areas of cortical destruction around the Sylvan fissure in the area of the central sulcus, together with severe retrograde degeneration in the medial geniculate nuclei of the midbrain. The authors conclude that "Language function therefore appears to have been subserved by pathways other than the primary auditory thalamo-cortical projection system." I am citing this case here to illustrate the far-reaching plasticity of the human brain (or lack of cortical specialization) with respect to language during the early years of life. There is clinical evidence that similar lesions in a mature individual would have produced severe and irreversible defects in reception and production of speech and language. The implication is that the brain at birth and during the subsequent maturation process may be influenced in its normal course of organization, which usually results in the specialization of areas.

Postnatal cerebral organization and reorganization have been demonstrated by several workers for a variety of mammals (Benjamin and Thompson, 1959; Brooks and Peck, 1940; Doty, 1953; Harlow, Akert and Schiltz, 1964; Scharlock, Tucker, and Strominger, 1963: and others). Various kinds of postnatal cortical ablations leave no or very minor deficit, whereas comparable ablations in later stages of development result in irreversible symptoms.

Arrest of Language Development in the Retarded

The material reviewed might give the impression that the age limitation is primarily due to better recovery from disease in childhood and that the language limitations are only a secondary effect. This is probably not so. In a study by Lenneberg, Nichols, and Rosenberger (1964), fifty-four mongoloids (all raised at home) were seen two to three times per year over a three-year period. The age range was from six months to twenty-two years. The appearance of motor milestones and the onset of speech differed considerably from individual to individual, but all made some progress—even though very slow in many cases—before they reached their early teens. This was true of motor development as well as of speech. In all children seen in this study, stance, gait, and fine coordination of hands and fingers were acquired before the end of the first decade. At the close of the study 75 percent had reached at least the first stage of language development; they had a small vocabulary and could execute simple spoken commands. But interestingly enough, progress in language development was recorded only in children younger than fourteen. Cases in their later teens were the same in terms of their language development at the beginning as at the end of the study. The observation seems to indicate that even in the absence of structural brain lesions, progress in language-learning comes to a standstill after maturity. Figure 3 is a graphic illustration of the empirical findings.

The Effect of Deafness on Language at Various Ages

The study of acquired deafness during childhood and later life gives further insight into the importance of age in language acquisition. The most common cause of sudden and total loss of hearing is meningitis.

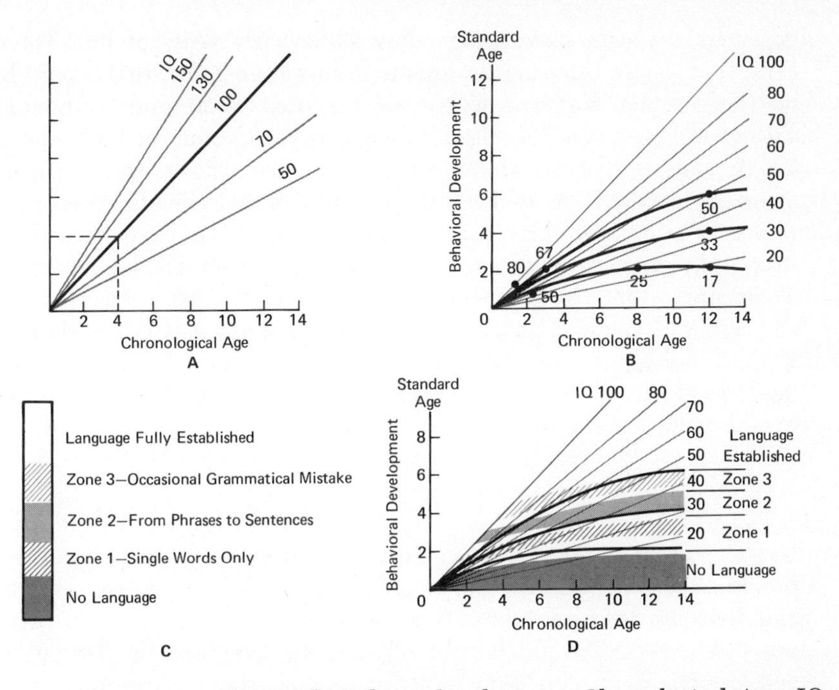

Figure 3 The Triple Relationship between Chronological Age, IQ as Measured by the Merrill-Palmer Test, and the Attainment of Stages in Language Development. A. Theoretical Representation of Constant IQ's; B. Predictable Tendency of IQ to Decay in Most Forms of Mental Retardation; C. Traversal of Certain Language Stages at Specific Ages in Normal Development; D. Empirical Determination of the Relationship between the Three Parameters in the Mentally Retarded.

The virulence of the disease is such that many a child falls ill and is left without hearing practically overnight. Throughout childhood sudden acquisition of deafness has an immediate effect upon voice and speech, and, before the age of six, also on language habits. Within a year or less the small child, say up to about four years of age, will have lost the ability to control his voice and articulatory mechanisms for ordinary speech sounds and will develop noises and habits very similar or even indistinguishable from those heard and seen in the congenitally deaf. His education has to be relegated to special teachers in the schools for the deaf. Both these populations, those who become deaf before and those after the onset of speech, sound and behave like the congenitally deaf children. But those who lose hearing after having been exposed to the experience of speech, even for as short a period as one year, can be trained much more easily in all language arts, even if formal training begins some years after they have become deaf. On the other hand, children deafened before completion of the second year do not have any facilitation in comparison with the congenitally deaf (based on work in progress). It seems as if even a short exposure to language, a brief moment during which the curtain has been lifted and oral communication established, is sufficient to give a child some foundation on which much later language training may be based. The effect of deafness thus complements our knowledge obtained from the effects of acquired aphasia. While the prognosis for recovery from aphasia gets worse and worse with advancing age after about ten, the prognosis for speech habilitation in the deaf improves directly with the advance of age at onset of the disorder.

Fry presents material at this conference that stresses the paramount importance of age in the establishment of optimal speech habits. He has recorded utterances of children whose audiograms would indicate profound hearing loss, but whose quality of voice, intonation patterns, and articulation are far superior to anything that is achieved either in America or the European continent. Fry's explanation is that these children were provided with hearing aids during earliest infancy and were given intensive "sound training" long before school started. In America children are also given hearing aids and sound training, but the latter does not begin seriously until age four or even later, and the hearing aids are often given little attention until school begins. If these findings can be verified on a larger scale, it would indicate an even shorter span of the critical age for optimal language acquisition than advocated in this paper.

Concomitants of Physical Maturation

Language cannot begin to develop until a certain level of physical maturation and growth has been attained. Between the ages of two and three years, language emerges by an interaction of maturation and self-programmed learning. Between the ages of three and the early teens, the possibility for primary language acquisition continues to be good; the individual appears to be most sensitive to stimuli at this time and to preserve some innate flexibility for the "organization of brain functions" to carry out the complex integration of subprocesses necessary for the smooth elaboration of speech and language. After puberty the ability for self-organization and adjustment to the physiological demands of verbal behavior quickly declines. The brain behaves as if it had become set in its ways, and primary, basic language skills not acquired by that time, except for articulation, usually remain deficient for life. (New words may be acquired throughout life because the basic skill of naming has been learned at the very beginning of language development.)

I shall now make a few comments on the state of the brain during the initial period for language acquisition. I must stress, however, that this is not an attempt to discover the specific anatomical or biochemical basis of language development *per se*. The specific neurophysiology of language is unknown, and therefore it would be futile to look for any specific growth process that would explain language acquisition. Nevertheless it may be interesting to know in what way the brain, particularly the cerebral cortex, is different before the onset of language and after primary language acquisition is inhibited. The answer to such a question does not point to the cause of language development but tells us something about its substrate and its limiting or prerequisite conditions.

In a separate paper (Lenneberg, 1966) I have collected anatomical, histological, biochemical, and electrophysiological data on the maturation of the human brain. I have presented the material there in the form of a series of maturation curves, one example of which is shown in Figure 4. While the curves differ one from the other, they are all intercorrelated, as might be expected. Their significance for a study of language acquisition is simply this: the curves define what is meant by maturation of the brain. All of the parameters of brain maturation studied show that the first year of life is characterized by a very rapid maturation rate. By the time language begins to make its appearance, about 60 per cent of the adult values of maturation are reached. Then maturation rate slows down and reaches an asymptote at just about the same time that trauma to the left hemisphere begins to have permanent consequences. Thus, by the time primary language acquisition comes to be inhibited, the brain can also be shown to have reached its mature state, and cerebral lateralization is irreversibly established.

The question remains of what the significance is of the coincidence between these brain-maturation phases and the onset and gradual decline of the capacity for language-learning. Could they be entirely spurious? The infant's obvious incapacity to learn all but the most primitive beginnings of language during his first fifteen months is, at least intuitively, attributable to a general state of cerebral immaturity. The maturational data for the end of the critical period is more difficult to interpret. If it were not for the consequence of different types of evidence that language acquisition is indeed inhibited at this time, the maturational data alone would lack all interest. As it is, however, we may think of these data as contributing to the diverse circumstantial evidence that puberty marks a milestone for both the facility in language acquisition and a number of directly and indirectly related processes in the brain. I am, therefore, suggesting as a working hypothesis that the general, nonspecific states of maturation of the brain constitute prerequisites and limiting factors for language development. They are not its specific cause.

This hypothesis leads to a rather revealing generalization. Because the various aspects of cerebral maturation are so highly correlated, we may think of maturation of the brain as a single variable (perhaps as

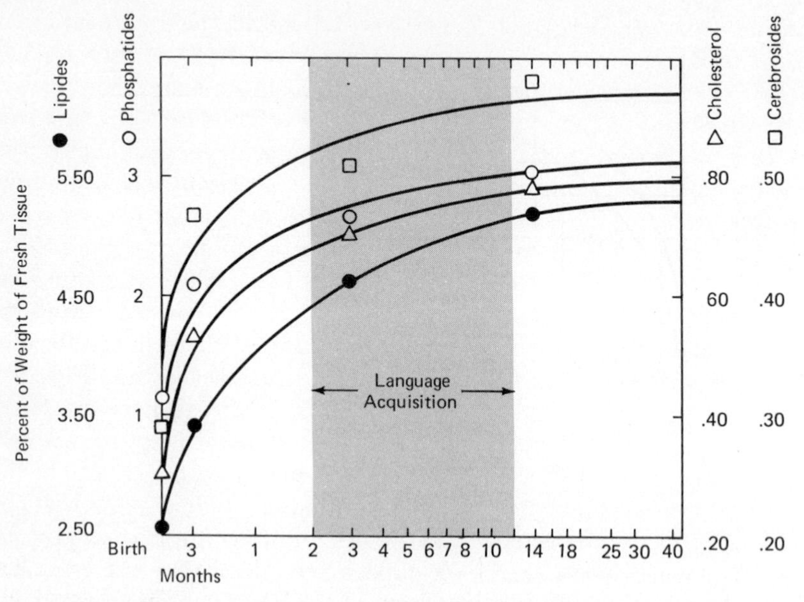

Figure 4 Chemical Composition of the Human Cerebral Cortex Plotted as a Function of Chronological Age. [Based on Data by G. Brante, "Studies on Liquids in the Nervous System; with Special Reference to Quantitative Chemical Determination and Topical Distribution," *Acta Physiologica Scandinavica* (18), 1949; and Also Based on Data by J. Folch Pi, "Composition of the Brain in Relation to Maturation," H. Waelsch, ed., *Proceedings of the First International Neurochemical Symposium.* This Is One of a Family of Similar Curves Published in E. H. Lenneberg, "Speech Development: Its Anatomical and Physiological Concomitants," in *Speech, Language, and Communication. Brain and Behavior,* from UCLA Forum in Medical Science, 3, no. 4, edited by Edward C. Carterette. Berkeley: University of California Press, 1966. Reprinted by Permission of the Regents of the University of California.]

a shorthand sign for the sake of the following demonstration). As the brain matures, the growing infant successively attains various developmental milestones, such as sitting, walking, and joining words into phrases. In Figure 5, we see these milestones as "developmental horizons," signifying thereby the breadth of the maturational accomplishments; that is to say, sitting or walking are not the only developmental achievements of these various periods; at the same time there is a whole spectrum of sensory and motor development, and sitting or walking are merely their most outstanding characteristic. Figure 5 shows that if the normal maturation function is slowed down, developmental horizons are reached later

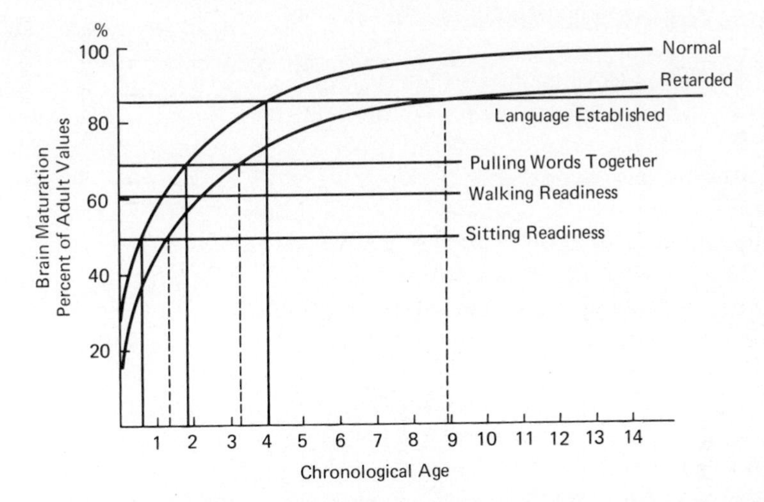

Figure 5 Relationship between General Rate of Maturation of the Brain, in Per Cent of Mature Values, to Chronological Age. The Parameters Are Normal Individuals and a Typical Case of a Mentally Defective Individual. In the Defective, Developmental Milestones Fall Farther and Farther Behind.

and, most importantly, the spacing between the milestones becomes more prolonged without altering the order of sequence. Normally twelve to fourteen months elapse between sitting and putting words together, and language is fully established within another twenty months. But in the retarded the lapse between sitting and putting words together may be twenty-four months, and language may not be established fully for as long a period as another sixty months. This is precisely what is found in generally retarded children. Their earliest milestones seem delayed by just a few months, but the delay is increased with advancing age, and the lag behind the norm becomes worse and worse even though the retarding disease may be completely stationary and maturation is progressing steadily but slowly. The autopsy findings of brain weight of retarded individuals conforms fairly well to this picture, though retardation may be caused by so many different factors that it is not surprising that the correlation is not perfect.

The working hypothesis expounded here does not postulate cerebral "rubicons" or any absolute values of brain weight or composition as the *sine qua non* for language. It is not so much one or the other specific aspect of the brain that must be held responsible for the capacity of language acquisition but the way the many parts of the brain interact. Thus it is mode of function rather than specific structure that must be regarded as the proper neurological correlate of language.

Growth Characteristics of the Human Brain

The irrelevance of absolute values, such as brain weight, for the capacity of language is discussed in Lenneberg (1964). The example is cited of nanocephalic dwarfism where brain weight and brain-body weight ratio may temporarily be identical with the corresponding values of a three-year-old chimpanzee. It is only when we look at some aspect of the ontogenetic histories of the two organisms that differences appear and the nanocephalic dwarf follows the developmental history of man, whereas the chimpanzee has a curve of its own. Any parameter of maturation must be studied in relation to the natural history of the organism's life cycle.

Species differ in their embryological and postnatal developmental courses. Much more important for an understanding of speciation are the ubiquitous failures of ontogeny to repeat phylogeny in specific detail, the species-characteristic deviations from the hypothetical course of a "general" history of evolvement. Consider the ratio of the weight of the brain to the weight of the entire body: let us call this the *brain-weight index,* a number that does not change appreciably after maturity, although it cannot remain completely constant because the body weight of a young adult slowly increases with advancing age. At birth, however, the index is about six to seven times the value of adulthood, and it decreases gradually throughout infancy and childhood.

Man's brain-maturation history is unique among primates. All lower forms approach the adult condition at a relatively quicker pace than man (Schultz, 1940, 1956). On the other hand, except for man's first six months, there is no brain-weight index value that is unique to him, as shown in Figure 6. His adult index value is about 2.2; in chimpanzee that value is attained shortly before midinfancy. At the end of the first childhood quarter the chimpanzee's index is 3.5, which is the same as man's at the close of the third quarter. It is not the measurements themselves that are different but the developmental stage at which these values are attained. When we discussed the nanocephalic dwarf, we were comparing a form near puberty with a form shortly before midinfancy, and hence the values for both brain and body could be matched. But the present material points to a fundamental difference between dwarf and chimpanzee, namely, their developmental history. The dwarf quickly approaches the human curve and merges with it by puberty.

In comparison with lower primates, man's brain remains large throughout life. Yet we obtain a different view if we compare the growth rate of the brain *per se.* Man's body increases at a slightly faster rate than chimpanzee's, but the difference is not very startling. The growth curves of the brains are very different, however. During the first quarter of childhood man's brain gains about 800 grams as against 110 in the chimpanzee.

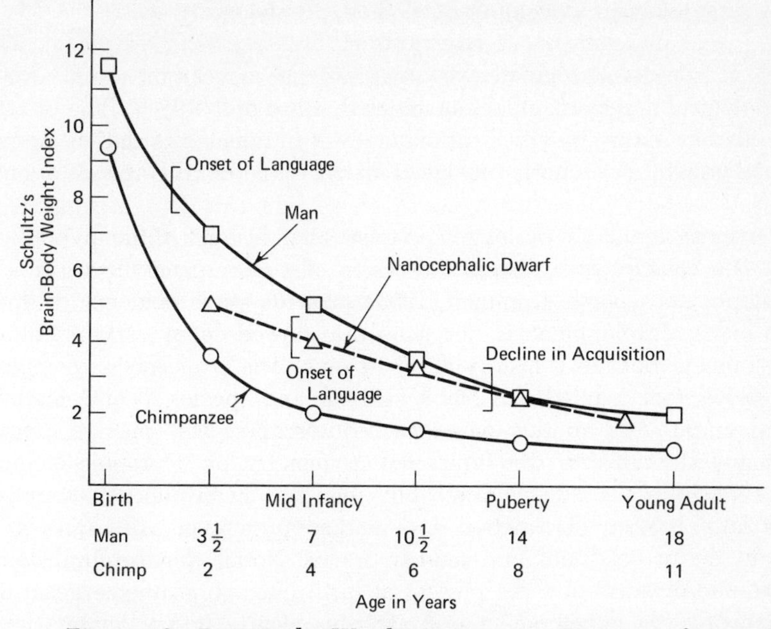

Figure 6 Brain-Body Weight Ratio as a Function of Age. The Developmental Curve of the Nanocephalic Dwarf Resembles That of a Normally Growing Human, Whereas That of the Chimpanzee Always Remains Different.

In terms of relative increase man's brain weight at birth is only 24 per cent of the adult weight, while chimpanzee starts life with a brain that already weighs 60 per cent of its final value. During the first quarter of childhood the chimpanzee's brain gains only 30 per cent, whereas man's gains 60 per cent.

What conclusions may be drawn from these differences? We do not have the maturational data for chimpanzee brains that we have for human brains, but it is interesting to note that in man growth characteristics of the brain weight are closely paralleled in time by growth of microscopic structures, by growth in chemical compounds, and by growth in electrophysiological parameters. There is always rapid acceleration in the first quarter and attainment of the adult condition by puberty. By extrapolation we may assume that the maturational events of the chimpanzee brain during childhood differ from those in man in that at birth his brain is probably much more mature and all parameters are probably more stabilized than in man. This would indicate that the facilitation for language-learning is not only tied to "a state of flux" but to a maturational history that is characteristic for man alone.

We should not suppose that one might be able to train a chimpanzee to use a natural language, such as English, simply by delaying the animal's physiological development. When physiological delay occurs in man (as in mongolism), it also protracts his speech development. The onset of speech is regulated by maturational development of certain physiological and perceptual capacities that are probably lacking or take on different forms in lower primates. Maturational retardation cannot induce growth of the basic biological matrix for cerebral language organization.

Pertinent to man's prolonged maturational history is the hypothesis that man constitutes a "fetalized" version of a more generalized primate developmental course. Kummer (1953) presents much evidence to show that man's development is not simply a slowed-down version of ape development but has a history all of its own. The cross-species comparison shows that man's brain has a peculiar and species-specific maturational curve. Add to this our earlier observation that man is unique among vertebrates in the functional asymmetry of neurophysiological process within the adult brain. Only man has hemispheric dominance with lateralization of function and marked preference with respect to side in the use of limbs and sensory organs. Notice that cerebral dominance and dexterity are not present at birth but regularly emerge in the course of early development and are thus clearly tied to maturational processes. I have, indeed, cited evidence that laterality is a process of innate organization and polarization that is inevitable in the normal course of development but may be blocked, so to speak, within certain age limits by destruction of tissue in either hemisphere. I would like to propose that lateralization is a phenomenon of growth and development in man.

The development of language, also a species-specific phenomenon, is related physiologically, structurally, and developmentally to the other two typically human characteristics, cerebral dominance and maturational history. Language is not an arbitrarily adopted behavior, facilitated by accidentally fortunate anatomical arrangements in the oral cavity and larynx, but an activity that develops harmoniously by necessary integration of neuronal and skeletal structures and by reciprocal adaptation of various physiological processes.

Summary of the Argument

The notion "need" explains nothing because (*a*) of its subjective nature and (*b*) if the infant's needs change in the course of the first two years of life, they do so because of his own growth and maturation and not because of arbitrary extrinsic factors.

We must assume that the child's capacity to learn language is a consequence of maturation because (*a*) the milestones of language acquisition are normally interlocked with other milestones that are clearly attributable to physical maturation, particularly stance, gait, and motor coordination; (*b*) this synchrony is frequently preserved even if the whole maturational schedule is dramatically slowed down, as in several forms of mental retardation; (*c*) there is no evidence that intensive training procedures can produce higher stages of language development, i.e., advance language in a child who is maturationally still, say, a toddling infant. However, the development of language is not caused by maturation of motor processes because it can, in certain rare conditions, evolve faster or slower than motor development.

Primary language cannot be acquired with equal facility within the period from childhood to senescence. At the same time that cerebral lateralization becomes firmly established (about puberty), the symptoms of acquired aphasia tend to become irreversible within about three to six months after their onset. Prognosis for complete recovery rapidly deteriorates with advancing age after the early teens. Limitation to the acquisition of primary language around puberty is further demonstrated by the mentally retarded who can frequently make slow and modest beginnings in the acquisition of language until their early teens, at which time their speech and language status becomes permanently consolidated. Further, according to Fry, the profoundly deaf must receive sound training and prosthetic aid as close to age two as possible to develop good speech habits. The reverse is seen in acquired deafness where even short exposure to language before the onset of the disease improves prognosis for speech and language, the outlook becoming better in proportion to the length of time during which the patient had been in command of verbal skills.

Thus we may speak of a critical period for language acquisition. At the beginning it is limited by lack of maturation. Its close seems to be related to a loss of adaptability and inability for reorganization in the brain, particularly with respect to the topographical extent of neurophysiological processes. (Similar infantile plasticity with eventual irreversible topographical representation in the brain has been demonstrated for many higher mammals.) The limitations in man may well be connected with the peculiar phenomenon of cerebral lateralization of function, which only becomes irreversible after cerebral growth phenomena have come to a conclusion.

The specific neurophysiological correlates of speech and language are completely unknown. Therefore emergence of the capacity for language acquisition cannot be attributed directly to any one maturational process studied so far. But it is important to know what the physical states of the brain are before, during, and after the critical period for language acquisition. This is the prerequisite for the eventual discovery of more specific neural phenomena underlying language behavior. One finds that

Table 3

SUMMARY SURVEY

Age	Usual Language Development	Effects of Acquired, Lateralized Lesions	Physical Maturation of CNS	Lateralization of Function	Equipotentiality of Hemispheres	Explanation
Months 0 to 3	Emergence of cooing.	No effect on onset of language in half of all cases; other half has delayed onset but normal development.	About 60 to 70 per cent of developmental course accomplished.	None: symptoms and prognosis identical for either hemisphere.	Perfect equipotentiality.	Neuroanatomical and physiological prerequisites become established.
4 to 20	From babbling to words.					
21 to 36	Acquisition of language.	All language accomplishments disappear; language is reacquired with repetition of all stages.	Rate of maturation slowed down.	Hand preference emerges.	Right hemisphere can easily adopt sole responsibility for language.	Language appears to involve entire brain; little cortical specialization with regard to language though left hemisphere beginning to become dominant toward end of this period.
Years 3 to 10	Some grammatical refinement; expansion of vocabulary.	Emergence of aphasic symptoms; all disorders recover without residual language deficits (except in reading or writing). During recovery period two processes active: diminishing aphasic interference and further acquisition of language.	Very slow completion of maturational processes.	Cerebral dominance established between 3 to 5 years, but evidence that right hemisphere may often still be involved in speech and language functions. About one quarter of early childhood aphasias due to right hemisphere lesions.	In cases where language is already predominantly localized in left hemisphere and aphasia ensues with left lesion, it is possible to re-establish language presumably by reactivating language functions in right hemisphere.	A process of physiological organization takes place in which functional lateralization of language to left is prominent. "Physiological redundancy" is gradually reduced and polarization of activities between right and left hemisphere is established. As long as maturational processes have not stopped, reorganization is still possible.
11 to 14	Foreign accents emerge.	Some aphasic symptoms become irreversible (particularly when acquired lesion was traumatic).	An asymptote is reached on almost all parameters. Exceptions are myelinization and EEG spectrum.	Apparently firmly established, but definitive statistics not available.	Marked signs of reduction in equipotentiality.	Language markedly lateralized and internal organization established irreversibly for life. Language-free parts of brain cannot take over except where lateralization had been blocked by pathology during childhood.
Mid-teens to senium	Acquisition of second language becomes increasingly difficult.	Symptoms present after 3 to 5 months post insult are irreversible.	None.	In about 97 per cent of the population language is definitely lateralized to the left.	None for language.	

in almost all aspects of cerebral growth investigated, about 60 per cent of the mature values are attained before the onset of speech (roughly at two years of age, when speech and language become rapidly perfected), while the critical period comes to a close at a time when 100 per cent of the values are reached. This statement must not be mistaken for a demonstration of causal relationship between the variables involved. It merely suggests what structural and physiological substrates there might be that limit the capacity for cerebral organization and reorganization.

Species differ in their embryological and ontogenetic histories. Brain-maturation curves of *Homo sapiens* are different from those of other primates. Man's brain matures much more slowly, and there is evidence that the difference is not merely one of a stretched time scale but that there are intrinsic differences. Thus man is not born as a fetalized version of other primates; the developmental events in his natural history are *sui generis.* The hypothesis is advanced that the capacity for language acquisition is intimately related to man's peculiar maturational history and the unique degree of lateralization of function. Table 3 presents the argument in tabular form.

ON COGNITIVE GROWTH II*

Jerome S. Bruner

In this discussion we take the view that symbolic representation stems from a form of primitive and innate symbolic activity that, through acculturation, gradually became specialized into different systems. The most specialized "natural" system of symbolic activity is, of course, language. But images, as we have already noted, can be infused with the properties of symbolic functioning, as can tool-using involving action. It can also be said that skilled motor activity can be shaped to the requirements of such a system, as in the articulatory aspects of speech, in the eye movements of readers, and the like.

It is a legend that during the fourth quarter of the nineteenth century the *Cercle linguistique de Paris* forbade the presentation of papers about the origins of language. Today it is a topic that is coming back into vogue, stimulated by inquiries into language universals (e.g., Greenberg, 1963). In trying to understand the origin of language not as a phylogenetic

* From Jerome S. Bruner, "On Cognitive Growth," *Studies in Cognitive Growth*. New York: John Wiley & Sons, Inc. Pp. 30–67.

matter but as an ontogenetic one, one might follow a similar line and examine the universal properties of earliest infant language, regardless of what language community the child belongs to. Presumably there will be certain features of this early language that are general enough to suggest something about the basic or stock phenomenon of symbolic activity before it becomes specialized into language. One could also examine in the same way the ontogenesis of such symbolically relevant activity as tool-using with the same objective in mind, but unfortunately we can find no scientific literature that throws light on the subject.

If in the following paragraphs it may at times appear as if our search is for the nature of *language* growth, let the reader be assured that we are searching instead for the nature of the protosymbolic activity that supports language and all other forms of symbolization. We shall set down the properties of early language as "instances" and then try to attain the concept. We begin with reference.

Symbol Reference

The idea that there is a name that goes with things and that the name is arbitrary is generally taken as the essence of symbolism, and indeed White (1949) makes such "symboling" the unique basis of all human behavior. One cannot tell what a symbol stands for by sensing it: the color appropriate to mourning may be yellow, black, white, or irrelevant, depending upon what has been decided. It is apparent that a fully developed use of symbolic reference in this sense is not immediately available to the child who begins to talk. For one thing, the child first learns words as signs rather than as symbols, standing for a thing present before him (see Brown and Berko, 1960) and conceives of the word rather as an aspect of the thing. As Vygotsky puts it (1962, p. 129):

> The word, to the child, is an integral part of the object it denotes. Such a conception seems to be characteristic of primitive linguistic consciousness. We all know the old story about the rustic who said he wasn't surprised that savants with all their instruments could figure out the size of the stars and their course—what baffled him was how they found out their names. Simple experiments show that preschool children "explain" the names of objects by their attributes. According to them, an animal is called a "cow" because it has horns, "calf" because its horns are still small, "dog" because it is small and has no horns; an object is called "car" because it is not an animal. When asked whether one could interchange the names of objects, for instance, call a cow "ink" and ink "cow," children will answer no, "because ink is used for writing and the cow gives milk." An exchange of names would mean an exchange of characteristic features, so inseparable is the connection between them in the child's mind . . . . We can see how difficult it is for children to separate the name of an object from its attributes, which cling to the name when it is transferred, like possessions following their owner.

Yet, for all that, the young child, like the young Helen Keller, comes early to the notion that things have names that they go by and that the name cannot be found by inspecting its referent.

We should like to propose, in line with a suggestion by McNeill (1966a), that the learning of reference, that is, the "semantic function" of language, is a slow process (unlike the learning of syntax, discussed below) largely because it involves a basic type of cumulative process that has been very interestingly described by Katz and Fodor (1963). It involves learning the semantic markers of a word—the senses that it has or the contexts into which it fits, much as the word *build* fits into contexts having to do with construction, with the shape of a body, with piling blocks on top of one another, and so on. In learning how to speak or to recognize whether what he hears is semantically sensible or anomalous, the child is learning to match the semantic markers of some words he has learned to the selection requirements of others that he is using in a sentence. It is not surprising that young children have more difficulty in distinguishing between a semantically anomalous sentence like "The flower ate the cheese" and one that is not anomalous like "The mouse ate the cheese." This conception of semantic learning is a far cry from the "learning by pointing" that is a stock in the trade of the older textbooks. The "original word game" may be characteristic of earliest semantic learning, but it is less and less frequent over the first years of language learning. It takes no elaborate investigation to know that we do not have to take five-year-olds into the presence of each "thing" for which they learn a word. Rather, their "What's that?" is increasingly directed to unfamiliar words that are being used and to the unfamiliar senses in which familiar words are being used. It is probably this feature of semantic learning that puts the children with strong imagery in Kuhlman's experiment (1960) at a disadvantage, for to learn the senses or semantic markers of words is to learn the constraints on their conceptual range—an intellectual task, not a perceptual one. Leopold (1949) in his classic observations of his own children also comments on this gradual progression "from coarse to fine" semantic distinction, and we shall return later to the point in discussing the relation of semantic and syntactic learning in the early acquisition of speech.

Categoriality

It is universally the case that, as Roger Brown (1956) has most persuasively put it, the child's *use* of language is categorial. Words cover classes of things, and these classes are, on close examination, found to be rule-governed so that new members can be added. Indeed, even morphological and syntactical rules, of which more is said in a moment, have a categorial use very nearly from the outset. In short, the type-token distinction is universal in all language, including early language.

Grammaticalness

All human language, once it passes the first stage of the comprehensive "one-word" utterance, the holophrase, is characterized by grammar. This is not the place to set forth a capsule characterization of what is meant by grammar, and indeed linguists are not in complete agreement as to how to characterize it. All languages, we can say, have a base grammar or structure that provides for at least three fundamental properties to sentences: *verb-object, subject-predicate,* and *modification.* There are no human languages whose sentences do not contain rules for these three basic sentential structures, and there are no nonhuman languages that have them. Languages use many devices, such as *order,* various kinds of *markers,* such as suffixes and function words, and different forms of prosody, as in intonation; but all reflect or respond to the basic grammar. One further universal of grammar, of the same fundamental status as the three basic relations of subject-predicate, verb-object, and modification, is *transformation.* All languages have rules for "rewriting" sentences: for example, rules for rewriting passive, interrogative, or negative sentences into basic-structures-plus transformation; that is, *The dog bit the boy* can be written with any one, two, or three of the optional transformations noted directly above—with all three of them yielding. *Was the boy not bitten by the dog?*

One can examine the nature of language more closely still by considering the classes of morphemes it contains, the minimum units that carry meaning, and the privileges of occurrence in sentences that they have. One can examine more carefully the small stock of phonemes in a language, combinations of which permit the production of a large stock of morphemes. Just as a morpheme is the minimum unit that carries meaning, a phoneme is the minimum sound element, a change in which alters meaning in a morpheme (or word) that contains it. These are matters which, though of great interest, need not concern us here.[1]

In general, when we speak of the grammar of a language we mean the set of rules that will generate any or all permissible utterances in that language and none that is impermissible. When a person speaks a language, he "knows" these rules in some fashion, though he cannot (like a linguist) recite them to you. In studying a child's grammar, we attempt to infer the rules that govern his utterances: we write a grammar for him much as we would write a grammar for some remote tribe. The difference is that, in the case of the child's language, we are interested in how he "moves toward" the model grammar of the adult linguistic community; therefore we study not only the differences between his grammar and

[1] The reader will quite properly sense that the present account is strongly influenced by the views of Noam Chomsky (1957, 1965) and George Miller (1965).

adult grammar but also the transitions between his grammar at one age and at another.

The rules of grammar are, in effect, combinatory and "productive." In the very nature of natural language these rules cannot be in the form of a device for "generating utterances from left to right," for all grammars ever studied have rules that permit imbeddings and forms of dependency that cannot be generated in this serial way. Chomsky (1957) illustrates this universal property with a set of declarative sentences, S_1, S_2, S_3 in some natural language, say English. It is possible in English or any language to generate these sentences:

(a) If S_1, then S_2.
(b) Either S_3, or S_4.
(c) The man who said that S_5 is arriving today.

In all three instances a simple serial rule of order is plainly inadequate. There must be a basic grammar that can govern the over-all sentence rather than only what word can follow a preceding word. This matter, perhaps a little abstruse on first encounter, becomes important shortly.

Let it be clear at the outset that the kind of grammar we are discussing differs strikingly from other forms of communication that come before it in the child's life. Mere gesture or vocalization, for example, is agrammatical. Dorothea McCarthy (1954, p. 521) in her exhaustive review of the literature on speech development has this to say:

> It is quite generally agreed that the child understands gestures before he understands words and, in fact, that he uses gestures himself long before he uses language proper. He looks for objects he has dropped, he reaches for objects, etc., long before he can ask for them. These and other overt bodily movements which are used as means of early expression and communication are often accompanied by early vocalizations. It has been claimed that words constitute substitutes for actual gross motor activity.

Such substitution of vocalization for gesture is often interpreted as the route from prelinguistic to properly linguistic or symbolical reference. This is the view of the Sterns in their *Kindersprache* (1907): first there is gesture, then gesture and vocalization, and then gesture drops out and vocalization carries the main burden. Such a statement fails to notice the enormous discontinuity between mere "gestural vocalizations" and the complex and the highly constrained grammar of language.

As Meumann originally suggested a half century ago (1908), the first functions of vocalization are affective and volitional, expressive of inner states, on the one hand, and of demands on the environment, on the other. How does the child make the step toward a properly combinatory grammar capable of productivity? There is consensus in the literature that first sentential utterances usually contain one common restricted morpheme and a second open or free-varying one. Such sentences as *that truck, that baby,* and *where dolly, where truck* illustrate this. The unique thing about this first step toward sentence formation is a forming rule that permits the generation of new sentences, like *where baby* and *that mommy,* when *where* had previously been used only with *dolly* and *truck* and *that* with *truck* and *baby.* The bound or pivotal forms are combined with other words in a manner that is strikingly rule-bound. Braine (1963b, p. 13)[2] gives the most succinct account of the process.

> The simplest account of the phenomena of the first phase of development seems to be as follows: out of the moderately large vocabulary at his disposal the child learns, one at a time, that each of a small number of words belongs in a particular position in an utterance. He therefore places them there, and, since he has not learned anything else about what goes where in an utterance, the complementary position is taken by any single-word utterance in his vocabulary, the choice determined only by the physical and social stimuli that elicit the utterance. As a consequence of this learning, the word combinations that are uttered have a characteristic structure containing two parts of speech. One part of speech, here called pivot, comprises the small number of words whose position has been learned. The other, here called X-class, is a part of speech mainly in a residual sense, and consists of the entire vocabulary, except for some of the pivots. During this first phase the language grows structurally by the formation of new pivot words, i.e., by the child learning the position of new words. The language grows in vocabulary by adding to the X-class.

To illustrate the rule-bound nature of the combinations, Braine considers the problem of rare word combinations not likely to be found in parental speech (p. 10).

> An objection which has been raised against the assumption that any pivot can occur with any x-word is that it puts into the children's mouths some implausible expressions which seem highly foreign to English. For example, the formulae would allow Gregory to generate *more hot, big dirty, allgone hot*; Andrew to say *see read, other fix, I short,* and Steven to say *that do, there up, high do.* This objection is sometimes based on the idea that the children's utterances are a recall or delayed imitation of things they heard adults say. As against this, there are a number of expressions in the corpora which are sufficiently strange to render it most unlikely that the children had heard them, e.g., *see cold, byebye dirty, allgone lettuce, no down, more high, want do pon* ("put on"), *there high* . . . . Moreover,

[2] M. D. S. Braine, "The Ontogeny of English Phrase Structure: The First Phase," *Language,* 39 (1963), pp. 1–31.

> several "strange" combinations, similar or identical to utterances generated by the formulae, appear in the fifth and sixth months; [after the beginning of the two-word sentences] examples are *more wet, allgone sticky* (Gregory after washing his hands) . . . etc. Manifestly, the strangeness of an utterance is no criterion of its grammaticality at this age.

How rapidly the child avails himself of this new system of making combinations in his utterances can be grasped from sampling the record of one child studied by Braine during the first six months after the appearance of the first two-word utterance: for each successive month the count was 14, 24, 54, 89, 350, 1400, 2500 plus.

Brown and Fraser (1964, p. 79) conclude their report on the acquisition of grammar with the following suggestion:

> For the present, then, we are working with the hypothesis that child speech is a systematic reduction of adult speech largely accomplished by omitting function words that carry little information. From this corpus of reduced sentences we suggest that the child induces general rules which govern the construction of new utterances. As a child becomes capable (through maturation and the learning of frequent sequences) of mastering more and more of the detail of adult speech, his original rules will have to be revised and supplemented. As the generative grammar grows more complicated and more like the adult grammar, the child's speech will become capable of expressing a greater variety of meanings.

None of the above accounts for the *first* appearance of grammar, why for example, the child shifts to the $P(x)$ form. We shall return to the point later. For the while it suffices to say that grammar does come forth and is typical of the child's utterances from about the second year on. It is an obvious point, but a powerfully important one. What is less obvious is that from the very first appearance of grammar the child seems to be master of the use of categories (as in his use of pivot and open classes) and of hierarchies. The latter are found in the form of "nested constructions" in which a single class is replaced by a sequence of instances that are, in effect, an unfolding of the original class. Ervin-Trip and Slobin (1966) comment in their review of the literature on language acquisition that a "layering of constituents in adult phrase structure is found from the very beginning of grammar." We shall have occasion very shortly to return to these two features of grammaticality.

Effective Productivity

As we have already seen, the rule-bound nature of grammar assures that the child (once he has acquired it) will be able to produce an endless number of utterances of a syntactically legitimate nature and will, moreover, produces none that is not. In short, he learns rules. We know from the engaging study by Ruth Weir (1962) that children practice pushing these rules to the limit. But the critical point for the student of symbolic representation is that these powerful productive rules of grammar are linked to the semantic function as well—to the "real world"; that is to say, having translated or encoded a set of events into a rule-bound symbolic system, a human being is then able to transform that representation into an altered version that may but does not necessarily correspond to some possible set of events. It is this form of effective productivity that makes symbolic representation such a powerful tool for thinking or problem solving: the range it permits for experimental alteration of the environment without having, so to speak, to raise a finger by way of trial and error or to picture anything in the mind's eye by imagery. "What if there were never any apples?" a four-year-old asked upon finishing one with gusto!

But it is apparent that much language is used without seeming awareness of the semantic implications of what has been said or of what our transformations imply for the world to which our words refer (e.g., Inhelder and Piaget, 1964); that is, although words can potentially be used to turn reality on its head for hypothetical purposes they are not often used in that way. This fact has been treated in various ways—this failure of the Whorfian hypothesis to predict as much as it should about a speaker's world view on the basis of the language he speaks. The popular approach is to follow the lead of Vygotsky (1962) and to assume that language becomes part of our apparatus of thinking when it becomes "interiorized" or "internalized," the assumption being that language starts as dialogue and then becomes monologue and finally "inner speech." It is a view whose chief flaw is that one can say too little about the nature of inner speech. In any case, it may be that one can avoid this problem of nonspecificity with respect to inner speech by examining how language becomes related to reality.

Now, to begin with, the very use of language presupposes certain underlying cognitive processes required for its use. We have just discussed some of these processes in considering the universal grammaticality of language. It seems as if these concepts are first used and perfected in the sphere of language and are only gradually transferred to thinking in general, and often not well transferred. How a child comes to transfer such formidable concepts as hierarchy and transformation from the linguistic to the nonlinguistic sphere is obscure, but a conjecture is in order. Brown and Bellugi (1964b) speak of a contingent cycle between the child and the tutor in original language learning. Recall that the child makes an utterance in his own particularly grammatical form: the adult "tutor" idealizes and expands the child's utterance into a more model grammatical form and the child matches his next utterance selectively to the adult model (if he is "ready"). Brown's observations are confined principally to the second and third years of age. Some such process as this goes on throughout childhood—with parents, with teachers, with older children. This exchange is as much a matter of learning to organize one's thoughts in a certain way as it is of learning the rules of *grammar.* Brown (1966) provides an example from the speech of Eve, one of his young subjects in her fourth year of life:

> Eve: *I sit table.*
> Eve: *I get big.*
> Mother: *That's right, when you get big you can sit at the table.*
> Now we look back and find that there has always been a latent structure in the order of the simple sentences the mother produced, seemingly with a more logical texture in the two more educated families (conditional, causal, disjunctive, and a more concatenative structure in the one less educated family).

It is at this point that McNeill's (1965)[3] discussion is particularly useful. Recall that he comments upon the slow accretion of semantic markers in the child's grasp of the "meanings" of a language—the allowable contexts into which words are to fit. It is worth pursuing his argument in more detail, from the point at which the child first uses holophrases and then shifts to words.

> It is clear that children have some kind of semantic system at a very early point in their linguistic development. We have already mentioned that children first seem to use words holophrastically, a phenomenon that suggests that the *earliest semantic system consists of a dictionary in which single words are paired with several complete sentence-interpretations.* A holophrastic dictionary is burdensome for a child's memory and susceptible to ambiguity, two conditions that might lead to the creation of a sentence dictionary. Again, because of memory load, *a sentence dictionary is itself abandoned,* this time for the ultimate solution, a word dictionary. Each of these transitions effects a re-working of a child's semantic system. Of the two, however, the second transition is by far the most significant, and it is from this point that we can date the rudiments of a system that is basically similar to adult competence. In the case of the first change, from a holophrastic to a sentence dictionary, a child continues storing undifferentiated semantic information, in that the definition of one sentence is not related to the definition of any other. The transition from a sentence dictionary to a word dictionary, on the other hand, introduces a fundamental change in the format of the dictionary entries themselves, for in compiling a word dictionary, a child must begin to build up a system of semantic markers. The

[3] D. McNeill, Development of the Semantic System. Unpublished paper, Center for Cognitive Studies, Harvard University, 1965. Pp. 1–3.

evidence is that the accretion of semantic markers is, in contrast with the acquisition of syntactic competence, a slow process that is not completed until well into school age.

We can confidently place the first effort to compile a word dictionary as occurring not earlier than the use of base-structure rules in a child's grammar. *It is very difficult to conceive of a word dictionary that does not receive input from some sort of syntactic component.* Without such input, a word dictionary would constitute a retreat to a point even more primitive than the original holophrastic dictionary. A word dictionary without syntax would result in a loss of power to encode information that previously had been encoded, viz., sentence meaning, and the cause of the transition to a word dictionary, we assume, is the need to retain an ability to encode sentence meanings while reducing the load on memory.

It may well be that language provides a kind of temptation to form concepts of objects and events that have a structure comparable to those contained in words, form classes, and sentences—as Brown (1956) and Werner and Kaplan (1950) have suggested. What it takes to bring the two into correspondence is a topic we must postpone until a later section. These concepts refer to the properties of communication that are characteristic of human language, though they may also be characteristic of nonhuman communication (see Hockett, 1960).

Design Features

Some aspects of early language are so specifically specialized for the transmission side of communication that they need not concern us in our search for "protosymbolic" activity. There are others that, although "transmission specific," may nonetheless suggest some general properties of human symbolic functioning. Consider these briefly.

Hockett's (1960) list of design features of communication provides a guide for us. That all human spoken language *uses the vocal-auditory channel* probably tells very little. As Sapir (1921, p. 42) said more than forty years ago, "The mere phonetic framework of speech does not constitute the inner fact of language." Yet there may be some special significance in the fact that to produce language we must develop articulatory motor responses that are governed by rules as subtle and productive as those governing distinctive features, phonemes, and allophones. At very least, vocal production may provide a prototype of all "programmatic" motor skills, and one cannot help wondering whether at some future time in human history the skill of the human hand may be directed by some form of combinatorial program such as that involved in speech production. The performing musician may be a precursor. Closely related to vocal-auditory channeling are the *interchangeability* of heard and spoken signals in human speech and the *feedback* of the human voice to the speaker's ears. So too is the *specialization* of human speech in the sound channel—so that the sounds of speech are produced for communication alone and not as a by-product of some other activity. The *broadcast transmission* of speech and its *directional reception* are part of the same complex. Their significance concerning the general nature of protosymbolic activity is obscure, save in the sense that, like our first example, they connote a specialized development of skills.

We should probably take a moment to scrutinize the feature of *rapid fading* in human speech, for though it may not reflect anything "primitive" about a symbolic system it does suggest that symbolism is an extraordinarily swift system, particularly in contrast to the two rather sluggish modes of representation discussed earlier: action with its habitual sequences and imagery with its lag and its slow transformability. It may well be that some of the startlingly abrupt spurts in growth reported in later chapters are a function of the shift from slow to fast functioning. The processing time required to solve problems by ikonic or enactive means is prohibitively long for a child whose attention span is short.

Hockett's list then goes on to several features of greater importance for inferring the basic properties of symbolism in man. Several have already been mentioned in preceding pages: *semanticity*, the *arbitrariness* of human symbols and their lack of any resemblance to what they specify; the *productivity* of utterance made possible by grammatical rules; and the *traditional transmission* of the language rather than its environment-free transmission through the genes. This leaves three critical features still to be examined for their possible significance: *discreteness, displacement,* and *duality of patterning.*

Discreteness in language refers to the fact that at the sound level, as at the level of meaning, the material of human language is discontinuous: there is no intermediate step between *bin* and *pin* that produces a word: /*b*/ and /*p*/ are discontinuous phonemes, and, should one voice a word that uses a sound midway between, the hearer will interpret it as one phoneme or the other. So too with words or morphemes; they are neither organized by continuum with a range from *hat* to *helmet,* nor are they form classes, such that one goes imperceptibly from nominals (or "nouns") to, say, functors such as *to, by,* or *at.* What this imposes on the speaker of human language is the requirement that he analyze the domain of sound and of sense into discontinuous components that can then be constituted into a message. The rule of discreteness in symbolic representation contrasts sharply with the rule of representation by perceptual similarity: the analogue rule of ikonic representation. Analysis and synthesis are literally *forced* on anyone who would speak human language. Language, then, breaks up the natural unity of the perceptual world—or at least imposes another structure on it.

Displacement as a design feature of language refers to the fact that one can represent something in language, though it may be remote from the place of speaking (or writing) in place and time. Save for the odd and fascinating cul de sac of bee-dancing, one finds this feature in no other part of the animal kingdom save man. What is interesting about displacement is that it is only gradually acquired by young speakers who, as we have seen, begin their speaking careers by referring only to those things that are present. More interesting still is the fact that, in problem-solving behavior dominated by ikonic representation, the principal feature seems to be "ostensiveness" rather than "displacement"; that is to say, the behavior is dominated by the requirement that all features of the task to be dealt with be present, present even to be pointed at. As we shall see repeatedly in experiments reported in later chapters, there is a critical point at which the child is able to go beyond the immediate situation of the problem-solving task to deal with features that are "remote in space and time"; for example, in playing a game of "Twenty Questions." It may well be, then, that there is some primitive respect in which a capacity for symbolic activity permits a human being to get this "distance" from a task. Indeed, it may be that without such distance it would be impossible to develop the analytic approach necessary for decomposing tasks into the discrete features in order to encode them better in the categories to which a language system can be applied.

With respect to *duality of patterning*, again we are dealing with a feature that is unique to human communication. Hockett (1960, pp. 4–6) describes the feature succinctly:

> The meaningful elements in any language—"words" in everyday parlance, "morphemes" to the linguist—constitute an enormous stock. Yet they are represented by small arrangements of a relatively very small stock of distinguishable sounds [phonemes] which are in themselves wholly meaningless. This "duality of patterning" is illustrated by the English words "cat," "tack," and "act." They are totally distinct as to meaning, and yet are composed of just three basic meaningless sounds in different permutations.

Little indeed is known about the significance in the broader cognitive sense of a system of this sort for human symbolic activity. What it suggests is again an enormous rule-governed capacity for analysis-and-synthesis. Duality of pattern, or the capacity that makes possible our mastery of it, probably also makes possible our capacity to use other "artificial" systems for analysis that have many of the same properties as language—notably mathematics and logic.

After this survey, what shall we make of the basic nature of symbolic activity in humans? We have looked at what we choose to call its "specialized" manifestation in spoken language, in the conviction that, since there is a gap between the child's competence with language *per se* and his competence with "reality," the former cannot tell us the whole story about his symbolic representation of the latter. Yet we do well to try to sum up the inherent properties presupposed by the use of language.

We may begin with Roman Jakobson's account of the fundamentals of language. He says (1956, p. 58): "Speech implies a selection of certain linguistic entities and their combination into linguistic units of a higher degree of complexity." With respect to these two fundamentals, selection and combination, Jakobson then goes on to remark (p. 60):

> Any linguistic sign involves two modes of arrangement. (1) *Combination.* Any sign is made up of constituent signs and/or occurs only in combination with other signs. This means that any linguistic unit at one and the same time serves as a context for simpler units and/or finds its own context in a more complex linguistic unit. Hence any actual grouping of linguistic units binds them into a superior unit: combination and contexture are two faces of the same operation. (2) *Selection.* A selection between alternatives implies the possibility of substituting one for the other equivalent to the former in one respect, and different from it in another. Actually, selection and substitution are two faces of the same operation. The fundamental role which these two operations play in language was clearly recognized by Ferdinand de Saussure [1916].

To the psychologist, the two properties in question can be readily translated into *category* and *hierarchy*, the latter providing a system for nesting the former (See Bruner, Goodnow, and Austin [1956]). Let us assume for the time being that these are two basic features of symbolic representation as well as of language.

To stop at category and hierarchy in describing the underlying structure of language is to leave out of account the nature of sentences and grammar taken in the fine. We have already commented on three universal properties of base-structure grammar, properties as characteristic of the child's grammar as of the adult's: subject-predicate relations, verb-object relations, and modification. Each of these presupposes an underlying logical form that can be simply stated. The subject-predicate relation is a linguistic expression of the argument of a function, "x is a function of y," where x is the subject and y the predicate. For "S-P" of the grammatical jargon, we can write $x = f(y)$ in logical jargon. Where the verb-object relation is concerned, it is a linguistic rendering of cause and effect. Modification in its turn is an instance of the intersect of classes—*green hat* is the intersect of *green* things and *hats.* The sentence, *The man wore the green hat,* contains the underlying relations, then, of a man wearing something (or x being a function of y), a hat being acted upon (being worn, which is causing it to be worn), and a certain kind of hat at that (modification by green). And so we might add to our list of language "fundamentals," to go on with Jakobson's term, the three symbolic operations of *function, causation,* and *intersection* (or logical addition of classes, as it is often called).

Now reconsider effective productivity. If linguistic categories organized in hierarchies are to have relevance to the "real world," then experience itself must be organized into hierarchically organized categories. We know that the child's language is organized in this way, and we know equally well from experiments that his experience is not. I am prepared to believe that in the linguistic domain the capacities for categorization and hierarchical organization are innate and so, too, are predication, causation, and modification; I will attempt to justify this belief later, but it is also plain that the child's experience is not organized by these principles. Here I find myself in fundamental agreement with Piaget's discussion (1961) in his difficult but rewarding book, *Les Mécanismes Perceptifs.* In discussing the difference between perception and intelligent thought he comments, "La perception primaire ignore l'abstraction." It is precisely the initial unanalyzability of perception that strikes Piaget: "Lorsque, par example, devant les figures classique de Müller-Lyer, le sujet est prié de comparer les deux médianes horizontales, il n'y a pas de problème déductif, puisque la question est seulement de comparer pour 'voir' " (p. 358).[4] This immediacy and nonabstract quality of perception is precisely what makes difficult the analysis-and-synthesis that is necessary for adult thought to be brought to the task—adult thought that has been organized to correspond to the properties of symbolic activity.

I would argue that language itself is not what is "imposed" on experience—as already suggested in my rejection of Vygotsky's idea that language is internalized and becomes inner speech, which is tantamount to thought. Rather, language comes from the same basic root out of which symbolically organized experience grows. I tend to think of symbolic activity of some basic or primitive type that finds its first and fullest expression in language, then in tool-using and finally in the organizing of experience. It is by the interaction of language and the barely symbolically organized experience of the child of two or three that language gradually finds its way into the realm of experience.

Let us assume that at the outset, and concurrently with the earliest form of vocal or bodily gestural sign, the child has developed some working conception of identity: that an object now encountered is the same object previously encountered. Indeed, Bower's (1965b) finding, using the method of moving objects behind a screen and then making them reappear in an altered form, even suggests that a primitive grasp of

[4] "When, for example, a subject is presented with a pair of classical Müller-Lyer figures and asked to compare their horizontal lines, it is not like a deductive problem, since the question is only one of comparing the two visually."

identity may be present in the first month of life, if not at birth. Primitive identity is surely, then, a sufficient beginning for semanticity—although it lacks the arbitrary quality of symbolic reference in full. It can be argued that it is not until the child has, in Piaget's (1954) full sense, achieved the *concept* of an object that there is naming, even by the beginning of the holophrase; for the use of an arbitrary name or holophrase requires some freeing of the object from its immediate context of action.

As McNeill (1965) properly points out, the shift from holophrase to words is inconceivable without a syntactic component that would give words a productive benefit over either holophrases or nondecomposable "sentences" in the child's repertory. That first syntax is simple, to be sure, but it is clearly a step beyond what went before—as for example in the pivotal grammar $P(x)$, already discussed. What nourishes the emergence, at just about eighteen to twenty-four months, of a syntactical system in language that contains our basic properties is anything but apparent. Is it the new achievement of a certain amount of freedom from enactive representation that permits this quick unfolding of syntactical learning? The suggestion comes indirectly from Lashley's paper on serial order (1951) already considered in . . . [an earlier chapter of Bruner's book]. Lashley comments that serially ordered behavior, like grammatical language, requires that there be some atemporal principle of organization: "Syntax is not inherent in the words employed or in the idea to be expressed. It is a generalized pattern imposed upon the specific acts as they occur." Or, more specifically, "The facility with which different word orders may be utilized to express the same thought thus is further evidence that the temporal integration is not inherent in the preliminary organization of the idea."

We suggest that the strong motoric element in enactive representation that prevails during the child's first year of life—described so skillfully by Piaget (1954) and underlined by the work of Held and his associates (1963, 1965)—probably interferes with the development of this atemporal schema. Indeed, the speculation fits in well with the observations of Kagan (1966) that hypermotoricity interferes with "reflectiveness" in young children, seeming to delay language development and the development of linguistic ability. When the child cannot inhibit the motoric acting out of responses, he cannot organize a central pattern sufficient for language—or at least for any language more complicated than the immediate holophrase. It is important to bear in mind that there is a long period of delay between the appearance of the first holophrase and the first syntactically ordered sentence—in the case of Leopold's (1949) daughter, for example, from the ninth month to the twentieth. It may be a period for steadying and overlearning.

The maturing of the syntactical system, as we know, takes place by a process of reciprocal interaction between child and tutor—the expansion-idealization cycle to which Brown and Bellugi (1964b) refers. Whatever the mechanism is that operates in this learning—it is certainly *not* a form of simple imitation, as we know from the work of Brown, Fraser, and Bellugi (1964)—is not clear. Brown uses the expression "Original Word Game" for semantic learning, and it is plain that this is a different activity from the "Original Sentence Game," in which the child is learning some sort of base-structure grammar. It is rather intriguing to consider that while the child is learning the ostensive referent of words and their semantic markers the very same words are already embedded in some form of syntactic hierarchy. In short (and we shall encounter the fact repeatedly in the studies reported in later chapters), the child is learning to use his words and their semantic markers for the picturable or ikonic aspects of his world, but the words themselves are enmeshed in a highly abstract and hierarchical system of categories used formally to signal causation, predication, and modification through sentences that the child can "rewrite" according to transformational rules that he early masters (see Brown, 1966 and Slobin, 1963). However, as we have said, these powerful, only potentially semantic features of language are rarely used by the child for giving structure to his experience. He can neither use the superordinate rule of categorization consistently . . . nor can he organize what he knows in a hierarchical organization. . . . He has not, to use Herbert Simon's (1962) phrase, mastered the "architecture of complexity" for things, but only for words; and, as with Simon's mythical watchmakers, Hora and Tempus, he is never finished with his problem-solving tasks, though he is quite capable of finishing his sentences.[5] Nor can he deal adequately with causation (Piaget, 1930), functions (In-

[5] Simon's fable (1962, p. 470) is worth repeating: "There once were two watchmakers named Hora and Tempus who manufactured very fine watches. Both of them were highly regarded and the phones in their workshops rang frequently—new customers were constantly calling them. However, Hora prospered while Tempus became poorer and poorer and finally lost his shop. What was the reason?

"The watches the men made consisted of about 1000 parts each. Tempus had so constructed his that if he had one partly assembled and had to put it down—to answer the phone, say—it immediately fell to pieces and had to be reassembled from the elements. The better the customers liked his watches, the more they phoned him . . . .

"The watches that Hora made were no less complex than those of Tempus. But he had designed them so that he could put together subassemblies of about ten elements each. Ten of these subassemblies again, could be put together in a large subassembly; and a system of ten of the latter subassemblies constituted the whole watch. Hence, when Hora had to put down a partly assembled watch in order to answer the phone, he lost only a small part of his work, and he assembled his watches in only a fraction of the man-hours it took Tempus."

The five-year-old, very strikingly, goes about his manipulation of the real world in a fashion akin to that of Tempus, distraction being his undoing. But his language behavior prospers, like Hora.

helder and Piaget, 1958), nor modification (Piaget and Inhelder, 1962; Inhelder and Piaget, 1964).

Then, if the child lives in an advanced society such as our own, he becomes "operational" (to use the Genevan term for thinking symbolically), and by age five, six, or seven, given cultural supports he is able to apply the fundamental rules of category, hierarchy, function, and so forth, to the world as well as to his words. Let it be explicit, however, that if he is growing up in a native village of Senegal . . . among native Eskimos . . . , or in a rural *mestizo* village in Mexico . . . he may not achieve this "capacity." Instead, he may remain at a level of manipulation of the environment that is concretely ikonic and strikingly lacking in symbolic structures—though his language may be stunningly exquisite in these regards. (The reader is reminded, in Sapir's words [1921] that "we know of no people that is not possessed of a fully developed language. The lowliest South African Bushman speaks in the forms of a rich symbolic system that is in essence perfectly comparable to the speech of the cultivated Frenchman.") But Miller and Chomsky (1963, p. 488), after a searching analysis of the formal properties of language, conclude:

> An organism that is intricate and highly structured enough to perform the operations that we have seen to be involved in linguistic communication does not suddenly lose its intricacy and structure when it turns to non-linguistic activities. In particular, such an organism can form verbal plans to guide many of its non-verbal acts. The verbal machinery turns out sentences—and for civilized men, sentences have a compelling power to control both thought and action.

We shall consider next how sentences might come to have this power.

To sum up our discussion of symbolic representation, symbolic activity stems from some primitive or protosymbolic system that is species-specific to man. This system becomes specialized in expression in various domains of the life of a human being: in language, in tool-using, in various atemporally organized and skilled forms of serial behavior, and in the organization of experience itself. We have suggested some minimum properties of such a symbolic system: categoriality, hierarchy, predication, causation, and modification. We have suggested that *any* symbolic activity, and especially language, is logically and empirically unthinkable without these properties.

What is striking about language as one of the specialized expressions of symbolic activity is that in one of its aspects, the syntactic sphere, it reaches maturity very swiftly. The syntactical maturity of a five-year-old seems unconnected with his ability in other spheres. He can muster words and sentences with a swift and sure grasp of highly abstract rules, but he cannot, in a corresponding fashion, organize the things words and sentences "stand for." This asymmetry is reflected in the child's semantic activities, where his knowledge of the senses of words and the empirical implications of his sentence remain childish for many years, even after syntax has become fully developed.

One is thus led to believe that, in order for the child to use language as an instrument of thought, he must first bring the world of experience under the control of principles of organization that are in some degree isomorphic with the structural principles of syntax. Without special training in the symbolic *representation of experience,* the child grows to adulthood still depending in large measure on the enactive and ikonic modes of representing and organizing the world, no matter what language he speaks.

In view of the autonomy of the syntactic sphere from other modes of operating and of its partial disjunction from the semantic sphere, one is strongly tempted to give credence to the insistence of various modern writers on linguistics that language is an innate pattern, based on innate "ideas" that are gradually differentiated into the rules of grammar Chomsky 1965, Katz 1965, McNeill,1965).

Considering now the interaction of the three systems of representation, we quickly find ourselves dealing with the question of how experience is organized to correspond in some measure to the structure of language.

The Interaction of Systems

...[In Chapter 1 of Bruner's book] the distinction was drawn between "representation by" and "representation for," the former referring to the medium in which a representation is couched, the latter to the uses to which it is put; doing, sensing, or symbolizing. We have already explored briefly the various ways in which systems of representation are translatable (or partially translatable) into each other. What we must do now is examine more carefully the nature of this interaction and its history. For it must be apparent from the preceding discussion that there is much more to this matter than first meets the eye.

The first and most general issue is raised by Lashley (1951), commenting on a dilemma of memory. Much of memory, he remarks, is spatially organized. Yet for even simple reproduction to occur, it is necessary to translate the images of memory into a serial order of their reproduction (or recall). We are constantly translating atemporal images into serially organized actions—as in the diagram of the battery, lamp, and switch, should we have to use the diagram as a guide to the reconstruction of the simple circuit it portrays. Even at a more primitive level the maintenance of postural equilibrium consists in monitoring the relation between an enactive, postural system governed by the semicircular canals, and a visual system with spatial coordinates. The coordination of the two systems, visual and kinaesthetic, most surely was the condition for survival of our arboreal, primate ancestors. The system must be well organized by the time the human child is able to crawl, for by that age he already displays avoidance of the visual cliff (Walk, 1965). One is tempted to look exclusively to the phenomenon of reafference as the source of the integration of the sphere of action and of sensory perception. . . . Plainly there is reason to think that the feedback between the two systems provides the basis for much of sensorimotor integration. But what of the integration of those systems with a symbolic system?

Before launching into that issue, we would do well to consider briefly the relations that can exist between two systems of representation, using the relation of the visual system and action as a model. In abstract terms, there are three basic ways in which two systems can relate to each other: by *matching*, by *mismatching*, or by *independence* of each other. In visual and labyrinthine definitions of space, for example, there is a kind of point-to-point correspondence between the systems so that independence is not possible. Where the two systems match, the organism goes on with its ordinary operations without any special problem solving. When the two systems mismatch, as in the experiments of Witkin and his collaborators (1962), one or the other is suppressed or some sort of correction is made in the schema that coordinates the two. In short, mismatch creates the kind of "trouble" that requires handling. In the Witkin experiments it is plain that before children are in adolescence they have come to deal with "mismatch troubles" either by developing a general preference for the visual representation of space or for the kinaesthetic-labyrinthine. What is remarkable about the long series of experiments carried out by the Witkin group (using the rod-and-frame procedure, tilting rooms, etc.) is not only that there is considerable consistency in these preferences for one or the other modality but also that many other features of behavior become correlated with the preference. This type of clustering suggests that there is a more general schema of adaptation than relates specifically to the preference for one or another medium of representation.

The existence of such a general schema suggests that when, in the history of the organism, mismatches occur between two systems there is a rather widely ramifying problem solving that is set up. There develops a range of "devices" and "preparations" that have to do with dealing with mismatch on the next encounter: trust in one's body image or strong reliance on a visual framework and the world of objects, etc.

Consider now the question of independence, match, and mismatch between two systems in which one system is symbolic. We have already commented on the fact that the syntax of language is mastered by the young child before he seems able to use it to organize his experience or his actions. It is as if independence prevailed in a major degree, but as soon as there develops an organization of experience that is in some sense capable of "matching" the properties inherent in language the state of independence between the two systems recedes. Consider this issue in connection with two quite different matters: the first, Sapir's (1921) analysis of imagery and language, and the second, the series of experiments stimulated by the early experiment of Carmichael, Hogan, and Walter (1932).

Sapir says (p. 12):[6]

> The world of our experiences must be enormously simplified and generalized before it is possible to make a symbolic inventory of all our experiences of things and relations and this inventory is imperative before we can convey ideas. The elements of language, the symbols that ticket off experience, must therefore be associated with whole groups, delimited classes of experience, rather than with the simple experiences themselves. Only so is communication possible, for the single experience lodges in an individual consciousness and is, strictly speaking, incommunicable.

Pursuing the thought further, he says (p. 14):

> We have seen that the typical linguistic element labels a concept. It does not follow from this that the use to which language is put is always or even mainly conceptual. We are not in ordinary life so much concerned with concepts as with concrete particularities and specific relations. When I say, for instance, "I had a good breakfast this morning," it is clear that I am not in the throes of laborious thought, that what I have to transmit is hardly more than a pleasureable memory symbolically rendered in the grooves of habitual expression. Each element in the sentence defines a separate concept or conceptual relation or both combined, *but the sentence as a whole has no conceptual significance whatever.* It is somewhat as though a dynamo capable of generating enough power to run an elevator were operated almost exclusively to feed an electric doorbell. The parallel is more suggestive than at first sight appears. Language may be looked upon as an instrument capable of running a gamut of psychic uses. Its flow not only parallels that of the inner content of consciousness, but parallels it on different levels, ranging from the state of mind that is dominated by particular images to that in which abstract concepts and their relations are alone at the focus of attention and which is ordinarily termed reasoning.

[6] From *Language* by Edward Sapir, copyright 1921, by Harcourt Brace Jovanovich, Inc.; copyright, 1949, by Jean V. Sapir. Reprinted by permission of the publishers.

> *Thus the outward form only of language is constant; its inner meaning, its psychic value or intensity varies* freely with attention or the selective interest of the mind, also, needless to say, *with the mind's general development.* [Italics ours.]

The reader will perhaps bear with this lengthy quotation from Sapir, for in fact we could not put it better. He notes very properly that, although at the outset language may be used for labeling, it is not until there is a certain level in the development of mind that it can be fitted to thought, particularly language's sentential or syntactical structure. Again, Sapir says it better than we can (pp. 14–15):

> From the point of view of language, thought may be defined as the highest latent, or potential, content of speech, the content that is obtained by interpreting each of the elements in the flow of language as possessed of its very fullest conceptual value. From this it follows at once that language and thought are not strictly conterminous . . . To put our viewpoint somewhat differently, language is primarily a prerational function. It humbly works up to the thought that is latent in it, that may eventually be read into its classifications and forms.

How human beings work their way up to the point where they derive the "fullest conceptual value" from their language is left rather loosely formulated by Sapir, and his looseness of statement is probably responsible for the misinterpretation of his position, which often goes by the name of the Whorf-Sapir hypothesis. He says (p. 15):

> It is, indeed, in the highest degree likely that language is an instrument originally put to uses lower than the conceptual plane and that thought arises as a refined interpretation of its content. The product grows, in other words, with the instrument, and thought may be no more conceivable, in its genesis and daily practice, without speech than is mathematical reasoning practicable without the lever of an appropriate mathematical symbolism.

There is a curious contradiction in this last point: that thought and the structure of experience depend for their growth on a reflective consideration by the speaker of the language he uses. How, then, would he have been able to *apply* language to experience and thought had these not achieved a form that matched them to the properties of the linguistic elements and sentences to which Sapir earlier refers? He clearly senses this problem, for a few paragraphs later he remarks (pp. 16–17):

> One word more as to the relation of language and thought. The point of view we have developed does not by any means preclude the possibility of the growth of speech being in a high degree dependent on the

> development of thought . . . We see this complex process of the interaction of language and thought actually taking place under our eyes. The instrument makes possible the product, the product refines the instrument.

In sum, then, there is some need for the preparation of experience and mental operations before language can be used. Once language *is* applied, then it is possible, by using language as an instrument, to scale to higher levels. In essence, once we have coded experience in language, we can (but not necessarily *do*) read surplus meaning into the experience by pursuing the built-in implications of the rules of language.

Until the time that this "surplus meaning" is read off from our linguistic coding of experience, language and experience maintain an important independence from each other. A child can say of two quantities that one is greater than another, a moment later that it is less than the other, and then that they are the same—using his words as labels for segments of experience. . . . It is not until he inspects his *language* that he goes back to his experience to check on a mismatch between what he sees with his eyes and what he has just said. He must, in short, treat the utterance as a *sentence* and recognize contradiction at that level. He can *then* go back and reorder experience, literally *see* the world differently by virtue of symbolic processes reordering the nature of experience.

We do not wish to quibble over the point, but it is *not* language *per se* that provides the reordering of experience. Rather, it is a genuine restructuring of how we perceive. The child says, for example, "that one *looks* bigger, but they are *really* alike," and later there is a firm distinction drawn between "the looks of appearance" and "the looks of reality." There remain, to be sure, those anomalous instances, the *visual illusions* that persist in defying perceptual reorganization, though we "know better," but even they can be divided into those that decline with age and experience (like the size-weight illusion) and those that are intractable (like the Müller-Lyer illusion).

Sapir's striking image of language as the "dynamo . . . operated almost exclusively to feed an electric doorbell" brings us to the second matter at hand, illustrated by the experiment of Carmichael, Hogan, and Walter (1932). Note first that labeling a figure (even one that is "meaningless") is almost irresistible. The dynamo seems hitched up to the doorbell very early. In Chapter 5, [of Bruner's text] where the visual recognition of children of different ages is examined, . . . even the three-year-olds name ambiguous figures. The labeling, once it occurs, has a strong influence on the form in which the figure is recovered from memory, with the well-known assimilation of the form to the label that has been applied to it—an effect even greater than originally found when appropriate controls are introduced as in the study of Herman, Lawless, and Marshall (1957).

Indeed, if the figure is labeled *in advance* of its being presented tachistoscopically, it can be shown (Bruner, Minturn, and Busiek, 1955) that its immediate reproduction will also be affected. Certain features will be made to conform with the defining properties of the class denoted by the label. Note that *if* the experiment is kept entirely within the ikonic mode—pictures are shown, labeled or unlabeled, and the subject's task is to *recognize* the pictures from among a set presented to him visually as in Prentice's (1954) experiment—then the effect of labeling is canceled out. If one has only to *match* a memory image to a present percept, the intervening aid of a label is not needed, but once the task becomes complex enough and involves a serial task such as reproduction, language is necessary as an aid to reconstruction. It is much as in a remark of Sapir's (1921, p. 15): "No one believes that even the most difficult mathematical proposition is inherently dependent on an arbitrary set of symbols, but it is impossible to suppose that the human mind is capable of arriving at or holding such a proposition without the symbolism." When the limits of direct imagery are reached (and they are soon reached), it is necessary to use another means—symbolic representation or even enactive, postural orientation, as in the experiment by Drake (1964) cited earlier. . . .

It is interesting that Vygotsky (1962) in his evocative book, *Thought and Language*, takes a view very similar to the one expressed here with respect to how thought comes to conform to language—this in spite of his rather loosely formulated view that "thought is inner speech." He clearly recognizes the separateness of "the stream of language" and "the stream of thought" and is sensitive to the need for organizing thought in a fashion that corresponds to language. He remarks that "we must uncover the means by which man learns to organize and direct his behavior" (p. 56). His famous sorting test, in which the subject has to form groupings of a set of blocks that themselves vary in width, color, shape, and height, is designed to explore this organization process. The very young child forms his groupings as "heaps," a seemingly unorganized congeries. In this phase "word meaning denotes nothing more than a vague conglomeration of individual objects that somehow or other coalesce into an image in his mind." At first, the child forming heaps picks up pieces at random to find out whether they will fit; a kind of grouping by doing. Later he chooses objects on the basis of "some other more complex relationship [produced] by the child's immediate perception" (p. 60). The somewhat older child organizes his groups by a process Vygotsky calls "thinking in complexes." A complex is a "family-relationship," and it may take several specific forms—the sharing of a common attribute with a nucleus object or the grouping of them by a theme rather than a universal property—"this is the mother block, this the baby." As Vygotsky puts it (p. 65),

> A diffuse concept in the child's mind is a kind of family that has limitless powers to expand by adding more and more individuals to the original group. The child's generalizations in the nonpractical and nonperceptual areas of his thinking, which cannot be easily verified by perception or practical action, are the real-life parallel of the diffuse complexes observed in the experiment.[7]

Finally, the conventional superordinate category emerges governed by the logical rules of inclusion, exclusion, and overlap: "all tall, red figures here; all short, blue ones there," etc.

Interestingly enough, Vygotsky does not discuss the emergence of these grouping rules as an instance of external language becoming inner speech but rather as the effect of instruction—instruction, indeed, in "scientific thinking" and even in Marxist thinking. I take him to mean, although he is not clear on the point, that the contents of experience must be prepared and organized better to fit the requirements of being handled by language and that once this occurs there can also be a "reading off" of the surplus properties of the language in much the fashion that Sapir suggested. Indeed, it is only a little extreme to suggest that there are symbolic rules contained in the principles of grouping. Heaps are characterizable as isomorphic to the verb-object relation: "These are in the group because I was looking at them or had them in my hand"; or, in the case of complexes, the grouping conforms to a grammar of prediction . . . in which, "This block is the baby of that bigger one." Finally, the principle of superordination is the introduction of categorial hierarchy and of modification, where "red circles" are opposed to "green squares."

There remains a puzzling problem about the interaction of symbolic representation and action. Why is it impossible simply to *tell* somebody how to ride a bicycle? Consider some Soviet experiments on the regulation of action by speech. The work of Luria and his students (1961) is a good instance to begin with, in particular an experiment by Anokhin. It had been discovered that if a child of two were asked to press a bulb when a light came on the *instruction* itself led the child to press whether the light had come on or not. To quote Luria (1961, pp. 59–60):

> Some assumptions by Sechenov, later reproduced by Anokhin, helped us in tackling this problem. These scientists maintained that the inhibition of a given action usually results from conflict between two excitations, the one inhibiting the other. Is it then possible to make use of the impeding, initiating action which adult speech already has for the child [as demonstrated by the experiments of Yakovleva] and on this basis to

[7] For the role of such complexes in "intellectual pathology" in childhood, notably learning blocks, the reader may consult J. S. Bruner, *Toward a Theory of Instruction* (1966b), particularly the chapter "On Coping and Defending."

produce a conflict between two excitations which would result in the inhibition of the reaction already begun?

With this aim in mind, we performed a very simple experiment, the results of which fully came up to our expectations. . . . We asked [the child] to perform two simple actions in succession: to squeeze the ball at the flash of an electric light signal, and then move his hand away at once (e.g., put it on his knee). When he had obeyed this double act of starting on instructions (which did not present any difficulty to him), we gradually reduced the distance he had to move his hand after pressing the bulb. First he was told to put it not on his knee, but on the table by the [bulb]; then we reduced the distance still further, and at last after some time, we were able to cut out the second intermediary part of the experiment altogether. Having learned through performing the second action thus to inhibit the first, the child was now able to cope quite easily with a task which he had previously found impossible . . . . Verbal instructions, previously ineffective, could now produce the required effect, thanks to the preliminary influences prepared by the preliminary conflict between the two successive excitations.

In short, if the action itself could be organized first in a fashion that conformed to the instructions that were to come, the instruction could be effective. It is the same point made earlier about the structure of experience. Athletic coaches also recognize that *telling* an athlete something that is useful to him depends first upon his having the requisite motor behavior segmented or organized in a fashion that corresponds to words. Where no such preliminary correspondence exists, there is an independence between the two systems, and we have the ubiquitous case of "he who talks a better game than he plays."

In sum, then, if one is using symbolic representation to guide looking or to guide action, the success of the effort will depend upon the extent to which the sphere of experience or action has been prepared to bring it into some conformance with the requirements of language. If Lashley (1951) is correct about the atemporal, grammar-like control of most serially ordered behavior in human beings looking and acting alike—there should be a very considerable scope for the achievement of control by symbolic means in the growth of the child. So we would have to agree strongly with the view expressed earlier by Miller and Chomsky (1963) that "sentences have a compelling power to control both thought and action."

Growth, Culture, and Evolution

On the occasion of the one hundredth anniversary of the publication of Darwin's *The Origin of Species*, Washburn and Howell (1960) presented at the Chicago centennial celebration a paper containing the following passage (pp. 49f.):

> It would now appear . . . that the large size of the brain of certain hominids was a relatively late development and that the brain evolved due to new selection pressures *after* bipedalism and consequent upon the use of tools. The tool-using, ground-living, hunting way of life created the large human brain rather than a large-brained man discovering certain new ways of life. [We] believe this conclusion is the most important result of the recent fossil hominid discoveries and is one which carries far-reaching implications for the interpretation of human behavior and its origins . . . The important point is that size of brain, insofar as it can be measured by cranial capacity, has increased some threefold subsequent to the use and manufacture of implements. . . . The uniqueness of modern man is seen as the result of a technical-social life which tripled the size of the brain, reduced the face, and modified many other structures of the body.

This statement implies that the principal change in man over a long period of years (perhaps five hundred thousand) has been alloplastic rather than autoplastic. That is to say, he has changed by linking himself with new, external implementation systems rather than by any conspicuous change in morphology—"evolution-by-prosthesis," as Weston La Barre puts it. The implement systems seem to have been of three general kinds: (1) *amplifiers of human motor capacities* ranging from the cutting tool through the lever and wheel to the wide variety of modern devices; (2) *amplifiers of sensory capacities* that include primitive devices such as smoke signaling, and modern ones such as magnification and radar sensing, but are also likely to include such "software" as those conventionalized perceptual shortcuts that can be applied to the redundant sensory environment; and finally (3) *amplifiers of human ratiocinative capacities* of infinite variety, ranging from language to myth and theory and explanation. All these forms of amplification are in major or minor degree conventionalized and transmitted by the culture, the last of them probably the most so, since ratiocinative amplifiers involve symbol systems governed by rules that, for effective use, must be shared.

Any implement system to be effective must produce an appropriate internal counterpart, an appropriate skill necessary for organizing sensorimotor acts, for organizing percepts, and for organizing our thoughts in a way that matches them to the requirements of implement systems. These internal skills, represented genetically as capacities, are slowly selected in evolution. In the deepest sense, then, man can be described as a species that has become specialized by the use of technological implements. His selection and survival have depended on a morphology and set of capacities that could be linked with the alloplastic devices that have made his later evolution possible. We move, perceive, and think in a fashion that depends on techniques rather than on wired-in arrangements in our nervous system.

Where representation of the environment is concerned, it too depends upon learned techniques; and these are precisely the techniques that serve to amplify our motor acts, our perceptions, and our ratiocinative activities. We know and respond to recurrent regularities in our environment by skilled and patterned acts, by conventionalized spatioqualitative imagery and selective perceptual organization, and through linguistic encoding which, as so many writers have remarked, places a selective lattice between us and the physical environment. In short, the capacities that have been shaped by our evolution as tool-users are those we rely on in the primary task of representation.

The consequence of the development of such a representational system, as psychologists and anthropologists alike have pointed out, is to make possible a kind of integration over space and time that approaches the conditions necessary for dealing with past and future in the present and for dealing with the distant as if it were near. As Hallowell (1955) put it in his presidential address to the American Anthropological Association,

> The psychobiological structure that the hominid evolved is one in which intervening variables which mediate between immediate stimuli and overt behavior come to play a more primary role. Such intervening variables include unconscious processes such as dreams, as well as conscious operations like thinking and reasoning, "Whereby the remote as well as the immediate consequences of an impending overt action are brought into the psychological present, in full force, so to say, and balanced and compared" [Mowrer and Ullman, 1945].

In time, there develops within human culture the traditionally transmitted means for making these activities more easily learned and more powerful.

As Peter Medawar (1963) notes, the point in primate evolution at which adaptation occurs almost entirely by the development of techniques rather than by change in morphology is the point at which evolution becomes reversible and, in a figurative sense, Lamarckian. We depend for survival on the inheritance of acquired characteristics from the culture pool rather than from a gene pool. Culture then becomes the chief instrument for guaranteeing survival, with its techniques of transmission being of the highest order of importance.

Man's dependence on a cultural heritage is supported by the strikingly appropriate accidents of man's morphological evolution that has produced a long period of early helplessness. Bipedalism, with its need for a stronger pelvis, reduces the birth canal at a time in evolution when brain size is increasing. The morphological compromise is a neonate brain of marked immaturity and with few prepared response patterns. This circumstance and the marked dimorphism of the human sexes give ample time and woman-power for child-rearing and ample opportunity for enculturation.

What a culture "teaches" and how it goes about the task of doing so will concern us in much detail in later pages. What seems plain enough is that the most characteristic thing about the lessons that are imparted—whether with respect to matters of value, of existence, or of self—is their productive generality. Whatever is learned seems to be converted into general rules that are applicable to many contingencies never before encountered. In learning a culture, we learn rules, it would seem, and in this sense there is a great kinship to learning language. Obviously, "behaving culturally" is no more "caused" by the culture's "rules" than speaking a language is "caused" by the grammarian's rules. Kroeber and Kluckhohn (1952, p. 170) cite a letter by the present author directed to the section of their monograph dealing with implicit and explicit culture:

> The process by which the implicit culture is "acquired" by the individual (i.e., the way the person learns to respond in a manner congruent with expectation) is such that awareness and verbal formulation are intrinsically difficult. Even in laboratory situations where we set the subject the task of forming complex concepts, subjects typically begin to *respond* consistently in terms of a principle before they can verbalize (a) that they are operating on a principle, or (b) that the principle is thus-and-so. Culture learning, because so much of it takes place before very much verbal differentiation has occurred in the carrier and because it is learned along with the pattern of a language and as part of the language, is bound to result in difficulties of awareness. Thought ways inherent in a language are difficult to analyze by a person who speaks that language and no other, since there is no basis for discriminating an implicit thought way save by comparing it with a different thought way in another language.

Perhaps one should emphasize too that the "nonverbalizable" phase of concept learning, rule learning, and skill learning constitutes the *preparation* of motor, sensory, and intellectual life for assistance by language; the issue was discussed in the section just preceding. Once this learning occurs, the trait then emerges that is most characteristic of human behavior: its symbolic quality, through which we substitute words and sentences in place of events in order to have a vicarious trial run on reality.

We have also commented that the extent and shape of this intellectual development, since in its very nature it depends on assistance from a culture, will vary as a function of culture. Cultures, so goes the classic line of relativism, are all different. Yet to take this position in analyzing the impact of culture on growth dooms one to a study of what is different about growing up in different places, and that is surely a trivial pursuit in comparison with the study of the few powerful shaping forces in culture that produce enormous uniformities in growth and a few crucial differences. Consider now some of these differences.

We have recently had the opportunity to observe carefully on film the play of baboon juveniles and of Kung Bushman children in a similar African habitat.[8] The comparison extended, of course, to our own culture. It is typical of baboon juvenile play that it is virtually all interpersonal. The young males chase and mock-fight one another, developing the skills needed in filling the role of the adult dominant male. The females do a certain amount of chasing around, but soon center on the infants in the troop, whom they groom and try to carry or care for. The adults do no "instructing" and interfere only to place general limits on the juveniles. The limits are on such things as noise that gives away position, for one often sees an older male "policing" a great juvenile chase-about. Adults also herd the young animals into a proper position in the troop or in a tree when a predator is around. There is also intergeneration exchange of grooming. But in the juvenile group, generally, learning is mostly in the peer play that develops skills which, when orchestrated, make an adult repertory. The important thing about this play of the young baboons is that it exercises and develops skills that in rearranged pattern have a place in adult life.

Among the Bushmen there is also very little explicit teaching, but what is strikingly different is the amount of joint activity between the children and the adults. What the child knows, he learns from direct interaction with the adult community, whether it is learning to tell the age of the spoor left by a poisoned kudu buck, to straighten the shaft of an arrow, to build a fire, or to dig a spring hare out of its burrow. Yet in thousands of feet of film, one sees no *explicit* teaching in the sense of a "session" out of the context of action to teach the child a particular thing. It is all implicit. One does see children imitate in competition with one another what they have been participating in with the adults, as in one beautiful scene in which Bushmen children are shooting miniature arrows at beetles. The only exception is the well known teaching of rituals in some indigenous societies at the time of the *rites de passage*. But that ceremony, wherever it has been observed, is for teaching rituals, chants, myths, and ceremonies—never skills (see Spindler, 1959). In Bushman society, indeed, there is rather little of this kind of instruction. Usually one finds that the dances, games, and rituals are first encountered when the young baby is still being carried in his mother's *karosse* and miraculously held there while the mother goes through the complicated steps of, say, the centipede dance.

By far the most detailed field study of the learning process of children in an indigenous society is to be found in the monograph by Professor

[8] We are indebted to Professor Irven De Vore and to Educational Services, Inc., for the opportunity to use the baboon film footage, and to Mr. and Mrs. Laurance Marshall for the use of the Bushman footage. Their kindness and patience in discussing these matters are gratefully acknowledged.

Meyer Fortes (1938) of the University of Cambridge, *Social and Psychological Aspects of Education in Taleland.* This monograph, published as a supplement to the journal *Africa*, is not readily available, so it is worth examining its contents here in some detail. For one is struck by the extent to which certain of Fortes' observations correspond to those of others who have looked at the educational process in indigenous societies.[9] The first point that Fortes makes about the matter is crucial (pp. 8–9):[10]

> The process of education among the Tallensi, as among a great many other African peoples of analogous culture, is intelligible when it is recognized that the social sphere of adult and child is unitary and undivided. In our own society, the child's feeling and thinking and acting takes place largely in relation to a reality—to aims, responsibilities, compulsions, material objects and persons, and so forth—which differs completely from that of the adult, though sometimes overlapping it. This dichotomy is not only expressed in our customs, it comes out also in the psychological reactions which mark the individual's transition from the child's world to the adult's—the so-called negative phase of adolescent instability which has been alleged to be universal in our society. It is unknown in Tale society. As between adults and children, in Tale society, the social sphere is differentiated only in terms of relative capacity. All participate in the same culture, the same round of life, but in varying degrees, corresponding to the stage of physical and mental development, nothing in the universe of adult behavior is hidden from children or barred to them. They are actively and responsibly part of the . . . system. Psychological effects of fundamental importance for Tale education follow from this. For it means that the child is from the beginning oriented toward the same reality as its parents and has the same physical and social material upon which to direct its cognitive and instinctual endowment. The interests, motives, and purposes of children are identical with those of adults, but at a simpler level of organization.

The same point about continuity has been made about many indigenous societies and, indeed, an illustrative episode presented by Fortes (p. 11) finds an almost identical parallel among the Netsilik of Pelley Bay, so meticulously studied by the Danish anthropologist Rasmussen (1931).

[9] One does not find many careful studies of education in primitive society, with education viewed literally as leading the child into the views, beliefs, and skills of a society. For more than a quarter century, most American writing has been dominated by the "culture-and-personality" point of view that concerned itself with the handling of emotional crisis points in a child's life (bowel training, weaning, the arrival of new siblings, etc.). It is difficult in such studies to find much on the development of skills or the process for passing on attitudes and values. Spindler's (1957) book has rich references and one can find occasional, well-informed articles in the pages of *Africa*.

[10] M. Fortes, "Social and Psychological Aspects of Education in Taleland," Supplement to Africa, XI, No. 4, 1938; also Memorandum XVII of the International African Institute. *Africa* is published for the I.A.I. by the Oxford University Press.

The Tale episode involves the nine-year-old girl Maanyeya, who had eaten none of the meat of the previous night's sacrifice. Asked why not, she replied:

> When they sacrifice to Zukek, women don't eat of the meat. If they do, they will never bear any children, they become sterile. [What's that to you?] Am I not a woman? Who wants to be sterile?

So too with a Netsilik girl of about the same age, loaded down with amulets, who, when asked why she wore them, listed a catalogue of adult woes from which she was guarding herself.

Everybody instructs among the Tallensi: adult to child, older child to younger, peers among one another; but the instruction is always in the context of the action or endeavor in progress. There is virtually nothing by way of standardized and deliberate methods of training children. The system is premised on the "expectation of normal behavior. . . . In any given social situation everybody takes it for granted that any person participating either already knows, or wants to know, how to behave in a manner appropriate to the situation and in accordance with his level of maturity" (pp. 25–26). The child is given tasks commensurate with his capacity and it is taken for granted that he will participate in and contribute to the economic, social, and ritual tasks of adult life, in however small a way. Even play, of which there is a fair share, is marked by the same centering on themes that are as much part of adult as of children's lives.

But because so much of the "meaning" of what is being learned is intrinsic in the context in which the learning occurs, there is very little need for verbal formulation. Fortes remarks (p. 30),

> The natives say that small children frequently ask questions about people and things they see around them. However, listening to children's talk for "why" questions, I was surprised to note how rarely they occurred; and the few instances I recorded referred to objects or persons foreign to the normal routine of Tale life. It would seem that Tale children rarely have to ask "why" in regard to the people and things of their normal environment because so much of their learning occurs in real situations. . . .

Consider again the question raised earlier about why, under certain conditions, intellectual development moves toward an elaborated form of symbolic representation with all the powerful accompanying symbolic activity that becomes accessible for use. Five possible sources of this development were mentioned in the earlier discussion, and we are now ready to consider them in more detail. They were, in brief: (a) the use of words as invitations to form concepts; (b) contingent dialogue between adult and child; (c) the importance of "school" as an innovation;

(d) the development in a culture of "scientific" concepts; (e) the possibility of conflict between modes of representation.

Note first that when a society grows more complex in its technology and division of labor, there are two deep changes that must necessarily occur. First, the knowledge and skill within the culture comes increasingly to exceed the amount that any one individual can know. Almost inevitably, then, there develops a sharp disjunction between the worlds of the child and of the adult. The unity of the Tale world becomes impossible in more complex societies. Increasingly, then, there develops a new and moderately effective technique of instructing the young based heavily on *telling* out of context rather than on *showing* in context. The school, of course, becomes the prime instrument of this new technique but by no means the exclusive one. For, in fact, there is also a great increase in telling by parents, again out of the context of action, for there come to be fewer spheres in which such learning *in situ* can be practiced. It is probably by virtue of this development that the "why" question becomes so important a feature of the child's response to his environment. It serves to provide a verbal context in the absence of the context of action characteristic of technologically simpler indigenous societies. Indeed, it has even been remarked that the world of learning of the child in school becomes detached from life as lived in the greater society, and one hears the voice of the reformer asking that school be drawn closer to life.

Yet, as commented upon in detail elsewhere (Bruner, 1965), it may indeed be that the important thing about the school as now constituted is that it *is* removed from immediate context of socially relevant action. This very disengagement makes learning an act in itself and makes it possible to embed it in a context of language and symbolic activity. For now indeed it is the case that words are the major invitations to form concepts rather than the action contexts so aptly described by Fortes (1938) and others. *Verbal* understanding, the ability to *say* it and to enumerate instances becomes the criterion of learning in such a context to the Tale concept of "yam," which according to Fortes connotes wisdom and resourcefulness about how to *behave*—both practically and morally.[11]

In more evolved technical societies, then, the very nature of the learning situation *requires* a contingent dialogue between parent or tutor and

[11] It goes without saying that the separation of thought and action has not been an unblemished blessing in technically mature societies. Hamlet, "sicklied o'er with the pallid cast of thought" or Goethe's contrast, "Gray is all theory/Green grows the golden tree of life" bespeak the problem eloquently. Yet it is all too evident that the unity of "the noble savage" fares poorly when the powers of analytic thinking work their separatist way. As McGranahan (1963, p. 16) remarks, ". . . in viewing the human implications of technological change we [must] not become so fascinated by the bad as to forget the good, and so protective of the present cultures of underdeveloped areas as to wish to preserve these cultures against the very idea of progress which we embrace for ourselves."

the child, for once one is out of the task context in which learning occurs directly one can no longer point or "let the situation carry the meaning." We would predict, and the prediction is almost trivial because it is so obvious, that comparable tasks to be performed by a member of a more technical society and a member of a less technical society will always be more accessible to verbal description by the former. It is precisely in this application of symbolic recoding of "what one knows" that the process of translating action and experience into its symbolic, vicarious forms occurs. Little enough is known about the process, but at least a few studies (Crutchfield's study of increased inferential power in children [1964] and Saugstad's [1955] experiment on a solution of Maier-type problems) suggest the manner in which prior verbal analysis of a task can produce an increase in reasoning solutions. We need mention these matters only in passing, for they will concern us again in the discussion of the striking difference one finds among Wolof-speaking school children and their unschooled peers in Senegal. . . .

Finally, with respect to the problem of *conflict* between systems of representation as a source of impulsion to grow (the disequilibrium theory), there is striking evidence elsewhere in this volume for the efficacy of the conflict between verbal and visual formulations . . . and for visual and enactive modes. . . . However, there is a deeper problem that is worth a moment's exploration. It can be introduced by a comment in *A Study of Thinking* (Bruner, Goodnow, and Austin [1956], p. 50) to this effect:

> It is curiously difficult to recapture preconceptual innocence. Having learned a new language, it is almost impossible to recapture the undifferentiated flow of voiced sounds that one heard before one learned to sort the flow into words and phrases. Having mastered the distinction between odd and even numbers, it is a feat to remember what it was like in a mental world where there was no such distinction. In short, the attainment of a concept has about it something of a quantal character. It is as if the mastery of a conceptual distinction masked the preconceptual memory of the things now distinguished.

There is a dilemma here. In one sense it is a loss for the growing human being to "lose" his older, more innocent conceptions and skills, but at the same time it appears to be a necessary condition for acculturation (and, indeed, for cultural control of the individual) that there be this "childhood amnesia," as Schachtel (1947) has called it. It is perhaps Neisser (1962, pp. 63ff.)[12] who deals most directly with this problem from the point of view of the role that culture plays in shaping the growth of the child:

[12] U. Neisser, "Cultural and Cognitive Discontinuity," *Anthropology and Human Behavior.* (1962) pp. 54–71.

Essentially, the experiences of childhood are incompatible with the schemata of the adult. It is no wonder, then, that the adult cannot recall them. The events, activities, and emotions of childhood were assimilated in a way that is no longer open. Years of sophisticating accommodation have made it as impossible to remember our own childhood as to fully understand anyone else's. The early years are like a forgotten dream. The simile of the dream is appropriate: our inability to remember dreams is based on the same factors. The coherent schemata of waking life have no room for their childlike illogic. It is worth noting the great individual differences in memory for dreams, as in memory for childhood experiences: not all persons are equally accommodating to society's demand.

This means that the universal amnesia for childhood is not primarily the result of anxiety or guilt, and is not based on an active process of suppression. It is, instead, a necessary consequence of the discontinuities in cognitive functioning which accompany growth into adulthood. From preverbal to verbal, from naive to sophisticated, from carefree to responsible, from weak to powerful—the cognitive accommodations which accompany these transitions seem to make the past inaccessible. Schachtel finds the changes repugnant, preferring the spontaneity of the child to the stereotype of the adult. For him, both the amnesia and its cause are a matter for deep regret. Over and over again he stresses the value of what is lost in the process of acculturation and accommodation, as if adulthood was an essentially impoverished condition. To me, his view appears one-sided. Childhood is not so simple, nor maturity so barren. Indeed, it is in the prematurely rigid schemata of early and middle childhood that the roots of adult neurosis are found. Psychotherapy does not aim at regressing the patient to the conceptual innocence of childhood, but at permitting him to grow beyond childhood. The healthy person is not the one who refuses to assimilate, but the one whose schemata are adequate to reality.

As the mental apparatus of a growing child develops, and the information-handling processes become more intricate, thinking goes through a succession of stages. These are semi-stable states of accommodation, phases through which the cognitive mechanisms must evolve on the route to intellectual maturity. A child assimilates and "distorts" the world of his experience in ways characteristic of his age. As he grows up, constant necessity for accommodation results in cognitive change. This change can come about in three fundamentally different ways, which we must clarify.

The first mode of accommodation is *absorption*. Later forms of cognitive schema may absorb earlier ones completely. This is what usually happens with repeated exposure to a piece of music. The inharmonious jumble that was experienced the first time simply ceases to exist. It cannot be perceived again and cannot be remembered. The new schema has swallowed up all the elements and interrelations of the old. Absorption is a common experience in hidden-picture puzzles. When we finally find the outline of the squirrel that the artist has cleverly concealed in the bark of the tree, we cannot lose it again; it is impossible to imagine how we could have failed to see it before. The same thing tends to happen in successful rote learning. One who knows a poem by heart usually does not recall the individual trials on which he practiced it. In a sense, he has an amnesia for them. Are they "forgotten"? Yes and no: they have a continuing effect (because they established the schema which now exists) but they cannot be individually recalled.

A second mode of accommodation might be called *displacement*. Part of the cognitive apparatus does not evolve, but continues to exist side-by-side with a new schema which assimilates the same environmental events in a different way. A trivial example of such dual mental functioning is the "double-take." The double-take occurs when you suddenly realize that something heard or seen a few moments ago actually had quite a different significance from that which you had, perhaps inattentively, ascribed to it. One assimilation process interpreted the event as unimportant. A second process, occurring simultaneously but unconsciously, interpreted it very differently; the double-take occurs when the second assimilation becomes conscious. More sustained instances of displacement are common among social scientists, who are able to react to a social situation either "personally" or "professionally." The behavioral results of these two ways of assimilating racial discrimination (for instance) can be poles apart.

Adequate consideration of the consequences of "displacement," in this sense, would go far beyond the intent of this paper. It is possible to interpret the classical evidence for unconscious cognitive processes in these terms. Suppose that an adult has preserved the assimilative mechanisms of a four-year-old with respect to certain events, for example, situations involving sexuality. These schemata will be "unconscious" from the point of view of his organized adult awareness, but will continue to process information and control behavior. The results will be perceptual defense, forgetting of intentions, and perhaps other symptomatic phenomena. This interpretation is not far from that implied by the concept of "dissociation," but here the displacement is viewed as one possible outcome of a developmental process that will take *some* form in any case.

The third mode of accommodation is the *integrative* one. In many cases it is possible for a new schema to make use of an old one without destroying its integrity. Integration requires a step to another level of abstraction or understanding, in which outputs of the older modes of processing are only part of a more comprehensive whole. This hierarchical organization can be taken for granted in some aspects of perceptual development. We do not lose our ability to see figure and ground when we understand the three-dimensional permanence of figured objects. Moreover, their perceptual solidity is not impaired when we endow them with cultural or personal meaning. To be sure, wide individual differences exist here. Some persons see much more than others of the natural shapes and colors about them. The artist's world is filled with shapes and colors that go unnoticed by the less perceptive. He has somehow maintained—and developed!—the integrity of assimilative systems that are absorbed or displaced in the rest of us.

The foregoing analysis of the accommodative processes can be applied to the problem of childhood amnesia. Both absorption and displacement of earlier schemata must lead to "forgetting." Both types of change must almost inevitably occur as a baby becomes a child, and a child grows to adulthood.

Indeed, even integration leads to a certain loss; a childish mode of functioning somehow preserved in an adult structure cannot be identical with its unintegrated form. However understanding a parent (or teacher or a therapist) may be, he remains an adult. But this change of perspective is insignificant compared with the total amnesia for infancy which we all share, and which I attribute to absorption and displacement. What circumstances lead to one mode of accommodation rather than another? An adequate answer can hardly be given. It is very likely, however, that the manner in which different developmental stages are handled by environment and culture are particularly important. Displacement will tend to occur where cultural factors emphasize the discontinuity and incompatibility between different phases of development, while integration must be easier where several stages of assimilation are welcomed and used together in a consistent way. Thus we must expect a close relationship between the continuity or discontinuity of developmental patterns on the one hand, and the continuity of memory on the other.

Neisser's comments are worth quoting in such detail not simply for the light they shed on the adult's inability to reconstruct his own development but because, in a deep sense, they provide a justification for doing research on intellectual development in children! Growth has a way of minimizing the conflicted ways of knowing, making retrospective efforts at creating a developmental psychology worse than hazardous. With that much said, we can turn directly to processes we have explored by examination of how they manifest themselves in the behavior of young children.

THE PRELINGUISTIC CHILD*

Eleanor and George Kaplan

Introduction

Virtually all children begin to speak somewhere between 18 and 24 months. By the age of 3 they have mastered many of the basic syntactic and phonological components of their language. The incredible rapidity of this process in the face of the enormous complexity of linguistic systems has led many to speculate on the ontogeny of language. Various investigators have suggested that everyone acquires his native tongue in roughly the same way despite seemingly significant individual and cultural differences.

---

* This selection was written expressly for this volume.

In approaching the theoretical issues raised by the acquisition of language, psychologists and linguists realized that there was relatively little systematic data on language acquisition. To rectify this lack of data, psychologists and linguists over the last two decades have concentrated on experimental investigations of the child's knowledge of grammar. This research on grammatical development, stimulated by the important work of Chomsky (1957; 1965) has added much to our understanding of linguistic development. Concentrating upon grammar, however, necessarily limits one to developmental sequences commencing somewhere between 12 and 18 months. This, in turn, has led many to regard the emergence of grammatical speech at this time as the advent of "true" language (Ervin and Miller, 1963; McNeill, forthcoming, 1970.)

In what follows we will question the assertion that "true" language commences at 12 months. Specifically, are there psycholinguistically significant changes in receptive and productive abilities within the first year? Or more generally, is it appropriate to call any child "prelinguistic?" We will address ourselves to these questions by examining two theoretical approaches to the relationship between early and later language, and by presenting recent data bearing upon several questions raised by the main issue.

Different Approaches

Learning Theory Positions

The most popular view of language development is that the learning of a language can be accounted for by the same principles operating in any other learning situation. The importance of early vocal behavior is stressed. It is assumed that a child, in his initial vocalizations, produces all the sounds used in any language. Commenting on much of the early literature on the development of language, Latif (1934, p. 60) says, ". . . the random sounds first produced by an infant serve as the raw material for its later linguistic progress." According to this view, an infant begins by emitting all possible sounds only some of which are represented in his or any language.

> One cannot fail to hear all the vowels and consonants, dipthongs, aspirates, sub-vocals, nasals, german umlauts, and tongue trills, French throaty trills and grunts, and even the Welsh L. (Bean, 1932, p. 198).

Ultimately, through a process of imitation and differential reinforcement, the infant's repertoire of sounds comes progressively closer to adult speech in a particular language. Thorndike (1943) aptly termed this position the "Babble-Luck Theory" because it was supposed that as the infant randomly produced sounds he was lucky enough to be reinforced for only those sounds which are represented in his language.

A second position within the learning tradition was proposed by Mowrer (1958) and more recently by Winitz (1968). Mowrer did not argue that the child produces all possible sounds, but instead that the sounds adults make, when coupled with increased comfort, cause these adult sounds to acquire value as secondary reinforcers. The child imitates the sounds he hears because of their reinforcing, or "comforting," value. Mowrer called this the "Autism Theory" of language learning because imitation of appropriate sounds results in a pleasurable experience for the infant.

Both of these positions emphasize the importance of early vocal behavior for later language development. In addition, they concentrate upon environmental support for the language learner via imitation and reinforcement. Consequently, both would predict large variability in linguistic development depending upon variations in environment. In addition, they would predict a smooth transition from early to later verbal behavior.

Linguistic Position

Roman Jakobson, a linguist, was struck by the lack of continuity between early vocalizations and later speech. His pioneering work, *Child Language, Aphasia and Phonological Universals* (1941), serves as the main statement of his influential theory. He agrees with most other investigators that the young infant produces a random collection of the sounds used in many languages. But he notes that this facility with sound disappears toward the close of the first year. For example, in his early vocalizations, the young infant might produce a sound which is acoustically similar to /r/. However, 6 to 8 months later when the rudiments of the adult phonological system begin to appear, this same sound /r/ is absent. This later inability to produce previously articulated sounds is not due to any motor deficiency because "one often secures the 'parrot-like' repetition of single sounds and syllables from children, even though the very same sounds continue to be absent where they talk spontaneously" (Jakobson, 1968, p. 23). Thus, Jakobson makes an important distinction between the production of a sound and the systematic use of that sound in a phonological system.

In mapping out the changes which occur in vocalizations, Jakobson observes that the stage of indiscriminate production "merges unobtrusively" into the systematic use of a small repetoire of sounds. Although several interpreters of Jakobson's theory have argued that there is an abrupt shift from one stage to the next, Jakobson seems to argue only for a gradual transition. The important point for the present discussion is that there *is* a transition, and that vocalizations in the two stages differ in two significant respects. In addition to the fact that many fewer sounds are initially produced during the second stage, Jakobson argues that the first stage does not show ". . . any general sequence of acquisition," whereas in the second stage ". . . we observe a succession which is universally valid" (Jakobson, 1968, p. 28). Referring to this latter stage he says,

> Whether it is a question of French or Scandinavian children, of English or Slavic, of Indian or German, or of Estonian, Dutch or Japanese children, every description based on careful observation repeatedly confirms the striking fact that the relative chronological order of phonological acquisitions remains everywhere and at all times the same. (Jakobson, 1968, p. 46.)

According to Jakobson, this universal ordering is the result of a direct relationship between linguistic universals and language development. There are two universal principles of language governing the course of language acquisition. The first is a frequency principle. Not all features of language are evenly distributed. Some sounds, such as nasal consonants, are found in all languages; in accordance with the frequency principle, they appear rather early in development. Sounds which are relatively rare in languages, such as nasal vowels, occur rather late. The second principle is that secondary components of languages are never found in the absence of primary components, but primary components are found in the absence of secondary ones. According to Jakobson, primary components are always acquired before secondary components. For example, in all languages the fricatives do not exist unless stop consonants also exist, but the reverse is not true. This is manifested in human development by the fact that consonant stops always appear before fricatives. Unfortunately, he does not postulate any mechanism which could account for the relationship between these two principles and language development. He, instead, points out the important fact that these relationships do exist.

In sum, Jakobson argues that there are two distinct stages of early vocal behavior. The first shows no particular order of development and is unrelated to later language. The second follows a universal order of development and is the matrix out of which "true" language arises. Thus, he predicts that the characteristics of very early vocalizations will be insignificant with respect to later language acquisition. In addition he argues that the transition from early to later verbal behavior is marked by large qualitative and quantative changes.

The theories described above suggest opposite answers to our initial question, "Is there any such thing as a prelinguistic child?" For a consideration of these contrasting answers, we must turn to the relevant data on the following topics: the development of productive and receptive abilities, the role of the environment, and the functional and linguistic significance of early language development.

Productive and Receptive Abilities

Obviously the most important data are those which give us direct information about the development of productive and receptive abilities. Only recently have laboratory procedures been used to examine the early development of these processes. Although there is a limited number of laboratory studies, there exists a wealth of observational data on the longitudinal aspects of child language to which we will also refer.

Production

Chronology of Stages. Most observers (Lewis, 1951; Lenneberg, 1967; Murai, 1960 and 1963; Nakazima, 1962 and 1966; Wolff, 1966b; see also the host of "baby biographies" mentioned later in this selection and referenced in the bibliography) generally agree that the chronology of early vocalizations can be divided into several overlapping stages, although not all agree about the defining characteristics of each stage and transitions between stages.

Stage 1: Crying. The first stage begins with the "birth-cry" and various assorted coughs and gurgles. This period is usually defined by the presence of a basic pattern of crying along with several variations (Lewis, 1951; Lieberman, 1967; Wolff, 1966b). The basic pattern, according to Wolff, is a rhythmical one consisting of a cry, a rest, an inspiration, and a rest, the whole event lasting about a second. It has a rising-falling frequency contour extending over the entire exhalation phase of the breathing cycle (Lieberman, 1967; Wolff, 1966b). The cry is present in this basic form for at least 6 months. Several variants of the basic crying pattern can also be noted. One of these, identified by parents as the "mad" or "angry" cry, has the same basic temporal sequence but is extremely loud, its excess turbulance resulting in frequency distortion. A second variant is associated with physical pain and has a much longer crying portion (Wolff, 1966b).

Stage 2: Pseudocry and Noncry Vocalizations. These begin to appear at 3 weeks and show a greater variety of temporal and frequency patterns and use of the articulatory organs than crying (Wolff, 1966b). By the end of this period (approximately 4 to 5 months) there is a large variety of noncry utterances which become increasingly distinct from crying and from each other.

Stage 3: Babbling and Intonated Vocalizations. The babbling stage is an extension and further differentiation of the previous stage. The vocalizations become increasingly speechlike. Greater numbers of clearly articulated vowellike and consonantlike sounds appear and are combined into reduplicated syllabic constructions. For the first time, seemingly adult intonation patterns are heard. Many observers have remarked on the tendency of infants to begin imitating adult speech at this point. This phase generally lasts until the end of the first year.

Stage 4: Patterned Speech. This last phase, beginning at from 9 to 12 months, is considered by many to mark the close of the "prelinguistic" period and the onset of "true" speech. According to Jakobson (1968) and Shvahckin (1948) the utterances found within this period are quite distinct from all previous utterances. There is a decrease in the large variety of phonetic forms previously heard in the child's speech. Gradually the small repetoire of remaining sounds systematically differentiates into a larger number of sounds directly related to the phonological structure of language. The child's first words begin to appear.

Continuity versus Discontinuity of Stages. One of the major differences between the learning-theory approach (continuity) and the linguistic approach (discontinuity) is their conceptualization of the transition between different stages (particularly between Stages 3 and 4). Let us now look at some data on these transitional periods. There is little disagreement with Jakobson's description of the dynamics of Stage 4. Most investigators have found the uniform and lawful progression in the development of the phonological system, which Jakobson (1968) and Jakobson and Halle (1956) outlined from an initial vocalic-consonantal distinction through a progressively finer differentiation within each class (Leopold, 1947; Nakazima, 1962 and 1966; Shvachkin, 1948; Veltan, 1943). See Ervin and Miller's review (1963) for further details.

Can the notion of successive differentiation also be applied to the transition from Stage 3, the babbling period, to Stage 4, "true" speech? Here the evidence is ambiguous. The data and observations of Jakobson (1968), Shvachkin (1948), and Bever (cited in McNeill, forthcoming) all suggest that there is no continuity between babbling and later speech. On the other hand, data collected by other investigators (Irwin, 1952. Lewis, 1951; Murai, 1960, 1963; Nakazima, 1962, 1966) do not show strong quantitative or qualitative changes between stages.

There are two conceptual problems raised by this question. The first is in deciding what constitutes a significant discontinuity. The second is the possibility that there are some elements of language which show a continuous development and some which do not.

With respect to the first problem, the discontinuity position claims to be able to demonstrate a difference in the patterning of utterances occurring before and after the onset of Stage 4. Supporters of this position generally agree that the articulated sound segments common to Stages 1 through 3 (the "prelanguage" period) are unpatterned, randomly produced without voluntary control, and do not appear in any universal order. These investigators also point out that there is an observable decrease in the child's ability to produce many previously well-articulated sounds toward the end of the third stage. Depending on the particular sound involved, recovery of a previously well-articulated sound may take anywhere from several weeks to several years. In fact, many have observed the tendency of young children of even 5 years of age to continue to make certain regular sound substitutions. This seems to occur despite the fact that at an earlier age these children had no trouble in articulating these sounds. These observations have led Jakobson and Shvachkin to conclude that the young child has no voluntary control over his articulations and that it is only after the onset of Stage 4 ("true" language) that any voluntary control over the sound system is actually attained.

In contrast, the continuity approach claims that there is order and patterning in the utterances of the "prelanguage" period, particularly in the babbling stage, which forms the basis for later utterances. Despite Jakobson's statement to the contrary, it is our impression that there are a considerable number of developmental regularities prior to Stage 4. For example, virtually all accounts of babbling contain reference to an initial preoccupation with vocalic sounds (V), followed by consonantal sounds (C), and then CV and CVCV patterns produced with language-like intonation and rhythm. However, the fairly consistent observation that certain early sounds disappear only to later reappear in the child's systematic speech is surely troublesome for the continuity position.

With respect to the second problem, it is possible that segmental aspects (phonemes, or actual sound classes of a language) and suprasegmental information (prosodic information, that is, pitch, stress, and duration) show different developmental histories. Even if there is a significant discontinuity with respect to segmental information, there exists some information which suggests the opposite for intonation. For example, Lieberman (1967, p. 42), in a study of the basic respiration patterns involved in crying, suggests that the ". . . infant's hypothetical innate referential breath-group furnishes the basis for the universal acoustic properties of the normal breath-group that is used to segment speech into sentences in so many languages." In addition, Lewis (1951), Nakazima (1966), and even Shvachkin (1948), note the appearance and importance of intonation and its gradual transformation into adult patterns throughout the last three stages.

We would be remiss without a discussion of the distinction between production of a form and the systematic use of that form. Many would argue that in the babbling stage "sounds pronounce themselves" (Shvachkin, 1948) as opposed to being used in a systematic linguistic fashion. This argument is predicated on the assumption that there exists no such system in babbling. To the extent that no other systematic analysis of babbling besides Irwin's (1957) has been performed, we cannot eliminate the possibility that there is a highly structured system underlying babbling.

Although neither of the major theoretical positions we have discussed makes specific claims about the transitions between even earlier stages, it will be useful to examine the nature of these changes if we want to know when language begins. For example, in considering a restatement of our question ("When does language begin?") it is helpful to know whether language develops continuously from the initial crying stage. Two psychologists, Lenneberg (1967) and Wolff (1966b), have recently addressed themselves to this important question. Lenneberg distinguishes between two kinds of vocalizations: crying and cooing. The former is present at birth and appears to correspond with Wolff's description of the "angry" cry. With the exception of some minor maturational changes within this pattern, it remains virtually the same throughout an individual's life. Cooing emerges independently at 6 to 8 weeks, although it too is similar in origin to crying. It is cooing and not crying, according to Lenneberg, which develops into babbling and eventually speech. Lenneberg's main evidence for the independent genesis of these two kinds of vocalizations is his failure to observe any acoustic or articulatory similarities between them. On the other hand, Wolff's data suggest the origin of cooing or noncry vocalizations is found in crying. At 3 weeks he notes the emergence of what he calls the "fake" cry. It differs from the basic cry pattern in that it is longer in duration, different in fundamental frequency toward the middle of the utterance, and does not fall in frequency toward the end. Wolff observes that in its early stages the "fake" cry always precedes full crying, but becomes independent at approximately 6 to 8 weeks. (It is important to note that this is the age at which Lenneberg marks the onset of cooing.) It is the early temporal dependency between "fake" crying and "basic" crying that leads Wolff to believe that the former evolves from the latter. It is difficult for us to resolve the discrepancy between Wolff's and Lenneberg's findings at this point, since both researchers have made careful study of vocalizations over the same time period. Perhaps further study of the transition from the first to the second stage will clarify the issue.

An interesting source of data which gives indirect support to Wolff's assertion that cry and noncry vocalizations are closely related is provided by Cullen, *et al.* (1968); Cullen, Fargo, and Baker (1968); and Fargo, Mobley, and Goodman (1968). These authors find that the crying of 2-year-old infants and the noncrying sounds of 6- to 19-month-old infants are similarly affected by delayed auditory feedback (DAF). That is, modification of the normal time delay between speaking and hearing what one has just said affects cry and noncry vocalizations similarly. Although they did not actually examine infants of between 3 to 8 weeks of age, the data they did collect suggest that it would be informative to do so. If, in that age range, cry and noncry utterances are similarly affected by DAF it would suggest at least a partial relationship between the two.

The Basic Sound Repertoire. As we pointed out above, a commonly held position is that the young infant's sound repertoire contains at the very least all the sounds occurring in all languages of the world. Although this claim is not important for Jakobson's approach, it clearly is of the utmost importance for the learning-theory approaches of Latif (1934), Skinner (1957), and Thorndike (1943). If it is true, the infant might either (1) begin producing all sounds simultaneously, or (2) acquire them gradually so that he has all of them before 11 or 12 months of age. All available data suggest that there is a gradual increase in the frequency and variety of sounds produced during the first year—in short, that all the sounds do not appear simultaneously (Irwin, 1957; Lewis, 1951; Nakazima, 1962; and others). There is also some evidence that the young infant has not acquired all possible sounds by the end of the first year. For example Preston (cited in Moffitt, 1968) has found that infants 10 months of age, from a variety of language communities including English, do not produce the aspirated unvoiced stop /p/ as in *pie*. This observation leads one to ask whether there aren't other sounds absent from the child's repertoire throughout this period.

Again we meet the problem of the disappearance and ultimate reappearance of many speech sounds at the end of the first year of life. Any attempt to explain language acquisition which emphasizes the early appearance of sounds and their incorporation into the child's phonological system—such as the learning-theory approach—will have to account for this phenomenon. The problem is complicated by the fact that sounds which occur relatively early in one stage, such as /r/ [as uttered in the comfort sound *ra* (Lewis, 1951)] may emerge relatively late in the next stage. It is interesting to note that the absence of previously controlled sounds is restricted to the simple wordlike utterances which begin to appear toward the end of the first year. The child may still produce these sounds in spontaneous babbling (which continues for some time after the end of the first year) and according to Jakobson's (1968) and Shvachkin's (1948) observations, he also retains his ability to imitate these sounds. Superficially, there seem to be two different sound systems in operation at this point. This fact suggests that the information-processing abilities of infants are limited. It could be that a great deal of processing is required to control an immature articulatory system well enough to produce early wordlike forms. Or, as some researchers suggest (Shvachkin, 1948), it may reflect the child's increased involvement with semantic processing, leaving less processing available for the control of production. We will return to the involvement of semantic processing in early language in a later section.

Reception

In the previous section we traced some aspects of the development of speech in the first year of life. It is important not to confuse what the child says, his speech, with what he knows about his language. Lenneberg (1962) has documented the case of a boy who, because of a congenital inability to control articulation even minimally, was unable to learn to speak. Nevertheless the child exhibited normal comprehension of language. Thus, the vocal aspects of language should be taken as symptoms of an underlying language system, rather than as the system itself.

Theoretical approaches to language acquisition have not often made explicit their views on the development of receptive abilities. Early learning-theory approaches assumed that production preceded reception, but Lenneberg's data, just referred to, cast strong doubt upon that assumption. Later learning-theory and linguistic approaches generally assumed that the child must be able to appreciate differences before he could systematically produce them. Strikingly little data were collected, however, on the ontogeny of reception or its relation to production and the underlying language system. Many investigators believed that the relationship between speech and language was rather direct and therefore saw no reason to focus on language perception. In addition, experimental techniques which could be used to examine early receptive development were not available.

Chronology. The development of the child's discriminative system is not so easily divided into stages, but it is possible to sketch a rough picture on the basis of the kinds of auditory inputs that can be discriminated at various ages.

Stage 1: The infant at birth seems to respond to some auditory stimulation. For example, Wertheimer (1961) demonstrated that an infant only a few minutes old could successfully localize sound sources. There is, in addition, a growing body of evidence on the neonate's and young infant's abilities to distinguish among sounds differing in frequency, intensity, duration, temporal patterning, and location in space. (See Lipsitt, 1963; and Spears and Hohle, 1967 for a more complete description of this literature.) However, until 2 weeks of age there is little evidence that infants discriminate between voice and nonvoice inputs.

Stage 2. At approximately 2 weeks, according to Wolff (1966b), the infant begins to distinguish between voices and other sounds. His evidence is that the voice, as opposed to inanimate nonhuman sounds (for example, bells, whistles, rattles) is especially effective in arresting crying. Other investigators have noted the prepotency of the voice in eliciting smiling and vocalizations within the next month and a half (Champneys, 1881; Hetzer and Tudor-Hart, cited in Lewis, 1951; Lewis, 1951; Nakazima, 1966; Preyer, 1890; and Rheingold, 1961).

Stage 3. During a third stage beginning at the end of the second month and continuing into the third and fourth months, the infant seems to discriminate between affective qualities of utterances. Bühler and Hetzer (cited in Lewis, 1951) found that angry voices produced withdrawal responses at this age, while friendly voices elicited smiling and cooing. Similar observations are reported by Wolff (1963) and Lewis (1951). During the early part of this period we see the development of the ability to distinguish between familiar and unfamiliar voices (Wolff, 1963). The end of this stage is marked by discrimination between male and female voices (Kaplan, 1969).

Stage 4. In the fourth stage, which begins at approximately 5 to 6 months, the infant begins for the first time to show the ability to attend to and discriminate between several kinds of linguistic information. Many observers have commented on the increasing sensitivity of the infant at this stage to the intonation and rhythm of adult utterances. Kaczmarek (cited in Weir, 1966) reports that a 5-month-old Polish child made the same response to a series of utterances which all had the same overall intonation pattern whether composed of Polish or non-Polish words. In a similar experiment, Tappolet (cited in Lewis, 1951) trained his 8-month-old son to turn toward the window when he said, "Wo ist das Fenster?" Following this training he said, "Où est la fenêtre?" with the identical intonation, and the child looked at the window. Tappolet concluded that his child had been attending to the overall intonation pattern. There are many other reports emphasizing the importance of intonation to the young child.

Until quite recently it remained to be shown experimentally that young children could discriminate between alternate intonation patterns. E. L. Kaplan (1969) conducted an experiment utilizing newly developed experimental techniques for assessing perceptual development in infants.

Specifically, 4- and 8-month-old infants listened to repetitions of the sentence, "See the cat." One group of children of both ages heard half the repetitions pronounced with a rising intonation contour typical of some classes of questions—"See the cat?" On the remaining repetitions the sentence was pronounced with a falling intonation, a contour characteristic of statements and some classes of questions—"See the cat." Another group got the reverse order. In both cases the sentences contained the normal American-English stress patterns. A second group of children heard the same two intonation patterns, but without the normal terminal stress changes. Discrimination of the intonation patterns with degree of stress held constant was tested. Subjects heard a sequence of repetitions with intonation held constant, then the intonation was changed. The data were analyzed for significant changes in heart rate and behavioral orientation at the point of stimulus change. The results indicated that 8-month-old infants could discriminate the two intonation patterns when they contained normal stress assignments but not when normal stress was lacking. The 4-month-old infants could not discriminate in either case. The data seem to support the assertion that in Stage 4 infants develop the ability to discriminate normal intonation patterns.

With the exception of a study by Moffitt (1968) most of the literature suggests that at Stage 4 the child is primarily processing suprasegmental aspects of speech such as patterns of intonation, stress, and duration. Moffit's study used procedures essentially identical to those used by Kaplan (1969), but he was interested in a segmental feature of speech. He examined 5-month-olds' discrimination of the contrast between two phonemes, /b/ and /g/, embedded in a common vocalic environment (*bah* versus *gah*). The crucial acoustic basis for this discrimination for adults is an extremely short 60 msec. period in which the two sounds exhibit different spectral characteristics. (Properly speaking the difference between the two sounds is given by the differently shaped second formant transitions.) Moffit found clear evidence of the ability to make the discrimination. We will return to a discussion of this experiment in a later section.

Stage 5. Toward the end of the first year, the segmental aspects of language become more important. The child becomes able to distinguish among the various phonemes of his language. There is an astonishing lack of data on perception in Stage 5. The only systematic study of which we are aware is by Shvachkin (1948). In a rather ingeneous experiment, this researcher tested the ability of Russian children from 10 to 18 months of age to comprehend commands to retrieve certain objects. Because the names of various pairs of objects differed only in the initial phoneme, he was able to infer the kinds of phonological contrasts present in the child's system. He found that the development of the child's phonological system proceeded very much in the order suggested by Jakobson (1968) and Jakobson and Halle (1956). These researchers appear to have independently derived this same order.

The Relevance of Early Discrimination to Later Language. There is little disagreement with the chronology sketched above. However, there is considerable disagreement about the significance of early receptive abilities in later language acquisition. The disagreement is usually focused on the transition from Stage 4 to Stage 5. Let us consider the role of intonation as an example. Some writers (Bever, Fodor, and Weksel, 1965, and McNeill, 1966b) have held that appreciation of the role of intonation can come only after the mastery of syntax. McNeill (1966b) notes that

> A widely accepted generalization about languages is that there is a close connection between phonology and syntax, especially in the imposition of intonation contours. The existence of this connection has caused some psycholinguists to suggest that intonation—which is observable in speech—might be the vehicle on which children arrive at rudiments of syntax. At first glance this is a plausible view. . . . [However] it is difficult . . . to see how intonation could guide a child to syntax; for no matter how strong the tendency is for children to imitate speech they receive from their parents, they will not imitate the appropriate feature unless important parts of the syntax have already been acquired. (1966b, pp. 52–53)

An alternative view has been presented elsewhere by Kaplan (1969). He argues that McNeill's position, and those similar to it, are based on a misinterpretation of an earlier study by Lieberman (1965). Lieberman found that when linguists were asked to identify the intonation and stress patterns in utterances using a rather complex notational system, their accuracy and reliability varied as a function of their knowledge of the structure and content of the utterances. A second finding often neglected by interpreters of this study is that increased accuracy and reliability resulted when a simpler notational system was used to transcribe the same set of utterances. This finding casts strong doubts on the arguments of Bever, Fodor, and Weksel; and McNeill.

The alternative position is that intonation could be used to provide some minimal kinds of semantic and syntactic information. For example, Lieberman (1967) has documented how intonational signals are used to define major semantic-syntactic boundaries. Speakers seem to adjust their speech rates and respiratory cycles in such a way that the onsets and offsets of intonation contours (or, in Lieberman's term, Breath-Groups) occur primarily at sentence boundaries or major phrase boundaries within sentences. In addition, speakers seem to exploit the use of intonational signals at points of potential ambiguity in sentences. Lieberman's data suggests to us that listeners could rely on such information in decoding language. Conceivably an infant tuned to detecting intonation patterns could begin to isolate semantically and syntactically important aspects of messages on the basis of intonation. [See Kaplan (1969) for more complete documentation of this argument.] Braine (1963a) and others take a stronger position, apparently suggesting that intonation is a vehicle by which a child might learn some rather subtle syntactic distinctions in his language. All we mean to suggest is that the early appreciation of intonation might provide infants with a primitive way of segmenting an otherwise continuous stream of sound. We are inclined to believe that the attention to intonation in Stage 4 marks a significant stage in language acquisition.

Continuities and Discontinuities in Receptive Development. Information on continuity between early stages of receptive development does not distinguish between the two major approaches to language acquisition as easily as did information on the stages of production. This is primarily because, as we noted earlier, the advocates of neither approach concern themselves with reception. We can speculate, however, that if elaborated to include reception the theories would emphasize either continuity (learning theory) or discontinuity (linguistic approach) between stages.

With respect to this issue it is interesting to consider the transition between the first three stages and Stage 4. We noted earlier that the second stage was characterized by attention to the voice, the third by discrimination of affective tone, and the fourth by discrimination of suprasegmental patterning. Let us look into the possibility that there is some uniform characteristic linking these three stages.

For reasons which we do not fully understand the young infant at around 2 weeks of age (Stage 2) shows more interest in the human voice than in other sounds. It could be, as Lewis (1951) and others before him have suggested, that this interest in the voice is innate; the infant may come equipped at birth with the ability to discriminate species-specific sound patterns (Marler and Hamilton, 1966). Or it may be, as Mowrer (1958) suggested, that by 2 weeks the infant has learned to associate the presence of a nurtural caretaker with the human voice. Be that as it may, discrimination of human from nonhuman sounds alerts the child to the specific affective properties of the voice. It is likely that there is at least some overlap between the acoustic properties which differentiate between, for example, angry and pleasant expressions and those which underlie the suprasegmental system. If this is the case, then detection of these properties in the context of different affective states will point the child toward, and perhaps even sensitize him to, the systematic use of suprasegmental features and all the subsequent information which he is directed to through intonation. Thus, we are arguing that with respect to at least some receptive aspects of the child's language system it is possible to find antecedent events as early as 2 weeks. This argument, in many respects, agrees with Lewis' (1951) approach to the ontogeny of language.

Lieberman (1967) also argues that the roots of the intonation system are present at birth with the "first" cry. However, according to him, attention to intonation does not differentiate out of attention to affective qualities of speech; instead, intonation is salient to the child because it is present in his own earliest vocalizations, being determined by properties of the efferent nervous system.

> . . . during the first minutes of life children employ "meaningful" intonational signals. The cries are at first meaningful only in that they have a physiological reference. We believe that these signals, which appear to be innately determined, provide the basis for the linguistic function of intonation in adult speech. (1967, p. 41)

It is clear that Lieberman believes that the development at least of intonation is a continuous process. Although he does not take a strict "motor theory" approach, it is not clear what he feels the consequences for reception would be of not being able to produce normal intonation patterns, or any speech for that matter. It would be interesting to know how Mongoloid children perceive intonation patterns. According to Lieberman they do not exhibit the cries characteristic of normal neonates which ultimately lead to normal adult intonation.

There do not seem to be any data or observations suggesting discontinuities between Stages 3 and 4. The difficulty is that most investigators of receptive development have started their discussion with Stage 4. Generally speaking, observers note that receptive control over the suprasegmental system (Stage 4) emerges before the child gains equivalent control over the segmental system (Stage 5), (Fry, 1966; Kaplan, 1969; Lenneberg, 1967; Lewis, 1951; Shvachkin, 1948; Zhurova, 1963). For example, Lewis distinguishes the following 3 stages:

> 1. At an early stage the child shows discrimination in a broad way, between different patterns of expression in intonation.
> 2. When the total pattern—the phonetic form together with intonational form—is made effective by training, at first the intonational rather than the phonetic form dominates the child's response.
> 3. Then the phonetic pattern becomes the dominant feature in evoking the specific response; but while the function of the intonational pattern may be considerably subordinated, it certainly does not vanish. (1951, pp. 115–116).

However, Moffitt's (1968) data, referred to earlier, are an apparent exception to this trend. In light of all the observations to the contrary, it is surprising that Moffitt's 5-month-old subjects were able to discriminate the acoustically subtle distinction between *bah* and *gah.* It is possible to argue that Moffitt's subjects were making a phonetic rather than a phonemic discrimination—that is, discriminating on the basis of continuous differences along the sound spectrum rather than classifying the two sounds as categorically different. Stating this rather simply, it is possible that the discrimination was not a language-specific one. In order to demonstrate that the discrimination was phonemic it would be necessary to show that allophonic variations (specifically, acceptable variations of a phoneme) are not perceived as being different from that phoneme. For example, it would be necessary to demonstrate a discrimination between the two phonemes /b/ as in *bah* and /g/ as in *gah* but not between the /b/ in *bah* and its allophone /b/ in *tab.* A similar observation can be made about Kaplan's (1969) study of intonation. However, since the results are consistent with most other observations the question does not seem as critical.

What preliminary conclusions can we draw concerning the two main approaches to early language development outlined earlier? There is evidence for some continuity of development in the discrimination and production of suprasegmental aspects of language, and this evidence would seem to favor the learning-theory approach. However, it seems hard to reconcile the evidence that there are ordered sequences of development with such an approach. This latter evidence is of course consistent with the linguistic approach associated with Jakobson, although there are suggestions of even more regularity in development than he would have predicted. The disappearance and subsequent reappearance of certain sounds toward the end of the first year cast doubt upon all the learning-theory approaches discussed. Other observations concerning the presence of language in certain clinically interesting patients and the apparent fact that infants do not produce all sounds further question the validity of most learning-theory approaches. It seems unlikely that either approach can adequately account for the data on production and especially reception. However, there remain two other important issues which require discussion before we can confidently complete our evaluation of these two theoretical approaches.

Role of the Environment

A complete understanding of the course and nature of language acquisition must include a description of the role of the environment. It plays a role, of course, for at some point the child acquires the ability to communicate in his "mother tongue." Theorists disagree widely, however, on how the child manages to "lock in" on his language community. The traditional learning theorist assumes that all the requisite linguistic information is made available to him by the speakers in his environment. The child begins with no knowledge of his language and by means of a combination of imitation, selective reinforcement, and generalization gradually grasps the substance of his language. Others have been less convinced of the efficacy of the environment in shaping the acquisition process. Recently, many linguists and psycholinguists, have pointed out that there are both empirical and theoretical problems surrounding such a position (Chomsky, 1959; Lenneberg, 1967; McNeill, 1968). They have suggested that children are born with a biologically based, innate capacity for language acquisition which includes considerable knowledge about universal aspects of language. The role of the environment, then, is to provide information about particular languages.

In evaluating these positions we will discuss evidence relating to the influence of the linguistic context and the behavioral environment in which early language acquisition takes place.

The Role of the Linguistic Context

In discussing the influence of the linguistic environment on the early acquisition process, it is important to focus on aspects of language which characterize particular languages. The absence of such data on this suggests that observers have noted no significant early developmental control over these aspects. There are only two recent studies which suggest the appearance of characteristics specific to individual languages in the first year of life. Weir (1966) examined the babbling of 6-month-old infants from monolingual American-English, Chinese (Cantonese), and Russian families for evidence of differences. Her impression was that the vocalizations of Russian and American infants differed from those of the Chinese infants, especially in intonation and stress patterns. Although there seem to be some technical problems in the design of this experiment and the subsequent scoring, a similar experiment by Tervoort (cited in Weir) lends support to this major finding. Tervoort found that the babblings of American and Dutch infants were discriminably different to naïve observers. Unfortunately, it is unclear on what basis his judges were able to make this discrimination.

The suggestion that productive control of intonation is one of the earliest appearing characteristics of a particular language is confirmed by Nakazima (1966). He compared spectrograms of the vocalizations of American-English and Japanese infants from early infancy to around 8 months of age. The earliest difference he could detect was in intonation at about 1 year.

Failure to find evidence for the acquisition of features which characterize particular languages does not indicate that the linguistic context has had no effect. It is entirely possible that the early role of the linguistic context is to transmit information about universal aspects of language, or, if the child already possesses this knowledge, to serve as a release of such information. Some of the data on production and perception which we reviewed earlier (Kaplan, 1969; Lieberman, 1967; Moffitt, 1968; Shvachkin, 1948) suggest, in accordance with Jakobson (1968), that it is indeed those universal aspects of language which emerge earliest.

The Role of the Behavioral Environment

While it is clear that the people as well as the voices that surround a child must be important, it is not at all clear of what relevance they are to language acquisition. Learning-theory approaches to language acquisition have placed much emphasis on the way others respond to sounds the child makes. According to one group (Latif, 1934; Skinner, 1957; among others) the infant emits vocal responses which are selectively reinforced according to their similarity to adult speech. A second approach (Mowrer, 1958; Winitz, 1968) claims that the infant imitates the speech sounds he hears and his caretakers positively reinforce any infant vocalizations similar to adult speech. Both of these theories rely heavily upon the principles of imitation and reinforcement. Let us now consider some aspects of these notions, as applied to early language acquisition. A study by Rheingold, Gewirtz, and Ross (1959) found that it was possible to increase the frequency with which vocalizations occurred in 3-month-old infants by the use of social reinforcement. Other studies (Routh, 1969; Wahler, 1969; Weisberg, 1963) have attempted to extend this analysis. Generally these studies demonstrate only that it is possible to increase or decrease the frequency with which a given type of already present vocalization occurs. They cannot account for the occurrence of new forms. In addition, as Wahler points out in discussing his results, there is evidence that in a natural situation mothers reinforce vocalizations indiscriminately, not differentially. Indeed, parents may sometimes reinforce vocalizations which will be inappropriate at a later date. This certainly seems to be the case with "baby talk." Thus, there are at least empirical reasons for rejecting the selective-reinforcement approach.

There are even more serious theoretical reasons. For one thing, how can this approach explain the universal ordering of development in productive and receptive systems? It seems unlikely that any parent is familiar enough with the distribution of linguistic features across languages to construct the appropriate reinforcement schedule. What about the fact that, despite widespread variation in socialization practices, the initial stages of language acquisition seem to proceed relatively uniformly (Irvin, 1957; Lenneberg, 1967)? Finally, the rapidity and complexity of even the initial stages of a language-acquisition process seem to us to argue that reinforcement alone is not a sufficient condition for language development.

There are also empirical and theoretical objections to the idea that imitation is essential for language acquisition. Lenneberg and his colleagues (1967; 1965; 1964; 1962) recorded the vocalizations of children of deaf and hearing parents from birth to the end of the third month. If imitation were important during this period the vocalizations of the children with deaf parents should be considerably different from the vocalizations of children born to hearing parents, but the crying and cooing sounds of the two groups were identical. The result was the same in the case of a deaf child born to deaf parents. Lenneberg's report of the congenitally inarticulate boy who showed normal comprehension is another indication that example of imitation is unnecessary for language acquisition.

Furthermore, it seems hard to account for the side-by-side existence of two phonological systems toward the end of the first year within the framework of a theory of imitation. Presumably the language environment from which the child selects material to imitate remains constant while his productive system is undergoing significant reorganization. There are also changes in the receptive system and, short of any elaborate mediational approach, it is hard to see how imitation and reinforcement could account for these changes. Finally, it is often pointed out that the principles of imitation and reinforcement are not very useful when trying to deal with the generative aspects of language, be it phonology, syntax, or semantics.

Summarizing the above discussion, it appears that the traditional learning-theory approach to the role of the environment in language acquisition is inadequate. In our opinion a revision of the concept of reinforcement and imitation will not help; a whole new set of explanatory concepts is needed. Unfortunately, there do not seem to be any psycholinguistic or learning theoretic approaches rich enough to deal with the phenomena at hand.

The Functional and Linguistic Significance of Early Language Development

Traditionally the infant's first year has been seen as a laborious, hard-fought battle to adapt to the environment. Perceptually his world was assumed to be, as William James put it, "a blooming buzzing confusion." We have reason to believe that this is not at all an adequate representation of the infant's perceptual and cognitive abilities (Bower, 1966b). In our opinion there is evidence that it is not an accurate representation of his language abilities either. In the preceding sections we presented data indicating that the process of language acquisition, at least with respect to some features, begins perhaps as early as the first few weeks of life. However, our discussion was almost entirely restricted to the phonological system and, to a very limited extent, the semantic-syntactic systems.

In this section we want to speculate further on the development of the semantic system. Many investigators (for example, Ervin and Miller, 1963; McNeill, forthcoming; Mowrer, 1958) mark the occurrence of the first word as the beginning of this system. This would have a certain methodological value in that we could then look at distributional data on the use of these early words and infer from this data what the child's semantic system amounts to. However, in our view this would result in a severe underestimation of the child's semantic structure. For example, if the young infant's semantic system were more highly developed than his phonological system, then estimates of his semantic system which were based on his phonological system would certainly be inaccurate. Putting it somewhat differently, although we can infer facts about the child's semantic system from the way he uses words, the absence of conventional words does not necessarily indicate the absence of an underlying semantic system.

How are we to gather information about the child's early semantic system? One possible way is to consider the development of communicative functions in the child and try to infer possible semantic organization. We do not mean to suggest that the semantic and communicative systems are identical; rather they share some properties. We will sketch an extremely tentative chronology of the development of communicative functions, based primarily on the observations of a number of "baby biographers" (Darwin, 1877; Kirkpatrick, 1910; Leopold, 1947; Lewis, 1951; Major, 1906; Moore, 1896; Preyer, 1890; Stern, 1924).

By *communicative functions* we mean the systems by which information is transmitted between the child and his environment, including everything from crying to pointing to speaking. The crucial aspect of such systems is that they have a set of differentiated "messages." From birth we find such differentiation of messages in crying. Wolff's study (1966b) demonstrated that the child has a repertoire of different cries which communicate distinct messages about the infant's state. What is more, he confirmed the common observation that mothers can discriminate between the various cries. From about 2 weeks old, observers note, the infant is able to discriminate speech from nonspeech sounds. Somewhat later what appears to be voluntary control over the frequency and type of vocalization emerges. The frequency of vocalization increases in the presence of other people, and the child appears to vocally interact with others in his environment. At the same time the child seems to distinguish between familiar and nonfamiliar faces and voices, and responds angrily to removal of a favorite toy. During this period babbling-type sounds which appear to be contextually defined emerge. The context consists of the momentary environmental situation as well as gestures and intonation. During the final period, we see a progression of wordlike forms which begin to phonologically approximate adult words. Semantically, these initial words may be very different from adult usage and appear to be rudimentary sentences or "sentence-words."

This rough sketch suggests that there is a progressive differentiation and elaboration of the communicative system. It might be argued that it cannot represent the beginnings of a true, symbolic semantic system. Since many of the early "messages" are tied to the affective state of the child, one might conclude that they do not possess the requisite symbolic aspects which characterize semantic systems. But we are inclined to believe that only a very parochial view of semantics would allow that conclusion. Most current linguistic theories recognize that an adequate semantic-syntactic theory must include distinctions which reflect universal properties of the environment and events in that environment, including affect (Chomsky, 1968b; Fillmore, 1968; Katz and Fodor, 1963). We assume that if we can show that important distinctions about the environment and the events within it are present early in the development of the child's communicative system, there is every reason to believe that those distinctions are also part of the child's emerging linguistic competence.

Although we are severely limited by lack of data, we want to present a few examples of features likely to be present in the early semantic system. Our identification of these features, their developmental ordering and the relationships between them should be taken as highly provisional. A similar approach has been suggested by Ingram (1969) who has attempted to adapt Fillmore's (1968) "case system" to descriptions of child language. He is interested in an analysis of early syntactic development, whereas we are at present primarily concerned with an analysis of early semantic development.

One of the early distinctions seems to be between human and nonhuman. We might represent it as "± human." Evidence for this discrimination is the differential response of the infant to voices versus other sounds. An initial differentiation of self versus other, which could be marked "± ego," is suggested by the very early effect of DAF on crying, indicating that the infant can distinguish between his own voice and other sounds. The ability of infants to manipulate the meaning of what superficially appears to be the same phonetic form by the application of different intonations and gestures suggests that the infant realizes his communicative control; this generative use of intonation and gesture could be taken as evidence for further differentiation of the "± ego" distinction. The child's realization that he can affect his environment by his speech must surely mark an important step in his cognitive and linguistic development. A highly related development probably occurs slightly later when the infant becomes able to differentiate vocally between requesting and rejecting objects not directly relevant to his "state." We can represent this as "± request."

At the same time it is possible that there are similar developments with respect to the child's knowledge of object properties. Recently Wolff (1966b) has reported that 2-month-old infants seem to respond specially to the presence of favored toys. We could take this, together with other effects of familiarity, as evidence for the marking of "± presence" of an object. It is interesting to note that at approximately the same age Bower (1967) reports that infants discriminate between ecologically possible disappearances of objects versus nonecologically possible disappearances. Thus we might have evidence for a distinction "± presence" which is joined with "± existence."

There is other evidence which suggests the existence of marked distinctions such as "± caretaker," "± female," and "± past." There is reason to suspect the existence of many more such contrasts. For example, it would be interesting to know when the distinction "± agent" appears. Ingram (1969) suggests that this distinction appears around 15 months. This is an important distinction in the realm of both semantics and syntactics, because it requires the synthesis of object and event distinctions. In general we expect to find continuous changes in the ordering, priority, and interrelationships of various semantic features. Observation of such changes can tell us much about the principles underlying the evolution of semantic systems.

Of course this is all highly speculative. Considerably more experimental work is needed before we can conclusively say when the child possesses control over these distinctions. Such work would need to pay careful attention to the possible use of intonation, gesture, context, and so forth in early semantic systems. Earlier we discussed the use of intonation and gesture as devices for marking semantic differences. A recent unpublished study by Charrow, Ingram, and Dil (1969) suggests that context is also very important for understanding what children mean. They selected twenty-five 2- and 3-word utterances from Leopold's biography of his daughter's language between 19 and 23 months. Leopold had noted the context in which each utterance was spoken so that it was possible to assign a "correct" interpretation. Charrow, *et al.*, gave the twenty-five utterances, out of context, to three groups of judges—linguists, mothers, and graduate students—who were asked to interpret them. For only 36 percent of the utterances did the judges assign the correct interpretation. Presumably the effect of context is even greater with "1-word" utterances.

In the absence of further work, it seems reasonable at the present time to hypothesize that the child's semantic system develops continuously from early distinctions present in his communication system. Thus we again find some evidence for very early beginnings of child language. Neither of the two major approaches to early language acquisition considered that the child's semantic system began this early. Although the evidence does not allow us to choose between the two positions, it seems likely that analyses of early semantic development will put significant constraints on any new approach to language acquisition.

Concluding Remarks

We began by asking, "Is there such a thing as a prelinguistic child?" In answering this question we considered two major approaches to early developments in the child's language system. What can we conclude about the relative merits of these two approaches? The assumptions, characteristics, and mechanisms of the learning-theory approaches do not seem able to accommodate the bulk of the available data. On the other hand, though the linguistic approach illuminates many important and empirically documented trends in the early development of language, it does not provide us with any hypotheses about the underlying mechanisms which could account for language acquisition. In addition, it posits a strong discontinuity in development for which there is only marginal empirical support.

In answering our question we have also examined the evidence that language functions develop continuously. In our opinion there is evidence for significantly more continuity than has previously been supposed. In any event, it has become increasingly clear that the first year is a significant period for language acquisition. Of course, much more work is needed. We need more analyses of the changes in receptive abilities which take place during the first year. It would be helpful to have better analyses of early vocalizations, including babbling, with particular emphasis on the role of suprasegmental and contextual features. Additional data on language development in children from different language communities as well as children suffering from congenital abnormalities is needed to give us information on the interplay between universal and environmentally determined aspects of language.

Finally, a crucial but often neglected domain is the early development of the communication and semantic systems. Aside from the relevance of semantic developments to the "continuity" question, this issue may be important in answering the interesting question of what motivates changes in the developing linguistic system. For example, Shvachkin (1948) has suggested that semantic development motivates phonological change. It is even possible to speculate that the semantic system is the most highly developed language system during the first year of life. Attention to concurrent changes in early semantic systems and other cognitive systems is likely to be crucial to any real understanding of language acquisition. Although all these questions must ultimately be answered, it seems reasonable to conclude at this point that there is probably no such thing as a prelinguistic child.

THE STATUS OF DEVELOPMENTAL STUDIES OF LANGUAGE*

Jacqueline Sachs

Much of the current research on language development focuses on the question: "Why does the child's form of expression change?" For example, a young child may say, *no push* rather than *don't push me*. What can account for the change from the initial sentence structure, through various stages, to the ultimate adult form?

Is Language Learned by Imitation?

One commonly held idea about how the child learns language is that he imitates what he hears and practices these forms until he can say them correctly. The child does imitate, to be sure, but there are strong arguments that imitation is not necessary for language learning, including the fact that some children learn language without being able to speak at all. (See Kaplan and Kaplan in this volume.)

Furthermore, an examination of the speech of normal children suggests that their sentences are not merely attempts at imitating adults' sentences. Children say many things that are very unlikely in adult speech and that they have, in all probability, never heard before. Utterances like *allgone outside* (said when the door was shut, apparently meaning the outside is all gone) or *more page* (meaning the mother should read some more) (Braine, 1963b) cannot even be considered shortened versions of adult sentences because those structures and orders

* This selection was written expressly for this volume.

are not found in the adult grammar. Nor are such utterances unpredictable, random combinations of words. If the child says *allgone outside,* he is also likely to say *allgone daddy* and *allgone car.*

Language Is a System of Rules

Most current psycholinguistic theories hold that the child creates a system of rules or grammar which governs his speech. By *grammar* we do not mean a set of prescriptive rules like the ones learned in school (such as "do not end a sentence with a preposition"). We are referring to the regularities that can be found in the language, that are adhered to by speakers although they are usually not even aware of them. Some examples will help clarify the concept of grammar and how we determine what grammatical rules the child is using.

The past tense of most verbs in English is formed by adding a suffix. In writing, we add "-ed." In speech, however, the actual sound added varies according to the last sound in the verb. For verbs that end in /t/ or /d/, we add /-əd/ (for example, *melted*).[1] If the verb ends in a phoneme other than /d/ that is accompanied by vocal chord movement (such as a vowel, /b/, /n/), we add /-d/ (*played, learned*). If the final phoneme in the verb is an unvoiced consonant (/p/, /k/), we add /-t/ (*stopped*). Verbs that form their past tenses this way are called regular because the tense change can be exactly specified by rule. To determine whether a child is following these rules we could, of course, notice whether he was using the past tense of verbs correctly.

However, he might conceivably have learned each of the past forms individually. To be certain that the child knows the rules for past tense, we can teach him a nonsense verb. If he can produce the appropriate past form of this verb, we can assume that he is using the rule correctly. Berko (1958) tested preschool and first-grade children for their knowledge of past tense, plurality, and other grammatical distinctions. In her experiment the child was, for example, shown a picture of a man swinging something and told "This is a man who knows how to rick. He is ricking. He did the same thing yesterday. What did he do yesterday? Yesterday he ________." Most of the children supplied the correct past tense form. Berko concluded that:

> If knowledge of English consisted of no more than the storing up of many memorized words, the child might be expected to refuse to answer our question. . . . This was decidedly not the case. The children answered

[1] The / / sign means that we are indicating the pronunciation rather than the spelling of a sound. The sound within the slashes is called a *phoneme.*

> the questions. . . . The answers were . . . consistent and orderly . . . , and they demonstrated that there can be no doubt that children in this age range operate with clearly delimited . . . rules (p. 371).

Although he can use it in a new situation, the child cannot tell you what the rule is. Even adults cannot usually state many of their grammatical rules explicitly. It is the job of the linguist to discover the regularities of the language that we use. People can, however, tell you that something is wrong if a rule is broken. For example, almost anyone will recognize *Does he wanted to go?* or *The book was had by John* as wrong.

Aside from tests of productive rule use, there is much evidence in the child's actual speech that he is following rules, not just imitating the speech he hears. For example, Ervin (1964) has noted a very interesting aspect of the development of correct use of verb tenses. In English there are irregular verbs that do not follow the rules given above. For example, *go* becomes *went* in the past. In fact, many of the most common verbs in English are irregular (*go, give, break, fall*). Since the irregular verbs are so frequent, the child will probably say *we went, I fell,* and so on at an early point in his language development, having learned the words as separate vocabulary items. When the child eventually learns the rules for forming the regular past tense, he will temporarily stop using the correct, highly practiced irregular past forms and apply his new rule to all verbs. Thus, although he said *went* and *fell* earlier, he now says *goed* and *falled.* This phenomenon gives evidence of the powerful motivation to impose order on the language one perceives.

Other examples of grammatical rules are those which describe the form of negation. The negative element must be inserted into the sentence at a certain point, and this insertion often necessitates other changes in the sentence. To negate the sentence:

The boy is running,

we must insert the word *not* after the first auxiliary verb in the sentence:

The boy is not running.

However, to negate:

The boy ran,

we cannot simply insert *not* because there is no apparent auxiliary verb.

The boy not ran.

A marker for the first auxiliary verb is needed to carry the negation; namely, the appropriate form of *do*. Note that *do* in its simple form is not appropriate:

The boy do not ran.

The tense marking is carried by the auxiliary element. In this case *do* becomes *did* and *ran* reverts to the infinitive *run*.
Finally we produce the correct negation:

The boy did not run.

Though these rules for negation are not all inclusive, they are basic:

1. Insert *not* after the first auxiliary verb.
2. If no auxiliary verb is present, insert *do* before inserting *not*.
3. Mark the tense of the sentence by the appropriate form of *do*.

How could we determine what rules a child is using for negation? Bellugi (1967) studied the language development of three children from the time they first formed two word utterances until they had mastered the basic grammatical rules of English. The earliest form of negation was the simple addition of a negative word (*no* or *not*) to the sentence. The child would say *no the sun shining* or *wear mitten no*. A little later, many new forms were added:

I can't see you.
We can't talk.

I don't sit on Cromer coffee.
I don't want it.

Don't leave me.
Don't wait for me.

That no fish school.
There no squirrels.

No rusty hat.
Touch the snow no.
This a radiator no.

I want not envelope

Quite a few of these sentences could have been spoken by an adult. Can we conclude that the child is using the same rules for the formation of negatives that an adult would? Bellugi argues that the child, at this point, has learned several different ways to negate but does not yet have the unifying rules of negative placement and insertion of *do.* At this stage, the child does not yet say affirmative sentences containing *do* or *can*; so, although *don't* and *can't* appear, the auxiliary system does not seem to be present. He has apparently learned *can't* and *don't* as separate vocabulary items. Furthermore, the child still uses the earlier forms of negation like *Touch the snow no.* We must conclude that although the negations are more complex at this stage, they are not formed by the rules of the adult's system. Bellugi feels the child has a number of different rules, each applicable to a certain type of sentence.

By the next stage in development, however, the earlier forms of negation have disappeared:

Paul can't have one.
I can't see it.
We can't make another broom.

I am not a doctor.
This is no good.
Paul not tired.
That not a clown.

I not crying.
He not taking the walls down.

You don't want some supper,
You didn't eat supper with us.
I didn't see something.

Don't put the two wings on.
Don't touch the fish.

I not see you any more.
Ask me if I not make mistake.
Fraser not see him.

I gave him some so he won't cry.

For the first time auxiliaries are found in affirmative sentences (*I can see it*). We have reason to believe that the child may be using something like the adult rule of adding negation to the auxiliary, although there are still deviations from the adult form. For example, the verb *be* seems to be optional, so that some sentences will be like the adult form (*I am not a doctor*) but others not (*I not crying*). This is true in affirmative sentences also. Other, more complex rules of negation are also absent from the child's system at this point, but we could conclude that he has rules for inserting the negative element and adding the appropriate form of *do* if it is missing. In other words, he has the basic rules of negation in English.

The Role of Imitation in Language Development

Although the child's imitation of the speech he hears cannot fully account for language development, it may play some part. Certainly imitation is an obvious and pervasive part of the child's behavior. If imitation does serve to improve the child's grammar, we might expect that the imitations would be more advanced grammatically than the sentences he says spontaneously. Several investigators have sought to determine whether the child can and does say advanced sentences when imitating. Fraser, Bellugi, and Brown (1963) tested children between the ages of 37 and 43 months for their ability to produce, comprehend, and imitate sentences of varying grammatical structure. They found that these children could imitate sentences they could neither produce on their own nor comprehend.

Ervin (1964) recorded the spontaneous imitations of five children, from the time they were 22 months until they were 34 months. She found that the child's imitation was never more complex than his spontaneous speech. In fact, the imitations could be predicted from the same grammatical rules that she had written to account for the child's production. For instance, *I'll make a cup for her to drink* produced *cup drink.* Ervin concluded that "there is not a shred of evidence supporting the view that progress toward adult norms of grammar arises merely from practice in overt imitation of adult sentences" (p. 172).

The differences in the results of these two studies might be accounted for by two differences in the methods. Fraser, Bellugi, and Brown analyzed elicited, not spontaneous, imitations. The study of spontaneous imitation probably gives a better indication of the significance of imitation in development. Also, Fraser, Bellugi, and Brown tested an older group of children who presumably had a longer memory span. Ervin's subjects were at an age when the ability to repeat rote material (such as nonsense words) is quite limited. Thus their imitations were actually paraphrases of the meaning that they had understood, using their own grammatical systems. Menyuk (1963) has also found that young children, when repeating adult sentences, modify them by their own grammatical systems. Therefore, at the age when speech imitation is most frequent in the child's development, it would appear to play little part in improving the child's grammar.

Slobin (1968) has suggested another way in which imitation might play a role in language development. It had been pointed out (Brown and Bellugi, 1964a) that adults very frequently "expand" what the child has said (that is, correct it and add to the grammar). Slobin noticed that after the parent's expansion, the child sometimes imitated the expanded sentence. This imitation could be less complex than his original sentence, the same, or more complex, as illustrated:

Less complex	*Child:*	*Play piano.*
	Adult:	*Playing the piano.*
	Child:	*Piano.*
Same	*Child:*	*Just like cowboy.*
	Adult:	*Oh, just like the cowboy's.*
	Child:	*Just like cowboy.*
More complex	*Child:*	*Pick 'mato.*
	Adult:	*Picking tomatoes up.*
	Child:	*Pick 'mato up.*

Analyzing transcripts of child-mother interactions, Slobin found that about half of the imitations of expansions were more complex than the original utterance. When the child imitates the parent's version of his original utterance, we have an ideal instructional situation, since the parent's expansion will ordinarily express the child's intention perfectly.

Slobin has suggested that perhaps there is a critical age for expansions, when the child is very likely to imitate what an adult says. As he grows older he imitates less, and the adult expands his speech less often. At about 30 months, much of the interchange between mother and child seems to consist of attempts to clarify what the other means. By about 36 months there is more real conversation. Slobin suggests that the mother stops expanding when the child's utterances become clearer. In a transcript made at 18 months, 13 percent of the child's utterances were imitations. The percentage dropped steadily thereafter, until at 34 months the rate of imitation was only 2 percent.

We do not know yet whether Slobin's suggestions regarding the role of imitation in language development are correct. At this point we know only that imitation of expansions is a frequent phenomenon, not that it is necessary or important in the development of grammar.

The Initial Stages of Language Acquisition

Any theory of language acquisition must allow us to account for several important aspects of the initial stages of language development. All over the world language learning starts at about the same age even though different cultures have very different attitudes toward the emergence of speech. Some cultures allow each child to progress at his own rate. Others attempt to train the child and anticipate each new developmental stage eagerly. Such differences seem to have little effect on the pattern of language acquisition, as contrasted with the effects on learning of, for example, reading or arithmetic. (For further arguments concerning this point, see essay 16 by Lenneberg in this volume.)

A child has mastered most of the basic grammatical structure of his language by the time he is about 4 years old. When one compares the learning of a system as complex as language with the other types of intellectual tasks the child can master at this age, the speed of development is truly astounding. Moreover, it seems that the earliest grammatical structures children develop are very similar for all languages studied so far, although the structure of the adult languages may be very different (Slobin, in press). The studies Slobin reviews describe a similar rapid grammatical development during which the basic relations and categories are acquired. "For almost all children for whom sufficient data were available, the earliest stages of two-word utterances could be characterized by a definite structure" (p. 11). There are two classes of words: a small class called "pivot-words" (Braine, 1963b) and a large open class. In such utterances as *bandage on, blanket on, fix on*, "on" is the pivot word, and a large variety of other words can be combined with it. The similarity in types of pivot constructions found across languages is striking and Slobin concludes that a "classification of their conceptual content seems to reflect basic universal categories of syntax and semantics."

How can we explain the uniformity in the age at which language starts, the speed of acquisition, and the universality of the pattern of acquisition? Many researchers feel it is useful to think of the child as having certain characteristics from birth which predispose him toward language acquisition. There is in psycholinguistics today considerable controversy over the sorts of properties thought to be part of the child's innate mental structure. Some theorists propose that certain basic, universal aspects of the language are inborn. Others claim that some sort of special process for language acquisition is innate. These two types of theories are sometimes referred to as the "content approach" and the "process approach" (Slobin, 1966). There have, thus far, been few specific suggestions of what sorts of innate properties there may be, but we will briefly consider some hypotheses.

Slobin (1966) supports the "process approach." "It seems to me that the child is born not with a set of linguistic categories but with some sort of process mechanism—a set of procedures and inference rules, if you will—that he uses to process linguistic data" (p. 87). He points out that general cognitive abilities account for a great part of language acquisition. The child can learn certain types of semantic and conceptual categories and organize experience on the basis of abstract criteria. Although language acquisition takes place rapidly, so do many other aspects of cognitive development.

Fodor (1966) is also a "process theorist." He has suggested that the child comes equipped with a set of inference rules which function specifically in the creation of grammars from language input. These rules allow the child to process heard language with considerable foreknowledge about the kinds of grammatical rules that are worth trying. Fodor notes that of many errors that could appear in child speech, only a very limited set actually occur. The child may say *John eat lunch,* but one never finds *John has eating lunch.*

McNeill (1966b), a "content" theorist, proposes that essentially all universal aspects of language are "prewired." Among these universals are the "basic grammatical relations":

1. "subject of a sentence—predicate of a sentence" (NV) (for example, *Bambi go*)
2. "main verb of a predicate phrase—object of a predicate phrase" (VN) (*change diaper*)
3. "modifier of a noun phrase—head noun of a noun phrase" (mN) (*that coat* or *Adam coat*).

Essentially this means that the child does not have to discover, for example, that a sentence consists of a subject and a verb. He already "knows" that for a noun to be operated on by a verb, they must be in the same sentence. Since these relations are universal, they are appropriate bases for the acquisition of any language.

While analyzing two- and three-word utterances from one child, McNeill found that all of the utterances corresponded with the permissible universal patterns of nouns, verbs, and modifiers. For example, noun followed by verb followed by noun (NVN) is permitted (such as, *Adam change diaper*), but VVN is not. McNeill notes that in the speech the child hears, such patterns (VVN) do occasionally occur (*Come and eat your Pablum*). Such a sentence is not formed by the basic rules, but is produced by more complex rules of sentence conjunction and deletion that the child does not have at first. Such sentences, McNeill claims, are not imitated or used by the child because they do not conform to the basic grammatical relations.

Later Rule Acquisition

Whether or not some linguistic structure is innate, many rules are specific to the particular language that the child is learning and, therefore, must be learned by each child. One of the most perplexing characteristics of this learning is that the forms to be learned are not very well represented in the language the child hears. Bellugi (1967) has pointed out that the child hears only a very limited set of examples of some language structures, yet he is somehow able to reconstruct the underlying system from these examples. Learning to use the auxiliary verb *will* is a good example.

Bellugi (1967) analyzed 30 recorded hours of speech between mothers and children, and tabulated the frequency of occurrence of the various grammatical forms of the verb *will* and their negations (*I will, he will, I will not,* and so on). Although there are many forms, Bellugi found no examples of most of them in the sample the mothers produced. Moreover, the child must learn that some forms are only alternative expressions of the same meaning (such as *I will* and *I'll*). The mothers, however, hardly ever used the uncontracted forms. The child is able to reconstruct the entire system underlying the use of the verb *will* from the fragments he has heard. In fact, when he first learns the rules, he uses the uncontracted form—a form that he has practically never heard. Thus the young child will say *I will* instead of *I'll*. The child has probably used some of the contracted forms earlier as separately learned words, but when a rule is learned, it replaces previous usage, as we saw in the case of regular verbs. We can call this acquisition process a kind of induction of the system from the examples, but this label does not explain how the induction is accomplished when the examples are as scattered and incomplete as they often are in natural conversation.

Another important characteristic of the acquisition of rules is that, when a rule is learned, it has a great impact on the whole system the child is presently using. According to Slobin (1966), there are many examples from the Russian child language literature illustrating the fact that the emergence of a grammatical principle affects several aspects of the child's language simultaneously. In one child's speech, no words had endings to signal grammatical categories until about 20 months. In the period of one month, previously unmarked nouns became marked for: (1) number, (2) nominative, accusative, and genitive cases, (3) past tense, and (4) present tense. Somewhat later in the same child's development, gender agreement appeared simultaneously in two aspects of the language. Slobin states that "apparently once the principles of inflection and derivation are acquired . . . [they are] immediately applied over a wide range of types" (p. 136). Bellugi (1967) found this same aspect of development striking in all three English-speaking children she studied. For example, when the auxiliary verb system was learned, it appeared in the appropriate places in declaratives, negations, and questions at the same time.

Language and Cognitive Development

We have discussed the fact that language development occurs because of the acquisition of new grammatical rules. But why are the rules acquired? There are two aspects of this problem. First, the child is developing new ways to say old things. (At 24 months he might say

No throw ball and at 30 months, *Don't throw the ball.*) Beyond this he is developing the ability to say more complex things. The first aspect of grammatical change involves the invention of new rules that do not add to or change the semantic content of the child's sentences. McNeill (1966b) has suggested a reason for this type of development. The child invents the simplest possible grammar compatible with the language he has heard, but as he learns more he is forced to set up a number of special rules to handle things that his grammar cannot account for. For example, at one stage during the development of negation (Bellugi, 1967), the child has separate rules for negative questions, imperatives, and so forth. Eventually the child seems to discover regularities that simplify the system and create negatives more economically. McNeill calls the earlier, awkward set of rules "cognitive clutter" and suggests that grammatical advances come about to reduce this clutter.

The development of the conceptual, or semantic, system is another aspect of language change. Not only does the child begin with a grammar that is very different from the adult's; his whole system of structuring the world is different. The child must learn or invent all the categories and distinctions that he will eventually make as an adult member of his culture.

Though research into this interesting problem has begun only recently, the emerging pattern of results suggests that there is a complex interaction between conceptual development and language development. On the one hand, the child may learn some concepts because the language draws his attention to them. For example, English-speaking people distinguish between singular and plural. Yet the language of another culture may have three categories: singular, double, and more than two. Children in that culture would learn to think in terms of those categories. Slobin (1966) feels it is

> reasonable to suggest that it is language that plays a role in drawing the child's attention to the possibility of dividing nouns on the basis of animation; or verbs on the basis of duration, or determinacy, or validity; or pronouns on the basis of social status, and the like (p. 89).

On the other hand, the child has to be ready to discover the category before he notices it in the language. Ways in which the acquisition of distinctions in language depend on conceptual development have been shown in several recent studies.

In a review of Russian data on language development in children, Slobin (1966) suggested that the order of emergence of various syntactic categories depends on their relative semantic difficulty rather than on their grammatical complexity. The first grammatical distinctions to appear are those with concrete reference. For example, the singular-plural distinction comes very early and so do concrete noun suffixes. Categories based on relational criteria, such as cases, tenses, and persons of the verb, emerge later. Abstract categories continue to be added as late as 7 years. Conditional forms (such as "if-then" statements) are learned quite late, as in English. In Russia, however, their grammatical structure is very simple. This indicates that semantic complexity rather than grammatical difficulty is responsible for the time of development. Grammatical gender is the most difficult of all the categories for the child to master. It has almost no semantic correlates; it is arbitrary. There are no rules the child can discover to make the learning easier. Words referring to certain semantic categories emerge at the same time as the grammatical expression of those categories. For example, the word for *much* appears at the same time as the singular-plural distinction, and *right away* and *soon* at the same time as future tense. "The semantic and conceptual aspects of grammatical classes thus clearly play an important role in determining the order of their development and subdivision" (p. 142).

Bloom (1968) studied the development of language forms and semantic function from the age of 19 months in three children. She found that their negative statements could be divided into three semantic categories on the basis of the child's behavior and the situation at the time of the statement; these she called Nonexistence, Rejection, and Denial. In the Nonexistence category the referent did not exist in the context of the speech event. For example, *no truck* meant that there was no truck there. In the Rejection category the referent existed in the context but was rejected or opposed. *No truck* meant that the child did not want the truck. In the Denial category, the child asserted that some proposition was not the case. *No truck* meant that it was not a truck but a car. On the basis of relative frequency of occurrence and the progressive development of syntactic complexity, Bloom concluded that the order of emergence of these three classes was the order given above and was almost identical for each of the three children she studied.

A very similar pattern was revealed in an analysis of the development of negation in a Japanese child (McNeill and McNeill, 1967). The negative form is very simple syntactically in Japanese, since all negatives are formed by adding one of several words to the end of the sentence, depending on its meaning. According to the McNeills' analysis, negation falls into three semantic classes. The first is called "Existence-Truth" and involves the existence of things (for example, *There is no candy, It's not raining*) or the truth of propositions (*I'm not hungry*). This category corresponds roughly with Bloom's Nonexistence. The second category is "Internal-External." It refers to statements of individual desire (*I don't want to*) and is like Bloom's Rejection. The third category is "Entailment-Nonentailment" and refers to the ability to contrast two propositions with one another (*That is not yours; it's mine*). It is like Bloom's Denial.

The child studied acquired the categories in the above order, even though no category was any more difficult than another from a syntactic standpoint. These results, when combined with Bloom's, support the idea that conceptual development is crucial in determining the order of emergence of grammatical categories.

Cromer (1968) has supplied impressive evidence of the role of cognitive abilities in determining the language forms the child can use. He studied the development of temporal reference in two children over a four-year period. Several new types of reference to points in time began to occur regularly at about the same age for each child—4 to 4½ years. When viewed together, these new forms seemed to indicate that the child had a greatly expanded range of temporal reference and an increased sense of the possible relations between times.

Among other changes, Cromer noticed that the child could reverse the actual order of events, such as *And then sleep after I have lunch.* He could make statements of possibility (in addition to the earlier learned convention of "let's pretend") like *You could stay and drink with us,* or *Oh, I might get silver instead of black.* He began to use the present perfect tense, expressing the relevance of one time to another such as, *I wonder if Mommy has gone.* A clause in the sentence could be put in a tense other than the present, to express ability, possibility, and so on: *I thought you were coming with your car* (future reference in past); *How you won't be able to get it out?* (ability in future); *Maybe a man painted it* (possibility in past); and *You couldn't see me* (ability in past).

Cromer hypothesized that one factor in cognitive development was responsible for all the linguistic changes he observed in the expression of temporal relations. The child was freed from the immediate situation and the actual order of events. He could imagine himself at other points in time and view events as related to that time.

What can explain why certain forms are not acquired earlier? There is no lack of opportunity to learn. Cromer's children were exposed to many words and structures at an early age that they did not acquire or even imitate. For example, an analysis of the mothers' speech showed that they used the perfect tense frequently even when the child was very young. In fact, there was little variation in the grammatical forms the mothers used with children of different ages.

> Given that the child has the materials that make up the perfect tense and the capacity to produce utterances of a sufficient length to combine these, and given further that he has the model of such utterances in the adult speech around him, it is difficult to see why he is not using this form—until we examine its meaning. And when the data are analyzed from this point of view, it becomes apparent that the ability to properly use the perfect tense rests on a late-developing ability to consider the relevance of another time to the time of the utterance (p. 122).

The development of new structures took place in two ways. Sometimes the child would combine forms he already knew in a new way to express a new idea. For example, *before* and *after* were used spatially before they were used temporally. *Have* was used as a kind of auxiliary verb in the construction *have to* from an early age. No new linguistic forms were required for clauses to be placed in times other than the present. In other cases, the child did not have a form available to express a new idea. Often there was evidence in the child's speech of the search for a form to express an idea. One child sought to express ability in the past by saying *Uh . . . why did Batman couldn't go fishin'?*

> The development of a new cognitive ability leads the child to an active search for new forms to express the relations he is attempting to formulate. His acquisition from the surrounding "inputs" will not be a passive one. He will be actively directing his linguistic attention into a search for or heightened awareness of particular forms or structures used by adult speakers to express those newly understood relationships. There is directed activity, then, both in production and acquisition (p. 168).

Cromer concluded that certain cognitive changes occurring somewhat independently from language have a great effect on the language acquisition process. He suggests that language acquisition research needs to move away from describing the learning of linguistic rules, and toward an investigation of the cognitive abilities that are needed before the child can learn and use those rules.

Theories of language acquisition that consider only the linguistic aspect will not be able to explain why the child learns new forms when he does, or in fact why he ever changes his form of expression. It is only through more research on the complex relationship between cognitive development and language acquisition that we will have a full understanding of either. Hopefully in the future we will find more studies of this type, and a closer communication between psycholinguists and psychologists studying other aspects of child development.

part four

THINKING

The ability to think or to conceptualize distinguishes man from most other animals in the animal kingdom. No normal human is without this capacity, and no other species is known to possess it to the same degree. As William Perry has observed, "it seems evident that the higher animals gained their relative freedom and mastery through developing the ability to form concepts—that is, to think. Man has gone on to his own greater freedom—and bewilderment—by learning how to conceptualize about concepts, to think about his thoughts. Man is distinguished from the ape not by his reason, at which the ape is often no slouch, but by his meta-reason, which is a blessing with which the ape is presumably uncursed." (Perry, 1969)

Our possession of meta-reason, or the ability to conceptualize about concepts, carries with it some inherent limitations, however. The act of thinking, like the act of reading, continues to defy formal analysis. While we may be able to learn to "read" music, mathematics, and various languages, and while we may be able deliberately to increase our reading skill, we have only vague notions as to what the reading process itself entails. Similarly, while we may learn to think about a variety of topics, and while we may acquire great skill in thinking logically, the act of thinking remains a mystery.

Good mysteries provoke speculation, and the literature is full of speculations about the nature of thinking. "Knowledge: Concepts" by John Hospers begins Part Four with a philosophical and historical review of attempts to define concepts and the process of acquiring them. His essay parallels Ausubel's in that he considers various definitions of concepts as they relate to the influence of heredity and environment.

The second topic in Part Four consists of two complementary essays. Bourne and Kendler consider efforts by psychologists to translate philosophical concerns into a workable research framework. Bourne, in his introduction to *Human Conceptual Behavior*, presents a behaviorist's definition of conceptual behavior and then describes a general scheme for interpreting research in conceptual behavior. His essay provides many useful definitions for those unfamiliar with the literature, and his analysis of conceptual behavior may remind readers of the essays by Kessen, Gagné, Sachs, and others in this book. Kendler's "The Concept of the Concept" concentrates upon the efforts of learning theorists to extend relatively simple stimulus-response models of behavior to account for such higher-order behaviors as thinking. Like Kessen, he points out that our study of conceptual behavior is shaped by our preconceptions about human nature. Like Wohlwill, he concludes his essay with three dimensions of conceptual behavior in terms of stimulus-response language. Bourne and Kendler, in other words, demonstrate the difficulties involved in describing and analyzing the act of conceptualization or thinking.

Sam Glucksberg's "Thinking: A Phylogenetic Perspective" is the third essay in Part Four. Glucksberg considers the problem of conceptualization in terms of various categories and functions of representational processes. In particular, he points out differences between animals and man with respect to symbolic control as it develops through the use of stimuli as cues.

The fourth essay in Part Four is David Mouton's "The Principal Elements of Thinking: A Philosophical Examination." This essay directs our attention to the issues which confound our study of conceptualization when we attempt a logical analysis. Specifically, Mouton discusses three issues: the origins of thinking, the states of thinking, and the content or products of thinking. Mouton concludes his essay by sketching traditions in speculation about thinking in terms of Ausubel's heredity-environment continuum.

William Charlesworth's essay, "Cognition in Infancy: Where Do We Stand in the Mid-Sixties?" redirects our attention to the arena of research activity. He begins by discussing the relationship between infant testing and theories of human development and then considers the impact of current research upon possible thought about infant thought.

David Elkind, in "Cognition in Infancy and Early Childhood," the final essay of Part Four, contends that research in cognitive development has followed two paths of reasoning: that knowledge and the capacity to acquire it exist in some amount and can be measured; that knowledge and the processes of its acquisition change with age and need to be explained. Both paths of reasoning have led to some tentative conclusions which may have relevance to education. Piaget's theory is emphasized in Elkind's discussion of the second path of reasoning about cognitive development.

Some Topical Readings

SURVEY OF THEORETICAL POSITIONS

Berlyne, D. E., *Structure and Direction in Thinking*. New York: Wiley, 1965.
Harre, Rom, "The Formal Analysis of Concepts," in H. J. Klausmeier and C. W. Harris, *Analysis of Concept Learning*. New York: Academic Press, 1966.

COMPLEXITY OF PROCESSES

Deese, J., "Behavior and Fact," *American Psychologist*, 24 (1969).
Guilford, J. P., *The Nature of Human Intelligence*. New York: McGraw-Hill, 1967.
Pikas, Anatol, *Abstraction and Concept Formation*. Cambridge, Mass.: Harvard University Press, 1966.

COMPARATIVE STATUS

Hebb, D. O., *The Organization of Behavior*. New York: Wiley, 1948.

INTERRELATIONSHIPS

Werner, Heinz, and Bernard Kaplan, *Symbol Formation*. New York: Wiley, 1964.

INFANT RESEARCH

Fowler, Harry, "Effect of Early Stimulation in the Emergence of Cognitive Processes," in Hess and Bear, *Early Education*. London: Aldine, 1968.
Wohlwill, Joachim, "The Mystery of the Prelogical Child," *Psychology Today*, 1968.

LATER RESEARCH

White, Sheldon, "Evidence for a Hierarchical Arrangement of Learning Processes," in L. Lipsitt and C. Spiker, *Advances in Child Development and Behavior*. New York: Academic Press, 1965.

Additional Readings in Thinking and Conceptual Behavior

Adams, J. A., *Human Memory*. New York: McGraw-Hill, 1967.

Anderson, R., and D. Ausubel, *Readings in the Psychology of Cognition*. Holt, Rinehart and Winston, Inc., 1966, Part III.

Atkinson, R. and R. M. Shiffrin, "Storage and Retrieval Processes in Long-Term Memory," *Psychol. Rev.*, 76 (1969), p. 179.

Baldwin, J. M., *Thought and Things*, Volume I–III. New York: Macmillan, 1911.

Bartlett, F. C., *Remembering*. New York: Cambridge University Press, 1932.

Bartlett, F. C., *Thinking, An Experimental and Social Study*. London: G. Allen, 1958.

Berlyne, D., *Conflict, Arousal, and Curiosity*. New York: McGraw-Hill, 1960.

Bourne, L., *Human Conceptual Behavior*. Boston: Allyn and Bacon, 1966.

Brown, T., *Lectures on the Philosophy of the Human Mind*. Masters Smith, 1854.

Bruner, J., J. Goodnow, and G. Austin, *Study of Thinking*. New York: Wiley, 1956.

Donaldson, M., *Children's Thinking*. London: Tavistock, 1963.

Duncan, C., *Thinking: Current Experimental Studies*. Philadelphia: Lippincott, 1967.

Elkind, D., *Six Psychological Studies*. New York: Random House, Inc., 1968.

Flavell, J., *Developmental Psychology of Jean Piaget*. Princeton, N.J.: Van Nostrand, 1963.

Flavell, J., *et al., Development of Role-Taking and Communication Skills in Children*. New York: Wiley, 1968.

Fraisse, P. and J. Piaget, *Intelligence*. New York: Basic Books, 1968.

Gagné, R., *Conditions of Learning*. New York: Holt, Rinehart and Winston, Inc., 1965.

Gardner, R., P. S. Holzman, G. S. Klein, H. Linton, and D. P. Spence, *Cognitive Control*. International Universities, 1959.

Getzels, J. W., and P. W. Jackson, *Creativity and Intelligence*. New York: Wiley, 1962.

Goldstein, K., and M. Scheerer, "Tests of Abstract and Concrete Thinking," in A. Weider, ed., *Contributions Toward Medical Psychology*, Volume II. New York: Ronald, 1953.

Guilford, J., *Intelligence, Creativity and Their Educational Implications*, E.I.T.E., California, 1968.

Harms, E., "Fundamentals of Psychology: The Psychology of Thinking," *Annals of the New York Academy of Sciences*, 91 (1960).

Heidbreder, E., "The Attainment of Concepts," *Journal of Psychology* (1949).

Hellmuth, J., *Cognitive Studies*. New York: Bruner/Mazel, 1970.

Henle, P., "On Relations Between Logic and Thinking," *Psychol. Rev.*, 69 (1962), pp. 366–378.

Hess, R. D., and R. Bear, *Early Education: Current Theory, Research, and Practice*. London: Aldine, 1968.

Humphrey, G., *Thinking*. London: Methuen, 1951.

Hunt, E., *Concept Learning: An Information Processing Problem*. New York: Wiley, 1962.

Hunter, W. S., "The Symbolic Process," *Psychol. Rev.*, 31 (1924), pp. 478–497.

Inhelder, B., *Growth of Logical Thinking from Childhood to Adolescence*. New York: Basic Books, 1958.

Inhelder, B., *Diagnosis of Reasoning in the Mentally Retarded*. New York: John Day, 1968.

Inhelder, B., and J. Piaget, *Early Growth of Logic in the Child*. New York: Harper & Row, 1954.

Jensen, A., "How Much Can We Boost IQ and Scholastic Achievement," *Harvard Educational Review* (Winter and Spring 1969).

Johnson, D. M., *The Psychology of Thought and Judgment*. New York: Harper & Row, 1955.

Jordan, N., *Themes in Speculative Psychology*. London: Tavistock, 1968.

Kleinmuntz, B., *Problem Solving: Research, Method and Theory*. New York: Wiley, 1966.

Kleinmuntz, B., *Concepts and the Structure of Memory*. New York: Wiley, 1967.

Koffka, K., *Growth of the Mind*. New York: Harcourt, 1924.

Kogan, N., and M. Wallach, *Risk-taking: A Study in Cognition and Personality*. New York: Holt, Rinehart and Winston, Inc., 1964.

Langer, S., *Philosophy in a New Key*. New York: New American Library, 1964.

Lashley, K. S., *Brain Mechanisms and Intelligence*. Chicago: University of Chicago Press, 1929.

Leeper, R. W., "Cognitive Process," in S. Stevens, *Handbook of Experimental Psychology*. New York: Wiley, 1951.

Lunzer, E. A., *Development in Learning*. Amsterdam: Elsevier, 1968.

Luria, A. R., *Higher Cortical Functions*. New York: Basic Books, 1966.

McKellar, P., *Imagination and Thinking*. New York: Basic Books, 1957.

Morris, C. W., *Six Theories of Mind*. Chicago: University of Chicago Press, 1932.

Mussen, P., "European Research in Cognitive Development," *Child Development Monograph*, Serial 100 (1965).

National Foundation of Education in England and Wales, *Concept Growth and Education* (1966).

Norman, D., *Memory and Attention*. New York: Wiley, 1969.

Palermo, D., and L. Lipsitt, *Research Readings in Child Psychology*. New York: Holt, Rinehart and Winston, Inc., 1963, Chapter 12.

Piaget, J., *Development of Memory and Identity*. Worcester, Mass.: Clark University Press, 1968.

Piaget, J., *Psychology of Intelligence*. Totowa, N.J.: Littlefield, Adams & Company, 1960.

Pribram, K., "Neurological Notes on the Art of Teaching," *NSSE Yearbook*, 1964.

Price, K., *Thinking and Experience*. Cambridge, Mass.: Harvard University Press, 1953.

Rappaport, D., *Organization and Pathology of Thought.* New York: Columbia University Press, 1951.
Restle, F., "Cognitive Functions," *Annual Review of Psychology.* Palo Alto (1970).
Royce, J., *Psychology and the Symbol.* New York: Random House, Inc., 1965.
Russell, D., *Children's Thinking.* Boston: Ginn, 1956, Chapters 2 and 8.
Ryle, G., *Concept of Mind.* London: Hutchinson, 1949.
Saarinen, P., *Abstract and Concrete Thinking at Different Ages.* Helsinki, 1961.
Scheerer, M., *Cognition: Theory, Research, Promise.* New York: Harper & Row, 1964.
Scher, J., ed., *Theories of Mind.* New York: Free Press, 1962.
Sigel, I., "Concept Attainment," in Martin L. and Lois W. Hoffman, eds., *Review of Child Development Research.* New York: Russell Sage, 1964.
Sigel, I., and F. Hooper, *Logical Thinking in Children.* New York: Holt, Rinehart and Winston, Inc., 1968.
Smoke, K. L., "An Objective Study of Concept Formation," *Psychological Monographs,* 42 (1932).
Spence, K. W., and J. T. Spence, *Psychology of Learning and Motivation.* New York: Academic Press, 1968.
Stevens, S., *Handbook of Experimental Psychology.* New York: Wiley, 1951.
Thomson, R., *Psychology of Thinking.* New Orleans, La.: Pelican, 1959.
Thurstone, L., *Differential Growth of Mental Abilities.* Chapel Hill, N.C.: University of North Carolina Press, 1955.
Tolman, E., "Cognitive Maps in Rats and Men," in *Psychol. Rev.,* 55 (1948), pp. 189–208.
Vernon, M. D., *Intelligence and Cultural Environment.* London: Methuen, 1969.
Voss, J., *Approaches to Thought.* Columbus, Ohio: Merrill, 1969.
Wallace, Mc., *Concept Growth and the Education of the Child.* New York: Teachers College, 1962.
Wallach, M., "Research on Children's Thinking," *NSSE Yearbook,* 1963.
Wann, K., *Fostering Intellectual Development in Young Children.* New York: Teachers College, 1962.
Wertheimer, M., *Productive Thinking.* New York: Harper & Row, 1959.
White, J. D., *Philosophy of Mind.* Baltimore: Penguin.
Wild, K. W., *Intuition.* New York: Cambridge University Press, 1938.
Wiseman, S., *Intelligence and Ability.* Baltimore: Penguin. 1967.
Witkin, H., *Psychological Differentiation: Studies of Development.* New York: Wiley, 1962.
Wright, J., and J. Kagan, "Basic Cognitive Processes in Children," *Child Development Monograph,* Serial 86 (1963).
Vinache, E., *The Psychology of Thinking.* New York: McGraw-Hill, 1952.

KNOWLEDGE: CONCEPTS*

John Hospers

Our primary purpose is to examine human knowledge—its sources, its nature, and the various kinds of it that there may be. This is the principal task of the branch of philosophy called "epistemology" (from the Greek *episteme,* "knowledge"), or "theory of knowledge." But before we come to this task, we must briefly examine another matter which is preliminary to it: the nature of *concepts.*

Knowledge is expressed in propositions: "I know that I am now reading a book," "I know that 2 plus 2 equals 4," and so on. But before we can understand any propositions at all, even false ones, we must first have concepts. I cannot understand what is meant by the sentence "Ice melts" before I have the concept of ice and melting. We might well express this otherwise by saying that in order to understand what is meant by "Ice melts" we must understand the meanings of the *words* "ice" and "melt." But to understand the meanings of words, we must have concepts: to understand the meaning of a word already involves having a concept.

How do we acquire the concepts that we have? It was once thought that at least some of our concepts are *innate*—that they are, so to speak, "wired into us." Suppose that the concept of redness (or being red) were innate: then we would have it without having to experience any instances of it—that is, without ever having to see anything red. A person born blind could have the concept just as well as a man who could see. It seems so obvious that a person born blind does not possess the concept of redness, or of any other color, that no one has held that this concept, or the concept of any other sensory property, is innate. *Some* concepts, however, have been believed to be innate: for example, the concept of cause and the concept of God. If the concept of cause is innate, then we could know what the word means, and be in full possession of the concept, without ever having seen causes operating. This too seems implausible to us. . . . Perhaps the God example seems more plausible, since God, if one exists, is not seen or otherwise perceived, and yet we do seem to possess the concept (though this too has been denied). If we cannot perceive God and we nevertheless have the concept, how, it might be asked, do we come by it? May it not be innate? We shall try to answer this question when we consider the alternative theory, that concepts are derived from experience. Meanwhile, it is worth noting that the theory of innate concepts is no longer held. The rise of modern psychology has dealt it a death-blow. No evidence has ever arisen to show that any concept that people have is innate; perhaps they don't have certain concepts that they claim to have, but when they do have a concept, it is derived in some way from experience—that is to say, they would not be able to have the concept unless they first had certain experiences.

The obvious next step, then, is to say that all concepts are acquired through *experience*. (This view is sometimes called "concept empiricism," and the view that some concepts are innate is called "concept rationalism," but these names are liable to be misleading because they become confused with the far more important sense of "rationalism" and "empiricism" . . . which has to do with propositions and not with concepts.) This view was defended and made famous by three British philosophers: John Locke (1632–1704), George Berkeley (1685–1753), and David Hume (1711–1776).

Instead of the word "concepts," these philosophers all used the word "ideas," and the problem they undertook to answer was: "How do we come by the ideas we have?" All the ideas we have or ever shall have, they said, come from *experience*: (1) some through the "outer" senses, such as sight, hearing, and touch, and from these all our concepts involving the physical world are drawn; and (2) some from the "inner" senses, such as experiences of pain and pleasure, feelings of love and hate, pride and remorse, experiences of thinking and willing—from these we get all the ideas about our inner life. All our concepts are derived from these two kinds of experience. (Locke called the first "ideas of sensation" and the second "ideas of reflection.")

The use of the word "idea" was so general in the 17th and 18th centuries as to include all experiences, of whatever kind; but Hume made a clear distinction among experiences, between "impressions" and "ideas." Neither word was used in the 20th century sense, in which we say "I have the impression that someone is watching me" and "The idea of human progress is a delusion." Hume's use of these words can be illustrated as follows: If I see a green tree, I have a green *impression* (sense-impression), and then if I close my eyes and imagine something green, I have an *idea* of green—an idea being a kind of weak copy of an impression.

You have the impression when your eyes are open, but you can have the idea of something whenever you care to imagine it. Hume's main thesis in connection with these terms was *"No ideas without impressions."* If you have never seen anything green—that is, if you have never had a green sense-impression—it is impossible for you ever to have any *idea* of green. You must first have the impression in order to have the idea, and a man born blind could never have any idea of green or any other color, because he has never had any sense-impressions of colors. Similarly, a man born deaf could have no idea of tones, nor would a man born without a sense of smell have any idea of odors, and so on. For every idea X′, there is a corresponding sense-impression X; and without first having the sense-impression X, we cannot have the corresponding idea X′. The same considerations apply to the ideas gleaned from the "inner" senses: a man who has never experienced pain can have no idea of pain; a man who has never experienced fear can have no idea of fear; and so on. And a child who has not yet experienced sexual love can have no idea of love: he can observe how people having this experience *behave,* but he does not yet have any idea of what the feeling is like that impels them to behave in this way.

So much for the outlines of the theory. But as it stands it will not do, as Locke and Hume were well aware. For can't we have ideas of lots of things of which we have never had any impressions? We can imagine a golden mountain even though we have never seen one; and we can imagine a creature that is half man and half horse. True, we have seen pictures of centaurs, mythical creatures that are half man and half horse, but we could imagine these without ever having seen the pictures, and the persons who first drew such pictures must have been able to imagine them before they drew the pictures. And we can imagine (have an idea of) black roses even though the only roses we have ever seen are red, yellow, pink, and white. We can have ideas of all these things before we have ever had sense-experiences of them, and even if we never experience them at all.

Thus Locke was led to distinguish between *simple* ideas and *complex* ideas. We can imagine golden mountains and black roses without ever seeing them because, after all, we have seen the colors gold and black in other things. The idea of golden mountains and black roses are complex ideas: we simply take ideas we have *already* acquired through other experiences and put them together in our imagination in new combinations. The human mind can create all sorts of complex ideas from simple ideas already gleaned from experience; but the human mind cannot create a single simple idea. If we have never seen red, we cannot imagine red; and if we have never felt a pain, we cannot imagine pain. Red and pain are simple ideas. It is true that we might well be able to imagine a mountain or a rose without ever having seen one, but that is only because the ideas of mountain and rose are themselves complex ideas. If we have seen a hill and also have the idea of height from having seen some things higher than others, we can then form the idea of something higher and steeper than a hill, namely a mountain, even if we have never seen one. Similarly, we can have an idea of God, because we can combine certain ideas we derive from our experience of human beings, such as power, intelligence, kindness, and so on, and imagine these as being present to a greater degree than in any person we have ever encountered. . . . It is sufficient to note here that the idea of God, whatever it is, is a *complex* idea. (Having the idea of God, of course, does not prove that anything exists corresponding to the idea, any more than in the case of the golden mountain or the unicorn.)

The relation of simple to complex ideas is somewhat like the relation of atoms to molecules. Without atoms, you cannot have molecules; and atoms can be combined in different ways to form different molecules. Without simple ideas, you cannot form complex ideas; but once you have a number of simple ideas, you can combine them in your imagination in all sorts of different ways to form the ideas of countless things that never existed on land or sea.

> Nothing, at first view, may seem more unbounded than the thought of man, which not only escapes all human power and authority but is not even restrained within the limits of nature and reality. To form monsters, and join incongruous shapes and appearances, costs the imagination no more trouble than to conceive the most natural and familiar objects. And while the body is confined to one planet, along which it creeps with pain and difficulty, the thought can in an instant transport us into the most distant regions of the universe. . . .
>
> But though our thought seems to possess this unbounded liberty, we shall find, upon a nearer examination, that it is really confined within very narrow limits, and that all this creative power of the mind amounts to no more than the faculty of compounding, transposing, augmenting, or diminishing the materials afforded us by the senses and experience. When we think of a golden mountain, we only join two consistent ideas, gold and mountain, with which we were formerly acquainted. . . . In short, all the materials of thinking are derived either from our outward or inward sentiment; the mixture and composition of these belongs alone to the mind and will.[1]

It has never been made quite clear *which* of the ideas we have are simple ideas and which are complex; no complete list of them has ever

[1] David Hume, *An Enquiry Concerning Human Understanding*, Section II, paragraphs 4 and 5, 1751.

been offered, but only scattered examples. In general, the ideas of sensory qualities have been the stock examples of simple ideas: red, sweet, hard, pungent; pain, pleasure, fear, anger; thinking, wondering, doubting, believing. It may not be of great importance to decide in every case which ideas are simple ideas, but there is a problem about them nevertheless. "Simple ideas are those that cannot be broken down, or analyzed, into other ideas"—so runs the suggested criterion. But this does not always help us in trying to determine which ideas can, and which cannot, be further analyzed.

There is even a problem in the case of color-ideas, such as red, which are usually taken as the standard case of simple ideas: it is doubtless true that if you have never seen *any* shade of red, you cannot imagine any; but what if you have seen two or three shades of red? Can you then imagine only those shades but not the others? Or can you imagine (have an idea of) *any* shade or red after you have experienced (had an impression of) a few samples? Hume discussed such a case: Suppose that you have seen all the shades of blue there are except one, but you are told where that missing shade of blue lies in relation to the other shades on a scale ranging from the lightest to the darkest. Is it really impossible for you to imagine that shade without ever having seen it? Many persons would say that you *can* imagine it; or at least—what is not the same thing—that whether you can imagine it prior to seeing it or not, that you can *recognize* it as the missing shade after you *have* seen it. But if you *can* have an idea of it before you have had an impression of it, what happens to the view that "for every (simple) idea there must be a corresponding impression?" Is the idea of it then not a simple idea? Or if it is a simple idea, must there be perhaps a million simple ideas of blue corresponding to each of the million or more specific shades of blue? If the idea of each of these million shades is a simple idea, then it should be impossible to imagine the missing shade without having first seen it. On the other hand, if the simple idea is only blue-in-general (not any specific shade of blue), then presumably you *could* imagine the missing shade; but then you would have to say that the idea of this missing shade is a complex idea, composed of (1) the idea of blue-in-general and (2) the idea of being darker than, or lighter than, some other shade.

Here problems multiply: If you have seen only primary red, can you imagine scarlet, crimson, magenta? Are the ideas of these simple or complex ideas? If you have seen many shades of yellow and many shades of red, but no orange, can you imagine orange without ever having seen it? (Might some people be able to, but not others? Could an idea then be simple for some people but complex for others?) And if you answer "Yes" to the previous question, try this one: If you have seen blue and yellow, but no green, can you imagine green? (It is most important not to confuse the physical and the psychological here. Orange, we say, is a mixture of red and yellow, and green is a mixture of yellow and blue. But green doesn't *look* like a mixture of yellow and blue the way orange looks like a mixture of red and yellow. What color you get when you mix different paints, or combine different lights, has nothing to do with the question of what colors you can imagine [without having seen them] on the basis of other ones.)

Whatever the outcome of these speculations, it seems to be quite clear that without some impressions we cannot have certain ideas. A man born blind can have no idea of colors. And if we had never experienced shapes of *any* kind, we could have no idea of shape—not of triangular, rectangular, circular, or any other—though it may well be that if we had experiences (impressions) of *some* shapes, say a triangle and a pentagon, we could form the idea of other shapes, such as a rectangle and a hexagon, without having seen them. Clearly *some* impressions are indispensable before we can form *some* ideas, though it may be a matter for legitimate dispute just which ones these are.[2]

In general, the Lockean simple ideas are those whose names can be defined only ostensively. The reason we can define "red," "sweet," "pain," and many other sensory words *only* ostensively—by confronting the learner with the kind of experience of which these words are the labels—is that there is *no other way* of communicating to others what these words mean. These ideas are simple—that is, unanalyzable into other ones—so there is no way of acquainting people with what these words mean than to confront them with the relevant sense-experiences (impressions). By contrast, it *is* possible to provide someone with a set of instructions so that he will recognize a horse, a chair, or a tablecloth even though he has never experienced them, provided that he already has certain *other* ideas (of shape, size, solidity, and so on) derived from sense-experience. That is to say, the ideas of horse, chair, and tablecloth are complex ideas.[3] But there is no set of instructions we could give for shape, color, or solidity (or, for that matter, certain types of color such as red, or types of shape such as round) such that he would be able to form an idea of them with-

[2] Ideas of shape are different from those of color, in that shape is accessible to us through both sight and touch, whereas colors can be experienced through sight only. Thus, one may say, a man born blind can have an idea of shape—gleaned from the sense of touch—but no idea of color.

But we must be careful about this. Ideas of visual shape are not the same as ideas of tactual shape. (We use the word "shape" to cover both of them, forgetting that two very different kinds of ideas are involved here.) A man born blind can have, through the sense of touch, ideas of tactual shape, but he could have no more idea of visual shape than he could have of color.

[3] We probably *did* learn the meanings of all these words ostensively, by being confronted with instances of their application; but we didn't *have* to. A set of careful instructions about what a horse looks like could enable us to imagine a horse before we ever saw one, as well as to recognize one after we did see it.

out first having experienced them through his senses. This, then, is the connection between verbally indefinable *words* and simple *ideas.*

Concept v. Image

At this point, however, it is most important to expose an ambiguity in the word "idea" of which the philosophers we have just been discussing seem to have been unaware: In using the word "idea" one can be talking about either a *concept* or an *image.* Most of the time, they appear to have been talking about images, but sometimes the discussion of "ideas" shifted in such a way that it would be more appropriate to a discussion of concepts. Without having seen red we cannot form in our minds red *images*; but from this we cannot conclude that we can have no *concept* of red. To illustrate this point, let us take the example of ultraviolet. No human being can have ultraviolet images, since the human eye is not sensitive to that part of the spectrum; bees and certain other creatures can see it, but we cannot. Since we have no ultraviolet impressions, we can have no ultraviolet images. But we do appear to have a *concept* of ultraviolet. Physicists speak of ultraviolet light, and can identify it and relate it to other parts of the spectrum; indeed, they can talk about ultraviolet just as easily as they talk about red. Similarly, human beings do not have any sense that acquaints them with the presence of radioactivity the way they have senses like sight and hearing and touch that acquaint them with the sensible properties of physical objects. ("Sensible" in philosophy means "capable of being sensed.") We can't see, hear, smell, or touch radioactivity; we have to rely on instruments like Geiger counters to detect its presence. If any creature did have a sense acquainting him directly with the presence of radioactivity, we would have not the faintest conception of what it would be like; we simply have no "image" of radioactivity. (Remember that images need not be visual: there are auditory images, tactile images, olfactory images, and so on. When you imagine the smell of ammonia or the taste of scalloped potatoes, you are having olfactory images and gustatory images respectively.) Yet we do, it seems, have the concepts—at any rate physicists do—and physicists work as easily and familiarly with this concept as they do with concepts of which they *do* have sense-impressions (and consequently images). Hume's dictum "If no impressions, then no ideas" applies to images; it does not seem to apply to concepts.

Indeed, we can go further: A man born blind might become a physicist and specialize in the physics of color; this would be a somewhat peculiar choice, no doubt, but it would be a possible one. Such a man would never have seen any colors, and therefore he would have no color-images. But he might well know more *facts* about colors than you or I: he could tell us more about the light-waves and other physical properties of colored objects, and more about the physical conditions under which colors are seen, than most people can. He would in fact be able to tell us what the color of every object is; not by looking at it as we do, but by reading in Braille the pointer readings on instruments that record the wave-lengths of light emanating from the objects. He would be able to impart to us a great deal of knowledge about color and colored objects; and how could he do this if he did not have the concept of color? If he did not possess this, how could he know what he was talking about? Of course, he could correctly identify colors only as long as the correlation held between the seen color and the wave-lengths of light; if this correlation were no longer to hold, he would start making mistakes in color identification because he could not see the colors but had only the indirect evidence of the instruments recording light-waves. Still, must we not admit that he has the concept of color, even though he is unable to experience any color-images? How could he use the word, and even impart to us new knowledge that presupposes knowledge of what the word means, unless he had the concept?

What Is a Concept?

We are thus led to the all-important question: What *is* a concept? It is clear enough that there is something different from images, something that we have called concepts. But what is a concept? How do we tell when we have one?

Let us try one possible answer to these questions: (1) We have a concept of X when we know the definition of the word "X." But this answer is far too narrow: we do know the meanings of countless words—"cat," "run," "above"—and use them every day without being able to state a definition for them. We observed . . . why this is so. Whatever having a concept involves, it does not require being able to state a definition—something that even the compilers of dictionaries often have a hard time doing. And in the case of words like "red" that are not verbally definable at all, we can never state a definition—from which we would have to conclude, according to this view, that we can never have a concept of red.[4]

So let us try again: (2) We have a concept of X when we can apply the word "X" correctly; we have a concept of redness and orangeness when we can correctly apply the words "red" and "orange" in all cases. This criterion does not require us to give a definition but only to use the

[4] Strictly speaking, we should say a concept of red*ness*. Redness is a property—the property that all red things have in common—and the property is not red but red*ness*. By contrast, we have red images; a red image is a particular instance of the property redness.

word with uniform correctness. It is also much more in line with our actual use of the term "concept": we do say, for example, "He must have some concept of what a cat is, for he always uses the word 'cat' in the right situations—he never applies the word 'cats' to dogs or anything else."

There is, however, one way in which this criterion is still too restrictive: it assumes that in order to have a concept we must first be acquainted with a *word.* Doubtless this is usually the case, but it is not *always* the case. A person may have something in mind for which no word yet exists, and he may then *invent* a word for it; or he may use an old word in a new sense, giving it a meaning it never had before. In either case, it seems plausible to say that he had the concept *prior* to the existence (or new usage) of the word. When the first physicists adapted the use of the common word "energy" for their own special purposes, they had in mind a highly abstract concept, and they presumably had this concept in mind before they had a word for it. Doubtless there are many concepts that one cannot have without much prior acquaintance with language, but this cannot be the case for all concepts, else how would language have got started? Using a word correctly seems to be a *consequence* of having the concept, but not a precondition of having it: that is, if you have a concept, *and* know the word for it, you will then be able to use the word correctly; but having the concept is not the same thing as being able to use the word.

Let us, then, try once more, so as not to involve the acquaintance with a word in the having of a concept. (3) We have a concept of X (of X-ness) when we are able to distinguish X's from Y's and Z's and indeed from everything that is not an X. We might well do this whether we had a word for X or not, though of course it would be most convenient if we did have a word, and normally we would have. Thus if a child can distinguish cats from dogs and pigs and all other things, he has a concept of what is a cat, even though he cannot state a definition and even though he has never heard the word "cat" and connected the word with the thing by way of ostensive definition.

We have now specified what a concept is in such a way as to make it possible to have a concept without knowing any words. A dog that can distinguish cats from birds can be said to have these concepts, although it knows no words. Even this definition might be objected to, however, on the ground that being able to distinguish X's from Y's is, once again, a *consequence* of having the concept of X, but not what having the concept consists in. One is tempted to say that if you have a concept of X, you can, *as a result,* distinguish X's from other things; but you have to have the concept first. But what then would having the concept be? Moreover, we can devise machines that can effectively differentiate some things from others; do we wish to say that these machines have concepts?

In reply to such objects, we might say (4) that to have a concept of X is simply to have some *criterion-in-mind.* It would consist in some kind of "mental content" quite independent of words and quite independent of distinguishing X's from Y's and Z's. But it is not easy to state what such a criterion-in-mind would be like, or how one would know, through introspection alone, whether one possessed such a criterion. Surely the way one would know whether one had a criterion for X would be whether it would enable one to distinguish X's from Y's and Z's. A criterion for identifying X's would (it would seem) automatically be a criterion for distinguishing X's from non-X's. And so we would be back with our third criterion after all.

I can, of course, have a concept of X even though there are no X's in the world at all. I may have a concept of a sort of thing that is a reptile, larger than an elephant, and flies through the air. I could easily identify such a creature if it existed, and the fact that it does not exist does not prevent me from having the concept of such a creature. I have such a concept, then, although no such creature exists and there is no word to designate this peculiar combination of characteristics. (Let us take care to note that I can have the concept even if I cannot state any characteristics at all; for in the case of verbally indefinable words, no characteristics can be stated. I cannot state in words what distinguishes red from orange, though I know how to make the distinction in practice, and therefore I have a concept of these two colors.)

It is clear, then, that we can have a concept without having an image. If scientists can have a concept of ultraviolet without being able to visualize ultraviolet, surely a blind man can have a concept of red without being able to visualize red. True, for both the scientist and the blind man have a criterion for distinguishing X (ultraviolet, red) from non-X. But, we can now say, the blind man, though he has *a* criterion for distinguishing X from non-X, does not have *the same concept* as seeing men do, for he does not have the same criterion for distinguishing red from non-red that seeing men have. The blind man must use wave-lengths as his criterion, whereas we use (as men from time immemorial have used) the easily distinguishable (but not verbally describable) difference in the way red *looks.* Both have a concept, and there is a high degree of correlation between the concepts; but they are not *the same* concept, for there is not the same means of distinguishing red from non-red. (We, of course, *can* use both ways of distinguishing red from non-red, whereas the blind man can use only one.) Similarly, a person who could see ultraviolet would have a concept of it over and above the one we have, for he would be able to distinguish that color from others by means of direct inspection, without having to resort (as we do) to instruments for distinguishing it.

Are All Concepts Based on Experience?

Let us return, finally, to our main question: what are we to say of the view that all concepts are based on experience—which means that in the case of simple "ideas" a concept of X is impossible without a prior experience of X, and in the case of "complex ideas" that the concept of X is impossible without a prior experience of the simple ideas of which it is constituted? The view seems not only plausible but inevitable, for what is the alternative? We are not born with concepts, nor do we (as Plato thought) remember them from a state of existence prior to our birth; so how else could we acquire them except through experience?

The difficulty lies in showing in each case how the concept was actually derived from experience. With sensory concepts like redness, the case is relatively easy: as small children we had various red things pointed out to us, and by acts of successive abstraction . . . we came to recognize the characteristic, redness, that all the cases pointed to had in common. But how did we derive through experience the concept of liberty, of honesty, of marginal utility, of four, of logical implication? We do have these concepts, and, let us quickly note, we have them *without* any corresponding images. When we think of liberty, we may imagine the Statue of Liberty, and when we think of slavery, we may imagine African slaves being whipped; but neither of these images constitutes the meaning of the words "liberty" and "slavery"—others may imagine something very different when they think of liberty or slavery, and still others may have no images at all. There is no image *of* liberty or slavery the way there is an image of red or sweet. These are abstract concepts, to which there are no corresponding images. If we have images, they are not *of* liberty but of particular things or situations that may or may not exemplify liberty. We can all understand the same concept, liberty, even though we all have different images (or none at all) when we think about it. What we *think* of when we think of liberty is very different from what we *imagine* when we think of it; what we imagine, if anything, is only an incidental accompaniment.

This is not to say that we could have the concept of liberty if we had never had any sense-experiences at all: our having the concept is in some way or other dependent on experience; but it is far from easy to say how. Perhaps if we had always lived under a tyranny and never seen or heard of people who could express their opinions without fear of punishment, we would not be able to form the concept of liberty—though even this is doubtful, for as long as we were aware of restraints upon our behavior, we could conceive of a state-of-affairs in which these restraints were absent. It is, indeed, very difficult to know upon *what* experiences our concept of liberty is dependent. At any rate, the relation between the concept and sense-experience is very indirect: there is no particular sense-experience, or even any single kind of sense-experience, that we must have had before we can have this concept. Whatever the connection is between the concept and experience, it is sufficiently indirect that no one has given a clear account of exactly what this connection is in every case.

Let us consider another type of concept, those of arithmetic. Since we can distinguish between two things and three things, where did we get our concepts of two and three? "From experience," Hume would say. But exactly how? Arithmetic, we might say, studies the *quantitative* aspects of things; when we consider the sum of two and three, we don't care whether it's three apples, three boats, or three bales of hay. The concept of three (or three-ness) is formed through abstraction from many cases. What three apples, three boats, and three bales of hay have in common is their numerical *quantity*; that there are three *of* them is relevant to mathematics, not what it is that they are three *of*. The concepts of arithmetic are all quantitative—that is what defines them as arithmetical; and they are abstracted from experience, from our experience of things in the world. Without any experience of quantities of things, we would have no arithmetical concepts. So far so good. The problem comes when we realize that we have a concept of 12,038,468 just as much as we have of 3. Yet we have probably never observed exactly that number of things, and did not know it even if we did. What, we can then ask, is the relation between that number and our sense-experiences?

Or consider the meanings of such terms as "equality," "infinity," "implication," "deduction." We have concepts of all of these, for we can distinguish cases of the application of these words from cases of their non-application. Yet they do not seem to correspond to anything that confronts us in experience. If they are derived from experience in some remote way, it is not clear how, or what exactly the steps are.

Perhaps the experience we must have gone through in order to derive our concepts of numerical equality and of 12,038,468 is simply our experience of learning mathematics, or the experience of learning to use these words. But if so, this is a broader sense of the word "experience" than we have thus far been using, namely sense-experience (or sense-impressions).

More puzzling still, there are words that we can use with systematic accuracy that do not seem to be connected with experience, even by way of abstraction: consider connective words like "and" and "about," which have a function in a sentence but do not correspond to any distinguishable items in the world:

> One must not only know the meanings of nouns, verbs, and adjectives, one must also understand the significance of the syntactical form of the sentence; and for many sentences, one must understand various kinds

> of words that serve to connect nouns, adjectives, and verbs into sentences so as to affect the meaning of the sentence as a whole. One must be able to distinguish semantically between "John hit Jim," "Jim hit John," "Did John hit Jim?," "John, hit Jim!," and "John, please don't hit Jim." This means that before one can engage in conversation one must be able to handle and understand such factors as word order; "auxiliaries" like "do," "shall," and "is"; and connectives like "is," "that," and "and." These elements can neither get their meaning by association with distinguishable items in experience nor be defined in terms of items that can. Where could we look in our sense-perception for the object of word-order patterns, pauses, or words like "is" and "that"? And as for defining these elements in terms of words like "blue" and "table," the prospect has seemed so remote that no one has so much as attempted it.[5]

The acquiring of such concepts as these seems to require experiences of a different kind: learning a language, understanding sentences and sets of sentences, and the operations or performances with such symbols that are governed by linguistic rules.

In the light of such difficulties as these, Hume's requirement "If no impressions, then no ideas" seems thus far to be little more than a promissory note.

Traceability to Impressions as a Criterion of Meaning

At any rate, this discussion has confronted us with yet another criterion of meaning (though it has more to do with the meanings of individual words and phrases than with the meaning of sentences). According to this criterion, every word or phrase that we use must be traceable back to sense-experience in some way, whether the route be short (as with "red") or long (as with "liberty"). Or to put it in a different way, every word, to have meaning, must be either capable of ostensive definition itself or defined by means of other words, and these perhaps by still others, which are ultimately definable by ostensive definition. If this cannot be done, then the word or phrase is meaningless. If anyone claims, says Hume, that he has some idea (concept), we need only ask him, "From what impression is that supposed idea derived? And if it be impossible to assign any, this will serve to confirm our suspicion. . . . Commit it then to the flames: for it can contain nothing but sophistry and illusion."

Is this criterion satisfactory? That depends on whether Hume's thesis "If no impressions, then no ideas" is acceptable. And is it acceptable?

[5] William Alston, *Philosophy of Language*, 1964, p. 68.

We must wait a while before judging it. If we cannot trace a concept back to a sense-impression, this may be because we have not tried hard enough; it may be that the relation is very tenuous and difficult to trace. But, on the other hand, it may be because there is no such connection, and in that case it will be fatal to Hume's criterion. We . . . [sometimes] encounter some concepts—or at any rate alleged concepts—that seem impossible to trace back to any sense experiences whatever. If we can nevertheless satisfy ourselves that they are genuine concepts, and not meaningless strings of words, we shall then, and only then, have satisfied ourselves that Hume's criterion is inadequate.

HUMAN CONCEPTUAL BEHAVIOR*

Lyle Bourne

Introduction to the Area

The term "concept" has a multitude of meanings. Most of us have used or applied it in a myriad of ways, and among these uses there may not be a great deal of obvious similarity. For example, "concept" is commonly used as a synonym for idea, as when we say, "Now he seems to have the concept," in reference to someone who has finally caught onto a message. Or we may talk of an abstract state of affairs, such as freedom, and call it a concept. On other occasions, a concept seems to be akin to a mental image, as in the case of trying to conceptualize (visualize) an unfamiliar object or event from a verbal description. Undoubtedly, each of these examples captures in part the meaning of "concept." But clearly, it would be difficult (or impossible) to formulate an unambiguous definition from them.

In experimental psychology the term "concept" has come to have a rather specialized meaning, which may not encompass all its various ordinary uses. Psychology is the scientific investigation of the behavior of organisms, which includes as a subarea the study of how organisms (human beings and lower animals) learn and use concepts. In such an undertaking, explicit, communicable definitions of terms are an absolute necessity. "Concept" is no exception.

---

* Lyle E. Bourne, Jr., *Human Conceptual Behavior,* pp. 1–23.

"Concept" in Experimental Psychology

As a working definition we may say that a concept exists whenever two or more distinguishable objects or events have been grouped or classified together and set apart from other objects on the basis of some common feature or property characteristic of each. Consider the class of "things" called dogs. Not all dogs are alike. We can easily tell our favorite Basset from the neighbor's Great Dane. Still all dogs have certain features in common, and these serve as the basis for a conceptual grouping. Furthermore, this grouping is so familiar and so well defined that few of us have any difficulty calling a dog (even an unfamiliar dog) by that name when we encounter one. There is then the concept "dog"; similarly, the class of all things called "house" is a concept, and the class of things called "religion."

Each of us carries around a fairly large number of concepts. Most of them we have learned at some earlier time and use in everyday behavior, but we do continue to learn new concepts when the occasion demands. It is probably true that much, if not most, of the interaction between an individual organism and his environment involves dealing with classes or categories of things rather than with unique events or objects. This is fortunate. If an individual were to utilize fully his capacity for distinguishing between things and were to respond to each event as unique, he would shortly be overcome by the complexity and unpredictability of his environment. Categorizing is not only an easy way but also a necessary way of dealing with the tremendous diversity one encounters in everyday life. Concepts code things into a smaller number of categories and thus simplify the environment to some degree.

The bases upon which things may be grouped as a concept are legion. Perhaps the simplest is sheer physical identity or similarity among the instances. For example, we may think of all "red houses" as members of a concept. Here redness and the various observable characteristics of inhabitable dwellings serve as common features. On a different plane common or similar function may provide the basis for grouping. To illustrate, consider the concept "food." There is little physical similarity between a grapefruit and a beefsteak, but the use to which each is put links them together. Other bases . . . are even less obvious or more complex. Whatever the underlying principle is, however, it is usually logical, rational, and understandable.

We may note at this point that grouping things together means, in a certain sense, that all members of the group are responded to in the same way. For human beings the nature of this response is often wholly or in part verbal. Thus, most concepts are associated with a general descriptive name or label—as must be obvious for the foregoing examples.

Conceptual Behavior

Psychologists are primarily interested in the ways in which organisms acquire and use concepts rather than in any deep philosophical analysis of the nature and meaning of particular concepts. We may dub such activities of an organism as conceptual behavior. What are some of these activities? First of all, concepts just do not come into existence suddenly and spontaneously. Although the bases for a concept may exist in the environment, in the form of things which illustrate it, and although the organism may have the intellectual capacity to "understand" the concept, some learning process has to take place before the concept exists for the organism. Most concepts, if not all, are acquired. Many concepts, such as "roundness," are so simple and familiar that it is sometimes difficult for adults to imagine that learning was ever necessary. But empirical studies clearly demonstrate that even the simplest of groupings are often difficult for the young or naïve organism. The process of learning a new concept is one important form of conceptual behavior and is usually termed "concept formation."

Equally important, and perhaps more common for the adult human being, is a task or problem which requires the use of concepts which are already known. For example, the subject may have to search for, identify, and use in any given situation one from a collection of several alternative, familiar concepts. This aspect of conceptual behavior, which as we shall see has been given somewhat more attention in psychological research, is called here "concept utilization."

Some Helpful Definitions

The task before us is to outline the current state of knowledge about the psychological aspects of conceptual behavior. To undertake this review in a systematic fashion requires familiarity with basic terms and ideas, so we introduce at this point several of the most common.

Pertaining to the Stimulus

A concept is a category of things. Ordinarily, these so-called things are perceptible and have a real existence in the organism's environment. We refer to them as stimuli or stimulus objects. It is clear from earlier examples that not all stimuli belonging to the same concept are alike. Stimuli vary along *dimensions,* some of which are simple and obvious, like the dimensions of size or color, and some of which are exceedingly complex. An illustrative complex dimension of stimulus patterns, on which there has been some important research, is abstractness (or concreteness). Fortunately, these complex dimensions can often be analyzed into a set of simpler underlying dimensions, but we will consider that problem in a more appropriate place.

Not all of the dimensions on which the stimuli belonging to a certain conceptual class vary are important in defining the concept. Consider the concept "red triangle." What sets it apart from other concepts? Obviously, a particular color, redness, and a particular form, triangularity. Surely neither the size of the stimulus nor any of its contextual features make any difference. In fact, there is a large number of dimensions of variation which are unimportant in delineating the concept "red triangle." In the technical language, we refer to those dimensions which are important as *relevant* and those which are not as *irrelevant.*

A dimension has, by definition, at least two and usually many more discriminably different *values* or *attributes.* For example, red, orange, yellow, and so forth, are clearly different values within the dimension of color (or more properly, hue). Corresponding to the relevant dimension(s), then, we generally use the term "relevant value" or "relevant attribute" to refer to the particular level on a dimension which is involved in specifying a concept. Thus, both redness and triangularity are relevant attributes in the foregoing example.

One final point about the stimulus. Some things illustrate the concept and others do not. To say that there exists a particular definable class of things implies that there are other things which do not belong to the class. Technically, we refer to those stimuli which illustrate or exemplify the concept, for example, a particular red triangle, as *positive instances* (sometimes also as exemplars) of the concept, and those which do not, for example, a green square, as *negative instances* (or as nonexemplars).

Pertaining to the Response

Knowledge of a concept may be indexed in either of two ways. First, to know a concept is to be able to use it. Concepts are used to categorize. Therefore, knowledge of a concept may be evidenced by its use in categorizing a set of stimuli to which it properly applies. In an experiment, then, a subject may be asked to categorize a set of stimuli according to some arbitrary concept (or set of concepts) or to make a *category response* to each of a series of stimuli. If he performs the task accurately we conclude that he knows the concept.

Obviously, the number of different category responses in any experimental problem is always smaller than the number of unique stimulus patterns. Indeed, the number of categories is often two—positive and negative instances of a single concept. These responses are to be associated with stimulus patterns by the subject. However, the process is more than a matter of simple rote memorization of the individual stimulus-response contingencies. The subject assumedly acquires a more general principle which permits the categorization of novel or unfamiliar stimuli as well as those to which he has responded earlier. Commonly, the number of different stimuli used in an experiment is so large that the subject never is exposed twice to exactly the same one.

Second, usually, but not always, we would expect anyone who knows a concept to be able to provide a reasonably explicit description or name for it. Formally, such a description ought to include all the attributes which are relevant to setting apart those stimuli which are positive instances from those which are negative. Often, the subject in an experiment will be asked to demonstrate his knowledge of a concept by producing a description. Rather than categorizing stimuli as positive and negative instances, he may be required to name the concept which a set of positive (and/or negative) instances illustrates. These tentative guesses by S are called here *hypotheses.*

A problem is a problem, obviously, only if there is a goal or a solution to be learned or discovered. In conceptual problems, the solution is a concept or set of concepts—either a new concept to be learned (concept formation) or a familiar one which must be identified and/or put to use in some way (concept utilization). The attainment of solution is ordinarily a guided process. That is, the problem solver receives clues from his environment which, if properly interpreted and used, can keep him on the road toward the correct solution. Conventionally, these clues are referred to as "informative feedback." The environment feeds (gives) back to the performer information about the correctness of one or more of his responses.

So much for technical vocabulary. We proceed next to a description of some frequently used experimental paradigms.

Experimental Paradigms

A paradigm is a general plan or method for conducting research. It provides a useful and valid set of operations for "getting at" a particular phenomenon. It is not, however, an inflexible prescription for what an experimenter should do. On the contrary, its operations are typically quite modifiable and adaptable to the requirements of each new experimental problem. Further, it consists of little more than a skeleton, to which the particular manipulations and measurements of variables unique to the experiment may be attached.

Although the methodological differences among studies of conceptual behavior are large in number, there are two basic paradigms which are worth singling out for description here. Most other experimental procedures can be viewed as variations on these; significant departures are best noted where relevant later in the text.

Paradigm I—The Reception Paradigm

Most studies of conceptual behavior have employed one or another variation of what has been called the *reception paradigm*. We can trace the use of this technique to the first experiments in the area performed by Clark L. Hull in 1920. The following description of the reception paradigm is stated in general terms and does not represent the exact procedure used by Hull or in any other single investigation.

The experiment begins with a set of general instructions to the subject about the nature of his task. The degree of detail in instructions depends on the particular purpose of the experiment. However, the subject is typically told that his task involves learning how to categorize a group of stimulus patterns. Further, the manner in which patterns will be shown and the kind of response that must be made are outlined. Often, but not always, the stimulus dimensions are described for the subject so that he knows from the beginning what range of variation must be dealt with in the problem.

For simplicity, we shall assume that the patterns are to be divided into two categories and that the subject makes one of only two possible responses to each; for example, either he calls it a positive or a negative instance of the concept. Not all experimental tasks are quite this simple and well structured. Sometimes the subject is required to learn several concepts concurrently; that is, he may be asked to sort the patterns into 4, 6, 12, or any number of different categories, each representing a different concept. Further, the subject may be required to learn unique labels or names, such as DAX or VEC, for each of these categories. Obviously, this complicates the basic problem, not only for the subject but also for purposes of exposition here. Later, we shall discuss these complications more fully.

Geometric designs are commonly used as stimulus materials in laboratory studies of conceptual behavior. The simplicity, familiarity, and highly dimensionalized nature of these stimuli makes them ideal for many purposes. A set of such materials is shown in Figure 1. This population embodies three dimensions of variation; namely, Color, Form, and Size. Each dimension has three values: for Color—red, green, and blue; for Form—square, triangle, and circle; and for Size—large, medium, and small. The population contains designs showing all combinations of values on these dimensions and thus consists of a total of 27 distinctly different patterns. Many different concepts can be illustrated with this population, one of which will be selected for the subject to learn or identify. Take, for

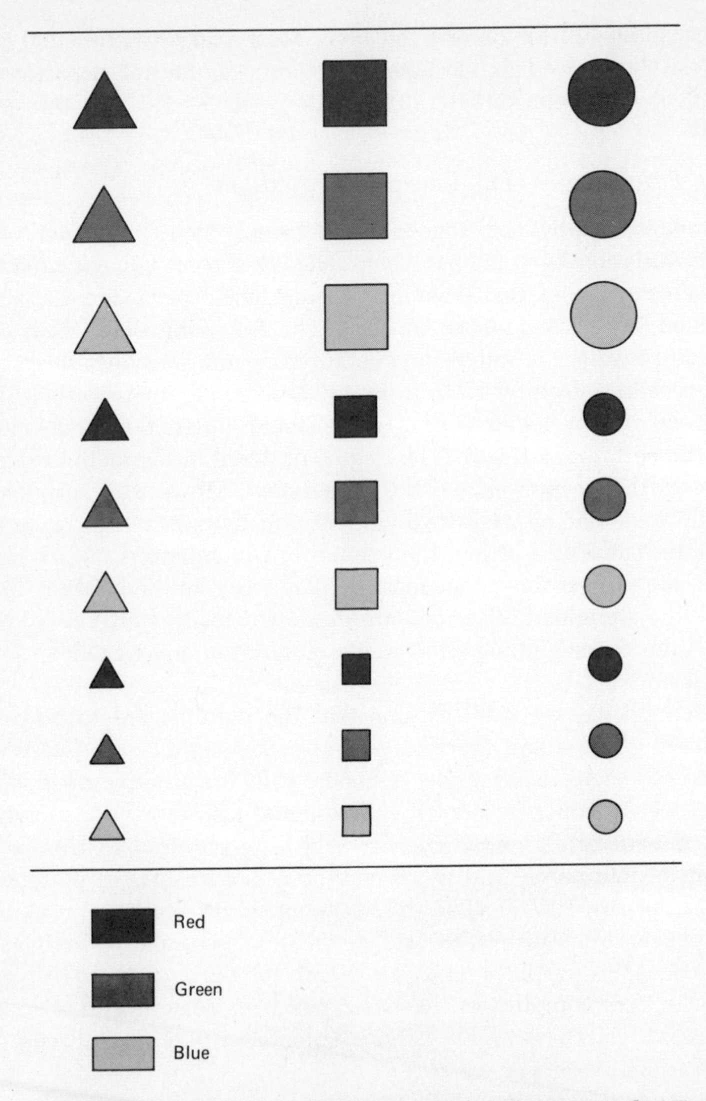

Figure 1 A Set of Geometric Designs Illustrating the Type and Dimensionality of Stimulus Materials Commonly Used in Experimental Work. The Size of This Population Can Be Extended Either by Increasing the Number of Dimensions (Here, Three) and/or the Number of Levels on Each Dimension (Here, Three).

example, the concept "red square." All stimulus patterns which embody both the redness and the squareness attribute, of which there are exactly three, are positive instances, and all which do not—24 in number—are negative instances.

The problem begins with the presentation to the subject of some randomly (or arbitrarily) chosen stimulus. The subject is required to respond to the stimulus by placing it in one of the two available categories. Informative feedback is then provided to indicate whether the response was or was not correct. After this the stimulus is removed, and a new one is presented for the subject's inspection and response. The three critical events—stimulus, response, and informative feedback—constitute one trial on the problem. These trials continue until the subject is able to demonstrate that he knows what the correct concept, or grouping of stimuli, is.

There are several points to be noted with respect to this paradigm. First, the stimuli are usually presented one at a time, or by the *successive presentation method.* Because no single trial ever provides enough information to solve a problem, the subject must keep track (in memory) of the events over a series of trials. Less frequently experimenters have used a *simultaneous presentation method* which permits the subject to see all stimulus patterns at once. This procedure can generally be expected to simplify the subject's task.

Second, the subject usually responds to each pattern by placing it in a category. Some experiments have required the subject to give, in addition, a reason (hypothesis) for the particular category chosen on each trial. For example, the subject may say, "I put the last stimulus (a large red circle) in the positive category because I think the concept is *large red* figure." In still other cases, the experimenter may tell the subject what the correct category for each successive stimulus is, and require only the hypothesis as a response.

The particular criterion of problem solution used in an experiment will depend on the actual response required. When category responses are made, the trial series usually continues until the subject makes a fairly large number (say, 10 to 20) of correct responses in a row. When hypotheses are given, a statement by the subject of the correct concept usually terminates the task.

Paradigm II—The Selection Paradigm

The *selection paradigm* represents a somewhat more recent methodological development than the reception paradigm, owing largely to the work of Bruner, Goodnow, and Austin (1956). Preliminary instructions to the subject are, in most respects, the same as those used in the reception paradigm. In this case, however, the stimulus population is presented in full at the outset. The problem begins when the experimenter designates one member of the population as a positive instance of the concept which must be discovered. On the basis of this information, the subject guesses what the concept is; that is, he states some hypothesis about the solution. If the guess is wrong the subject himself is allowed to select an instance from the population and to ask whether it is positive or negative. Once this question has been answered by the experimenter, the subject states his new or revised hypothesis. This process continues—another instance is selected by the subject and categorized by the experimenter—until the subject states the correct hypothesis; that is, the solution.

Because it allows the subject to select his own instances in a problem, this technique provides an interesting measure of performance which the reception paradigm does not. One can determine from stimulus selections (and corresponding hypotheses) whether or not the subject is using any systematic plan of attack or *strategy* in the problem. In reception experiments the subject is in a sense at the mercy of the experimenter for information about the solution. In a selection experiment the subject gathers information on his own. If he knows how to go about it he can use very efficient strategies and acquire the necessary information in a minimal number of trials.

There are many variations on the selection paradigm which have appeared in different experiments. For example, once the subject knows the stimulus population, he may be required to solve a problem "in his head"; that is, under a condition wherein the stimulus population is absent and selections must be made from memory. Under some circumstances the subject may be required to name the category for each selected stimulus as well as a hypothesis. Here corrective feedback ordinarily would be provided after each category response. Modifications of this sort will, of course, be discussed in the context of appropriate experiments.

The Relationships among Critical Events in a Conceptual Problem

In all conceptual problems the subject is required to learn or discover some arbitrary scheme for grouping stimuli through an inductive process based on the observation of a set of positive and negative instances. Information about the correct concept is presented in bits and pieces, on a trial-by-trial basis, until the subject can demonstrate that he knows the solution.

Each trial, as has been noted earlier, consists of three critical, objective events; namely, the stimulus, the response, and the informative feedback, occurring in that order. Temporal relationships among these events may be particularly crucial to the problem-solving process. It takes time, no doubt, to assimilate and use the information available on any single trial, and if time is limited performance may suffer. Consider the schematic representation of the critical events of a single trial in the reception

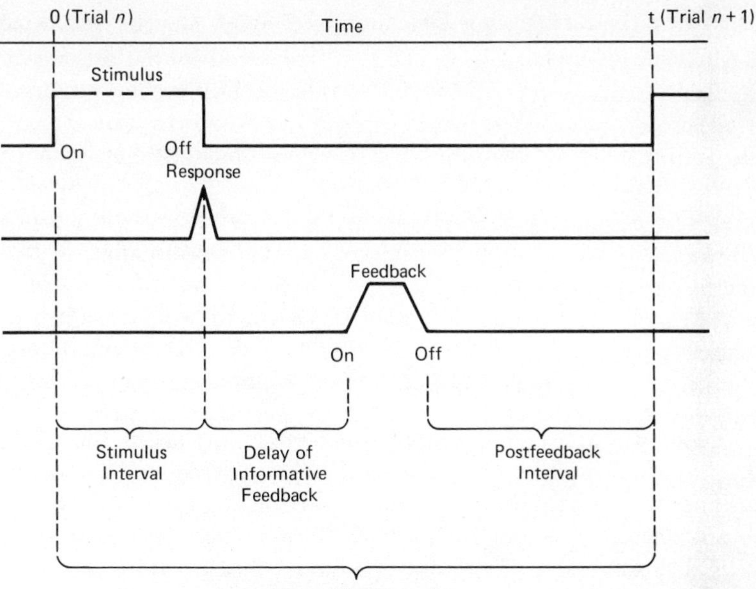

Figure 2 Order and Timing of Critical Events—Stimulus, Response, and Informative Feedback—within a Single Trial of a Conceptual Task.

paradigm as shown in Figure 2. The over-all length of a trial may be broken down into three major components: (1) Stimulus interval, or the period of time during which the stimulus is available to the subject for inspection. Typically, the subject will be required to make his response within this interval. (2) Delay of informative feedback, or the period of time between the subject's response and the presentation of feedback. Experimental procedures typically minimize this interval on the assumption that delaying feedback will have adverse consequences for performance, just as delay of reward has often been shown to slow down the acquisition of simple habits. (3) The postfeedback interval, or the time passing between the presentation of feedback and the occurrence of the next stimulus pattern. The length of this interval is particularly critical to solving conceptual problems efficiently, probably because it provides some time to "mull over" all information provided by the preceding critical events of the trial.

The determination of these time intervals varies from experiment to experiment. In many cases the subject is allowed to take as much time as he wishes to inspect each stimulus and/or to make his response (indicated by the dashing in Figure 2). In others the experimenter sets the interval, forcing the subject to respond in some brief period. Those studies . . . which have been designed to explore temporal relationships indicate that their effects on conceptual behavior may be pronounced. They suggest, moreover, that close attention be given to the *control* of these intervals in studies where the major purpose is to explore other variables.

Our purpose . . . has been to present a small technical vocabulary and a set of general empirical operations that are used in the analysis and specification of conceptual behavior. The availability of these permits us to proceed directly to a more detailed psychological analysis of the problem.

Analysis of the Problem

Most analyses of conceptual behavior emphasize the stimulus properties which delimit the class of positive instances. To understand a concept, it is clear that one must recognize all (or most) of the attributes which are important in its definition. Less obvious perhaps is the fact that some sort of *rule* for combining and/or using the relevant attributes is also involved in the definition of any concept. Consider, for example, the concept "red barn." This consists of the class of all things which are both red *and* barn. White barns and red houses do not fit the concept, for both lack one of the two relevant attributes. Members of the concept are denoted by the joint presence of both attributes. Joint presence or *conjunction* is the rule for combining the attributes in defining this concept.

It is fair to ask if the rule for forming concepts is not always the same; namely, conjunction. The answer is no. We need only look at those primitive concepts involving a relevant single feature, such as redness. Here the rule is simple *affirmation* of one specific attribute. But consider something a little more complicated. Let us probe more deeply into the concept "barn." One might say that a barn is a farm structure used for housing cows *and/or* for storing hay and other feed crops. There may be other attributes, but we need only consider two in order to illustrate the point. The existence of either or both of these attributes is sufficient to designate any instance in question as positive with respect to the concept, barn. Formally, the and/or rule is called a *disjunction* of attributes.

The number of different rules that serve to define concepts is large. Many rules depend on the definite presence (or absence) of certain specific attributes. Others are probabilistic, involving only the frequent occurrence of certain characteristics, such as a symptomatic description of a particular category of medical diagnosis. Some concepts are static in that the defining attributes are identified with a single unchanging object. Others might be labeled dynamic or sequential because a temporal order of events, such as in certain weather patterns, defines the concept.

There is no need here to provide an exhaustive listing of the possible rules for forming concepts. More detailed distinctions, such as between probabilistic and deterministic concepts, are more easily discussed later in the context of empirical work. The point is that both rules and attributes are involved in all cases. This is an important point for both experimental and practical purposes; both aspects of a concept may affect individual behavior in or outside the laboratory.

The majority of empirical studies has been concerned with the learning and the use of relevant attributes, commonly in problems where the rule has been spelled out beforehand or at least is simple and familiar to the subject, such as would be the case for conjunctive solutions. In circumstances, however, where the rule is more complicated and/or unfamiliar to the subject, its contribution to the difficulty of the problem may be substantial.

The Different Aspects of Conceptual Behavior

If the foregoing structural analysis of concepts is meaningful, psychologically, it should be possible to identify unique behaviors of an organism which are associated with the attribute and rule components of any problem. Such a possibility is implicit in the fact that concepts differ in difficulty depending on whether or not the rule is familiar to the subject. We shall make an attempt next to show specifically how attributes and rules are implicated in conceptual behavior.

The Role of Attributes

Most stimulus dimensions, for example, color or size, are continuously variable. Physical values on such a dimension are, of course, infinite and merge imperceptibly from one to the next. The attributes corresponding to a physical dimension are often best thought of as an arbitrary or conventional category (discrete or noncontinuous) scale, which all or most members of a given culture have learned to superimpose on the dimension and to use for sake of convenience. Each of these attributes is usually labeled with a distinctive name. The number of distinct gradations on this category scale depends on many factors including both the discriminative capacity of the appropriate sensory system and the importance and utility of fine discriminations to the individual (or the culture).

Although we often speak of an attribute as if it were a unitary characteristic of things, in most cases it would be more realistic to think of it as a range or category of physical variation. Indeed it is in this sense that attributes such as redness or largeness are primitive concepts from which more complex groupings are derived.

Because the dimension underlying a set of attributes may be continuous there is often a problem fixing the boundaries between one attribute and the next. Where does redness stop and orangeness begin, and how are the requisite discriminations made? As in the case of so many other behavioral phenomena, both inborn and learned factors make a contribution (Hochberg, 1962). No one needs to learn the different feelings that accompany contact with hot and cold objects. Basic aspects of discrimination like this are built into the normal organism. It would be misleading, however, to conclude that experience with the environment is inconsequential. There are at least two learning processes—perceptual learning and labeling—which are important to the discriminations we make among attributes.

Perceptual Learning. Relatively permanent and consistent changes in the way in which a stimulus array is perceived, solely as a result of experience with the array, have been reported. One example of this is the case wherein initially confusable stimulus objects become discriminable with practice. De Rivera (1959) required subjects to associate alphabetical letters (as responses) to different fingerprint patterns. One group of subjects learned a different letter for each pattern. A second group learned only two unique letter responses (each letter being associated with half the patterns) under instructions to look for common features among the patterns with the same letter response. Transfer to a second task, which required associating new responses with each pattern, was significantly poorer for the group which looked for common features, suggesting that these subjects *learned* less about the discriminable aspects of individual stimulus patterns (especially those assigned to the same response).

A related study was performed by Rasmussen and Archer (1961). They gave several kinds of pretraining with a set of unfamiliar nonsense shapes to subjects who eventually were required to categorize the shapes. For some subjects pretraining consisted of learning labels for particular stimulus characteristics while other subjects were asked to inspect and make an aesthetic judgment of the shapes. Those subjects who judged the shapes performed better in the later categorization task. In all probability this result was due to the fact that judgment pretraining led subjects to attend to and to discriminate more effectively among many of the attributes and dimensions of the stimulus shapes.

Experiments like these (Gibson, 1963b) point out that it is possible to learn to detect features of an object (or class of objects) which distinguish it from others. There is some evidence of a much weaker sort that the actual sensitivity of a sensory system may be enhanced or that detectable differences between two stimuli varying on the same continuum may be reduced in magnitude through practice (Engen, 1960). Learning to detect previously unnoticed features probably involves learning to use discriminations of which the sensory system is already capable.

Changes in sensitivity, however, imply a real difference due to practice in the capacity to detect. Both types of observations have been referred to as "perceptual learning" because they do involve a relatively permanent change in the way an organism perceives stimulation.

Labeling. A second contribution of learning, which is not entirely unrelated to the first, is *labeling*. "Labeling" is a term used to describe the process of associating distinctive names (or responses) with discriminable attributes (or more complex groupings). The fact that we customarily refer to objects of a certain color as "red" implies a prior association between the label "red" and those objects and/or similar ones. There is a fair amount of empirical evidence that these distinctive labels add to the discriminability of stimulus objects and their attributes (Rasmussen and Archer, 1961; Goss and Moylan, 1958).

It is probably true that everyday experience permits the organism to attain many of the discriminations among and labels for attributes or primitive sensory concepts. To some degree the young child's perception of sensory qualities is undifferentiated. For example, twoness, threeness, fourness, and so forth, may all be confused with manyness for the child. Further, the values on the number dimension may be confounded with correlated experiences of largeness, spaciousness, or heaviness. The conceptual system associated uniquely with the number dimension must be to some extent acquired. That much of this process involves learning is best demonstrated by the fact that certain cultures emphasize or impose somewhat different conceptual systems. If, for example, a child were born into a society which paid little attention to the difference between square and rectangle (or parallelogram) his concept of "squareness" may remain relatively undifferentiated even as an adult (Brown and Lenneberg, 1954).

Utilization of Attributes. It appears from the foregoing discussion that learning is involved both in the enhancing of the level of discriminability among stimuli and in the process of labeling attributes. Insofar as the conceptual problem requires finer differentiation among attributes or the acquisition of new labels, (discrimination) learning may be involved. Some conceptual problems, however, are better described as necessitating the utilization of previously learned discriminations and labels, rather than the learning of new ones. For example, when an adult human subject is given the task of discovering the correct sorting of geometric designs into two categories, for example, according to their color, little or no learning may be involved. The task amounts to identifying, on the basis of information provided by the experimenter, which of the already discriminated and labeled attributes are relevant to defining the category of positive instances. In the sense that "new" labels (positive and negative instances) are used in the task, one may think of some associative learning taking place. But ordinarily this requirement constitutes only a small fraction of performance, since it is mediated by already learned labels. The more important requirement is to discover and to identify the relevant attributes among the many that may vary from stimulus to stimulus. To denote this, the task is often called concept identification. More consistent with the terminology used in this discussion would be the phrase "attribute identification," because in the typical problem of this type the rule for classifying stimulus patterns is known to the subject.

The Role of Rules

Conceptual rules are rules for grouping. They specify how the relevant attributes are combined for use in classifying stimuli. For various reasons it is not clear that every concept embodies a rule. But even in the case of primitive concepts, wherein a single attribute provides the basis, there is a rule—either the attribute is present (positive instance) or absent (negative instance)—to implement the sorting of stimuli.

The independence of rules and attributes is worth emphasizing. Some combination of rule and attributes defines every specific concept. "Red and square" is a specific concept; "small and/or tilted" object is a specific concept. Both of these expressions could be used to classify an appropriate stimulus population. But the rules "and" and "and/or" are not bound to the attributes they combine in these cases. Other combinations of the same rules and attributes—for example, "red and/or square" or "small and tilted"—are allowable and meaningful within the same domain. Thus, it is entirely reasonable to think of rules and attributes as independent components of a concept.

Rule Learning. It is probably true that all sorting rules are learned. Through experience with a series of concepts each of which is based on the joint presence of two or more relevant attributes, the individual acquires the conjunctive rule. He learns a set of specific concepts, each of which is based on certain attributes combined by conjunction, but above and beyond this he also learns the conjunctive principle itself.

A simple example of what is involved here is provided by experimental studies of learning sets (Harlow, 1949, 1959). Consider the "oddity" problem. In this task the subject is confronted with a set of (three or more) stimulus objects one of which is different in some specifiable way from the rest. Suppose the stimulus array consists of one square block and two round ones. The subject is allowed to choose one of the objects. If he chooses the "correct" one, in this case the square, he is rewarded with food, candy, or some other incentive. If he chooses incorrectly (one of the circles), all objects are removed and no reward is given. On Trial 2, the same objects are presented in some different positional arrangement. Once again the subject is allowed to choose one and is rewarded for a correct choice. The problem continues for several trials, over which one can usually expect some observable increase in the probability of a correct choice.

A new problem is begun soon after the last trial on the first. Problem 2 differs only with respect to the stimulus objects involved. Suppose here one block is smaller than the other two. Once again, the odd (that is, small) object is associated with reward. The subject is given several trials with this problem.

The series may continue for tens, even hundreds, of problems. In each successive problem the stimulus objects are different and the particular critical cue or attribute, such as shape, size, color, and so on, changes. But the rule governing reward—associated always with the odd member of the stimulus set—remains the same. Figure 3 shows a schematic representation of several oddity problems, along with sample data.

Two aspects of the outcome of such an experiment are to be noted. First of all, within each problem there is typically a gradual increase in the percentage of correct responses which indicates that at least some subjects learn to respond to a particular stimulus in order to obtain reward. This observation hardly qualifies as rule learning, for the association is specific to a single stimulus. The second, more interesting, facet of the data is the interproblem improvement. In general, performance functions in later problems are higher than the initial ones. Subjects show general improvement from problem to problem—a general positive transfer effect. Subjects become better and better at learning (or solving) oddity problems, and this learning is based on something other than a simple stimulus-response association, for the stimulus attributes change from problem to problem.

Such an observation is probably not surprising to many readers. Improvement over a series of problems of the same type is a rather familiar process. We get better with practice at working crossword puzzles, at writing letters, at memorizing poetry, at solving anagram problems, and so on. Obvious or not, it is an important observation for each of these examples involves, in part, learning rules—ways of behaving which are not bound to particular stimulus features but which can be used in an entire class of tasks.

In the illustrative experiment subjects learned the oddity principle or rule. In later problems subjects chose almost unerringly the odd member of each new stimulus set—the bright object in a group of dull ones, the

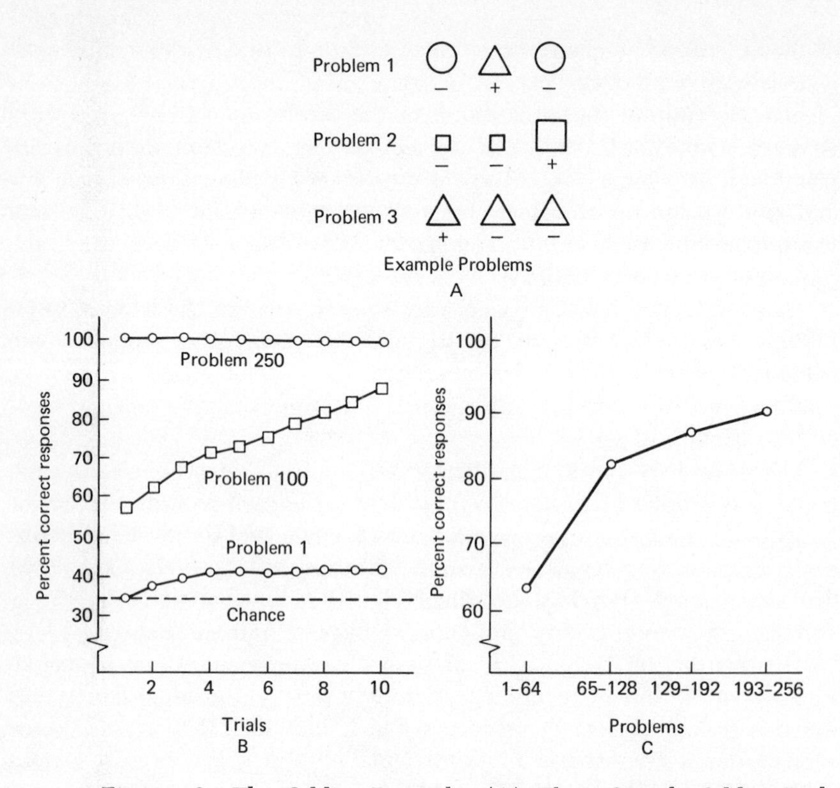

Figure 3 The Oddity Principle. (A) Three Sample Oddity Problems. The Subject Is Rewarded on Each Trial Only If He Chooses or Responds to the Odd Member, Marked +, of Each Stimulus Set. (B) Hypothetical Data Showing Performance Changes over Trials within Single Oddity Problems. The Learning Function Is Elevated with Successive Problems until, after Very Many Problems, Performance Is Errorless. (C) Actual Data from A Study of Oddity Learning in Monkeys (L. E. Moon and H. F. Harlow, "Analysis of Oddity Learning in Rhesus Monkeys," *Journal of Comparative Physiological Psychology*, vol. 48 (1955). Copyright 1955 by the American Psychological Association, and Reproduced by Permission.) The Score Plotted on the Ordinate Is the Percentage of Responses to the Odd Stimulus on Trial 1 of Each Problem. After 250 Problems These Subjects Make about 90% Correct Trial 1 Responses.

green in a group of reds, the symmetrical in a group of asymmetrical objects. While stimulus attributes provide the cue, only knowledge of the rule can produce 100% (or any level in excess of chance) correct choices on Trial 1 of each new problem.

The oddity rule is a simple one, perhaps too simple and obvious for most adults to appreciate fully. Actually, it is a simple elaboration of affirmation in the sense that the critical objects are distinguished usually by a single attribute. Yet, the data suggest it is not obvious for the young or naïve subject. It must be learned and it is used without error only after extensive practice.

Demonstrations of rule learning in the adult human being are more difficult to construct. Because of his vast backlog of experience it is almost impossible to find a rule which is completely new to this subject. But interproblem improvement has been shown even for the adult with some grouping rules, such as the disjunction. While an individual may have had some experience with this rule, most people require a certain degree of training to use it perfectly. Ability to form concepts with the disjunction does improve with practice. Data on this point constitute evidence for genuine rule learning in human adults.

Rule Utilization. Grouping operations, once learned, provide the organism with powerful conceptual tools. A repertoire of rules permits rapid acquisition of unfamiliar stimulus classifications based on it, increases the range of concepts that can be formed with any particular stimulus population, and enhances the flexibility of the subject's conceptual behavior, in general. Like other learned rules, such as mathematical formulas, grammar, chess moves, study habits, and so on, conceptual rules implement problem solving, thinking, and other acts of so-called complex or higher-order behavior which are characteristic of the mature, sophisticated organism.

Experimental studies of rule utilization are unfortunately rare. Experiments on problem solving by human adults provide perhaps the best examples of the use of rules. Maier (1930) developed a number of such experimental tasks, one of which is called the "pendulum problem." This task requires the subject to find a way to tie together the ends of two strings suspended from the ceiling of a room. The strings are too far apart to allow the subject simply to carry one end over to the other. To solve the problem, the subject has to attach a weight to the end of one string, set it in pendulum-like motion, take hold of the free string, and then catch the swinging string when its arc carries it close by. The critical element in problem solution is the pendulum principle (or rule). That the principle is not quite so obvious as it may sound is documented by the fact that in Maier's experiment no subject attained solution without some guidance from the experimenter.

A General Scheme for Describing Conceptual Behavior

Thus far we have discussed in general terms the various facets of conceptual behavior as it is studied in experimental psychology. Deeper penetration requires a greater concern with the actual theory and research. We pause at this point to provide a brief summary of these facets, to indicate their interrelationships and organization, and to emphasize certain methodological and terminological problems.

Conceptual behavior covers all the activities of an organism which in any way involve concepts. These include forming new and/or using already known concepts in some way. A concept is analyzable into two structural components, rule and attributes. To the subject, whose task it is to learn or discover a concept, either or both components may be unknown.

These considerations produce rather directly a scheme, portrayed in Table 1, for classifying research and theory in the area. The scheme amounts to a simple 2 x 2 representation with problem elements, that is, rules for grouping and attributes, providing the basis for one distinction and the nature of behavior, that is, learning and utilization, providing the other. According to this scheme there are four fundamental types of conceptual tasks or problems; namely, attribute learning, attribute utilization, rule learning, and rule utilization.

Table 1

Schematic Representation and Examples of Tasks Involving Conceptual Behavior

	Type of Behavior	
	Learning	Utilization
Attribute	Perceptual learning, labeling	Concept identification, sorting tasks
Problem Element Rule	Formation of learning sets, positive transfer across conceptual problems based on the same rule	Rule identification, problem solving

Examples of each type of task are shown in Table 1. In the attribute-learning cell we include those tasks wherein the perceptual characteristics of a stimulus array are changed, either through enhanced sensitivity to the underlying stimulus dimensions or through learning to detect distinguishing features of stimuli. Under attribute utilization fall those tasks which require the discovery and/or use of already discriminable and labeled attributes, such as concept-identification problems. Rule-learning tasks are those wherein new principles of grouping are acquired; many studies of learning set formation nicely illustrate this category. And finally, rule utilization is meant to encompass those tasks which require the selection and use of known principles, such as many problem-solving situations.

Some Notable Difficulties

A Pure Conceptual Task. One may think of the examples of tasks falling into each of the cells of Table 1 as "pure" in the sense that presumably only a single behavioral requirement, dealing either with attributes or rule, is imposed. It is clear, however, that not all experiments or theories in the area of conceptual behavior deal exclusively with "pure" cases. Actually many conceptual problems embody several requirements. For example, some experiments employ a task in which the subject must both identify relevant attributes and learn an unfamiliar rule for grouping. An experimenter need not use pure tasks to gain useful and informative data on conceptual behavior. But it is important to recognize the existing components of any task and to evaluate the results of the experiment accordingly. In this respect the portrayal in Table 1 will assist in the interpretation of later experiments.

The Distinction between Learning and Utilization. Any attempt to draw a firm line between learning and utilization is fraught with difficulties. Suppose we provide some opportunity for a subject to practice a new response or response sequence or to form some new association(s) between stimulus features of his environment and a response. How can we tell when learning is complete, and that the subject is "ready" to use his newly acquired behavior? While it is true that all learning necessarily entails a modification of behavior and that continuing practice tends to yield smaller and smaller behavioral changes, it is not clear that even the simplest of learning processes ever terminates. Furthermore, complete learning is probably no prerequisite to the *utilization* of acquired behavior. Even weakly established associations, for example, may be called into use in a different situation.

The distinction between learning and utilization in Table 1 is then arbitrary. In general, we use the term "learning" for those tasks and experiments wherein the emphasis and central interest lies in the acquisition of differential responses for formerly confusable attributes or of some complex behavior strategy which implements a formerly unfamiliar rule for grouping. When it is clear from strictly formal considerations that learning has taken place and when the task demands some use of that prior learning, we use the term "utilization." It may be noted here that the vast majority of utilization tasks are essentially identification problems; that is, problems wherein the primary requirement is to discover and specify which of several known alternatives (either rules or attributes) is or are relevant to problem solution.

The Distinction between Rules and Attributes. Concept formation and concept utilization entail dealing *both* with attributes and with rules. Indeed, as we have noted, discriminations among attributes and the behavior required to implement a rule are typically acquired within the context of specific conceptual problems. As a result the difference between rule and attributes is not always clear; both are bound up in specific groupings of stimulation which are learned. Certainly, the typical subject does analyze a learned concept into its component parts without external prodding and direction. The properties of rules actually may be thought of in terms of certain relationships among attributes, such as "different from" (oddity) or "joint presence of" (conjunction), and may appear not to carry any information other than in the context of specific concepts.

It can be shown, however, that the rule-attribute distinction is an important one to make. It is possible, in both a formal and a methodological sense, to define a rule, independent of any attributes. Further, it can be shown that the acquisition of knowledge about a rule and of behavior associated with it proceeds separately of attribute learning or utilization. It is on this basis that we draw a strong distinction which we find to have considerable utility in the interpretation of behavior in conceptual problems.

Overlap with Other Areas of Psychological Research. Efforts to categorize behaviors are both common and useful. While behavior is basically continuous and multidimensional, for the sake of efficient communication and the organization of knowledge it is handy to have a finite set of categories and category labels at our disposal. Conceptual behavior is such a category.

We should recognize, however, that categories of behavior overlap; they are not mutually exclusive. In part, conceptual behavior involves *perception*; environmental stimulation is received, transformed, and in most cases, organized before we respond to it overtly. Clearly, basic *learning* processes are involved; discriminations are in part acquired, verbal and other labels are acquired, learning sets are acquired. On another dimension conceptual behavior impinges on *thinking* and *problem solving*, for certainly adequate performance in conceptual tasks depends on internally organized symbolic activities and complex behavioral outputs.

Other examples of overlap could be cited. The point, however, is that conceptual behavior as an area of inquiry within psychology is not an isolated body of knowledge. Rather it is a part of an integrated, but incomplete, network of facts and theories which defines the whole of psychology. We shall treat it in this way, calling upon the information available in other categorical areas of psychology which help to make a coherent and intelligible picture of what is known about the conceptual behavior of human organisms.

Words and Concepts

For human beings words and concepts are inextricably bound. It is difficult even to think about any known concept without the immediate intrusion of its verbal associate(s) or description. Not only is it the case that most conceptual groupings have meaningful verbal labels but also some concepts are learned and used almost exclusively in a verbal context. For example, at least some people use and understand concepts such as "electron," "gene," "economic recession," and "nuclear holocaust" without much of any experience with their empirical referents. This state of affairs has led at least one psychologist who has worked extensively in the area to the conclusion that concepts ". . . are meaningful words which label classes of otherwise dissimilar stimuli" (Archer, 1964, p. 238). He goes on to point out the indisputable fact that a shortcoming in ability to use words is the single most important factor in the slow acquisition of concepts by preverbal human subjects and lower organisms.

While the importance of language in conceptual behavior cannot be denied, there is no reason to assume a strict identity of words and concepts. It has long been known that people can learn concepts or stimulus groupings without being able to verbalize or put them into words (Hull, 1920). Still there are at least two fundamental functions of words in conceptual behavior—first as symbols, and second as cues or signs. Both functions are based on an assumption that language is a learned representation or code for events, actions, objects, and relationships in the real world. Words are a cultural convenience which in many behavioral processes "stand for" other things. Clearly, this is a vastly oversimplified description of language, but it is suitable for present purposes.

As symbols, words are internal (or cognitive) mediators of behavior. They are the tools or elements of (most) thought processes. They allow us to think about concepts in the absence of any examples thereof. They are the basis of much covert activity which may or may not be accompanied or followed by overt action.

As cues or signs, words carry information about concepts from speaker to listener. In this sense, words implement communication. Because they are conventionally established representatives of something else which may be awkward or difficult to produce or point out in any situation, they provide an efficient and convenient means for the transfer of information.

Words, then, are responses which have been associated with states of the world. When multiple associations—between a response and several dissimilar stimuli—exist for a subject, the word is a label for a concept and can be used symbolically. Insofar as these associations are the same for two or more people, the word may function as a sign in communication. When verbal labels do not exist, either because the concept is arbitrary and unnatural or because the stimulus attributes do not lend themselves easily to verbal associations, the subject may have difficulty describing the basis for his category response. This merely shows that words and concepts are independent, though often tied in natural circumstances.

This position toward the relationship between language and thinking in general is similar to that developed by Luria (1957) and Kendler (1961). It stems from a belief that as a child matures his behavior is more and more influenced by self-generated stimuli. His own verbal behavior is the most important source of self-stimulation. Verbal responses, whether overt or implicit, mediate and regulate other overt behaviors. Words as symbols govern much of what we do.

The aforementioned functions are in reality rather primitive in comparison to what human beings actually can accomplish with language. For example, we have said nothing about the syntax and rules of language which specify the meaningful ordering and organization of words into larger linguistic units. But we shall not delve any more deeply into the complexity of language. There is just one further point about the relation of words and concepts that needs to be indicated. Words as signs are external, public, and objective stimulus events. Like any other stimuli, they can be grouped. They serve as the basis for "verbal concepts." They (or their meanings) may be combined, moreover, in new ways, which in some circumstances generate concepts with no known or previously experienced referents. Thus, through words and language the range of conceptual behavior of human beings becomes vastly expanded. So much so that, as Archer implied, it becomes almost impossible in many circumstances to distinguish the concept from its verbal label.

THE CONCEPT OF THE CONCEPT*

Howard H. Kendler

The writer could not help but speculate, after receiving the invitation to participate in this symposium, about whether his assigned topic would have been included if everyday language were without a word such as *concept.* Being so common and possessing such apparent psychological significance forces the psychologist to treat this topic as a fundamental one.

This observation about the linguistic import of the term concept emphasizes the point that the present subject, as distinguished from the others in this symposium, with the exception of problem solving, is not a technical term having its origin in a clear-cut experimental methodology. Of course, many respectable scientific concepts do have their roots in common parlance. They, however, achieve respectability and importance only after these original roots have withered and their place and function have been taken over by a technical term, or terms, that have the advantages not only of being less ambiguous but also, in the experimental sense, more meaningful. Although there are signs that this healthy course of development is beginning for the concept of the concept, the fact is that it is still vague and amorphous. This point is made to impress the reader that because of the scientific infancy of his topic, the writer is in that unenviable but not uncommon position of a psychologist who is not quite sure about what he is writing. This predicament nevertheless does have one advantage. It frees one from being constrained by any orderly array of facts and theories that demand a particular kind of systematic treatment. It allows for flights of fancy and broad generalizations without much fear of being embarrassed by any clear-cut contradictory evidence.

With this prologue concluded the general approach that will be made in the analysis of conceptual behavior can now be described. Initially the methods used to investigate concept learning will be examined, particularly in terms of how the empirical attacks upon this problem have been shaped by theoretical preconceptions. An attempt will be made to tease out significant characteristics of these approaches so that some overview,

---

* Howard H. Kendler, "The Concept of the Concept," in Melton, *Categories of Human Learning,* Academic Press, 1964, pp. 211–237. The writer wishes to express his great appreciation to the Office of Naval Research and to the National Science Foundation for their support of his research efforts in the field of classificatory and problem-solving behavior. Thanks are also due to Tracy S. Kendler for her helpful reading of the manuscript.

although no doubt an incomplete one, will be offered of the current scene of research and theory in the field of classificatory behavior. Finally, the problems of conceptual behavior will be systematized in terms of stimulus-response language. For the most part this analysis will be concerned not with criticism but with clarification, not with truth but with meaning.

Models of Concepts

Typically, models of behavior have been extended *to* concept learning. As a result there is often a greater interest in demonstrating how concept learning is like something else than it is like itself. There is, of course, nothing wrong in approaching a research area with certain preconceptions, or what may be more appropriately called a model (Lachman, 1960). Such biases—and they are biases—sometimes lead to important theoretical generalizations (not to mention the comfort they can provide during lengthy periods of ignorance). But until the models pay off, it should be remembered that they are neither demanded by, nor an outgrowth of, the data. These preconceptions are brought in from the *outside*. Researchers must guard against the tendency to be overconcerned with demonstrating the appropriateness of the model at the expense of coping with problems, both empirical and theoretical, that characterize the behavior that is being investigated.

The possible prejudicial effects of pretheoretical models are sometimes minimized because of the prevailing conviction that they can always be disentangled from empirical results. Although the writer has no desire to enter the debate about whether ideally scientific language can be divided nicely and neatly into a theoretical and empirical component (Hanson, 1958), he does wish to emphasize that it is often far more difficult than scientists like to believe. Theories and models often persist simply because they are impregnated with facts. As long as the facts stand, the theory and model survive. These comments, it should be understood, are not directed at attacking the *use* of pretheoretical models; instead a warning is being issued against their possible *misuse*.

S–R Conceptions of Concepts

The initial sortie into the realm of models of concept learning will be in the direction of S–R conceptions. There are two reasons for this choice. The first is that S–R conceptions have probably influenced more workers in the field of concept learning than any other view. Second, they serve as a frame of reference to explicate and analyze other formulations.

S–R models emerge from three different orienting attitudes: (1) Behavior can be represented in terms of S–R associations; (2) the science of behavior is investigated best with a sophisticated behavioristic methodological orientation, and (3) S–R theories of behavior of simple phenomena such as conditioning and discrimination learning are fruitful sources of models for more complex behavior.

While no detailed analysis of these three components will be offered at present, it will be helpful to refer to certain points, particularly in relation to the meaning of stimulus-response associationism as a language system *representing* psychological events.

The essence of stimulus-response language is contained in the S–R paradigm. Accordingly, there are three important characteristics of behavioral events. These are the stimulus or stimuli, the response or responses, and the association (the—) between them. On the simplest level possible this means that these terribly complex phenomena which we call *behavior* can be represented advantageously in language of stimulus-response associationism. Behavioral events can be described by specifying some features of the environment (stimulus) and some component of the total behavior pattern (response), and the relationship between the two.

Much of the confusion surrounding stimulus-response language has resulted from overburdening the descriptive linguistic system with theoretical properties. Stimulus-response language need not be committed to any physiological conception of behavior, nor need it be committed to any view of how a functional relationship (Robinson, 1932) between a stimulus-response association is established or strengthened. It simply is a way—and not the only way—of describing behavior.

When this S–R language system is combined with a behavioristic methodological orientation, which can be succinctly characterized as physicalistic, operational, and experimental (Estes, 1959), it is not surprising that early S–R psychologists clutched to their bosoms conditioning procedures and phenomena. The consequences of this action should not be ignored. Not only was it decided to investigate and theorize about conditioning, but also to perceive other behavioral phenomena as if they were in some way (Hilgard & Marquis, 1940) like conditioning.

The first and perhaps the most important extension of the conditioning model was to the realm of discrimination learning. Pavlov (1927), in his investigations of differentiation, indicated how this phenomenon could be explained as resulting from the combined operation of excitation and internal inhibition. But more important in shaping the history of the psychology of learning was Spence's (1936) classical theory of discrimination learning. Conceptually it was an extension of the view that conditioning processes (conditioning, generalization, and extinction) were operating in discrimination learning. It, in turn, extended this conditioning-generalization-extinction model to new empirical realms. It may not appear much of a leap from Pavlov's differentiation to Spence's discrimination. Nevertheless, it proved sizable, as witnessed by the numerous problems and controversies it generated.

Perhaps the major effect of taking this apparently small step from conditioning to discrimination learning was the lessening of the control by the experimenter over the subject's (S's) reception of the critical stimulus. In classical conditioning the S's behavior is so restricted and the environment is so unchanging that the presentation of the conditioned (positive) or differentiated (negative) stimulus practically guarantees stimulus reception. In contrast, in the typical discrimination problem conventionally used with rats, the S is moving constantly about and the critical stimulus does not occupy such a perceptually dominant position in the S's environment. In addition there are two critical stimuli to which the organism must respond. This usually complicates the stimulus reception process by creating special problems of stimulus patterning and receptor orienting responses.

The differences between conditioned differentiation and discrimination learning can be reduced by arranging the discrimination apparatus to minimize problems of stimulus reception (e.g., use black and white alleys in which no matter what the animal does, except closing his eyes, he will see black or white). The use of a successive discrimination problem instead of a simultaneous one also has the effect of making the discrimination problem more similar to the differentiation one.

It can be argued that it is neither necessary nor desirable to try to investigate discrimination learning in an experimental situation arranged so that it resembles the classically conditioned differentiation. If, however, one is particularly interested in extending the conditioning-generalization-extinction model, there are obvious strategic advantages in initially selecting discrimination problems that appear similar to classical conditioning.

The relationship between conditioning and discrimination learning, however, was not perceived only in terms of research strategy. It was also considered in terms of basic theoretical issues. With the help of sharp pens and thin skins, theoretical discussions produced arguments and acrimony, disagreements and distortions, confusions and confabulations. An Olympian sage, looking at these controversies in the tolerant light of history, might conclude they revolved about two related problem areas: (1) whether the principles of conditioning, generalization, and extinction could support a general theory of discrimination learning, and (2) whether independent empirical stimulus variables could potentially be coordinated to the independent theoretical stimulus variable of any S–R theory (Koch, 1954).

In analyzing the first problem it is helpful to view the conditioning-generalization-extinction model in both a weak and strong form. In a weak form it is very little more than a simple description of discrimination phenomena expressed in terms of S–R associations. A successful discrimination between two features of the environment results from learning to respond to the positive one and eliminating the generalized tendency to respond to the negative one. The strong form of S–R discrimination theories specifies important additional properties: the nature of the stimuli and responses as well as the associative process; the reinforcing mechanisms that form and strengthen associations; the principles of stimulus generalization; the manner in which one habit gains ascendancy over another, etc. The weak and strong forms of the theory can be considered as end points on a dimension of theoretical development. Quite obviously contemporary S–R discrimination theories are past the weak point but equally obviously, they have yet to reach the strong point. If this appraisal is correct, then one cannot as yet decide whether or not the conditioning-generalization-extinction model has the potential to develop into an adequate general theory of discrimination. The proof of the pudding is in the eating and the pudding is still in the oven. In order to prevent this remark from being considered as an attempt to shield S–R discrimination theories from possible criticism, it may be appropriate to state flatly that S–R discrimination theories are much weaker than they are made to appear. A visit to the laboratory with an intention to test S–R discrimination theories will reveal that fairly exact predictions can be made in only the simplest of situations.

The second problem, which is that of defining a stimulus, is one that is constantly with us. The view that it is insoluble, or that it lends itself to only one possible solution seems premature. At the same time no amount of turning one's back on this problem will make it disappear. (This symposium, with its frequent reference to the problem of the stimulus, testifies to its existence.) In classical conditioning and operant conditioning with discrimination stimuli, neat relationships are obtained between a physically defined stimulus and behavior. No definitional problem seems to exist. Even when a simple physical definition of the stimulus fails to produce a tidy empirical relationship, it is possible to cope with such a problem, as did Hovland (1937), by redefining the stimulus dimension in terms of different mathematical properties. Although such derived measures can assume many forms requiring both ingenuity and mathematical sophistication, epistemologically they share the common property of defining a stimulus as a physical event. Looking at the problem of stimulus definition within the limited scope of conditioning phenomena, it is easy to see how it is possible to conclude mistakenly that the problem is nonexistent.

When one leaves the security of conditioning and enters the uncertainty of discrimination learning, it becomes apparent that some special mechanisms are required to select out the appropriate stimulus. Such concepts as receptor orienting or observing responses, attention, stimulus patterning, and response-produced cues all function as selective associative mechanisms.

In experimental situations in which the relevant stimuli do not "surround" the S, he must first learn to make appropriate receptor orienting acts (e.g., Ehrenfreund, 1948). The problem of attention is brought home in the physiological work (Hernández-Péon, Scherrer, & Jouvet, 1956) showing that a distinctive and obvious sound can be prevented from eliciting neural impulses in the auditory nerve when the organism is attending to something else. Since Pavlov (1927) it has been recognized that all elements of a stimulus compound are not equally important in associative formation. Whether an explanation of stimulus patterning can be deduced from fundamental conditioning principles as Hull (1943) had hoped, or whether it depends on independent perceptual principles as Lashley (1938) and Krechevsky (1938) indicated, or whether a learning-perceptual orientation can be combined as Dodwell (1961b) suggests, is at present a moot question. But it is quite clear that the stimulus patterning problem must be solved if the S–R discrimination theory is to become "stronger."

The final line of evidence that focuses attention upon the problem of defining a stimulus is the one that will concern us most. It has to do with defining the stimulus situation in terms of the cues resulting from responses, both explicit and implicit, of the organism. There has been a definite trend in many S–R theories (e.g., Spence, 1956) for large parts of the environment to which the S is responding to be introjected into the organism. Historically this trend bears some relationship to Gestalt psychologists' distinction (e.g., Koffka, 1935) between the behavioral (psychological) and geographical (physical) environments, a distinction that was offered to illustrate the right and wrong way to proceed in analyzing behavior in relation to the environment. Bergmann neatly gets to the heart of the matter. "But even so, what is the predictive value of the suggestive metaphor 'psychological environment'? Is it not the business of science to ascertain which objective factors in the past and the present states of the organism and its environment account for the difference in response, so that we can actually predict it instead of attributing it, merely descriptively and after it has happened, to a difference in the psychological environment?" (Bergmann, 1943, p. 133).

This sort of criticism, it should be realized, is not basically critical of any construction of an environment that intervenes between the physical environment and overt behavior. The weight of the criticism is directed at ad hoc constructions. Postulating some "intervening" environment is permissible and potentially useful if: (1) its conceptual properties can be described in an a priori manner, and (2) its influence on behavior can be specified. These are the goals that S–R formulations have sought to reach when postulating mechanisms like the anticipatory goal response (Hull, 1930) which in effect shifts attention from an external to an internal environment. These hypothetical intervening environments have resulted in part from the prodding of, and reaction to, cognitive (Tolman, 1932) and Gestalt-type formulations of an internal environment, even though it should be recognized, as Goss has shown (1961), that the earliest behaviorists (e.g., Max Meyer, John B. Watson) assumed an intervening environment in the form of mediating verbal responses.

The analysis of the expansion of the stimulus construct serves to lay the groundwork for beginning our analysis of conceptual behavior. You recall that the extension of conditioning theory to discrimination behavior, an empirical gap that intuitively does not appear very wide, was executed with some difficulty, particularly in relation to coordinating the independent theoretical stimulus concept with new experimental operations. While this extension from conditioning to discrimination learning was taking place, discrimination theory was simultaneously trying to bridge a much wider gap existing between discrimination learning and concept learning. Without intending to be critical, this theoretical strategy may be likened to a battle plan in which an attack on an enemy's position is planned from a position not yet won. In any case it is necessary to recognize that important defining characteristics of concept learning tasks have emerged from the methods of discrimination learning and the theory of conditioning. The main difference between the typical experimental methodology used by S–R psychologists to study discrimination learning and concept learning is that in the former situation single stimulus events are discriminated from each other, while in the latter situation classes of stimuli are discriminated. This has led T. S. Kendler to define the present area of inquiry in the following manner: "Concept formation is taken to imply the acquisition or utilization, or both, of a common response to dissimilar stimuli. It is the problem of those who study concept formation to analyze the process and determine which variables influence it" (T. S. Kendler, 1961, p. 447).

This definition seems sensible and appropriate. It should be recognized, however, that such a definition has its roots in the conditioning-generalization-extinction model of discrimination learning. It is so much a function of these orienting attitudes that it practically forces one to consider concept learning as continuous with discrimination learning. By this definition the simplest kind of concept learning would be exemplified by stimulus generalization, the phenomenon in which a common response is made to *different* stimuli on the same physical dimension.

The definition under consideration seems so intuitively obvious that it becomes difficult to think of any alternative. If, however, one shifts his frame of reference to that of Piaget's work with concepts (1953), it soon becomes evident that a definition that emphasizes "a common response to dissimilar stimuli" scarcely seems relevant. Admittedly the complexity, or perhaps the obscurity, of Piaget's work precludes any satisfactory working definition, but the fact remains that such concepts as *object, space, time, coordination,* and *causality,* concepts which interest Piaget, do not fall comfortably into the aforementioned definition.

Now that the orienting attitudes and prejudgments of S–R psychologists who have adopted a conditioning-generalization-extinction model have been explicated, it becomes possible to examine their implications for the study of concepts. For the present the analysis will be limited to only two points.

The first is that major attention has been paid to the learning of concepts, or what has been more commonly referred to as *concept formation, concept attainment,* or *concept identification.* This problem has been mainly investigated with an experimental methodology similar in design to that used in discrimination studies, except for convenient modifications borrowed from rote learning procedures (e.g., Oseas & Underwood, 1952). As a result such problems as concept utilization, concept modification, level of abstraction of concepts, to name a few, have largely been ignored.

The second problem raised is that of the relationship between concept behavior and other forms of responses. Operationally specifying a concept as a common response to dissimilar stimuli fails, as already mentioned, to distinguish it from ordinary discriminations and responses to generalized stimuli. Osgood (1953) has tried to make such a distinction, however, not on empirical grounds, but instead in terms of a theoretical principle. He raises the question whether Hull (1920), in his classical study in which he found concept attainment resembling ordinary discrimination learning, was ". . . actually studying *concept* formation." His argument is that Hull's Ss were required to discriminate a *common* element in a group of Chinese characters. These elements were the same in all the characters the group to which the Ss had to learn a common response. Does not a dog, asks Osgood, go through the same task when he selects out the relevant tone from a multiple-stimulus environment? Osgood prefers to describe Hull's concept learning experiment as a study in labeling, believing that *concepts* require some "abstraction" process. His position is expressed in the following quotation from his analysis of Fields' (1932) work on the development of the "concept" of triangularity by the white rat.

"Patterning his work directly on Hull's approach, he showed that rats could jump toward a triangular form, when paired with other forms in a discrimination situation, despite large variations in size, shading, position, amount of outline, and so forth. Yet should we conclude that the rat can understand the *abstract* concept of triangularity? Would the rat respond positively to three dots in a triangular arrangement versus four dots in a square? Or react positively to three people, three places on a map, a three cornered block, as 'triangles'?" (Osgood, 1953, p. 667)

The distinctive feature of a concept for Osgood is that it depends on a mediational process. That is, instead of a common stimulus in different stimulus patterns becoming associated with a common response, as occurred in Hull's experiment, Osgood believes that a *true* concept emerges when varying stimulus patterns, *not necessarily containing any features in common,* elicit a common mediating response which serves as the cue for conceptual behavior.

This apparently meaningful distinction becomes less meaningful under close scrutiny. Although it is easy to specifiy the similarity between two stimulus patterns, it is extremely difficult to decide when two particular patterns fail to possess a feature in common. A coffee can and doughnut possess roundness. What in this world is completely different from either? A shoe? It is definitely not round but it does possess physical characteristics in common with the coffee can and doughnut. They all have weight and substance. They all are *something.* If one argues this way, then only nothing is uniquely different from something. But even here an objection might be raised by some philosopher that nothing is something, or at least the concept of nothing depends on the concept of something.

Disregarding the complex semantic overtones of Osgood's distinction, it is necessary to realize that it has served as a bifurcation representing the two directions that investigation of concept attainment has taken.

One direction has been the discovery of systematic relationships between stimulus events (words, geometrical designs, etc.) and a common response. Attention in this sort of work is focused primarily on the stimulus-response relationship. The second direction is represented by the major interest being directed at the mediational mechanism responsible (at least in the theoretical sense of "responsibility") for concept behavior. The internal cue, instead of the association, becomes the main focus of attention. An example of the former would be the work of Archer, Bourne, and Brown (1955), Grant, Jones, and Tallantis (1949), and Hovland (1952), who report that certain stimulus features of the experimental task (e.g., complexity, distinctive feature, information) are related to the speed of concept attainment. The second orientation is represented by the efforts of Goss (1961) and Kendler and Kendler (1962), who postulate mediating mechanisms and try to determine variables to which they are related.

It would be misleading to suggest that the conditioning-generalization-extinction model is the only S–R conceptualization of concept behavior, although as indicated historically it has been the most dominant one. In recent years two kinds of phenomena, both of which are represented in S–R language, have been reported that promise to exert influence on the development of research techniques in conceptual behavior. One is an outgrowth of operant conditioning techniques, and the other is that of clustering effects in free recall.

Operant Conditioning. Concept learning tasks that emerge from operant conditioning techniques appear not too different from those which derive from classical conditioning. Green (1955) instructed college students to tap a key when a correct discriminative pattern was present, and not to respond when the incorrect pattern was shown. The procedure differed mainly from conventional concept learning methodology involving discrete trials, in that fixed-ratio schedules were used in which discriminative stimuli were shown for different periods of time, allowing the S to respond repeatedly during one stimulus exposure. The results showed that, "The extent to which Ss discriminated was related inversely to the ratio of responses to reinforcement (1:1, 15:1, and 30:1) and directly to the length of time the discriminative stimuli were presented to them in conditioning (3, 30, and 60 sec.)" (Green, 1955, p. 180). Of greater interests, and probably of more importance than either the operant conditioning procedure or the results it produced, was the introduction of a verbalization technique which required the S to state what he believed to be the characteristics of the correct card. This made it possible to study simultaneously the performance of two operant responses, one verbal, the other instrumental key tapping. The results showed the correct verbalizations were ". . . related to the parameters of the experiment in the same way as was the key-tapping discrimination" (Green, 1955, p. 180).

This dual operant response technique was refined and extended later by Verplanck and Oskamp (1956) in a card-sorting task. The S was required to state a hypothesis before each card placement, with either the hypothesis statement or card-placement responses being reinforced. One of the most interesting findings was that a discrepancy often existed between the S's stated hypothesis and where he actually placed a card. This dissociation between words and action is also noted by Kendler and Kendler (1962), who postulate that several "horizontal" chains of S–R associations may be operating simultaneously in concept-learning tasks. These chains can function independently of each other, or they can become integrated by a stimulus from one becoming associated with a response from the other. Presumably, when S's verbalizations do not match his actions, the verbal and card-placement chains function in a parallel fashion, i.e., they do not interact. Combining developmental studies with a mediational S–R model, the Kendlers have sought to understand how the acquisition of verbal processes influences conceptual behavior, i.e., how verbalizations interact with instrumental responses.

Although Verplanck and the Kendlers share interest in a common problem, their methods of coping with it are decidedly different. Functioning within a Skinnerian factual, atheoretical framework tends to encourage treating "verbal hypotheses" and "card placements" as independent events. Each being a response in its own right, the problem is to discover how their rate of responding is influenced by various schedules of reinforcement. The problem of the relationship between the two is gingerly touched upon by showing how they covary. Such an approach can easily degenerate into a kind of psychophysical parallelism in which the occurrence of each response is scrutinized carefully, while the basic problem of *how* and *why* they interact, or fail to interact, is ignored. It must be obvious that the writer's preference is to deal with this problem in a theoretical manner by postulating the mechanisms responsible for the interaction between verbalizations and discrimination responses and then test their implications.

Clustering. Although the phenomenon of clustering has emerged from studies in memory, its implications are relevant to conceptual behavior. Bousfield (1953) found that when S is instructed to recall a list of randomized words in the order that it occurs in memory, he typically reports clusters of related words. For example, a list of 40 randomly distributed words comprising 10 names of animals, 10 vegetables, 10 professions, and 10 articles of clothing will not be recalled in a random manner. In free recall there will be a tendency for words from the same category to cluster together. A name of an animal will have a greater probability of being followed by another name of an animal than by the name of a vegetable, profession, or an article of clothing.

Explanations of clustering have varied from those attributing it to the operation of simple associative connections (Jenkins, Mink, & Russell, 1958) to those postulating high level mediated superordinate responses (Bousfield & Cohen, 1956). Whatever the explanation may be, the phenomenon of clustering illustrates concepts in action. Words are spontaneously organized into conceptual groups in the absence of any specific instructions. Thus clustering would seem to be an important factor in problem solving when the selection of an appropriate word or words has functional significance. Judging from the work of Maltzman (1960) on originality as mediated by word associations, work on clustering should ultimately reveal important information on how concepts are utilized in intellectual behavior.

Classificatory behavior has been approached in ways other than those already described. In fact, there is practically no limit to the number of pretheoretical models and orienting attitudes that can be used to investigate conceptual behavior. In judging what appears to be the most influential approaches, not so much in terms of their achievements, but rather in their hopes for the future, consideration of Piaget, "computer simulation" and mathematical models seems to be in order. These will now be discussed in a brief, and hopefully, a cohesive manner, so that the current scene does not appear, as it so often does, as a group of isolated efforts.

Piaget

If we dispense with evaluation and analyze Piaget's method of investigating concepts, we find an approach that is markedly different from that used by most S–R psychologists. Piaget (1953) does not have any preconceived notion that concept formation is like any other behavior, as was the case for the preconceptions of S–R psychologists of the similarity between classificatory behavior and conditioning and discrimination learning. Piaget nevertheless prejudges issues by using a frame of reference to describe behavior that emanates mainly from logic and to a lesser extent from biology. As a result he sees conceptual behavior largely in terms of logical operations.

Piaget superimposes his logical model on genetic data. He attempts to distinguish in a normal child's development successive stages of thought, each of which is characterized by different kinds of logical operations. Because of this genetic orientation concepts are considered to emerge at particular ages (just like teeth), with little or no consideration being given to specific learning experiences. The major attention then is paid not to the acquisition of concepts but rather to their utilization.

Piaget is a psychologist in the European tradition (although there are signs that, like everything else, he is becoming more "American"), and his studies and those of his co-workers, dealing with physical and mathematical problems, do not achieve standards of experimental control and statistical sophistication demanded by that new breed of behavior scientist, the psychonomist. Great trust is placed in the verbal responses of children to the questions of the experimenter when it is not always clear that agreement exists about meaning of terms.

If substantive issues are put aside, we find that for the most part Piaget's orientation is not contradictory, but is instead supplementary to that of the S–R psychologists. The principal area in which this is particularly true concerns their respective treatment of developmental and learning processes. A common criticism, and a justified one, that is often directed at Piaget is that he ignores the influences of learning in the development of conceptual behavior. It would be equally appropriate to direct the same kind of criticism, in a reverse form, at the S–R psychologists. With comparable indifference they have ignored developmental factors. This equal but different neglect stems from a tragic trend throughout the history of psychology to view developmental changes and learning as antithetical processes, a consequence, no doubt, of all the confusion surrounding the heredity-environment controversies. What actually is needed to break down this wall dividing these two research areas is some representation of behavior that can be applied to results of developmental as well as learning studies. One such attempt that is now being made in a very modest fashion is the analysis in S–R terms of the classificatory and problem-solving behavior of children of different ages (Kendler & Kendler, 1962). Needless to say, it would be helpful to the entire study of psychology if additional attempts were made with different pretheoretical models.

Piaget's work also supplements the S–R approach by providing interesting areas of investigation for possible exploitation. Although problems of studying such concepts as transitivity, probability, and causality are technically more difficult, there is no reason to believe that they cannot be studied in a manner consistent with behavioristic tradition.

Computer Simulation of Cognitive Processes

Perhaps because psychology is reaching its adolescence, it is succumbing more and more to changes in styles of research as well as theorizing. Witness the impact of such recent innovations as information theory, mathematical models, and computer simulation of behavior. One of several unfortunate consequences of an atomic holocaust would be that we would be deprived of learning which approach was a fad and which had permanent value. One can be spared from any doubts about the contributions of computers if one is willing to accept Newell and Simon's (1961a, 1961b) enthusiastic appraisal of simulation of human thinking by computers. A behaviorist might find it somewhat difficult to accept these authors' omnisscience. He will discover that in spite of being awarded their seal of approval for his rigorous methodology, their opinion is that he is interested in insignificant questions.

If computer models are to achieve the goals that their enthusiasts have set for them, much greater attention will have to be paid to the behavior that is being simulated and the generality of computer programs. The writer is reminded of the reaction of a minister to a lengthy account of the kinds of human behavior that computers have simulated. His comment was, "I'll take this stuff seriously when they get a computer that starts worrying after its parts begin to wear out." This anecdote nicely illustrates the point that in evaluating computer simulation one must also consider the behavior that is being simulated. The data that come from the continuous verbal reports of S as he is trying to solve a problem may not, as is assumed, mirror the process. For example, the reporting might conflict with the thinking and thus change it. According to Watt and Ach (Boring, 1929), a person's thinking occurs *before* he knows about what he is thinking. If this is true, then at its least a continuous verbal report slows thinking down, and at the very most, actually distorts it. Obviously much more information is needed about the relationship between utilizing concepts in problem solving under "normal" conditions and those of continuous verbal report. But even if a great difference were found between these two kinds of problem solving, the ability of computers to explain thinking while talking would still be a major contribution. Although it would not explain all forms of thinking, it would at least provide insights into the behavior of that small segment of our population who are required to talk while they think (e.g., professors who are unprepared for their lectures and wives who never shut up).

More important, perhaps in the long run, is the manner in which these verbal reports are gathered. It is awfully easy to shape verbal behavior (e.g., Greenspoon, 1955). Unless the most rigorous experimental techniques and controls are used, it is quite possible that experimenters can shape the S's introspective report to fit the mold of some preconceived hypothesis. "When it is demonstrated that human problem solving resembles that of the computer, it would be wise to ask whether such results are due to the computer simulating human behavior, or whether the experimental technique shaped the subject's behavior so that it resembled that of the computer" (H. H. Kendler, 1961, p. 190).

Of course, there is no reason why computer programs have to be applied to verbal reports. They can and have been applied to experimental results such as those from partial reinforcement and rote learning studies (Newell & Simon, 1961b). The important point is that to apply a program to a single experimental result is only the beginning of effective theorizing. What are needed are general theories—or programs—that can be applied successfully to a variety of empirical data. To return to Newell and Simon's General Problem Solver, a computer program that fits some logical manipulations (the fit is not perfect since the human leaves out some of the steps of the computer), the giant step forward will come when it, or some other program, is expanded to fit a variety of demonstrated phenomena such as functional fixedness, the relative effectiveness of reversal and nonreversal shifts in conceptualization for different age groups, the effects of verbal pretraining on subsequent concept behavior, etc.

Lest these comments be interpreted as completely negative, reference is made to Hovland (1960), who appropriately points out that computers can function as potential *aids* ". . . in sharpening our formulations concerning mental processes and phenomena," in encouraging ". . . theories that have both descriptive and predictive power," and for coping with the complex problem of dealing with a multitude of interacting variables.

Mathematical Models

Formally speaking, there is no difference between representing behavior by computer programs or by mathematical models. A computer program ". . . used as a theory has the same epistemological status as a set of differential equations or difference equations used as a theory" (Newell & Simon, 1961b, p. 2013). But in other respects they are different. Although it is difficult to generalize from all the various mathematical models, it is safe to say that there is at least one difference that is relevant to this discussion. To an appreciable extent mathematical models, unlike much of the computer simulation work, have emerged from the S–R language and methodological traditions.

One effort to deal with concepts has taken the form of a concept identification model (Bourne & Restle, 1959) which is an elementary extension of a mathematical model for discrimination learning (Restle, 1955). The reason that this transition is made with relative ease, aside from the clarity of the model, is that Bourne and Restle restrict their attention to relatively simple conceptual behavior having its roots in empirical relationships between clearly defined stimulus variables (e.g., number of relevant dimensions, number of irrelevant dimensions, delayed feedback) and choice behavior. Their work can be located conceptually along the "nonmediational" path emanating from the bifurcation that has been associated with Osgood's distinction between labeling (nonmediational) and concept behavior (mediational).

Mathematical models lend themselves to this sort of single-stage S–R analysis. Problems that have required a mediational analysis tend to be avoided. When they have been dealt with, as is the case of Restle's analysis of learning set data (1958), mediating stimuli resulting from responses of the organism are postulated. It would be most interesting if attempts are made to deal with "mediational-type" phenomena with a single-stage S–R mathematical model. For example, can Restle's discrimination theory (1955) with the assistance of his "adaptation" mechanism represent the changes in the relative rapidity in which reversal and nonreversal shifts are executed by children of different ages (Kendler & Kendler, 1959; Kendler, Kendler, & Wells, 1960)?

There is little doubt that mathematical models will play a more and more important role in attempts to investigate and interpret classificatory behavior as research moves ahead. Disagreements exist as to what exactly this role will be. Will it be an innovator that leads to the discovery of new phenomena while systematizing the old, or will it make its main contribution only in terms of the latter function? In either case, the warning of Skinner that comes from the wings, or from above, depending how one views the current scene in psychology, should not be *completely* ignored; "What is needed is not a mathematical model, constructed with little regard for the fundamental dimensions of behavior, but a mathematical treatment of experimental data" (Skinner, 1961, p. 62).

The Nature of Concepts

The impression has no doubt been given of talking around the concept of the concept. Any reader, hoping to discover what a concept really is, would no doubt have his curiosity unsatisfied, if not downright frustrated. Admittedly, the writer or psychology may be partly at fault, but it can be argued, perhaps in self-defense, that the major difficulty lies with this hypothetical reader's uncritical curiosity. After all, a concept is not a single thing or event whose attributes and functions can be simply listed and described. A concept is a complicated psychological phenomenon, a complete description of which will be contained someday in a theory capable of explaining the numerous empirical laws involving the term *concept.* That day will only come about when knoweldge about *all forms* of behavior is greatly increased.

Although the question of the discontented reader cannot now be satisfactorily answered, the direction which such an answer will take can be suggested. In order to do this, some method of conceptualizing behavior in general, and classificatory behavior in particular, must be offered. Stimulus-response language can serve this function. As a *first approximation,* concepts have three conceptual properties: they are associations, they function as cues (stimuli), and they are responses. This classification scheme, whose divisions are not as ironbound as they should be, is useful in ordering some of the major empirical problems existing in conceptual behavior.

Concepts as Associations

The greatest experimental effort has been directed toward the learning of concepts. Typically this has meant the association between dissimilar stimuli and a common response. The experimental methodology used to investigate this process resembles in many ways the conventional discrimination procedures used with animals; a single stimulus event is presented, whether it be in a card-sorting task or a rote-learning-type experiment, and the S is required to make a response which in turn is either reinforced or not. One important problem that has been raised is whether the concept learning that goes on in experiments of this sort really represents a single kind of concept learning or instead certain stages of concept learning. The terms *concept identification* and *concept acquisition* (or attainment) highlight a possible distinction. In a conventional card-sorting task the S, who is usually a college student, does not "acquire" a concept such as number or shape or color. Instead he identifies one. Prior to training the S knows these concepts; he merely has to learn appropriate responses to the appropriate stimuli. But the basic associations have been formed. His task is different from that of a child who does not know what shape or color or number is at the beginning of the experiment. The child has to learn, for example, the *associations* between square shapes and a common response and circles and another response. This distinction is operationally meaningful if we define concept identification tasks as occurring when instructions could produce the same behavior (e.g., sorting in terms of number) as the conventional training procedures. Concept acquisition would be restricted to the situation in which a simple set of instructions would not suffice. The S would have to acquire a concept from the "very beginning."

This distinction does not rest only upon the ages of the Ss or their verbal ability. College students who could learn to *identify* a number concept would have to learn to *acquire* such concepts as *disinhibition* and *spontaneous recovery*. And correct verbalizations of children in simple concept learning tasks are not always a guarantee that their choice behavior will be appropriate (Kendler & Kendler, 1962).

It would seem that acquisition occurs prior to identification, although there is no reason to suggest that both processes could not occur simultaneously. The basic problem is to discover how, if at all, these processes differ. Is only the acquisition process truly associative, while the identification process is a relatively simple case of transfer, in which a new response is associated to an old class of stimuli? Or are these processes extreme points on a single dimension? If comparable studies could be done with concept identification and concept acquisition tasks, then the differences between the two processes would be better understood.

The selection of a particular experimental procedure would be a problem for the investigator who decided to try to answer the questions just raised. Concept learning experiments have barely tapped the wide variety of possible conditions—as well as questions to ask. Answers to the simple practical question of what are the optimal conditions for concept learning can reap a rich harvest. This problem has not been ignored, as indicated by interest shown in the relative importance of positive and negative instances (Hovland, 1952), in the relative difficulty of identifying conjunctive and disjunctive concepts (Bruner, Goodnow, & Austin, 1956), and the retarding effects of irrelevant information (Archer, Bourne, &

Brown, 1955). But obviously much more has to be done. A particularly important problem is suggested by a study of Marx, Murphy, and Brownstein, who in seeking to train Ss to recognize classes of complex stimuli find that training with ". . . certain kinds of abstracted patterns results in greater recognition than presentation of the fully drawn stimuli" (1961, p. 459). Although their study is primarily concerned with what may be properly described as principles of perceptual organization, the problem touches upon one that is central to conceptual behavior, whether it be concerned with classifying geometrical patterns, words, ideas, or what have you. This problem is what kind of learning allows for the most effective transfer from the abstract to the specific and from the specific to the abstract.

In discussing associative problems of conceptual behavior the topic of retention cannot be ignored. We know concept identification occurs more rapidly if the S is freed from the task of remembering previous correct and incorrect instances (Bruner, Goodnow, & Austin, 1956). But if learning a concept is dependent to some extent on retention, as it usually is, what are those principles of retention that can be exploited to facilitate concept learning? It is customary when studying concept learning with memory drums or card-sorting tests to present all the positive and negative instances in successive mixed sequences. Although these sequences are necessary for experimental control, the order itself may be an important variable. Perhaps an instance should be repeated until the correct response is forthcoming. The problem being raised is related to that old and seemingly forgotten one of "part" versus "whole" learning. Previous attacks on this topic failed to provide any definitive answer. Perhaps psychologists did not know enough about learning and methodology at that time. But it might be noted that problems of conceptual behavior were largely ignored when this topic was originally investigated. An approach that considers conceptual behavior might make sense out of what has appeared for many years to be nonsense.

It is possible to make the transition between associative and stimulus and response problems of conceptual behavior by raising questions related to developmental factors in concept learning. What is the psychological basis of the developmental changes occurring in various kinds of concept learning? To attribute these differences to intelligence, at best, is offering an incomplete answer and, at worst, is begging the question. Is the locus of these differences in associative functioning or stimulus or response mechanisms? This question cannot be answered now (a number of partial answers are being offered), but if stimulus-response language is useful, some answer should be forthcoming after appropriate research is completed. And these answers will represent giant strides forward in understanding classificatory behavior.

Concepts as Cues

One of the major differences between S–R correlationists and mediational theorists is that the former are interested in concepts as associations and the latter in concepts as cues. The result is that the former have been more concerned with concept formation, while the latter are primarily interested in concept utilization.

Although much theoretical use has been made of concepts as cues (Goss, 1961; Kendler & Kendler, 1962), there are some (Bruner, Goodnow, & Austin, 1956; Miller, Galanter, & Pribram, 1960; Shepard, Hovland, & Jenkins, 1961) who believe the mediational approach with its emphasis on the cue function of concepts is not enough. They argue that something like "ideas" or "principles" or "strategies" are needed in addition to what is referred to as S–R associations. This difference reminds one of what appeared to some to be the basic issue surrounding the latent learning controversy, namely, whether S–R associations or cognitions were learned. This distinction has always puzzled the writer (Kendler, 1952) because it seems to reflect personal preference for models and language systems adopted to represent behavior instead of fundamental theoretical assumptions. If the vast potentialities of S–R language resulting from chaining and mediated generalization of verbal behavior are considered, then what some prefer to call "principles" or "ideas" or "strategies" can be, although need not be, expressed in terms of cues arising from *chains* of S–R associations.

A somewhat more meaningful aspect of this problem was expressed in an exchange between Bousfield (1961) and Osgood (1961) concerning the justification of postulating a concept of "meaning" that is separate and independent of the cue functions of words. Bousfield is willing to dispense with "meaning" because it is unnecessary and it generates confusion. Meaning for him can be handled by word associations; a concept's meaning would be expressed by the word it elicits. Osgood disagrees, believing that responses to words depend on other things besides meaning (e.g., syntax). Meaning per se is better measured by the semantic differential. The reaction to this disagreement by the members of the symposium at which this exchange occurred suggested that this issue was not crystal-clear. It may be that when we expand our method of measuring the cue function of words to include chains of words (i.e., phrases and sentences), then this apparent difference will disappear. Or it may be, as Osgood argues, that meaning can never be reduced to the cue function of concepts. Whatever the outcome of this intratheory squabble, the results cannot help but have important repercussions for mediational theory in general and conceptual behavior in particular.

The mechanism basic to both these intertheoretical and intratheoretical differences can be categorized as problems of concept generalization. Having learned a conceptual response, what are the instances that will evoke it? A question like this is extremely relevant to the area of problem solving. Perhaps when programed learning is used more as an experimental tool (Gagné & Brown, 1961) than as a sales commodity, basic information about concept generalization will be obtained.

Concepts as Responses

When a child has acquired a concept, what response has he learned? According to current mediational theories (Goss, 1961) he has acquired some implicit response, usually, although not necessarily, verbal in nature, which serves as a cue for his overt behavior. This is obviously not the only possible interpretation. Zeaman takes a position more in line with Gibson's (1959) perceptual formulation, ". . . distinctiveness is acquired not by the addition of response produced interoceptive stimulation but rather by the acquisition of responses (observing or attention) whose function is to make effective certain features of *exteroceptive* stimulation not previously responded to" (Zedman, person communication).

This alternative to S–R mediational theory should—and no doubt will—be pursued. But certain problems must be recognized. First, it will not be sufficient to equate observing responses with attention. Although they are functionally equivalent in that they both operate to "select out" from the total pattern of stimulation those components that will become associated, it may be that principles governing their operation are different. Observing, or what some prefer to call receptor-orienting responses, will determine what part of the environment will strike the organism's sensorium. Attention, on the other hand, decides what stimulus component of a pattern of stimuli, falling within the "receptor gaze" will stand out and become associated. In short, both observing and attending will influence the stimuli that are to be associated, but their influence may operate through different mechanisms, the former through principles governing the learning and performance of instrumental responses, the latter through principles governing perceptual organization.

The second problem that must be recognized in developing the perceptual formulation of the sort Zeaman suggests is to understand how it differs from a mediational approach. There seem to be three possibilities. One is that perceptual and mediational responses can be parts of separate segments of the same behavioral chain (Kendler, Glucksberg, & Keston, 1961). Another possibility is that perceptual and mediational responses interact, thus influencing each other in a reciprocal manner. Finally, they may not be distinct events even though they have been assigned different names. It would surprise the writer if all three possibilities did not possess some degree of validity for at least some situations. For the moment it would be wise to consider the possibility that apparent differences could hide intrinsic similarities.

Disregarding the relationship between mediation and perception, it would seem reasonable to expect that many, although not necessarily all, developmental differences will be discovered to be localized in different response capabilities. This would be true for many of the differences noted in athletic skills, and there is probably some psychological similarity between athletic and intellectual skills. Response differences resulting from developmental changes in human intellectual functioning can be categorized in four different groups. First, the degree of abstraction can be increased. Second, partly as a result of language development, the child can learn to respond in a mediated manner. Third, the acquired verbal habits increase in complexity and uniqueness. One reason that Gagné and Brown (1961) find that self-discovery of a principle is superior to being taught the principle is that the former procedure allows the principle to fit better into the individual's existing verbal system. Fourth, as development proceeds, the verbal behavior which initially develops as an independent chain has increased opportunity to become integrated with other patterns of behavior that are constantly being learned.

Before terminating this section attention should be called to the importance of conceptual behavior in education. If researchers in the field of programed learning believe that the major response being learned is the one the student writes in the frame, they are sadly mistaken (Gagné & Brown, 1961; Kendler, 1959b; Silverman, 1961). What is being learned are concepts, and what is modified are large segments of the student's verbal repertoire. Unless we understand this process better than we do now, the hopes of programed learning will never be realized.

The Place of Classification Learning in the Psychology of Learning

In a symposium such as this one cannot avoid considering the interrelationships among the various categories of human learning. In fact, the present sequence of topics implies an organization of learning processes that is widely accepted. The underlying theme of this organization is that conditioning is the simplest form of learning. One can observe in the conditioning situation the formation and strengthening of associations in all their natural glory and simplicity. As one proceeds from conditioning to rote learning through probability learning and finally on to problem solving, the learning phenomena become more and more complex, more and more a function of a larger number of variables and the interrelationships among them. This organization of the psychology of learning from the apparently simple to the complex is so eminently reasonable and sensible that it is often accepted as valid. In fact, the writer (Kendler, 1959a) once directed a criticism toward statistical learning theorists for using such complex empirical phenomena as probability learning to arrive at their fundamental theoretical assumptions. It was suggested that the appeal of probability learning was not its psychological simplicity, but instead the quantitative simplicity of the data it produced. This criticism is justified if one accepts the idea that conditioning mirrored most clearly the fundamental processes of learning. Such a "conditioning-oriented" view is obviously not demanded by the facts, but instead represents a prejudgment as to how the facts will ultimately be ordered.

Realizing this point, it is as justified to organize all learning around the facts of probability learning as around the facts of conditioning. In short, the answer to questions concerning the interrelationships among various categories of learning must of necessity be a theoretical answer. This does not mean that one answer is as good as another. Instead it means that the best answer must be supported by the best theory.

Although it might be possible to argue that concept learning should be considered as the core of all learning, the writer would just as soon not defend this view. But it might be mentioned that an allied view proposed by Tolman (1932) that discrimination learning is basic to all learning be considered more seriously. It is impossible to consider any learning situation in which competition between responses is completely absent. In classical conditioning the competition between the conditioned response and other reactions (e.g., the "investigatory" response) are typically ignored. The conditioned response, as a result of a long series of acquisition trials, may become so powerful that evidence of competing responses is absent. But the introduction of experimental extinction procedures will usually show that hidden behind every dominant response is another response that can take over.

When viewing the history of the psychology of learning over the past few decades one can notice a shift of interest from the development of a single association to the competition between associations. Perhaps this shift represents a reorientation toward a more fundamental problem—or at least, to a problem that is equally fundamental. The implication of this is that if the various categories of learning are organized into some meaningful relationship, perhaps discrimination learning with its emphasis on the process of habit competition, should be assigned the central role.

An attempt has been made to show how investigations of conceptual behavior have been shaped by preconceptions of such behavior. Some of the more popular preconceptions have been reviewed and analyzed. These models tend to focus attention and obtain information in different problem areas. The analysis was terminated by attempting to systematize the problems of conceptual behavior in terms of three psychological functions of concepts resulting from a stimulus-response analysis, and pointing out that the location of conceptual behavior against a background of all psychology will ultimately depend on the theory of behavior which is accepted.

It is easy to get depressed about the current status of our knowledge of conceptual behavior. We know so little. We have so much to learn. Yet depression, or even disappointment, is not justified. Different empirical techniques have been developed, knowledge is being gathered, and some integrating ideas, even if they are limited in scope, are being offered. No doubt individual psychologists viewing the effort that is being expended by behavior scientists in this field could recommend and more economical and strategic approach. For example, the standardization of our research techniques could be improved. This will be achieved not by wishing or proclamation, but by the development of techniques that capture the imagination of the younger psychologists who enter this fascinating field. Criticism about our present concepts of conceptual behavior can be offered. They are both clumsy and confusing. But problems of definition will dissolve as the field matures. The great need is to get more researchers in the field of conceptual behavior. And this is exactly what seems to be happening.

THINKING: A PHYLOGENETIC PERSPECTIVE*

Sam Glucksberg

Psychology has been defined to generations of students as the science of behavior. In spite of this definition, and despite my own training in the "science of behavior," I find myself involved in the workings of the mind. An esteemed colleague whose research interests are in physiological psychology refers to his laboratory as the "brain research laboratory." Since he had access to a machine which produced very impressive and official-looking signs, he asked me what title I would like to post over my laboratory door. When I told him "Mind Research Labora-

* This selection was written expressly for this volume. Preparation of this essay was facilitated by Grant MH-10742 from the National Institute of Mental Health, and by Grant HD-01910 from the National Institute of Child Health and Human Development.

tory," he looked puzzled, finally decided that I was not serious, and dropped the subject.

Perhaps I would have been taken more seriously had I tried to explain what I meant by *mind,* and, more specifically, what I mean by the word *thinking.* If this essay convinces my colleague that one can, indeed, deal with thinking as a comparative psychologist then I may someday get that sign to post over my laboratory door. If that day comes, copies of this essay will be handed out to all who ask, "What do you mean, 'Mind Research Laboratory'?"

In barest outline, I use the word *thinking* to refer to an activity within an organism which (a) in some way represents some aspect of the external world, and (b) may influence behavior. If I am hungry, then the sight or smell of roast beef will make my mouth water. But I don't need the sight or smell of roast beef; I can *imagine* it. Whether I imagine the roast in the form of words, visual images, or olfactory images, my *thinking* of that succulent, steaming platter may make my mouth water. In its simplest form, thinking is a representation of an external event. In its more complex forms, thinking may be a series of representations which follow formal transformational rules, or it may involve representations of still other representations. In other words, we can think about our own thoughts. You can provide your own example of a more complex form of representation by doing some mental arithmetic. The numerals you start with are representations of abstract number concepts, and you will perform some formal operations according to certain rules and eventually end up with a numeral which represents the other numbers that you had been thinking of and the relationships between them.

Various forms of representation exist, and representational processes serve a number of important behavioral functions. I will deal first with representational behavior and processes in the normal adult human. This will provide a conceptual framework for our discussion of thinking in other animals.

Categories of Representational Behavior

Jerome Bruner, in the first of two essays on cognitive growth (Bruner, *et al.*, 1966), distinguishes between three major kinds of representation: enactive, ikonic, and symbolic. In the enactive category, the external world is represented by motor actions. A home plate umpire, for example, keeps count of balls and strikes with his fingers. A young child may be able to remember that a toy is hidden under a box by keeping an eye on that box. If he is diverted, even for a moment, he may not remember where it is, just as the umpire might not remember the count if he were to relax his fingers.

In the ikonic category, representation is accomplished by images. If asked how many windows a room has, many people can generate a visual image, and then literally scan the image and count the windows. Visual images are perhaps the most common, but they are not the only kind of imagery. Many people can generate auditory images, and scan those images for information, say, about a song or a symphony. Images of tastes, smells, pains, and touches can all occur.

The third of Bruner's categories of representation is the symbolic. The most prevalent form of symbolic representation is language in which words and sentences represent reality. Symbolic behavior is not, however, confined to humans. As we shall see later, animals can generate non-linguistic symbolic representations, although this symbolic activity is crude when compared to that of human adults.

Imagine the following task. You are shown three doors, with a light bulb over each. You are told that a $100 bill is behind one of the doors, and that the light over the correct door will flash on for 3 seconds. You are not allowed to move toward that door until 15 seconds after the light has gone out. When it is time for you to act, there will be no external cues to guide you. You must supply your own cue based upon memory of which light flashed.

How might you go about doing this? You might point your head or your arm or, indeed, your whole body toward the appropriate door, and when the time came, walk straight to it. You would then be engaging in an enactive kind of representation. Another way to get to the right door would be to form a visual image of the scene and use it as a basis for choice and action. This would be a form of ikonic representation. Finally, you might simply say to yourself, "It's the middle door" or "It's the door on the right." This would be a form of symbolic representation.

The three major categories of representation are of great interest from a developmental and evolutionary point of view. The categories differ not only in quality but also in complexity and sophistication, and thus are characteristic of different developmental and evolutionary levels. Bruner (1966c) deals with them in terms of developmental levels: Representations are:

> . . . the means by which growing human beings represent their experience of the world; and how they organize for future use what they have encountered. There are striking changes in emphasis that occur with the development of representation. At first the child's world is known to him principally by the habitual actions he uses for coping with it. In time there is added a technique of representation through imagery that is relatively free of action. Gradually there is added a new and powerful method of translating action and image into language, providing still a third system of representation. Each of the three modes of representation—enactive, ikonic, and symbolic—has its unique way of representing events. Each

places a powerful impress on the mental life of human beings at different ages, and their interplay persists as one of the major features of adult intellectual life. (p. 1)

Functions of Representational Behavior

As Bruner suggests, the capacity for representation is at the core of our ability to function. We do not simply respond to events in the external world; as Plato pointed out in his myth of the cave, we do not even perceive or experience the "real" world. Instead, we provide ourselves with an internal model of it, and we act and think largely on the basis of the model. The model, if adaptive, articulates with the external world in such a way as to organize it and free us from inflexible external stimulus control. Behavior that is based entirely upon a system of representation which does not articulate well with the real world would be maladaptive. It could be disastrous, and sometimes is—as in extreme cases of psychosis. The psychotic's representation and organization of the world prevents him from functioning in important and necessary ways; he may not eat, or talk with other people, or protect himself from injury. Even so, his functional capacity to develop an aberrant world view is illustrative of the general human capacity to develop and to organize symbolic world views.

There is a great range of possible adaptive organizations, and our choice between them is often arbitrary. Because most people in a culture share the same set of representations, we are usually unaware of other possible sets. Consider, for example, our representations of color. In physical terms, color varies continuously with wave length from about 380 to 720 millimicrons. There is no sharp break in the spectrum, but we may discriminate sharp breaks. The location of those breaks depends in part upon the structure of our visual system and in part upon our symbolic representation of the color spectrum. In the English language we divide the color continuum into six categories: blue, violet, green, yellow, orange, and red. This division, however, is no more natural or right than any other division. The speakers of Shona, a language of Rhodesia, divide the spectrum into three categories, and the breaks between the categories do not coincide precisely with our breaks. Though the Shona speakers have the same color vision that we have, they refer to the colors that we call orange, red, and violet with the same name. Our blue and some of our green have a second label, and the third label refers to some of our green and yellow. Speakers of Bassa, a Liberian language, use only two major color categories. Our violet, blue, and green have one label, and our yellow, orange, and red have another. This does not mean that they can only discriminate between two colors. Just as we can distinguish between two shades of green, so can they distinguish different colors described by the same word. But, like us, they would find it hard to recall which color was which without appropriate discriminative symbolic representations (in this case, words), and have trouble thinking about different colors in the same word category except in ikonic ways.

A representational system, then, provides an organism with the capacity to organize experience. Maier and Schneirla, in their classic work *Principles of Animal Psychology* (1935), consider this capacity the hallmark of higher mental processes:

> We have seen how learning releases an animal from the limited forms of behavior furnished by maturation. Learning likewise has its limitations. At best the animal (through learning) can respond to a range of equivalent stimuli. . . . But mere learning cannot fit an animal for meeting entirely new situations. . . .
>
> In the ability to reorganize experience, the animal is released from stereotyped forms of behavior which are laid down by learning. With the development and dominance of this ability, learning takes on a new role. Past experience ceases to furnish the patterns of response, and instead furnishes *the data from which new patterns may be formed* [italics added]. This ability frees the animal from a particular bit of learning and makes possible almost unlimited patterns of response. (p. 479)

The data from which new patterns may be formed are, in terms of the view of thinking we have adopted, representations of experience which may, under certain circumstances, be the material for thought. We turn now to a brief survey of the evidence that animals other than man think.

The Evolution of Representational Processes (thinking)[1]

The evolution of thinking can, as we have suggested, be viewed in terms of the evolution of representational behavior, a particular kind of internal activity which may influence and determine overt behavior. Overt behavior, of course, is determined jointly by external stimuli and cues and internal stimuli and cues. *Stimuli,* for the purposes of this discussion, elicit behavior directly. There is no decision involved. Sunlight, as a stimulus, directly elicits tropistic behavior in plants and lower animals. The organism turns toward or away from light. *Cues,* on the other hand, provide information which guides behavior. Sunlight, as a cue, may provide the occasion for a variety of human actions, such as putting on sunglasses, lowering window shades, or buying a bikini.

One trend in phylogenetic development is a decrease in the impor-

[1] Much of the material in this section was adapted from a section of *Symbolic Processes* (Glucksberg, 1966).

tance of stimuli as direct elicitors of behavior and an increase in the importance of stimuli as cues which guide behavior. Accompanying the increase in the importance of stimuli as cues is an increase in behavioral flexibility and adaptability and a decreased stereotypic response. The development of learning in the phylogenetic scale can be viewed as the development of the capacity to use external cues rather than relying upon fixed action patterns determined by the interaction of stimuli with organismic states.

A second trend in phylogenetic development is the evolution of the capacity to generate and use internal cues. The kinds of representation we have been talking about serve as cues for overt behavior. The process which generates these cues may be simple or complex, enactive, ikonic, or symbolic. In the case of the ikonic and symbolic activities, internal cue generation substitutes for overt behavior as well as for external cues.

In its most basic form, we have noted that an internal cue is a stored representation of some external cue. At the highest level, it is a stored representation of an entire system of symbolic material. Clearly, human thought covers the range between these two extremes. Because we are capable of the greatest use of stimuli as cues, both external and internal, our behavior is less stereotyped, less controlled by external stimulation, and less limited by our particular experiences than is the behavior of any other animal.

Symbolic Activity in Subhuman Animals: Representations of Cues

What evidence do we have of representational processes of various levels of complexity in animal behavior? Does an animal demonstrate a capacity for enactive, ikonic, or symbolic representation?

Are any animals able to do the three-door task described above and, if so, what form of representation might they use? An enactive form, which Bruner contends to be the simplest in developmental terms, is the most readily observed in animals. In a delayed-response task similar to the three-door problem, a dog points his head or body toward the correct door and tends to stay in that position until the delay period is over. Hunter, inventor of the delayed-response technique (1913), found that dogs can maintain a point for as long as 5 minutes. A dog, like the umpire who relaxes his fingers and loses track of the count of balls and strikes, loses track of which door is which after a little while, if prevented from maintaining the point.

A more sophisticated form of representation can also be found in animals. If white rats are not permitted to maintain an enactive representation, they can still tolerate a delay of 10 to 15 seconds. It is clear that these animals are generating some form of nonenactive—that is, ikonic or symbolic—representation which enables them to remember an external cue and to behave appropriately in the absence of that cue. Similar experiments have shown that dogs, cats, monkeys, and others share this ability.

What about the capacity to generate a representation of an internal cue, such as one's own past activity? Animals have been tested for this capacity by the delayed-alternation task. One form of this task employs a simple T maze. The animal is released from a starting compartment, runs down an alley to a choice point, and then either turns left or right to get food. If the animal has turned right on the first trial, then he must turn left to get the food on the second trial, then right again on the third trial, and so on. No external cues are available to the animal at the time of choice. To succeed on the task, he must be able to represent his own behavior. The animal must remember how he acted and then do the opposite the next time. As might be expected from their delayed-response performance, white rats cannot usually succeed with long delays between trials in a delayed-alternation task. In general, the more advanced an animal is in the phylogenetic scale, the longer the delays may be before the animal fails in these tasks. The delay periods with animals other than man are measured in seconds, minutes, and with some, such as chimpanzees, hours. An adult human, of course, can tolerate a delay of, conceivably, many years.[2]

It is quite clear, then, that nonverbal species can engage in some forms of representational activity, or thinking. Examples of rather complex symbolic activity could be culled from reports of observations of animals, particularly primates, in the field. Even under laboratory conditions, there is abundant evidence of higher-order symbolic behavior.

In a series of experiments performed at the University of Wisconsin, Harry and Margaret Kuenne Harlow (1949) trained monkeys to solve relatively difficult problems without trial and error. They started by giving the monkey a simple discrimination problem. He could pick up only one of two objects which differed in color, size, and shape. Food was under the correct one. After many trials and errors, the monkey learned to select the correct object every time. Specifically, he had simply learned to select one of two specific objects by trial and error.

The next step was to give the monkey the same type of problem with two brand-new objects. This time the monkey required fewer trials to learn. After working on several hundred of these problems, each one different, the monkey's speed of learning improved consistently, until rarely was more than one error made on each problem. The animal

[2] A distinction must be noted between the duration of delay that an animal can tolerate, and the length of time that a learned response or habit may be retained. Delay tolerances in these tasks are very short; the retention of a particular discrimination or learned response may easily last for the lifetime of an animal.

learned that if the first object picked at the start of a new problem did not have food under it, then the other object always did, and if the first did have the food, then the other did not. He had learned something which enabled him to solve any new two-choice problem in one trial. If the monkey were human, we would say that he had adopted a strategy of win-stay, lose-shift. Since we have no way of knowing how the monkey represented his learning strategy, we must simply acknowledge that his behavior was being guided, in some way, by a rather complex symbolic activity.

The Harlows then demonstrated that the monkey was capable of learning far more complex solution strategies. If, after the monkey has learned that the food is always under a certain object, a switch is made, he makes many errors before he learns to reverse his original choice. When the objects are again switched, the monkey makes fewer errors before reversing again. After many such reversals, the monkey again exemplifies the strategy learned in the simple discrimination problem: win-stay, lose-shift.

Still more complicated is the oddity problem. Here, the monkey is given three objects. Two of them are identical and the third is different. The food is always under the odd object. In a typical problem, the monkey is first shown two small red cubes and a large green pyramid. The pyramid is the correct choice. On the next trial, he finds two pyramids and one cube. The cube is the correct choice. The monkey cannot use as the basis for choice the position of the correct object. Nevertheless, the pattern of behavior displayed in a long series of oddity problems is similar to the pattern observed in the simpler two-choice discrimination problems. Initially, laborious trial and error progressively gives way to faster and faster learning until the monkey makes no errors at all, even when he is given problems with objects he has never seen before.

This is no trivial accomplishment. The monkey must pay attention to all three objects, make judgments on the basis of some concept of sameness and difference, and select an object which he represents somehow as "the different one." The monkey has developed a *learning set*, a pattern of activities, some of them symbolic representational, which leads to immediate success in any new oddity problem. He has learned to generate appropriate representational activity to guide his overt choice behavior. In a very concrete sense, he has learned to think.

Primates, although thus proved to be capable of engaging in symbolic activity, do not seem to have the advantage of the symbolic capacities provided by an organized language system. Their behavior is not qualitatively different from the prelinguistic human child. If an adult were given an oddity problem, he could be instructed to "always choose the different one." If he were not given that instruction, he would, in one way or another, give it to himself after one or two trials. Neither the monkey nor the very young child can do this. Pavlov expressed this profound difference between adult mankind and all other organisms by saying, "It is nothing other than words which has made us human . . ." (1941, p. 179). By words, he meant the representational activity, the thinking that can be expressed by words—for symbolic activity, as we have seen, does not consist entirely of language.

A chimpanzee can engage in symbolic activity and can even learn to communicate using the arbitrary gestures of the deaf-and-dumb sign language. The lowly white rat displays some behaviors which are typical of the young child (Kendler and Kendler, 1968). But we are, as humans, unique in our capacity for complex symbolic activity, although foreshadowings of that capacity can be found, not only in our primate ancestry, but virtually throughout the mammalian series.

THE PRINCIPAL ELEMENTS OF THOUGHT: A PHILOSOPHICAL EXAMINATION*

David L. Mouton

The development of human thought has brought with it a natural and enormous interest in the nature of thinking itself. In part, this has been a matter of simple curiosity. But for fields such as philosophy, theology, and creative writing, where thinking itself is the principal means of advancement, presuppositions about the nature of thinking can determine the entire theoretical or artistic result. For a field such as pedagogy, an understanding of the nature of the various modes of thinking can be an important means to practical success.

The purpose of this essay is to set forth the basic philosophic theories of thinking as they emerge from an examination of the principal philosophic issues concerning thinking. This essay is intended as a contribution to the dialogue between psychology and philosophy which has lately been badly neglected. There are many passages in psychological literature where psychologists encounter and then step around philosophic questions for lack of active interchange with philosophers of similar interests. The outline provided here is intended to provide psychologists with the conceptual framework within which their own empirical investigations fall and which establishes their limits.

---

* This selection was written expressly for this volume. See the Consolidated Bibliography for complete references.

The three principal aspects of thinking which have consistently puzzled philosophers and other thinkers down through the centuries concern: (1) the mind, or organ of thought; (2) the mental states which partially or completely constitute thinking; and (3) the content and/or object of thinking.

The Organ of Thought

Thinking is either an activity of man and other intelligent beings or a state or process which obtains or occurs in them, or both. (The distinction between *acts* and *mere events* will be discussed in the next section.) In either case the question naturally arises as to the location of this state, process or activity, and it is traditional to employ the word "mind" to designate this particular component of a person.

There are numerous theories as to the nature of the mind, but the principal watershed lies between those theorists who believe the mind is identical with the brain or central nervous system and those who raise what they believe to be insuperable objections to any claim that the mind is physical in nature. It will serve our purposes here to note the main reasons for holding to one or the other of these alternatives.

There is one overriding fact about human experience which forms the basis for this entire controversy and which may be illustrated by the following example. When a meal is being digested, it is sometimes possible for a person to hear signs of, and even to feel, what is going on. In such cases it is obvious, independently of all anatomical and biological knowledge, where the digestive process is taking place. But when a person is meditating, solving problems in his head, dreaming, or daydreaming, he is not *merely by virtue of being conscious* aware of the organ whose state that consciousness is nor even of its anatomical location. This fact of nature goes a long way toward explaining many of the philosophical issues concerning minds and self-knowledge. To be conscious is necessarily to be conscious of something whether a thought, pain, or a physical object—but not directly of the organ of consciousness itself. To clarify by analogy: if dreaming is like watching an internal movie, it involves only the awareness of the pictures and sounds themselves, not of the location of the theatre, its walls, screens, and so forth. For this reason it was easy for men to ascribe their thinking to a nonphysical soul or mind since thinking does not appear to the thinker as a bodily function nor as a function of any *thing* or *location*. Reflecting on this fact about consciousness, however, will not reveal the whole story about the notion of the mind, but it does constitute an important starting point.

This fact about consciousness and its substantial base makes clear that our knowledge of the mind and its nature must be based upon its perceived properties rather than any direct perception of the mind itself. This point may be clarified by a comparison with the case of ordinary material objects. When I see a physical object (say, a tree), I see such properties as its shape and color. And whenever I perceive the shape and color of a thing, I do *ipso facto* perceive the thing. Why, then, do I not also conclude that when a person is conscious (in other words, is perceiving dreams, thoughts, feelings, sensations, and so forth) he is *ipso facto* perceiving his mind? The reason for this may be explained by means of the following distinction between types of properties. Some properties, such as color, shape, and bulk, are such that to perceive them is to perceive the object of which they are properties—but certain other properties, such as temperature and odor, are not of this type. Thus, if I perceive a certain heat and a certain odor, I have no way of knowing merely from these properties whether they emanate from one or two objects. Nor can I form an idea of the object(s) having these properties merely from the properties alone. I suggest we regard conscious states in general as being analogous to these latter, nonsubstantial properties. Thus to be conscious is *not* to apprehend one's mind, and all that we can determine about the nature of the mind must be based on these conscious states.

Psychophysical Dualism

The basic thesis of psychophysical dualism is that the seat of consciousness is the mind and that the mind is a nonphysical component of a human being. What does it mean to say that an entity is nonphysical? René Descartes (1596–1650), the most renowned champion of this view, answered this question by holding that the mind is nonspatial, that is, it does not occupy any *extended* place or point in space.[1] This would explain why we are unable to apprehend directly either our own minds or the minds of others.

Now consider the role this postulated entity, the mind, plays in our lives. It is by means of my mind that I experience such sensations as pains and that I perform such tasks as making decisions. When a pin is stuck into my hand, this purely physical occurrence brings about a certain sensation or conscious state in my mind. Thus we are led to the conclusion that *the body affects the mind.* And when I decide to move my finger and then do so, a purely mental event of deciding or willing causes or initiates the movement of my finger, thus suggesting that *the mind affects the body.* These two conclusions form the philosophical theory of two-way interactionism.

[1] René Descartes, *Meditations on First Philosophy,* VI, in Haldane and Ross (eds.), *Philosophical Works of Descartes,* Vol. I, p. 190.

This theory, although appealing to common sense, is difficult to reconcile with the thesis that the mind is nonphysical. It raises the following questions: How can the mind and body affect one another if there is absolutely no contact between them? And how can there be contact between a physical entity and another which is not located in space at all? Descartes, although not convinced that these questions represent genuine difficulties, in effect presented an answer by holding that the mind is located *within* the space of the brain but occupies only a nonextended point.[2] Not only does this fail to meet the objection, it introduces new and equally difficult questions centering around the notion of a point.

There are at least two familiar kinds of points, physical and mathematical. Physical points are not points, strictly speaking. Each of them is in fact extended since all space, including even very minute points, is divisible and hence each physical point actually contains further points *ad infinitum*. There is therefore no reason to think that any actual physical point is indivisible. Mathematical or geometrical points are by definition indivisible just as mathematical lines have by definition no width. This enables mathematicians to theorize about the points and lines of a triangle with absolute precision because their models—those in their definitions and their thinking, not the actual drawings—are themselves absolutely perfect. But there is no reason to think that mathematical points exist outside the ideas and definitions of mathematics. The mind could actually occupy neither type of point. If it occupied a physical point it would have to be extended since to be located at a physical point is to be coextensive with it. It could occupy a mathematical point only if such points were real and thus available, whereas in fact they are mere fictions of mathematical thinking for which we have no reason to believe there corresponds anything in the actual world.[3] Descartes has no answer to this objection and indeed the very notion of interaction between a physical body and a totally nonphysical, nonspatial mind appears inconceivable. These considerations may not be decisive but they do constitute a primary source of dissatisfaction with the theory of psychophysical dualism and its corollary of two-way interactionism.

Physicalism

If we reject the view that the mind is nonphysical in nature, then the most prominent alternative is its contrary—that the mind is physical and hence a part of the body, the brain and/or the central nervous system

[2] Descartes, pp. 345–346.

[3] This objection was presented to Descartes by his contemporary, Pierre Gassendi (1592–1655), in a set of objections contained in *Philosophical Works of Descartes*.

being the obvious nominees for this role. This is the thesis philosophers call Physicalism. Its principal strength lies in its simplicity; it attempts to explain mental events without appealing to any entity that is not a component of the human body.

The claim that the mind is actually the brain (or central nervous system) is of ancient vintage and has encountered many objections. The contemporary version—the psychophysical identity theory—is a refinement of past versions in the light of contemporary logical and linguistic studies. In order to explain what this theory is and is not, and to exhibit its strengths and weaknesses, let us examine it against the background of a statement of reasons for rejecting it by Professor C. J. Ducasse.

> Let us consider first the assertion that "thought" or "consciousness," is but another name for . . . molecular processes in the tissues of the brain. As Paulsen and others have pointed out, no evidence ever is or can be offered to support that assertion, because it is in fact but a disguised proposal to make the words "thought," "feeling," "sensation," "desire," and so on, denote facts quite different from those which these words are commonly employed to denote. To say that those words are but other names for certain chemical or behavioral events is as grossly arbitrary as it would be to say that "wood" is but another name for glass, or "potato" but another name for cabbage. What thought, desire, sensation, and other mental states are like, each of us can observe directly by introspection; and what introspection reveals is that they do not in the least resemble muscular contraction, or glandular secretion, or any other known bodily events. No tampering with language can alter the observable fact that thinking is one thing and muttering quite another; that the feeling called anger has no resemblance to the bodily behavior which usually goes with it; or that an act of will is not in the least like anything we find when we open the skull and examine the brain. Certain mental events are doubtless connected in some way with certain bodily events, but they are not those bodily events themselves. The connection is not identity. (1964, pp. 223–224)

Scientific Evidence

Professor Ducasse makes several distinct points. Consider first his charge that "no evidence ever is or can be offered to support" the claim of mind-brain identity. If a scientist wanted to investigate mental events, he would avail himself of some means for monitoring the processes in the brain, for example an electroencephalograph. His findings would consist of specifications of electrochemical changes in the nervous system. *At the same time* he would record the first-person reports of the subject whose brain processes were being monitored. These findings would consist of reports of sensations, feelings, thoughts, images, desires, and so forth. If successful, this would enable the scientist to discover, for example, that whenever a certain electrochemical event occurs in a particular place in X's brain, X feels a sharp pain in his right knee. If and when scientists are permitted widespread experimentation with electrodes planted in human brains to stimulate specific brain processes, they will have a means of checking these findings. If these findings hold for all human beings, the scientists will have discovered universal laws. One can easily imagine that the ultimate result of such research would consist of a complete listing of each possible mental event and its correlated brain process after the manner of a dictionary or telephone directory.

There can be no doubt that such information would be of inestimable value in medicine and psychology, but the important point here is that what the scientist is doing is *correlating* two different findings, not discovering that they are identical. Correlation is possible only where there are at least two things to correlate. It would, for instance, be strange to correlate A and B and then say that A *is* B.

Compare the physicalist's suggestion that the mind is the brain with a genuine discovery of identity. The true claim that a cloud is identical with a set of water particles in suspension has as one corollary that observation of the cloud at close range will enable one to perceive the particles which constitute it. But as one psychologist has expressed it, "a close introspective scrutiny [of one's inner mental states] will never reveal the passage of nerve impulses over a thousand synapses in the way that a closer scrutiny of a cloud will reveal a mass of tiny particles in suspension." (Place, 1962) In other words, whereas closer scrutiny of a cloud reveals that it is really a mass of water particles in suspension, closer scrutiny of brain events will not reveal that they are really pains, thoughts, and so forth, nor will closer scrutiny of pains or thoughts reveal that they are really brain events. This fundamental difference between the observations of a neurologist and the "observations" which constitute each person's awareness of his own thoughts, pains, perceptions, and such implies that the detailed correlation of human mental states with neurological states will always remain just that—namely, correlations of two distinct phenomena. The very fact that two sets of events can be separately apprehended and correlated implies that they are distinct phenomena. Thus these correlations, far from being evidence *for* physicalism, in fact constitute by their very nature evidence *against* physicalism.

It must not be thought that the mind-body problem lingers on in modern times only because neurology is still in its infancy.[4] All philosophers of whatever persuasion have assumed that mental and neurological events are correlated and many have assumed (as noted previously)

[4] For a sketch of the perplexed attitudes and diverse reactions of some psychologists and physiologists to the mind/body problem, see Farrell (1962).

that they stand in distinct causal relations with one another; but however detailed the description of these correlations and causal connections becomes, it will never constitute proof of physicalism. On this Ducasse appears entirely correct.

Semantic Evidence

In the light of this fact, what is the physicalist claiming? Ducasse suggests two different answers. First, he attributes to physicalism the claim that "'consciousness' is but another name for . . . molecular processes in the tissues of the brain." Let us express this as thesis

(*A*) *'mind' = 'brain'*

by which we mean that the words 'mind' and 'brain' have the same meaning. The second thesis Ducasse ascribes to the physicalist may be expressed as follows:

(*B*) *mind = brain*

where this is understood as the claim that minds are identical with brains. Evidently Ducasse believes that the advocate of physicalism is committed to (A), whereas Ducasse himself holds both that (A) is obviously false and further, if (A) is false, then (B) is false: that is, if the word 'mind' does not mean what the word 'brain' means, then minds are not brains. He is of course correct in holding that (A) is false since people do commonly understand both 'brain' and 'mind' without suspecting that minds are brains—an obvious impossibility if these words had exactly the same meaning. Can one, for example, understand both the synonymous words 'nighthawk' and 'bullbat' without knowing they are the same bird? But Ducasse's assumption that the falsity of (A) implies the falsity of (B) is of the utmost importance in understanding the contemporary physicalist thesis.

One of the principal achievements of recent analytic or linguistic philosophy is the discovery of a principle of language which enables us to clarify a question of precisely this sort. The principle was first enunciated by the German philosopher Gottlob Frege (1848–1925). Frege raised the question as to how it could possibly be interesting to discover that A is identical with B if the expression 'A is B' means "'A' has the same meaning as 'B'." If it were true that 'A is B' means or involves "'A' has the same meaning as 'B,'" as Ducasse assumes, then we should need only to know the meaning of 'mind' and the meaning of 'brain' in order to see whether they designate one or two different entities. Thus no interesting, or nontrivial, discovery of identity could ever be made. Discoveries of identity would always be discoveries of the synonymy of meanings of words.

But discoveries of identity are not always, or even often, discoveries of the synonymy of meanings of words. Consider, Frege suggested, the discovery by the ancients that the morning star is identical with the evening star. There was a time when men understood their corresponding expressions for 'morning star' and an 'evening star,' and yet did not know these heavenly bodies were identical, that is that there was only one such heavenly body, not two. Frege argued that we can only make sense of this example if we recognize that the meaning of a word is not as simple as it has been assumed, and distinguished between the *sense* of a word and its *reference*. The sense of 'morning star' is that which we understand when we understand this expression, "a certain bright star seen in the morning." The referent is that to which the expression refers, namely, a certain heavenly body. In this way it is easy to understand how two expressions can differ in their senses while referring to the same entity. An *interesting* discovery of identity consists in the realization that two words or expressions differing in their senses have the same referent.

When this principle of meaning is applied to the mind-brain distinction, it becomes an open question whether 'mind' refers to the same physical entity referred to by 'brain.' It is clear that the falsity of (A), which concerns the senses of these words, does not imply the falsity of (B), which concerns their referent(s). The controversy cannot be resolved merely by noting the differences in meanings of 'mind' and 'brain': the properties of the mental and the properties of the physical must be examined.

Evidence of Properties

A = B implies that any property of A is also a property of B and vice versa; and conversely, if every property of A is also a property of B and vice versa, then A = B. The man I saw fleeing the bank at the time of the robbery is the bank robber, if and only if they have all their properties in common. Given the impossibility of scientific evidence (that is, beyond mere correlation) for physicalism, the question of mental and physical properties becomes the central consideration. Ducasse speaks of this issue also when he writes: "What thought, desire, sensation, and other mental states are like, each of us can observe directly by introspection; and what introspection reveals is that they do not in the least resemble muscular contraction, or glandular secretion, or any other known bodily events." The validity of Ducasse's point does not depend upon comparing mental events with the particular macrophysical examples he selects since the more inaccessible microevents such as the chemical and electronic processes in the nervous system appear also to have properties utterly incompatible with mental events. As one writer noted, the claim that mental events are in actuality physical events appears to force us into saying "that physical processes are dim or fading or nagging or false, and that mental phenomena such as after-images are publicly observable or physical or spatially located or swift." (Cornman, 1962) Thoughts, but not brain processes, are true or false; brain processes, but not thoughts, are spatially located and publically observable. Can these differences be reconciled?

Consider first the case of spatial location. We do not experience brain processes as such at all and we do not experience thinking processes as being located anatomically. On this basis, then, there can be no distinction between the mental and the physical, and the consideration of spatial location seems to end in a draw.

There are, however, two other mental properties which are crucial in this issue. The first is *privacy*. A phenomenon X is mentally private to Peter if only Peter can directly apprehend X. It seems obvious that all physical phenomena, whether particles or properties, are essentially public. Being located in space, they can be perceived by anyone who can bring the appropriate perceptual powers to bear on them. But are states of mind public? It is questionable whether it makes any sense even in principle to suggest that Peter could directly apprehend the content of Paul's mental state. Since every pain that Peter feels is his own pain, how could Peter feel a pain not his own? This tough or necessary sense of privacy has convinced many thinkers that the mental cannot be identified with anything physical. The most that the advocate of physicalism can hope for is a stalemate (that is, success in showing that if a purely physical being could think, his thoughts would be just as private as if his thinking had occurred in an incorporeal soul).

The second property is *intentionality*. To be conscious is always to be conscious *of something*, whether the mode of consciousness is thinking, dreaming, feeling, seeing, or whatever. It is frequently claimed that no physical entity can be intentional in this sense. A book, for instance, may be *about* politics only in the sense that the black marks on the book's pages stimulate mental states in the reader which constitute his thinking of politics. Independently of all minds the book is not about anything. The marks on its pages could have been produced by chance factors in nature. Does it follow from the fact that such physical objects as books, signs, films, pictures, and such are not intentional independently of minds, that no physical entity, not even a nervous system, can be intentional? This is an extremely difficult question, and there is widespread disagreement about the answer.

Consider the following situation: Peter has his eyes closed but "sees" in his mind's eye a yellow flower (that is, he entertains or experiences a mental image of a yellow flower which he then reports as the object of his consciousness). This object has various perceived properties, such as its shape and color. The question before us concerns the proper account of this appearance to Peter and thus we can safely concentrate for simplicity on one property, the yellow of the flower. Now since Peter professes to "see" (that is, in some sense to be conscious of) yellow and his eyes are closed, the yellow he is aware of must be located either *in his mind* or *in his brain.* Yet both of these alternatives are either false or impossible or both. It is a demonstrable fact that no yellow occurs within a person's brain. If a neurologist had been able to gaze into Peter's brain while he was "seeing" the yellow, the neurologist would have discovered nothing yellow. How then can this appearance of yellow be identified with a property of the brain? Yet, although physicalism appears on this point to be patently false, there is no comfort for the proponent of dualism either. For dualism usually involves the thesis that the mind is not spatially extended (how else can mind and body be distinguished?) whereas all colors or appearances of colors are essentially extended. No appearance of color is possible without being extended. How then can any color be literally represented in a nonphysical mind?

Having thus reached an impasse, we should retrace our steps and examine the presuppositions which have brought us to this point:

1. There is an object, perhaps a mental image, which is the literal object of Peter's consciousness.
2. This literal object is literally yellow.

The only reason for holding premise 2 to be true is that it is implied by premise 1. For if Peter is aware of some object and that object is a mental image, then that object must appear to him as he reports it (namely, yellow). But if consciousness is a *state* of mind, perhaps premise 1 is false; perhaps there is no literal object of Peter's consciousness. Attempting to locate a literal object in either mind or brain seems to be a search for a will-o'-the-wisp. Perhaps this is because his "seeing" yellow is simply a particular state of mind or brain which has no need of being literally yellow.

This suggestion of a solution may be clarified by the following contrast. Given that Peter "sees" yellow, we can ask either:

A. What is the object before Peter's mind which accounts for his report of "seeing" yellow? or

B. What are the properties of Peter's conscious state that constitute his "seeing" yellow?

If question A is the correct query to pose, then premises 1 and 2 may well be true. In that case we are simply faced with a problem for which we have no answer. However if question B is the correct one to pose, then premises 1 and 2 are both false. The proper account of the intentionality involved in our example would be along the following lines: to "see" yellow is to experience a conscious state the "yellow" of which is not literal but is explicable in terms of various properties or characteristics of the conscious state. And, of course, if this account were true then it would be plausible to hold that these properties of the conscious state were specifications of the electrochemical state of the brain.

Anyone inclined to think that this sketch presents an acceptable defense of physicalism is well advised to pause long enough to wonder whether it may not simply be a way to sweep the problem of intentionality under the rug. Is this really a successful way to reduce the truth of a thought or the intensity of a pain or the red of a mental image to the physical specifications of a physical state? At this point—far short of any clear solution to this classic issue—we must leave the problem. Of one point we can be certain: The final solution to this problem is a function of such conceptual or philosophical considerations as the above and not of any scientific findings which perforce consist of correlating mental and neurological phenomena.

The Conscious States in Thinking

Ideas as Images

Thinking is a mode of consciousness and as such to be thinking is to be in some state of mind. We have seen already how difficult it is to say just what the nature of that organ or entity is to which the word 'mind' refers. Now let us raise a parallel question: What is the nature of the conscious state which constitutes thinking? As with the question of the location of mind, this one has received a variety of answers during the history of intellectual pursuits.

Perhaps the simplest theory to characterize is the one whose most renowned champion was David Hume (1711–1776). Hume held that all conscious states which he called "perceptions" are of two types, "impressions" and "ideas." The former are the more fundamental in the sense that the nature of ideas is dependent upon the nature of impressions and thus Hume defines 'idea' in terms of 'impression.' By 'impression' he means "all our sensations, passions and emotions, as they make their first appearance in the soul" and by 'ideas' he means "the faint images of these [impressions] in thinking and reasoning."[5] Thus I may see a certain house and then subsequently think of it, the former state being an impression, the latter an idea. An idea is a mental picture, since it differs

[5] David Hume, *A Treatise of Human Nature,* L. A. Selby-Bigge (ed.), p. 1.

from the original visual picture solely in its diminished intensity or vividness. In all other respects it is an exact copy.

According to Hume, then, to think is to experience a mental image; and mental images constitute awareness of their objects by picturing them. There is one outstanding difficulty with this theory. Some words designate particular persons, objects, or properties such as 'Socrates,' 'this page,' and 'the whiteness of this page'. Certain other words designate general notions of things or properties, such as *'man,' 'tree,' 'red,'* and so forth. Since people are able to understand words whether particular or general in meaning, it is natural to assume that there is a thought which corresponds to each meaningful word. By identifying thoughts with determinate images, Hume appears to explain what is transpiring in one's mind when one is thinking of a particular object or quality. To think of Pablo Casals is to entertain a mental picture of that particular person. But by the same token, Hume makes it impossible to explain what is transpiring in one's mind when one is thinking of a general notion such as man, color, or tree, or a general thought such as "All men are mortal." This problem is grounded in the fact that all images are perfectly determinate whereas all general words are indeterminate. There can exist no image or picture of man: Either the image must be very unclear and to that extent not represent anything at all, or the man must be represented by it as being of a particular size, shape, and color. Similarly there can be no color or representation of color which is not either of a particular shade of red or green or blue or . . . and so forth. There is a basic incompatibility between the determinate nature of images and the indeterminate nature of many of our most ordinary meanings, concepts and, consequently, thoughts. How, then, can thoughts be identical with mental images?

Hume was of course aware of this difficulty and he tried hard to produce an account of abstract thinking within the framework of his own theory.[6] It should be noted that this theory derives from another thesis which Hume regarded as self-evident—namely, that everything which actually exists is particular, and thus any general or abstract thought must somehow be explicable in terms of the particular things which make up the universe. His proposed solution to this problem lay in the distinction between an image (or word) and the signification of the image (word). Thus to think (of) "all men are mortal" is to picture either one particular man or one particular set of linguistic signs (such as 'All men are mortal') and then for that particular image to *represent all men.* The image, in other words, is particular in its nature but general in its significance. There are certain phenomena for which we would find this formula a perfectly natural explanation. It is, for example, a familiar

[6] Hume, pp. 17–25.

fact that a portrait can represent both a particular person and humanity in general. We speak of this as though these two modes of representation were actually in the particular portrait. Upon reflection, however, it is apparent that the representation of the particular person is built into the portrait, whereas the universal representation is merely suggested by the portrait or abstracted from the portrait by the beholder. In neither case would Hume's solution be supported. The general principle seems to be this: A particular can be general in significance only if it appears so to some thinker, and this is possible only if the thinker is *aware of some general fact.* But how can the mind be aware of a general fact if its conscious state consists exclusively of a particular picture?

This same basic question regarding the nature of thinking states is dealt with in the *Meditations* of Descartes. Descartes concluded that common sense was correct in its opinion that there are at least two varieties of thinking states. His evidence went as follows:

> I may entertain a mental image of a triangle or I may conceive of the triangle via the definition of 'triangle.' Now consider the parallel case of a chiliagon, a thousand-sided plane figure. I may conceive of a chiliagon as anyone would do in reading this discussion or I may entertain a mental picture of a chiliagon. . . . The remarkable fact . . . is that I cannot actually distinguish a picture of a 1000-sided figure from that of a 950-sided figure merely by glimpsing it, yet I cannot fail to know whether I am thinking of the one or the other. I do not . . . need to scrutinize my thinking states in order to know what I am thinking of as I must scrutinize the image of a chiliagon in order to know what kind of polygon . . . [I am seeing]. Hence my thought in conceiving of a chiliagon must be different from my mental picture in imagining a chiliagon.[7]

Since both types of thinking clearly do occur, Hume is mistaken in trying to reduce the one to the other. Descartes accepted this duality of conscious states in his thinking.

This may not be quite as decisive as it appears, however. A possible counter to Descartes' claim has been developed in recent years by the Oxford philosopher, H. H. Price (1953). While there may be, as Descartes says, no noticeable difference between a mental image of a 950-sided figure and a 1000-sided figure, there certainly is a noticeable difference between an image of the words 'a 950-sided figure' and an image of the words 'a 1000-sided figure.' If we include images of linguistic expressions within the class of images we wish to identify with thinking states, the range of materials available for the attempt to explain thinking states by images is significantly broadened. Price effected an even greater extension of the notion of "image" when, following Hume, he recognized

[7] Descartes, *Meditations,* VI, pp. 185–186.

a distinct type of mental image corresponding to each of the five senses as well as to kinesthetic sensations. Thus in addition to seeing a triangle in the mind's eye, one may hear a tune in the mind's ear, seem to feel the touch of velvet, seem to feel in one's muscles the sensation of the swing of a golf club, and so forth. Once this plethora of very subtle types of mental states is acknowledged, then it becomes noticeable immediately that it would require extraordinary powers of introspection for anyone to determine for himself whether every moment of thinking is constituted in part or whole by one or more of these very subtle images. Recent discussion of this question by philosophers has yielded the conclusion that, whatever the answer may be, individuals differ markedly in the variety and clarity of the images (if any) which they experience. This is in fact a lesson which they could have learned in the last century if they had read Francis Galton's illuminating study of mental imagery, *Inquiries into Human Faculty.* Galton sent a questionnaire to a large number of fellow countrymen whose replies ranged from detailed accounts of their wondrously complex but clear imagery to flat denials that such things existed! There is, however, one final philosophical point to note: If all thinking were in fact constituted in some sense by mental imagery, these images could not be simply identical with the thinking states. The complete state of consciousness which is in part pictorial is not exclusively pictorial since the image must be recognized as an image of ______ where the blank is to be filled in by the expression reporting whatever the person is actually thinking (of). Thus either the notion of an image must be revised so as to build in the aspect of a thought, or the image factor is never more than an aspect of a complex conscious state.

Mental Acts

There is one further issue which should be introduced at this point, concerning mental acts. Hume once used the following metaphor to describe thinking:

> The mind is a kind of theatre, where several perceptions [thoughts] successively make their appearance; pass, re-pass, glide away, and mingle in an infinite variety of postures and situations.[8]

Again, these perceptions

> are link'd together by the relation of cause and effect, and mutually produce, destroy, influence, and modify each other.[9]

[8] Hume, *A Treatise of Human Nature,* Book I, Part IV, Section VI, "Of Personal Identity," p. 253.

[9] Hume, p. 261.

This suggests that thinking is not something which human beings do, but something which happens in them, and this is the basic notion behind the "stream of thought" model. Thinking is construed, roughly, as though it were a kind of daydreaming. Hume was committed to such a view since he did not think there was any organ of thought, the mind for him being simply the system of perceptions.

The contrary to this model holds that some thinking consists at least in part of mental acts which the thinker performs. In this view I may deliberately compare A with B in my mind. There is no question but that ordinary experience supports the action model for some, though not all, instances of thinking. The fact that we find it natural and appropriate to modify reports of thinking with adverbs such as 'deliberately' and 'carefully' implies that it is either something we do or at any rate something we experience as if we did it. What reason is there to doubt the existence of mental acts?

First, there is the puzzle as to our basis for saying we have done something when there is no movement or component event that we can single out as part of our act. This is obviously closely related to our ignorance as to the nature of the organ of thinking.

Second, the issue of mental acts is merely an aspect of the general problem concerning the nature of actions. The twentieth-century philosopher Ludwig Wittgenstein formulated this issue in the following question: "what is left over if I subtract the fact that my arm goes up from the fact that I raise my arm?" (1953) If my arm goes up involuntarily, this is a reflex movement, not an act. If I raise my arm, this is an act on my part and it consists of this same arm movement plus some additional factors which differentiates it from a reflex movement. Apparently, then, the difference lies in the initiating event. It is quite simple to analyze the movement of a finger back through the muscles and nervous impulses to some event in the brain. On this suggestion the initiating event in the case of reflex movement is simply a physical event and, in the case of actions, an act of will, a volition.

This conclusion presents many difficulties. First, it appears to force us back to mind-body dualism with all its complications. Second, it is very problematic whether such volitions occur at all, much less that they are always necessary to actions and what, if they do always occur, their nature is. Finally, there is the difficulty of tying actions to the agents who perform them. This last problem or puzzle may be formulated as follows: If when I move my finger anything whatsoever can be said to be the cause of that movement—whether a neurological event, a desire, volition, or whatever—then it is false to say that *I* moved my finger, and hence to say that in this case anyone has *done* anything (Taylor, 1966, p. 113). This makes it appear that although human beings seem to be capable of analysis into various components, the pursuit of such analysis in the case of an action must either fail or pay for its success by showing that the act was not an act after all! It has been suggested that if we ascribe actions to people when they do muscular tasks without understanding the nature of actions, it would be unreasonable to withhold the ascription of action in the case of thinking simply because we lack that same understanding (Taylor, 1966, pp. 163–164).

Third, the peculiar nature of thinking introduces a new dimension into the question of mental acts. We want to say that a person can control the thoughts he thinks, in part because we encounter difficulty in discovering any other aspect of thinking which could be the basis of our calling the thinking an action. If I compare A with B in my thoughts, I entertain the thought of both and then notice their similarities and dissimilarities. All of this could simply occur to me. To say that I do this deliberately suggests (a) that I can either bring about or avoid the entire experience and (b) that I can control the attention I give to the thoughts before my mind. I must be aware of my own thoughts, however they are brought to mind, but it is up to me whether I regard them with a view to a comparison, analysis, or whatever other concern or purpose I may have. There is, however, a peculiar problem attached to these notions. If I can choose whether and how to think a certain thought, I must select from possible alternatives in each case. I may decide which speaker in a room to look at because hearing the speakers is different from, and possible without, looking at them. But I cannot really select one of two possible thoughts to think since there is no difference between being aware of the possible thoughts between which I am to choose and my simply thinking the two thoughts. This impasse drives one towards Hume's associationistic theory of thinking, according to which my conscious states in thinking succeed one another by force of an inner momentum, causal connections, and external stimuli independently of my volition. This is far too cursory a discussion to achieve more than a glimpse of the tip of the intellectual iceberg which goes by the name of 'mental act.' Much needs to be done before we shall truly understand these aspects of thinking.

The Object of Thinking

Adverbial Theory

The preceding section discussed two versions, one based on images and one on pure thoughts, of what is today referred to as the adverbial theory of thinking. Now let us consider explicitly what is involved in this notion. Some verbs take nouns while others take adverbs. Thus we can say "He saw the bird" and "He felt miserable." In some cases, it is unclear on which model we are to construe such linguistic reports as

"He felt a pain." The grammar of this sentence suggests that a pain is the object of some feeling in much the same way that a bird may be the object of some visual perception. But this may strike one as absurd. Perhaps 'He felt a pain' really means '*He felt painfully.*' In other words, the apparent object is really a quality of the mental state and should be expressed by an adverb assuming that we want the form of our language to reflect the manner in which we understand the world. If one rejects this adverbial theory, one must show what kind of object or entity a pain is.

The imagistic theory of thinking discussed above is an adverbial theory, as is the Cartesian theory, according to which images are but one of the two types of thinking states. In each case the conscious state contains or constitutes a representation of the object, and our discussion so far has concentrated on the nature of the representation. In any event the representation is regarded as sufficient to its role in thinking even if the object of the thought does not at the moment exist.

Theory of Diaphanous States

One of the motivating factors behind adoption of the imagistic theory is the diaphanous character of nonimagistic thinking states. In watching a film I may be either so intensely involved in it that I am unaware of anything but the story, or I may be simultaneously or alternately aware of both the story and the fact that it is being portrayed on the screen of a theatre. In an analogous manner I may think of X by means of a mental picture, oblivious of the fact that X is being represented to my mind by a mental image, or I may be simultaneously aware of both X and that there is an image of X before my mind. The normal experience is of the former type, but the fact that introspection is possible as in the latter case removes an element of mystery from thinking and provides the comforting illusion that we know just what is going on. We make images the beast of burden of our theory of thinking because they are such familiar introspectively discriminable and repeatable items. The idea that there are nonimagistic "pure" thoughts is regarded as mystery mongering because, unlike images and motion pictures, we seem able only to grasp their content and never the means by which the content is represented. They are diaphanous, almost given to our introspection but not quite.

Classical Theory

One solution to this puzzle, the so-called Classical Theory of thinking, jettisons both the adverbial theory and the theory of diaphanous states in favor of thinking states which are *completely transparent* and through which we cognize real objects. The thinking state functions like a clear pane of glass through which the object of thought is directly apprehended. Thus Plato held that in thinking we are apprehending the denizens of a realm of forms which constitute the objects of thinking.

The relevant evidence for this analysis of thinking is not immediately apparent. In fact it seems obvious that when I think of a physical object at which I am also looking, the object of my thinking is the same as the object of my sight. There appears to be no room in such cases for Platonic forms. This would suggest that the Platonic view is either carelessly based on a limited and select set of cases and/or grounded on metaphysical considerations independent of the psychological phenomenon of thinking. Both these suggestions contain some truth.

The natural area for the Classical Theory is an abstract field such as mathematics. What am I thinking of when I think of $2 + 2 = 4$ or of a triangle? In each instance the object does not appear to be identical with anything in this world and it seems to have properties which go beyond our meanings, ideas, knowledge, and so forth. Descartes expressed the point this way:

> When I imagine a triangle, although there may nowhere in the world be such a figure outside my thought, or ever have been, there is nevertheless in this figure a certain determinate nature, form, or essence, which is immutable and eternal, which I have not invented, and which in no wise depends on my mind, as appears from the fact that diverse properties of that triangle can be demonstrated, viz. that its three angles are equal to two right angles, that the greatest side is subtended by the greatest angle, and the like, which now, whether I wish it or do not wish it, I recognize very clearly as pertaining to it, although I never thought of the matter at all when I imagined a triangle for the first time, and which therefore cannot be said to have been invented by me.[10]

According to this view, if there is nothing perfectly triangular in the actual world of space and time and if the notion of triangularity is not created by the thinking of intelligent beings, then there is something, call it Triangularity or the Triangular Form, which exists in some sense eternally and independently of both the physical world and the world of thought.

This theory derives from the conviction that there are many concepts, especially ideal ones, which have no instantiation in the physical universe. There are many bodies and physical lines which appear triangular in shape. A triangle, however, is a plane figure whose three angles total 180°. This is possible only if the lines forming the triangle are straight and, as Descartes said, "I do not think that any part of a line has touched

[10] Descartes, *Meditations,* V, p. 180.

our senses which was strictly straight."[11] Apparently straight lines do not appear straight when seen through a microscope which reveals them, presumably, as they really are. This same point extends to such concepts as justice, beauty, goodness, perfection, and so forth, which we understand without ever having seen a perfect instance. The thesis may then be made still more general by noting that each thought which is propositional in form contains, as Bertrand Russell emphasized (1912), at least one universal term or property term about which these same questions may be raised. In the thought that the house is yellow, 'the house' may refer to a particular entity in the world, but 'is yellow' refers to a universal property which has innumerable instantiations. By 'yellow' we mean neither this instance of yellow nor all the instances summed together but the property itself regardless, as with the triangle, of whether there are any instances in the world or not. Thus 'yellow' seems to refer to an ideal Form in Plato's sense.

Since we are usually thinking propositional thoughts of one form or another and each of them contains at least one universal, this last step brings us very close to the thesis that all thinking consists at least partly in cognizing a particular type of nonphysical object peculiar to thinking and rationality. Once this step is taken, it is an easy matter to account for the thoughts we entertain but which obviously have no actual instances, such as unicorns, mythological and fictional persons, and imaginary creations.

Here again we are up against a problem only the tip of which is visible and which is obviously a general metaphysical issue and not limited to philosophical psychology. If Plato were correct, the unique nature of thinking would consist of its peculiar conscious states of cognition *and* the peculiar class of entities which constitute the objects of that cognition. On the other hand, if the adverbial theory in any of its forms is correct, the unique nature of thinking consists solely in the peculiar conscious states wherein its various objects, existent or imaginary, are represented.

The Development of Thinking

Thinking is a human activity, or at least a human function, and thus questions which are raised about human development can also be raised about thinking. In a general way the various possible theories of development have been outlined for us by Ausubel. We may simplify a bit and say that there are basically four possibilities: (1) Preformationism, or, more directly, no development; (2) Predeterminism, or development determined endogenously; (3) Environmentalism, or development

[11] Quoted in Anthony Kenny, *Descartes, A Study of His Philosophy*, pp. 148–149.

determined exogenously; and, finally, (4) any combination of these, especially the theory that human thinking is the result of an interaction of the organ with the environment. My purpose in the following discussion is to relate these possible theories of development to the elements of thinking as we have analyzed them.

The question of the development of the organ of thinking is immaterial, however the ontological dispute may be finally resolved. If the psychophysical identity theory proves true, the question of development will be relegated to natural history and science. If the mind is nonphysical, we will never have direct access to it and can only infer about its nature from what we experience as its thoughts. One possible inference is that whatever is absolutely nonphysical and thus (spatially) simple has no parts and therefore cannot develop. From here it is only a small step to the classical theological thesis that God creates minds *ex nihilo*. On the other hand, if the seat of thinking is physical, then its development is that of the brain or central nervous system; this has a natural history and is properly a subject for scientific investigation.

Somewhat the same points must be made about the conscious states of thinking and the objects of our thoughts. What are immediately given as introspectively discriminable items are the thoughts, the actual content of our conscious states, whether the content be regarded as internal or external to the thinker. It is only through the contents, concepts, and propositions we entertain and the manner in which we entertain them that we can discover anything philosophically about the development of thinking. Now let us turn to the moot questions.

The Environmentalist Principle

Common sense would no doubt find most reasonable an explanation of the development of human thought in terms of human experience, especially sense experience. It seems axiomatic that such activities as travel and reading result in a fresh set of ideas. As civilizations develop and history is recorded, the wealth of information and thus of thoughts is overwhelming. Common sense is unavoidably environmentalistic in its orientation on this question, many philosophers have agreed. When Hume defines ideas (thoughts) as faint copies of sense impressions, it follows that no one can entertain the thought of X unless one has experienced (that is, perceived by the senses) at least one X. Either this thesis is false or some possible Xs (God, for instance) must be redefined so as to permit some natural experiences to constitute the sensory basis for the corresponding thoughts. This kind of redefinition has appealed to some thinkers (consider, for example, Freud's claim that the idea of God is the idea of an authoritative, fatherly figure with exaggerated properties or powers); to others it could only appear as question begging.

Descartes' statement of this thesis is as follows: There must be as much actual reality in the cause of an idea as there is representative reality in the idea itself.[12] If the color green is represented in one's thinking there must be something which is really green and which caused the idea. This led Descartes directly to the argument that God exists since we obviously do think of God. (Note, in connection with the previous discussion of conscious states, how difficult it is to say what one's idea of God really is.) For our purposes what needs to be recognized is that Descartes' extension of reality to cover the spiritual realm as well as the natural constitutes an application of the environmentalist principle: we think whatever we do because of what exists and can hence be the cause of our thoughts. Although it is often said that Descartes believed in innate ideas, this seems to amount to nothing more than a belief that men have some nature and that it is necessary for their nature to be such as to respond to the causes of its thoughts. Thus it is natural for man to think of God, but the necessary final cause of this thought is an actual constituent of the "environment"—namely, God.

On the surface Plato seems to have held a very different theory. He believed it was possible to demonstrate that a person who knew no geometry could be led into formulating geometrical insights in such a way that the only possible explanation is that the person is recollecting something known before (in a prior life) rather than learning something new.[13] This theory of recollection rests heavily upon the theory discussed above to the effect that many of our thoughts are of (or by means of) concepts having no instances in this world. And if there are no instances in this world of, say, a triangle, then we could not become acquainted with triangles by any experience in and of this world. Therefore, the origin of these thoughts lies outside this world.

But again the important point to notice is that Plato is merely extending, not denying, the environmentalist position. His account of how I can think of a triangle requires that I become acquainted with at least one real triangle in an earlier life or incarnation. Like Descartes, he is noting a lack of correspondence between our thoughts and the objects in *this* perceived world and arguing for a wider view of reality in order to account for our thoughts. An environment adequate to account for our thoughts must include both a world of forms and a world of previous instantiations of ourselves.

This leads us to the following provisional conclusion: If some thoughts

[12] Descartes, *Meditations*, III, p. 163.
[13] Plato, *Meno* (many editions).

cannot be explained in terms of the objects and properties actually in this world—because these thoughts are not about such natural objects and properties—then their explanation seems to require either a theory or innate ideas or by a theory of reality extending beyond the natural world. Those philosophers who accept this conclusion tend to opt for the expanded universe and thus for a form of philosophical environmentalism. For them the real controversy centers on the nature of the environment (specifically, of reality) and not on the nature of thinking.

The final aspect of thinking to relate to the question of development is the manner of our thinking. When we speak of learning to think, we usually mean learning to think consistently, logically, clearly, and systematically. People think unavoidably, but they can learn to think skillfully. This development may be related both to individual men and to man as a species. With regard to the latter, there is good reason to wonder what progress in skilled thinking, if any, has been made in the recorded history of man in spite of the enormous stimulation of accumulated knowledge and information. Has the marvelous and intricate subtlety of the philosophical arguments, analyses, and theories of the Greeks of the fourth century B.C. ever been surpassed? With the assistance of modern mathematics and computers, we can do many things the ancient Greeks could not do; but left to his own wits the modern thinker is unlikely to surpass the mastery of the ancients.

Development, then, is of more basic relevance to the individual than to the species. Every normal individual develops enormously during his lifetime, both in skill and in breadth and depth of subject matter. And this is true regardless of whether the mind is the brain and the brain is extremely highly developed prior to birth, or whether the mind is noncorporeal and is completely developed at conception.

COGNITION IN INFANCY: WHERE DO WE STAND IN THE MID-SIXTIES?*

William R. Charlesworth

When this conference was first instituted by Merrill-Palmer in 1962, research on the cognitive capacities of the infant had just entered a very significant phase. Since the latter part of the 19th Century the history of such research was characterized by relatively long periods of inactivity separated by brief phases of sporadic interest. Then came a sudden leap forword. Of over 100 research papers which I have so far selected from American and foreign journals dating back to 1914, roughly 85% of them were written in the last seven years. I have not yet completed my research, but it is safe to say that the bulk of research on cognition in infancy has been generated from about 1960 on and there is no sign of a decline.

Why this phenomenon is occurring right now in child psychology is a difficult question to answer. Perhaps it is a case of a self-fulfilling prophecy engineered by the secret powers of the Merrill-Palmer Institute. Whatever the reason, the infant is edging closer to the center of the stage in psychology and doing quite well in his role. Unlike his ancestors who were described by Preyer (1888) and William James (1890) as having limited and undeveloped sensory capacities, today's infant appears much more sophisticated and competent. At this conference in 1962 Kessen (1963) noted that the infant was not only passively reactive as well as spontaneously active, but an initiator of behaviors and in his own way, a rather successful little problem solver. Furthermore, the infant is also doing better perceptually. As Lipsitt (1966) pointed out at this conference

* From William R. Charlesworth, "Cognition in Infancy: Where Do We Stand in the Mid-Sixties," *The Merrill-Palmer Quarterly*, vol. 14 (1968), pp. 25–46. Reprinted with permission of the author and publisher. Presented at The Merrill-Palmer Institute Conference on Research and Teaching of Infant Development, February 9–11, 1967, directed by Irving E. Sigel, chairman of research. The conference was financially supported in part by the National Institute of Child Health and Human Development. This paper was written while the author was supported by Public Health Service Grant MH 10275, and was supported in part by grants to the University of Minnesota, Center for Research in Human Learning, from the National Science Foundation (GS 541), the National Institute of Child Health and Human Development (PO-1-HO-01136), and the Graduate School of the University of Minnesota. The author thanks John Flavell for his reading of the manuscript, Burton White for his encouraging and helpful remarks, and Miss Joy Yanez for her typing and bibliographic assistance.

three years after Kessen, the neonates he tested in his olfactory study with Engen were capable of discriminating "components of odorants from the compounds of which they are members." But not only that—"the infants lined the odors up in terms of similarity just as adult observers do." Recent literature, in short, suggests that the infant is more discriminating and preferential towards stimuli than originally thought, and is also more cognitively organized. Furthermore, he seems to function no differently in certain respects than older subjects when faced with learning situations—he shows habituation, spontaneous recovery (see Engen and Lipsitt, 1965), "contingency awareness," according to Watson (1966a), and a modifiability of behavior not qualitatively unlike those many years in advance of him (see Bridger, 1961; Bartoshuk, 1962). His secrets are gradually being revealed and if those interested in him persist in being so imaginative, we will know much more about his competencies in the next ten years.

The question posed today is where do we stand now in this special area of child psychology? Two years ago at the Concept of Development Conference commemorating the anniversary of the Institute of Child Development at Minnesota, Kessen (1966) noted that the psychology of cognitive development was in an era of openness, variation, and doubt. These qualities are still characteristic of the field, and have in the meantime become more intense. Today we are being inundated by increasingly more rigorous and sophisticated empirical studies, which are changing our conceptions about what infants can do when faced with researchers and what researchers can do when faced with infants. Furthermore, to make the current scene even more open and varied, test psychology, developmental cognition, ethology, and psychoanalysis—once separate disciplines as far as infants were concerned—are now strengthening their converging points of interest. These are good times for those who want excitement.

Let me attempt to present briefly what I think some of the trends and issues are that are currently making the field so exciting and what achievements have brought us to this point. Because of limitations of space and my lack of familiarity with portions of the literature, I will by necessity omit certain studies and perhaps certain areas. For purposes of simplicity, I have broken the issues and achievements down into two sections—one involving substantive and methodological trends and problems and one involving the impact of current research on possible conceptualizations of infant cognition and its development.

Methods and Substance

In the area of infant testing some significant and very encouraging steps have been taken to strengthen the assessment enterprise. Critique and verification of infant tests have finally begun in all seriousness. Stott and Ball (1965) in their excellent monograph take at least four well-needed steps forward: (1) towards succinctly and accurately reviewing the history of infant tests, as well as concepts of intelligence, starting with Darwin, Galton, and Binet and ending with Piaget and Guilford; (2) towards critically and constructively evaluating no less than the Stanford-Binet Scale, the Goodenough Draw-a-man test, the WISC, the Gesell Developmental Schedules, the Cattell Infant Scale, and the Merrill-Palmer Scale; (3) towards clarifying the content of infant tests through factor analysis, a much-needed achievement in light of the fact that infant tests have been predicting little or nothing in terms of later mental-test results, and (4) last but not least, they have taken a well-needed step forward by publishing the first three steps in one compact, readable volume. The substantive relevance of their work for cognitive processes in infancy in part is their finding that even at the earliest ages some test items formerly viewed as motoric in nature are best interpreted as intellectual or psychological items, items which are capable of instigating "thinking" processes in infants as young as 3 months. While their interpretation is still subject to a critical evaluation, it has great implications for our conceptualization of the cognitive abilities of infants as well as for testing itself.

Let me make an additional observation, however. Stott and Ball use the Guilford structure-of-intellect conception as a basis for labeling the factors isolated at 14 age levels. What strikes one immediately is the resemblance some of the factors have to Piaget's sensori-motor abilities. At first glance, at least, convergent production looks quite similar to secondary circular reactions and divergent production looks almost identical to tertiary circular reactions. Perhaps these similarities can help produce a bridging point between intelligence testing and theories of cognitive development. I should also note that infant mental testing in the sixties has also been rejuvenated by Bayley's 1958 revision of the California First-Year Scale, a revision which has made possible a number of valuable substantive and reliability studies (see Bayley, 1965; Werner and Bayley, 1966).

A point of intersection between part of Piaget's formulations of cognitive change and mental testing exists now in the work of Uzgiris and Hunt (1966), Escalona and Corman (1967), and to some extent Charlesworth (1966b). In response to the absence of instruments "sensitive to changes in developmental pattern produced by diverse experiences," Uzgiris and Hunt have developed an infant psychological development scale based on the Piagetian notion of ordinality which characterizes intellectual development. As far as I am informed, their scales have been developed on the basis of tests of close to a hundred Ss and while still considered provisional are without doubt the right step in the necessary direction. The authors' concern with numerical inter-observer reliability and test-retest reliability as a measure of stability make their scales more scientifically palatable than the "clinical" scales developed by Piaget.

Hopes of improving Piaget's tests as a measure of infant intelligence received a jolt as well as a boost from work recently carried out by Escalona and Corman (1967). While examining problems of validating Piaget's theory of sensori-motor intelligence and extending the utility of the theory in leading to techniques to measure "overall infant intelligence," Escalona and Corman constructed scales of object permanence and spatial relationship and applied a Guttman type scale analysis to them. The tests scaled well, thereby supporting Piaget's contention that the stages appear in an invariant sequence. In addition to tracing vertical developmental progression, they also tested for horizontal decalages by giving each infant a number of different items, all of which purportedly measured the same stage. Results indicated that individual differences with respect to horizontal decalages were greater than differences in vertical progression (there were no individual differences in sequence of stage progression, thank goodness). In other words, children varied significantly in the extent to which they straddled two stages when tested with a number of different items at the two stages. The authors concluded that "horizontal learning occurs in different forms and sequences, and also because different components of sensori-motor intelligence develop at different rates, items based on Piaget's stages are not suitable for the measurement and prediction of overall infant intelligence within the normal range." While it is too early to determine how widespread the implications of these findings are, future researchers should be well-warned when attempting to find a point of easy confluence between Piaget's notions of intelligence and the Anglo-Saxon concept of general intelligence during the initial stages of infant test construction.

On the positive side, Escalona and Corman point out their scales can provide valuable information on areas of infant performance where environmental factors produce retardation or acceleration. For example, lack of contact with certain materials would be detected by horizontal decalage items. Furthermore, the Piaget scales, they feel, provide a valuable assessment of cognitive structures which play a relevant role in the sphere of social and affective development.

To be different and to make the infants' test experience an unforgettable one, I devised a test of the infant's level of object concept development which does everything that the above tests do not (Charlesworth, 1966b). I deceive the subject. While Piaget, Hunt, and Escalona abide by respectable procedures that acknowledge the limits of physical laws, I abide by the procedures of the magician who acknowledges no law but that which produces surprise. By making a preferred object, which has been placed under a cover, disappear "mysteriously" from under the cover, it is possible to generate some surprise or surprise-related responses. The long-range goal is to set up test situations which will elicit expressive behavior or non-instrumental responses that will serve as indicators of the subject's level of cognitive development. So far, most tests of intelligence have relied primarily upon instrumental behaviors (in the broad sense of the term) and ignored facial expressions, expressive body movements, inadvertent delays of simple instrumental behaviors as measured by reaction time, and autonomic changes (heart rate changes, GSR, vasodilation, etc.). By employing such measures, we might be able to add our existing techniques as well as to get clues to the relationships between cognitive structures in developmental transition and dynamic-affective variables which are certainly not separated in the human even if they are in textbooks.

Additional work with the object concept has been carried out by Gouin Decarie (1965). Concern with the relationship between the development of the object concept, as viewed by Piaget, and object relations as accounted for by psychoanalytic theory, has led her to the difficult task of developing an objectal scale to be used along with the Griffiths Mental Developmental Scale. Various objectal criteria range from specific reactions to feeding—through automatic or differential smiling, ability to wait, signs of affection—to discriminations of signs of communication. While no firm points of intersection were reached between such criteria of objectal developmental and those used by Piaget to measure object concept, Gouin Decarie's attempt was an important one because (a) it points out a need for more rigorous theorizing on the part of psychoanalytically oriented theorists so that specific deductions concerning early cognitive representation can be tested, and (b) it succeeds in demonstrating the need for a solid, more empirically based approach to the problems concerning relationships between affective phenomena that parallel developmental changes in cognitive processes.

So much for infant testing. In general, as I see it, the field is becoming more concerned with the relationship between tests and theories of development and appears to be moving toward a fruitful coalescence between the two. Furthermore, there is strong tendency toward developing an analytically critical and statistically more sophisticated posture toward testing methods, such as Piaget's, which have been predominantly "clinical" in nature. Rather, however, than supplementing the values of clinical flexibility and intuition, the new concern for rigor appears so far to be complementing them.

One noticeable trend today in infancy research *not* having to do with testing per se is a growing self-conscious concern with methodology (including measurement techniques, apparatus, etc.). There are a number of recent studies, such as those by Fantz (1956, 1959, 1961a), Lipsitt and DeLucia (1960), and Rheingold (1962), which are specifically methods-oriented. And there are numerous others, such as those by Ames and Silfen (1965), Cantor and Meyers (1965), and Charlesworth (1966), which are indirectly concerned with methodology or concerned with it in terms of a particular empirical substantive problem. In light of past efforts in child research, this is a very encouraging sign. Early in the history of infancy research rigorous methods were not seriously and consciously stressed or clearly understood, and even if they had been there would have been little supporting technology to carry them out. As a consequence, misconceptions of the infant's abilities inevitably prevailed. Conclusions of high generality and seeming sophistication were frequently based on unsound empirical or inferential methods. A good example is William James' (1890) famous statement "the baby, assailed by eyes, ears, nose, skin, and entrails at once, feels it all as one great blooming, buzzing confusion." While sounding intuitively valid, this statement was actually based on a speculation derived most probably from casual observation and an inadequate act of *Verstehen.* James could not have done better because he simply did not have adequate techniques for measuring the infant's discriminative capacities.

The methods issue, however, goes even deeper than miscalculations of the infant's capacities. It is inextricably linked with the conceptions we have of the phenomenon the methods are aimed to reveal. For example, Fantz (1966) recently pointed out that Gesell's conception of infant perception was inextricably related to sensory-motor capacities. The infant was tested in terms of ocular-motor coordination. As could be expected, he did very poorly initially and then gradually improved. It was assumed that his perception followed the same course of development. While such an assumption may have been justified as regards the infant's ability to localize objects accurately, nothing about object discrimination or recognition could be assumed from such observations. Nevertheless, many thereafter believed, according to Fantz, that the young infant had no pattern vision, and that perception had to wait until muscles were better coordinated. An almost similar example can be given using Piaget's approach as an object of criticism. However, this anticipates an issue dealt with later. For now I wish to identify some important methodological steps being taken today.

On the stimulus end, there is no doubt that the spirit of greater and greater precision is at work. The stimuli and stimulus dimensions used in studying the distribution of attention provide us with an interesting case study in the evolution of a stimulus technique. The rough lines of adaptive radiation appear to begin with Fantz's well-known ancestral faces—faces schematized, photographed, and scrambled next to precise bull's eyes—then branching into various ecological niches, and resulting in Lewis' (1965) realistic and distorted faces, Hershenson's (1964) and Ames' (1966) checkerboards, the Hershenson-Munsinger-Kessen (1965) randomly-generated shapes, and finally to the most recent shapes developed by McCall and Kagan (1967) specifically to control mean contour length, a stimulus dimension which will be the most precise one for some time to come. Whether this trend from multi-dimensional to uni-dimensional (if that is possible) stimuli will be productive remains to be seen. As Hershenson (1965) points out, so far there has not been any great progress in the dimensionalization of stimuli and that it might be more fruitful to "let the newborn himself supply the dimension from which his experiential world might be reconstructed." While the design to accomplish this may be unwieldly, allowing the infant to choose the relevant dimensions appears to be a more promising strategy than letting adults do it with their rulers. However, this remains to be seen.

In addition to static stimuli, moving or apparently moving stimuli are also showing a tendency to vary according to the environmental niche. Cohen (1966) and Lewis, et al. (1966a) employed a clear-cut matrix of blinking lights. Charlesworth (1966a) used a schematic terrycloth face that peek-a-booed from behind a screen. Bower (1964), while testing for depth discrimination and going as far into the realm of ecological validity as he could respectably go in an experiment, used a peek-a-booing experimenter's head. Results, happily enough, were obtained by all.

Future developments of attention-controlling stimuli will most probably include greater use of motion pictures, three-dimensional stimuli with moving parts—faces or masks with movable jaws, mouths that smile and frown reliably and objectively to experimentals and controls alike, eyes that open and close—and a few unpredictable offspring of "op art." The latter, I should note, has already been anticipated by White and Held (1966) with their red-white bull's eye disc which falls rapidly toward the face.

From a broad perspective, it appears as if the stimulus inventors are operating on the assumption that if they construct the right stimulus situation, they will thus demonstrate that the infant is much more cognitively advanced than his clumsy gross motor behavior indicates (as if his true competencies have been so far concealed by bothersome performance factors). Does this suggest a latent trend toward a new nativism or preformationism? Perhaps. Hershenson (1965) in his provocative paper on the development of form presented a list of prerequisite functions for the perception of form—on the receptor end, the scotopic and photopic systems, deeper insider EEG activity indicating level or arousal, presence of ocular convergence and conjunctive accommodation. All were available quite early in one form or other in early infancy. Putting it more cautiously, there is no evidence which strongly challenges the possibility of form perception in early infancy, and there is presumptive evidence, at least, which suggests that the infant may be more quantitatively rather than qualitatively different in his perceptual capacities from an adult. We will only know whether this is so by carrying out, as Hershenson urges, more complex experimental analyses than done so far of the various components of the perceptual system. One thing is certain. Our knowledge of the infant's cognitive capacities will grow as a function of the growth rate in our techniques and strategies of measurement. Whether the infant will emerge even more perceptive and intelligent than he is now is not so certain.

What intrigues me from the subject's point of view is how he is going to take all the supernormal stimuli that we are exposing him to. Would it not be interesting if he did something other than orient, visually fixate, cardiacally decelerate, or habituate? After all, biology has shown that animals have preadapted structures which are "hidden" until a particular environment realizes them through selective pressure. Why couldn't humans have developed preadaptive responses which are not manifest unless there are environmental changes of sufficient magnitude to actualize them?

On the response side of the infant, things are just as stimulating as on the stimulus side. Currently, there is a clear concern for the "internal" significance of an overt response. Lewis, Kagan, Campbell, and Kalafat (1966) in a number of studies employed two response measures to novel and familiar stimuli. Using visual fixation as measured by head and eye orientation and the cardiac responses, these researchers found that fixation time alone may be misleading in indicating whether the subject is actively processing a stimulus. He may be merely staring at it with no involvement in it at all. Cardiac deceleration more frequently occurs with active attention and hence serves to discriminate between the two kinds of attentional behaviors. Using both responses as measures of intensity level of attention, they feel, will be more sensitive than fixation time alone. Furthermore, they point out the value of using both measures in studying auditory input since attentiveness to various properties of auditory stimuli cannot be measured by the postural cues which reliably reflect attention.

Other work points in a similar direction. Attentional responses which are not overly different from one another can be associated with different central events (see Lynn, 1966). For example, Grastyan (1961), makes a distinction between the startle reaction and the orientation reaction. Both are behaviorally very much alike, but have different adaptive functions—the former defensive or avoiding, the latter investigatory or approaching. And both have different electrophysiological concomitants to prove it—the startle reaction is characterized by hippocampal desynchronization and the OR is accompanied by hippocampal theta rhythms.

At a higher level of cognitive functioning, in the realm of object concept development, as mentioned earlier, I am trying to find "new" response variables, i.e., other than those conventionally used in measuring infant capacities. These variables are not historically new and have not been systematically examined and used in the testing of infant cognitive functioning. While these variables could be classified many ways, I guess viewing them as respondents or expressive behavior or as both, depending on one's level of desperation, is one way of trying to distinguish them from the instrumental behaviors required of subjects in most tasks. Whatever the case, as I noted already, spontaneous facial expressions, autonomic mediated responses such as the GSR, vasodilation-constriction, heart-rate changes, etc., would be included. Historically facial expressions have not fared well, but I predict a comeback with the clever use of motion picture techniques. I find it interesting as well as very important to discover just what it is about an infant's facial expression that makes his mother say, "He didn't recognize you"; or "He saw his old toys and recognized them immediately." Our present lack of concern for such variables, while partly due to failures in the past of physiognomy, is also due to a greater present concern for instrumental behaviors. This issue will be discussed in more detail later.

The gist of this section is that there is a concern today to know more about conventional as well as new response variables that may be significantly associated with cognitive functions which we are trying to assess and understand.

Let me now go inside the infant. It is well known that there are at least three major ways of measuring discrimination capacities or preferences in nonverbal subjects. One, we can note differential behaviors to two stimuli presented simultaneously (paired comparison). Two, we can establish a CR to a particular stimulus and then note the degree of response generalization following. Or, three, we can note the properties of a response to a stimulus which is different or discrepant from stimuli that were previously fed into the subject. Of the three, the third is currently in vogue in many different forms, and there are a number of good reasons for this. For one, the other techniques pose methodological problems which some researchers find too unwieldly to handle. For example, as Kagan and Henker (1965) point out, paired comparisons are subject to inter-trial interactions and position habits, both of which can be avoided when a single stimulus is presented in a temporal sequence.

There is a more profound reason, however, why the third technique is becoming the focus of attention. Expectancy theory, whether it deserves it or not, in its broadest sense is coming back into psychology from many different directions. There is a widespread acknowledgement that experience results in the building up of schemes or models in the organism (even though no one knows if this is true or not). These models generate expectancies of forthcoming events and the expectancies in turn serve as a basis against which the new sensory input can be compared. These notions go back at least to Tolman (1932) and were bolstered by Hebb (1949) in his notions of "cell assemblies" and "phase sequences," neurological events which serve in part as comparison mechanisms for sensory inputs which are classified as congruous or incongruous in terms of ongoing central activity patterns. In Russia, Anokhin (1958) and Solokov (1960) spoke of neuronal models while carefully avoiding the term expectancy. Berlyne (1960) developed the concept of the collative properties of stimuli to cover those properties of stimulus variables which involve a comparison of current stimuli with stimuli that have been previously experienced (novelty and change) or stimuli and expectancies (surprise). In 1963 Spitz spoke of the proleptic, i.e., the anticipatory function of emotions which has emerged in its original form outside of the psychoanalytic framework as a CR. One year later Cofer and Appley (1964), while discussing the possibility of a unified theory of motivation, proposed an anticipation-invigoration mechanism. Finally, a whole school of "younger" psychologists such as Lewis, et al., (1966b), Cantor and Cantor (1965), and Charlesworth (1967) find operationally fruitful the notion that stimulus inputs which have become stayputs can serve as a basis of comparison for new inputs.

One need not look far in the empirical literature to see how ubiquitous the notion involving the concordance of discrepancy of a stimulus from a given input has become. Berlyne's work on novelty and curiosity is familiar to all of you. Studies of response habituation to repeated stimulation and response recovery when the stimulus is altered (Bartoshuk, 1962; Engen and Lipsitt, 1965) are examples of where this particular technique has been employed in learning situations. In studying visual discrimination Saayman, Ames, and Moffett (1964) utilized response to novelty as an indicator of visual discrimination in 3-month-old human infants. Lewis, Goldberg, and Raush (1967) in a study of age differences in response to familiar and novel stimuli based their experimental manipulation on the notion that novelty can be seen in terms of a violation of expectancy. Bower (1965a) in a study of perceptual unity in infants 4 to 40 weeks old used surprise as manifest in cessation of sucking as an indicator if the infant's sensitivity to various kinds of stimulus transformations. As mentioned earlier, I also used surprise as an indicator of the absence or presence of the concept of object permanence in 4- to 12-month-old infants (Charlesworth, 1965).

In short, the notion that internal processes, or short-term dispositional states such as expectancies, must be known and controlled in order to understand the relationship between incoming stimuli and subsequent responses has become a useful notion in our attack on problems concerning cognitive changes in infancy. The range of phenomena covered by such a notion is immeasurably large. It includes (1) sensitivity to novel departures from immediately preceding (and hence) familiar stimulus sequences, (2) sensitivity to apparent violations of physical laws (as in the case of objects disappearing); and (3) stranger sensitivity. These three areas of subject-environment interaction cover a very wide range of infant experience.

So much for our current ingenuity in providing the infant with stimuli, and response opportunities and expectancies to go along with them. We have the techniques and it appears as if nothing is going to stop us, given enough computer hardware, from developing them beyond the realm of current imagination. But one thing is being overlooked if we would stop here. And that is nature's ingenuity in pre-programming the infant. After all, human infants manage to adapt to their environment even though early environments vary tremendously across cultures. As Kessen (1963) pointed out at this conference 5 years ago, there is evidence of early adaptation but moderately little evidence of early learning according to the meanings inherent in our conventional theoretical models. Current conceptualizations of the learning mechanisms controlling behavioral change, while probably adequate to account for modifications of behavior at later ages, so far do not account adequately for the momentous changes that take place at a rapid rate during infancy. Phylogenetically pre-programmed mechanisms, which structure behavior to some extent independently of reinforcement contingencies and guide behavioral change across varieties of different kinds of reinforcements and reinforcement schedules, appear to be operating in every infant. While conventional reinforcement situations are recognizable in the laboratory, they are not so recognizable outside the laboratory and when they are, they never approach the intensity or orderliness of those controlled by the experimenter. The probability of having a learning experience in the natural environment comparable to that which an infant receives in the laboratory at Brown University or at Yale is close to zero for all babies past, present, and future. But despite this, the infant survives such an unexpected ecological improbability and even learns now and then according to the stimulus-response model. The human is indeed adaptable.

Clues as to how we should reshape our models so they will account for controlled changes will come from careful observation of both nonrestrained behavior and the environment in which it takes place. The methodological moral of this is that we still need to observe and record behavior as it occurs in natural or free-response situations. Burton White (1967) made an appeal for just this approach a year ago at the 1966 Minnesota Symposium on Child Psychology, pointing out that the two outstanding men in child psychology, Freud and Piaget, ultimately rest their theses upon vast amounts of data collected under "naturalistic or near-naturalistic conditions." Obviously there have been others in the field who have done a significant amount of careful observation, if not with such alacrity and such success. But now we must go beyond observing; we have to record behavior permanently if we wish to abstract its structure, submit it to detailed analysis, and restudy it at leisure. Biologists have permanent specimens, so do chemists, geologists, anthropologists. Where are ours? There is no museum of natural behavior where one can take his children on Sundays to see what psychologists are studying. With the exception of a scattering of efforts—Gesell's work, work done in

Berne, Switzerland under Meili (see Lang, 1967), the start made by Kessen's group (Kessen, Williams, and Williams, 1961)—there has been relatively very little accomplished systematically in the way of getting permanent records (motion pictures, not photographs which tell us virtually nothing) of infant behavior in general and of cognitively guided behavior in infants in particular. In Göttingen, Germany there is an archive for scientific films which has a sizable film library of behavior. For the most part the behavior is of turtles, wasps, cyclids, monkeys, and other lower animals. The films made by cultural anthropologists deal with larger units of behavior than those of interest to the child psychologist. There are a few films dealing with startle and other simple sensory-motor behaviors, but that is about the extent of their collection, and it should be mentioned that members of the archive try to keep abreast of major filming enterprises all over the world. Small wonder that we are a long way from other natural sciences when it comes to having permanent records of the phenomenon that interests us most.

There are many factors which explain why the existing library on human behavior is not larger, and why a museum of natural behavior is still a long way off. One factor that I feel is deterring researchers (aside from the cost involved) is the fear of recording too much behavior and then finding most of it of no immediate theoretical significance as well as unbearably familiar and boring.

There are at least two things we can do to avoid getting into this trap. One solution is not new; we can be discriminating in what we photograph. We can select environment-behavioral interaction sequences that have a high degree of adaptive significance for the child or theoretical significance for us, such as sucking, reaching and grasping, manipulating barrier objects to obtain a goal, and various problem-oriented attentional behaviors. Once such interaction sequences are identified, we can then create conditions which raise the probability that such behaviors occur. All of the interesting cognitive behavior that Piaget has elicited in infants with a few exceptions, e.g., Uzgiris and Hunt (1966) and Charlesworth (1966), has yet to be put on film and catalogued.

The second solution is the use of the speed-up technique developed by Eibl-Eibesfeldt and Hans Hass (1966) to record patterned behavior at the individual or group level. The technique simply involves taking a few frames (one to seven) per second instead of the conventional 16, 24, or 48. The net result is an acceleration effect (frequently humorous) which nevertheless still preserves the pattern of the responses. For accurate analysis of high-speed events such as the startle reaction 48 frames per second, or even 400 if necessary, would be required, but the total footage would still probably be no different than slow-speed events.

Let me just note briefly some current attempts to obtain adequate ethograms of behavior and the implications that result from them. An analysis of the infant's repertoire of innate and learned behavioral patterns has already been accomplished in varying degrees of rigor and detail at various developmental levels with and without films. Gesell (1952) so far has done most of this on the basis of direct observation and film record, and Piaget (1952; 1954) has carefully observed highly structural behaviors in naturalistic and semi-naturalistic test situations. More recently, B. L. White and his co-workers (see White, Castle, and Held, 1964; White, 1967) have made extensive and detailed analyses of visually directed reaching and of how such reaching develops from swiping behavior at two months to a "top level" reach some months later. At a more micro level, Prechtl (1958) has analyzed head-turning and allied movements in the human infant. Currently Kessen (1967) and the Yale University group are carrying out detailed analyses of sucking and looking behavior on neonates (1½ hours to 12 days) using elaborate recording techniques. The rationale for the relatively heavy investment of time and equipment involved in the production of an accurate and precise description of organized and adaptive behaviors derives from some leading ideas that, according to Kessen, are guiding the Yale group—ideas which are in harmony with the constructionist theme of Piaget. According to this theme, the child builds his reality as a result of interaction between innate schemas and events which cannot be immediately assimilable to them. From this interaction, higher orders of cognitive constructions gradually evolve. At each point in time such constructions constitute the child's existing theory of reality. Kessen maintains that one of our first jobs is to learn how the newborn "orders environmental variability in the first place." Whether he orders it on the basis of unlearned guidance mechanisms, unlearned plus learned mechanisms, or just learned mechanisms is important, but what is more important right now is that we find out in what way such early behavior is ordered. No zoologist would ignore embryology when faced with the problem of analyzing morphological characteristics of mature animals. Why child psycholigists should be any different when faced with the morphology of cognitively guided behavior is difficult to understand.

There are some data coming in from other sources which bear on this point as well. These data are worth mentioning because they reflect a growing interest in close and careful observation as a means of finding out what is actually going on in infancy. Peter Wolff (1966a) in his monograph on the causes, controls, and organization of behavior in the neonate, presents data for 12 newborns during the first few days. These data were based on systematic observations and experiments which lasted a mere 24 hours per day. One possible weakness of his technique was that he could not get anyone else to observe with him, although it should be pointed out that high inter-observer reliabilities were observed during the pretest period. The significance of Wolff's study is that it is, as far as I know, one of the few of its kind, aimed at integrating the results into the theoretical systems of psychoanalysis and Piaget. Furthermore, it is a refreshing study in these days of attempts to exert super stimulus control over the infant's behavior. His infants acted as they wished. As a result, he was able to obtain much needed data on infant's states as viewed in terms of spontaneous behavior and motor responses to external and internal stimulation. In an earlier work (employing roughly the same technique) which attempted to study dynamic and structural determinants of attention (i.e., against a background arousal state which makes visual discrimination and preferences possible or impossible), Wolff (1965) observed that alertness in infants from 0-1 month of age was more prevalent when the infant was not hungry than when he was. The absence of visceral excitation appeared a necessary condition for the attentive state. Furthermore, spontaneous and deliberate attention occurred when the infant ceased to be bodily active—body movements were negatively related to attentiveness.

Another condition which influences visual alertness has recently been identified by Korner and Grobstein (1966). They observed that picking up a crying newborn and putting it to the shoulder not only stops crying but increases the frequency of eye opening, alertness, and scanning. Such visual behavior is apparently made possible by a change in state of arousal which is under the mother's control.

The finding by White and Held (1966) that extra handling by nurses of institutionalized infants had a significant, positive effect upon the growth of visual attention is by now well known to all of you and need not be gone into further.

Observationally based findings similar to these are invaluable for our understanding of the mechanisms controlling visual behavior. Furthermore, their implications for future maternal behavior will be extremely significant if it can be demonstrated that early visual behavior is a functional antecedent to latter attentiveness and curiosity and these in turn are significant contributions to general intelligence. It would be interesting to know whether actual time spent during infancy quietly attending stimulus surroundings is partially related to later behavior requiring mediation, or if such early attentional interest is a precursor for later attentional behaviors only. Those who argue for the lasting developmental effects (e.g., Fantz, 1966) of early sensory-perceptual stimulation should welcome research which deals with visual behavior in terms of raw quantity, especially since babies differ greatly in their capacity for alertness and active curiosity.

Let me abruptly conclude this section on methodology with the simple observation that our present concern with method in the broadest sense of the term is a good sign. In speaking for the Greeks, Aristotle allegedly said: "Our love of things of the mind does not make us soft." Neither does our present concern with the delicate phenomenon of infant cognition.

Theory

In this final section, I would like to set forth some general impressions concerning our present theories of infant cognition. Let me start with an attempt to answer the difficult question, "What has influenced the way we now look at the infant's cognitive abilities? The answer in a simple-minded form is (1) our theories about cognitive abilities in general, and (2) our dominant mode of measuring such abilities.

Our major current theories share some points in common—they are essentially behavioristic, consequentialistic, and peripheralistic. They are action-centered. Piaget's theory in particular celebrates the vast hegemony action has over all areas of the infant's existence, a hegemony which goes as far as the infant's most valuable adaptive tool—his cognitive structures. In development, perceptions or cognitions having any veridicality must, according to Piaget, await motor development and the consequences of action. This is a powerful working assumption that has great implications for both our image of infant cognition and the way we approach the infant when we wish to know more about him. The question is whether it is a correct assumption. It will be a long time before we can answer this question but one thing is certain—it is a very popular assumption and a monolithic one that fits very well into our current Zeitgeist. For this reason alone it should be examined with utmost care.

In a paper on prehistory, Lewis Mumford (1966–1967) notes that from a certain perspective man has a curious picture of himself, a picture based on a single-factor explanation of his phylogenetic development. According to popular anthropological theory, man developed to his present status because he *first* became a tool user. By hammering on the physical world, a mind eventually developed inside of him. The theme of Man, the Tool Maker, is pandemic in modern interpretations of man. As a result of these speculations, the 20th century, claiming its glorious liberation from past dogmas, has cultivated an interesting dogma of its own—that man is basically an industrially talented animal whose work, activities, and tools shape his very consciousness and consequently his symbol systems and social institutions. Behaviorism, capitalism, communism, socialism contribute their share to this belief that action is prime and generative, first in man's adaptive moves to cope with his environment and consequently most efficacious in the building of cognitive and social structures. How this dogma progressed as far as it did is a complex and probably insoluble problem. In anthropology and paleontology, however, its origin may be due simply to the fact that data on prehistorical events are limited mainly to those materials which were hard enough to withstand destructive bacterial and geological forces. Fossils and stone artifacts are harder than the more complex and delicate nervous systems that accompanied them and, in my opinion, controlled their very development. They are also harder, as Mumford points out, than the symbol systems that made cultures what they were. Those things that have left no material traces, namely early man's rituals, language, and social organization, Mumford argues, should not be excluded in the final description of man's evolution. To make up for their absence modern man must carefully deduce from the available facts the "unseen and unrecorded context" of man's existence. This context is one of consciousness and communication, which probably made tool-using and symbol making possible in the first place.

In the present context, if the analogy is not too strained, we must ask (1) whether it is plausible and valid to infer early intelligence from overt task-oriented behaviors exclusively, and (2) whether we should consider postulating an early capacity for primitive consciousness in the form of veridical perception and reflection which may not be manifested in early actions upon the world or in tasks requiring gross motor involvement. It is possible that like primitive cultures the infant's capacities are underestimated when we evaluate him in terms of one class of outputs and not others. Let me change the analogy. Harlow (1958), while discussing the fact that reptiles, despite their greater cerebral structures than fish, show no superior learning capabilities, notes that "It is possible that our behavior tests on the turtle have never done justice to it, and that a latent imagination has been obscured by an introverted personality." Maybe we should seriously assume the human infant's cognitive capacities are obscured by his slow motor development. Trying to prove this assumption will require some real methodological innovations. I mentioned some of them earlier and am sure there are others. Right now a rationale for such an assumption should be made.

The action-behavioristic theorists have a strong point, when they argue that cognitions or cognitive structures develop as a result of the consequences of gross motor interactions with the environment. My point is (1) that their theory has so far not been adequately tested, and (2) it is possible that a different theory-perceptual learning theory, a cognitive theory, call it what you wish, can also account for early cognitive development—and I mean cognitive, not perceptual development. There are a number of reasons why I believe this.

For one, there is a vast amount of information familiar to all of you which strongly indicates that early stimulus enrichment and deprivation significantly influence later behavior and learning. One implication of this is that early organizations of information in the form of knowledge of the environment may take place before trial-and-error motor involvement with the environment ever occurs. When motor activity does take place, it is assimilated to pre-existing structures laid down by perceptual learning processes. In other words, the infant's visual system provides the initial matrix upon which later acquisitions based on tactual contact and kinesthetic, proprioceptive feedbacks are mainly built.

Obviously I can learn a lot about you by looking at you and moving my eyes only slightly. Why isn't it possible for the infant to learn many of the significant variances and invariances of his environment by just looking or staring with a minimum of eye movement? Does he really have to scan the edges of an object many, many times in an active fashion until he perceives its form; and does he have to lift, push, pull, and manipulate things until he can understand their important characteristics? While the four-month-old probably does not learn that his mother has weight by just looking at her, he certainly learns enough to comprehend her more significant properties.

When discussing Piaget and his heavy emphasis upon the necessity of action for cognitive growth, John Flavell often brings up the question of whether the child paralyzed at birth from the neck down ever develops the object concept, the first and most primitive invariant of intelligence, as Piaget terms it. One answer to the question (which is based on no empirical data whatsoever) is that even without the assistance of big muscle involvement in instrumental behaviors the child would inevitably construct the object and the terminal construction would be no different from that of children with normal muscle development. Piaget, keep in mind, claims that object construction is dependent upon the coordination of polysensorial inputs—stimulations from visual, kinesthetic, proprioceptive systems. This claim, as far as I know, has not been subject to a critical test. Such a test would probably have to be in the form of selective deprivation which is very difficult to achieve if not impossible.

Secondly, there are presumptive, but relevant data such as Fantz's which indicate that the infant is processing more distant information than originally thought. It appears that relatively more of his total behavioral repertoire is organized by cognitive structures which have been constructed on the basis of visual inputs than we were led to believe by psychoanalysis, Werner, and to some extent, Piaget. These theorists advocate that near receptors—visceral, tactual, kinesthetic, and proprioceptive—were the important sensory input sources which controlled behavior in infancy and early childhood and then later gave way to greater dependency upon other sources of stimulation. These notions may have come about partially as a result of the nature of infant observation and tests. Infant observation and tests are directed primarily at motor behaviors, many of which are instrumental. For example, of the 13 factors selected by Stott and Ball (1965) as characterizing the Cattell Infant Intelligence Test and California First-Year Scale for the 3-month and 6-month level, only three factors could be designated as predominantly cognitive in the sense that few motoric instrumental responses were involved—visual attention, perceptual foresight-anticipation of feeding, and smiling at one's image in a mirror. The other ten clearly involve gross motor activity of arms, hands and eyes. I am not saying that those 10 do not involve visual-hand-arm coordination. But they do differ in a significant way from the first three.

The third point is that at older age levels there is a known significant lag between perceiving and actually performing. In language learning, for example, it is an accepted generality that children discriminate sounds correctly before they articulate them and they do this much more frequently than the other way around (see Berko and Brown, 1960). The same asymmetry between perception and performance is found in the young child's ability to distinguish various forms from each other long before they can draw them (see Ling, 1941). Maccoby and Bee (1965) have reported on this lag phenomenon and discussed three different hypotheses which attempt to account for it. There is no time to discuss them, but these hypotheses are relevant for the point here. It is difficult to see why such a lag in older children would not be characteristic of the infant.

The point here is that considering the notion that perception precedes actual instrumental behavior in establishing early cognitions is a very useful working hypothesis. It forces us to develop our stimulus techniques such as those mentioned earlier, and to develop our stimulus techniques to the point where use of the subject's instrumental behavior in a diagnostic test would not be necessary or would at least be minimized. According to my bias, the prime behaviors to focus upon are orienting responses, expressive behaviors and such events as the consequences of interrupting sucking, breathing, etc.

In addition to forcing us to expand our testing techniques, the notion that perceptual activities alone precede instrumental behavior in establishing early cognition allows us to consider the possibility that schemas are actually formed in stages—the first involving information from distant receptors alone; the second involving information from near receptors which increases in volume as a result of increased effector activity in the form of locomotion and manipulation. The kind of knowledge acquired at each stage would, of course, vary.

Finally, let me just note that if we emphasize the centralist position that many of the early cognitive structures are constructed to a great extent before motor learning takes place, we will be drawn to the old problem of determining what, if anything, of adaptive infant behavior is innate in the sense that it is derived phylogenetically and stored in the genome of the organism. The earlier in infancy we find adaptive functioning, the greater looms the possibility that gross motor behavior and instrumental conditioning plays a minor role during early ontogenesis. This is not to suggest that learning theorists will have little to do with young infants. To the contrary, such discoveries will be welcome since they would help to reveal what environmental manipulations actually have to work with or against. The selection of an adequate reinforcement depends upon the "deep" vagaries of the organism as well as his known reinforcement history. If these vagaries are pre-programmed by heredity, and we are aware of them, so much the better. They are probably more stable than most environmentally determined response dispositions.

Summary

To summarize briefly, today research into cognitive processes in infancy is at a very exciting point. There are no firm conclusions and few certainties. Our monolithic theories are with us and waiting for empirical verification; delicate young alternate theories are gathering strength. Our methodology is improving and though we have not yet reached respectability through replicability, as many of the other sciences have, we are working toward it as well as toward new and interesting conceptual niches to fill with experiments.

COGNITION IN INFANCY AND EARLY CHILDHOOD*

David Elkind

Cognition has to do with knowledge and with the processes by which it is acquired and utilized. Research on the development of cognition has gone in several different directions depending upon the theoretical orientation of the investigator in question. These orientations can be loosely grouped within two broad categories. On the one hand there is the orientation which starts from the assumption that knowledge and the capacity to acquire it exists in some amount and can be measured. This is the *mental test* approach. A second orientation starts from the premise that knowledge and the processes of acquisition change or develop with age and the task of psychology is to describe and explain this development. This second orientation might be called *developmental*. It should be said that these two approaches do not necessarily contradict

---

* Reprinted with permission of The Macmillan Company from "Cognition in Infancy and Early Childhood" by David Elkind, in *Infancy and Early Childhood*, by Yvonne Brackbill (ed.). Copyright ©1967 by The Free Press, a Division of The Macmillan Company.

one another. The mental test approach is concerned with assessing individual differences whereas the developmental approach concerns itself with normative trends. Yet individual differences can only be assessed with reference to norms while norms are always abstractions from individual variations. In fact, many of the tasks which appear on intelligence tests are also used in the study of the nature and content of cognitive processes. In short, the difference between the orientations is relative rather than absolute and is more a matter of differing emphasis rather than differences in kind.

Defined in this broad fashion, these two orientations with respect to cognition in infancy and early childhood encompass a tremendous amount of research. Selection is obviously necessary not only because of the sheer amount of material but also because few investigators abide by the age limits of our present concern, so that many studies deal with age groups which overlap the infancy and early childhood periods. In the present chapter no attempt has, therefore, been made to be encyclopedic. On the contrary, the aim has been to select issues and areas of research that seem important and whose examination might lead to reevaluation and/or new insights into problems of theory or fact. Such an approach will, of necessity, involve some new categorizations and terminology as well as a moderate amount of speculation. Hopefully it will be of more use and interest than a single compilation of the available literature.

The first section of the chapter will deal with the mental test approach to cognition in infancy and early childhood. Three issues will be taken up: the prediction of later mental ability on the basis of infant tests of intelligence, infancy as a critical period in intellectual development, and, finally, the many faces of causality with respect to intelligence in young children. The second section of the chapter will deal with the developmental approach and will take up problem solving, memory, and conceptualization during the early years of life.

Intelligence

The Problem of Infant Intelligence

If any psychological finding has a claim to being axiomatic it is the observation that so called infant tests of intelligence are poor predictors of later intellectual level. Virtually everyone who has reviewed the research in this area (e.g., Goodenough, 1949; Jones, 1954; Bayley, 1955; Cronbach, 1962; Landreth, 1962) agrees that the usefulness of infant tests as predictors of later intelligence varies as a joint function of (a) age of initial testing; (b) the time interval between initial examination and retest (including the events which occur within that interval). By and large the earlier the test is given, the lower the correlation with later tests of mental ability and the shorter the interval between test and retest, the larger the correlation. Tests, for example, given prior to the third year are of little predictive value with respect to intelligence scores attained in middle childhood (Bayley, 1940). On the other hand tests given during the third year correlate significantly with IQs attained at age six (Ebert & Simons, 1943; Honzik, *et al.*, 1948).

Reactions to this state of affairs have been of two sorts. On the one hand there are those who accept these findings as inevitable because of the different capacities assessed by infant as opposed to noninfant tests of intelligence. These writers reject the notion of intelligence as a fixed capacity or quantum of mental energy that remains relatively constant throughout life. As Bayley (1955) writes, "I see no reason to think of intelligence as an integrated entity or capacity which grows throughout childhood by steady accretions" (Bayley, 1955, p. 807). In a similar vein Goodenough suggests that it may not even be justified to speak of intelligence in infancy and speaks of "The unsettled question as to whether or not true intelligence may be said to have emerged before symbolic processes exemplified in speech have become established. Attempting to measure infantile intelligence may be like trying to measure a boy's beard at the age of three" (1949, p. 310).

Other writers, however, have reacted differently to this anomalous situation. While granting the validity of the findings regarding quantitative prediction of intellectual standing from infant tests, they claim that qualitative estimates of intelligence made in infancy may still be of value. That is to say, although one may not be able to predict the later quantitative scores on the bases of scores attained in infancy, one can make successful predictions from the infants' general level of functioning. Thus if infants were categorized in gross terms as mentally retarded, below average, average, above average, and superior with respect to intellectual ability and on the bases of infant tests, the predictive power of these tests for later intellectual standing would be considerably improved. From a practical standpoint, such predictions would be of value to institutions, as adoption agencies, which are in great need of infant predictive indices.

The evidence for the predictive validity of such qualitative evaluations, although not overwhelming, is sufficiently impressive to warrant further exploration. Since these studies are less well known than those which demonstrate the lack of quantitative relationship between infant and later intelligence, a few of them will be reviewed here. In one study, Illingworth (1961) sought to demonstrate that at the low end of the intelligence continuum a diagnosis of mental level in infancy will have considerable predictive value with respect to later ability. Illingworth found that in a sample of 122 infants given a diagnosis of mental inferiority in infancy, 30 died and 65 out of the 87 survivors had an IQ score of less than 70 when tested several years later. Despite these findings, however, Illingsworth does not believe that, with the exception of mental subnormality, there will ever be a high correlation between infant tests of intelligence and IQ scores attained at later age levels.

This pessimism is not entirely shared by Simon and Bass (1956) who studied 56 infants tested before the age of one year and again prior to school age. When the infant test scores and the scores attained at the preschool level were grouped according to three categories—dull normal and defective; average; above average and superior—a significant relationship between the two sets of categorizations was obtained. These writers found, however, that this relationship was largely a function of having included children at the two extremes of retardation and superiority in the sample. MacRae (1955) using more subjects (102) and more categories (superior, above average, average, below average, and mentally defective) obtained similar results. The children were initially tested before the age of three with the Gesell Schedule and were retested after the age of five with the WISC. Of the 102 cases examined, only five cases deviated more than one category, and there was not a single instance of a deviation of more than one category. Furthermore, in striking contrast to all of the findings using quantitative scores, the predictive value of the infant tests was affected neither by the age at which the infant test was given nor by the interval between test and retest.

Escalona and Moriarity (1961) introduced clinical appraisal of total test performance into their calculations of the predictive value of infant intelligence tests. The subjects were 58 infants selected for "normalcy" on the bases of medical, social, and developmental criteria. Infant measures were (a) Gesell Schedule scores, (b) Cattell IQ scores, and (c) clinical appraisal based on total test performance. The subjects were again tested between the age of six and nine years with the WISC. The results indicated that for this sample, no method of appraisal predicted later intelligence range when utilized prior to the age of twenty weeks. For tests administered between twenty and thirty-two weeks of age there was a positive but not significant relationship with later measures of IQ. When clinical appraisal of test scores made during this same age period (twenty to thirty-two weeks) were related to later intellectual standing, significant correlations were obtained. The authors conclude, "When infant assessments were examined for their ability to distinguish between subjects who would later be of average or above intelligence, clinical appraisal (but neither of the test scores) achieved these discriminations at a highly significant level" (1961, p. 604).

Knobloch *et al.* (1963) also found that clinical assessment was a successful way of predicting later intellectual level and argued for giving up the test-score method of infant evaluation.

Although these studies are not free from methodological defects—a statement which also holds true for the studies dealing with quantitative scores—the results do suggest that if the full range of intellectual variation is taken into account and this variation is dealt with categorically rather than numerically, then the assessment of infant intelligence can predict intellectual level at later ages. Put differently, one might say that infant examinations may be useful in predicting gross differences in later intellectual level and particularly at the extremes (superior and mentally retarded range). For practical purposes, such as advising adoptive parents, such gross discriminations are much better than nothing and seem to justify the continued use of infant assessment methods for predicting later intelligence.

Do these findings refute the axiom that infant tests are poor predictors of later ability? Not necessarily. If intelligence is conceived in strictly quantitative terms as a score on a particular test, then the axiom still holds true. Infant tests cannot apparently predict later intelligence test *scores.* If, on the other hand, intelligence is conceived as a *ranking* relative to other children of the same age, then infant tests do seem to be able to predict the child's later standing with respect to his peers. Considering the fact that all intelligence scores are in reality ranks since they do not represent units—recalling six digits does not mean that one has two units more of memory than the person who recalls only four—the question boils down to just how much precision one is willing to settle for. If one is adopting a child even some precision is better than none.

Infancy and Early Childhood as a Critical Period in Intellectual Growth

The concept of the "critical period" seems to derive mainly from the work of ethologists such as Konrad Lorenz (1957) who used the term to describe some extraordinary circumstances of animal behavior. What Lorenz and others have observed is that there is a period during infancy when social attachments need to be made if they are to be lasting. Apparently these attachments are made regardless of species and Lorenz and other ethologists tell of chicks who follow them around as if they were the mother hen. Extensive work on this phenomenon has been done by Scott (1963) with dogs. Scott has shown that for dogs the period of socialization begins at approximately twenty days of age and continues for a few weeks thereafter. If social relationships are not established during this period it becomes increasingly difficult to do so later. Dogs with no experience of humans during this period are "wild" whereas those handled by humans during the same period are "tame."

It seems likely that something similar is apt to hold for the human infant. Schaffer and Emerson (1964) have, for example, shown that evidences of social attachments among infants begin to appear during the third quarter of the first year. The signs of such attachment are evidences of distress when a familiar person leaves the room, or when a stranger enters it. Infancy is not only a period when the child establishes social attachments, it is also a period in which it establishes a fundamental feeling tone about the social world. Erikson (1963) describes this attitude as one of *basic trust*, the feeling that the social world is reliable and that one's needs will be met. This attitude derives from the normal experiences of infancy in which the baby is cared for on an unconditional acceptance basis. In the absence of this unconditional acceptance and care, the child develops a sense of mistrust, the feeling that the world is a dangerous, fearful, and unreliable place, which then undermines all his later attempts to establish healthy interpersonal relationships.

These conditions, the establishment of emotional attachments and basic trust, thus seem to have their critical periods during the first year of life. If this is true then cognitive development must also have a critical period in infancy. This follows because for the infant, much more than for the older child and adult, intellectual and affective functions are undifferentiated. That is to say, anything which affects the child's affective equilibrium also affects his cognitive functioning. For the adult, whose intellectual abilities are fully developed, even severe neuroses may only dampen intellectual functioning. But for the growing organism whose intellectual capacities are in the process of development, emotional disturbance can be catastrophic. As the work of Ribble (1943), Goldfarb (1945b), and Spitz (1945) suggests, lack of appropriate social and affective stimulation in infancy leads to devastating consequences in both the personality and intellectual spheres. It is for this reason that infancy can be regarded as a critical period in intellectual development. At this age social and emotional deprivation are equally intellectual deprivation.

It is not, however, simply deprivation which affects later personality and intellectual development. On the contrary, research is beginning to show that the nature and quality of the stimulation provided infants may have enduring effects. Although such research is only now gaining momentum, it promises to reveal much about the early influences on intellectual growth. To illustrate this line of research and some of the most interesting conclusions several representative studies will be reviewed in detail.

In one of the continuing reports from the Berkeley growth study Bayley and Schaefer report on the intercorrelations between maternal and child behaviors and intelligence over an age span of eighteen years. The patterns of correlations for the 61 subjects tested repeatedly during the first eighteen years of their lives were complex and varied with the age and the sex of the individuals involved. What emerged from the study was the importance of what might be called "parental emotional temperature"—from extreme warmth to extreme coldness—shown during the early years of the child's life for later intellectual development:

> Hostile mothers have sons who score high in intelligence in the first year or so, but have low IQs from age 4 through 18 years. The highly intelligent boys, in addition to having loving mothers, were characteristically happy, inactive and slow babies who grew into friendly, intellectually alert boys and well adjusted, extroverted adolescents. The girls who had loving, controlling mothers were happy, responsive babies who earned high mental scores. However, after three years, the girls' intelligence scores show little relation to either maternal or child behavior variables, with the exception of negative correlations with maternal intrusiveness. The girls' childhood IQs are correlated primarily with education of the parents and with estimates of the mother's IQ (1964, p. 71).

Bayley and Schaefer conclude that these results support the hypothesis of genetically determined sex differences. Their results suggest that the effect of the environment, particularly maternal behaviors, exerts a constant influence on the developing mental capacities of boys but not of girls. Apparently this study presents some empirical evidence for the proverbial hardiness of the female sex in comparison to males. If these results are accepted, the boy would seem to be much more susceptible to environmental influence than are girls. From a clinical point of view, this might explain why many more boys than girls have learning problems and get into trouble with the law.

Another study concerned with parent-child relationships has been reported by Kagan and Freeman (1963) on the basis of data from the Fels longitudinal research investigation. As in the Bayley and Schaefer report, the obtained relationships were complex and varied with age and sex. Kagan and Freeman, however, were concerned with a different parameter of parental behavior. This parameter, which might loosely be called *parental control,* involved such activities as restriction, coercion, protectiveness, criticism, acceptance, and affection. On the basis of their findings, Kagan and Freeman concluded that "Maternal justification of discipline during the ages 4 through 7 continued to be associated with higher IQ scores for both boys and girls even when mother's education was controlled. Moreover, for girls, early criticism was positively associated with IQ at ages 3½ ($r = .52$), 5½ ($r = .51$), and 9 ($r = .46$) with maternal education held constant" (1963, pp. 905–906).

Both Bayley and Schaefer and Kagan and Freeman are careful to point out the dangers of interpreting correlation as causation. It may be the case, for example, that the obtained correlations are mediated by other variables not directly studied. What does seem clear is that there is a host of parental behavior dimensions such as emotional temperature and control which may influence intellectual growth. When these dimensions begin to be combined in the same investigation we can expect the results to be even more involved and complex. And that, after all, is as it should be since human behavior is no simple matter.

In some cases, the parental parameters are bound to overlap. It is hard to imagine a warm mother who is also critical. But such conflicting, or apparently conflicting, patterns do seem to occur as in the cold, hostile mothers who breast feed their babies out of a sense of duty rather than affection (Heinstein, 1963). Children who receive this kind of double communication often develop serious emotional difficulties in later life. We are thus now only beginning to appreciate the variety of parental behavior dimensions which may influence intellectual growth. It is still too early to say with assurance that warmth or coldness, coercion or affection will have this or that effect upon the child until we can measure the whole spectrum of parental behaviors in combination. Although the two studies reported here are only a start in that direction, they do underscore the intricate chain of causation that underlies what were once thought to be simple and straightforward relationships between the intelligence of parents and their children.

Before closing this section on infancy as a critical period in the development of intelligence, it might be well to point out that some writers, namely Fowler (1962), have stressed the positive aspects of this hypothesis. Fowler's point is that infancy and early childhood have been neglected by educators and parents out of, to Fowler's view, a mistaken belief that education was not appropriate during the early years and that it might even interfere with the child's personality development. Quite the contrary, claims Fowler. He argues that gifted persons have routinely shown early cognitive skills such as reading or playing instruments prior to the age of three. Fowler claims that this is due to earlier intellectual stimulation and training experienced by such persons. To substantiate his claim, Fowler reports work with his daughter whom he taught to read at an early age and who at the age of eight had an IQ of 150 to 170. Fowler admits that the girl does manifest mild emotional problems although in general she gets along quite well. Fowler, like Watson (1928) and, apparently, Bruner (1960) seems to take an extreme environmentalist position which asserts that one can after all make a silk purse out of a sow's ear. The current revival of interest in the Montessori methods (1964) suggests that Fowler is not alone in his belief in the importance of early education.

The issues raised by Fowler are of extreme importance. Assuming that one can train children earlier than we have been accustomed to doing, what would be the purpose of such training? Is intellectual development purchased at the expense of something more valuable? And if it is not, is the educational system prepared to handle children who read and do mathematics when they enter kindergarten? In short the matter is not just a philosophical or scientific one, but an eminently practical issue. For the early education of children presupposes a fundamental change in the hierarchy of the educational system. What sense does it make to provide early education for children if they will then only be bored when they go to school. Those who wish to educate at an early age must face the fundamental fact of educational existence, to wit, that the wheels of change regarding educational practice grind exceedingly slowly.

We began this section with the statement that infancy and early childhood are a "critical period" in the development of intellectual functioning. Research dealing with emotional deprivation and with maternal interaction patterns does seem to indicate that intelligence is vulnerable in the early years just because it is not yet differentiated from affective components of personality. This is not to say that infants are mere passive lumps to be molded by experience and parental behaviors—far from it. Considering the variety of parental behaviors one has to assume that infants are surprisingly hardy critters who will develop relatively well under an amazingly wide variety of conditions. What we need to know are the lethal parental behavior combinations and dosages as well as the optimal ones. Research such as that of Bayley and Schaefer and of Kagan and Freeman is a start in the direction of attaining that knowledge.

The Many Faces of Causality

The environment of the infant and of the young child can vary in so many different ways that it is possible to find a multitude of factors which may have an effect upon his intellectual development. There is no point in going over all of these materials in detail inasmuch as they have been frequently summarized in books and articles. There does, however, seem to be some value in reviewing the research on those factors which are particularly pertinent to the infancy and early childhood period. The three factors to be briefly considered here are prematurity, nursery school experience, and nutrition.

Prematurity

Does the child who does not develop to full term prior to birth develop normally thereafter? This is an extremely pressing question for parents of premature infants. Unfortunately, there is no simple answer to it. Indeed, there is not even complete agreement on what constitutes prematurity. Although most writers agree on the international criterion of 2500 grams (5½ pounds) others regard this standard as being too exclusive, inexact, and incomplete to meet the actual situation. Not only is the definition of prematurity in dispute, but the effects of early birth are particularly difficult to evaluate since premature children differ in many other ways from their full term cohorts.

"Their families tend to be poor and their mothers have often failed to use ante-natal services, have worked late in pregnancy or have had no help with the housework. They are also more likely to be girls, to be first born or to have mothers who are either much younger or older than average" (Douglas & Blomfield, 1958, p. 133). Differences between premature infants and full term youngsters are thus difficult to interpret since so many sociocultural differences are confounded with low birth weight.

One finding that does seem to hold across most studies of prematurity (e.g., Alm, 1953; Knobloch *et al.*, 1956; Drillien, 1958, 1964; Lubchenco *et al.*, 1967) is the negative prognosis for children of low birth weight (under 4 pounds). Such children are least likely to survive and, if they do, show a high incidence of retarded growth, physical illness, and handicaps as well as a high incidence of mental retardation and emotional disorders. Children with relatively high birth weights (above 4½ pounds) seem to have a better chance for normal growth and development (Capper, 1928; Mohr & Bartelme, 1930; Melcher, 1937; Alm, 1953; Knobloch *et al.*, 1956; Harper, 1959). Even in the case of high birth weight, however, the possibility of residual neurological deficit seems to be a possibility, and Douglas and Blomfield (1958) in a carefully controlled study found a higher incidence of reading difficulties among premature children than among normal controls matched for socioeconomic level, maternal age, and birth order. These investigators found no correlation between the incidence of such reading difficulties and birth weight. Thus, even among those prematures who seem to catch up with their peers in weight, height, and overall intellectual prowess, the possibility of residual deficit in specific intellectual areas such as reading is a very real consideration.

Apparently then, the outlook for premature children is not entirely reassuring. Parents of such children should probably be counselled in a realistic fashion as to the possibilities of illness and handicaps so as to be better prepared for them. It would probably also be helpful if school personnel were alerted to the special difficulties encountered by premature youngsters.

Nursery School Experience

In retrospect, the effort expended in attempting to determine the effects of nursery school experience upon intelligence appears as so much wasted labor. Not only are the experimental difficulties enormous as Jones (1954) has pointed out, but the results have been for the most part negative. As Goodenough (1940) concluded "The attempts to demonstrate the differential effects of different kinds of school practice upon child achievement have been disappointingly meager when suitable controls have been employed" (1940, p. 330). The troubling fact about these studies is the apparent assumption that the nursery school environment and experience is necessarily more enriching than a home environment. Anyone who has spent time in nursery schools knows that the quality of nursery school teaching and practice is extraordinarily wide. The physical plants also vary tremendously. Given a finding that nursery school did improve IQ in a particular instance, one could hardly generalize from this to all nursery school experience. What is important is not going or not going to nursery school but rather what goes on within the nursery school or home setting.

In this connection it is interesting that no one has as yet attempted to compare children attending Montessori nursery schools with those attending schools with nonspecialized programs. In contrast to most nursery school programs the Montessori schools have a very structured regime (cf. Rambusch, 1962) which is quite directly aimed at facilitating cognitive development. With the revival of interest in the Montessori schools, research on the effects of nursery school experience may well be revived. The writer has heard of several studies of this kind in the planning stage. Such research would make sense, since it is reasonable to ask whether one system is more effective than another. Obviously many experimental difficulties are still involved, but the attempt to determine whether specific types of experience have an effect upon intellectual development seems more surely founded than an attempt to find whether a general difference in experience such as attendance or nonattendance at nursery school affects mental growth.

Nutritional Factors in Intellectual Development

The present-day concern with the underprivileged in America may well produce a spate of studies dealing with such factors as nutrition and intellectual development. Some are already available. As in all investigations dealing with IQ changes, they must be accepted with caution because of the numerous experimental difficulties such investigations must overcome. In an early study of this type Kugelmass (1944) attempted to determine the effect of nutritional improvement on Kuhlman-Binet scores of 182 children two to nine years of age, for a period of 14 years. Kugelmass reports that children undernourished at the time of first testing showed an average rise of ten points for those who were initially retarded and an average rise of ten points for those who were initially normal. For children well nourished initially and at the time of retesting, the mean gain in IQ was zero. Kugelmass stated that the chance of improvement of mental function is greatest if the child is young when nutritional therapy is instituted, since the IQ rise is insignificant if therapy is started after four years of age. Similarly, less well documented results were reported by Poull (1938) and Knobloch and Pasamanick (1953).

An interesting new tack in this type of research was recently taken by Stock and Smyth (1963) in Cape Town, South Africa. They explored the effects of undernutrition during infancy upon brain growth as measured by the circumference of the head. The subjects were two groups of 21 colored children. One group was severely undernourished during the first year and more or less so thereafter. The undernourished lived in abominable conditions, many were illegitimate and the mothers were neglectful and apathetic. The well nourished children came from homes where the mothers were much more adequate as persons. All of the children were followed from the first through the eighth year of life. Head circumference, height and weight, and IQ were all significantly less in the undernourished as opposed to the well nourished children. While the authors are aware that the IQ differential could well be accounted for in terms of the difference in emotional and cognitive climate experienced by the two groups, they nonetheless point out that the difference in head circumference could not reasonably be attributed to these factors, and that this head circumference difference does seem to support the hypothesis of a direct relationship between intelligence and nutrition. Obviously we need to know much more in this area before reaching any definite conclusions, but in a general way there does seem to be some relationship between nutritional status and intellectual development.

The Developmental Approach

Cognitive Activities in Infancy

Although the sphere of cognitive activities is much more limited and much less differentiated in infancy and early childhood than it is among older children and adults, it is still possible to distinguish—at least for heuristic purposes—such activities as problem solving, memory, and conceptualization. The difficulty with making such distinctions is already familiar from the earlier discussion of intelligence. What we mean by memory, problem solving, and conceptualization in the very young person may be operationally and factually different from what we mean by these terms for the adult. There is thus always the question as to whether we are really talking about similar or parallel phenomena or phenomena which are epigenetically distinct from one another. What are needed are criteria that bypass the particular differences at each developmental level and which demonstrate the continuity of mental development across age levels. Such criteria are still lacking so that when we speak of problem solving, memory, and conceptualization below, the reservation that we may be talking about qualitatively different phenomena going under the same name must be kept in mind.

Problem Solving in Infancy and Early Childhood

Generally speaking, problem solving situations are those in which the subject is desirous of attaining some goal, the direct access to which is blocked in some way. When, for example, a child sees a box of candy on the top shelf of a cupboard, he is confronted with a problem situation. He desires the candy but is prevented from reaching it because of its height. If he recognizes that he can utilize the kitchen stool to climb to the top of the cupboard, he has solved the problem.

Several things should be noted about this situation. First of all, the solution clearly depends upon the child's past experience and learning. Without having experienced the sweetness of candy, it would not have been desired and there would have been no problem to be solved. Likewise, without having previously learned that stools, chairs, and benches could be used not only for sitting but also for climbing and jumping, their relevance to the problem solution would probably not have been recognized. Second, the attainment of a solution amounts to a kind of learning (Gagné, 1964). Once the child has discovered how to use the stool to mount to the top of the cupboard, he is likely to continue to use this solution whenever he wants to get things outside of his immediate reach. His behavior has thus been modified by past experience, he has therefore *learned* in the broad sense of that term.

The third characteristic of problem solving is that the goal can be attained in any number of different ways. In lieu of using the stool, the child could have pulled out drawers and constructed a stairway, or he might have used a stick to knock the candy off its perch. This last point is very important because it distinguishes problem solving from other types of learning in which the goal can only be attained in a manner prescribed by the experimenter. Problem solving allows for innovation and this is why it is frequently used to study the higher mental processes.

One of the issues that dominated problem solving research for many years, was the question of trial and error versus insight. Do solutions to problems come suddenly as the cognitive field is reorganized or do the solutions emerge gradually as a result of trying out a variety of alternatives? An eclectic resolution of this conflict has been offered by Harlow (1949) who has described what he called "learning sets." After an animal or human subject has learned a great number of similar problems he becomes an expert with this type of problem and can solve them readily sometimes in a single trial. If the subject's past experience were unknown one might say that the solution was insightful when in fact it was based upon considerable previous experience. The data on problem solving in infancy seem to support a learning set hypothesis. Both trial-and-error and sudden insightful solutions have been observed in the very young and often are observed in one and the same child. In short, trial and error and insight are, in all likelihood, not mutually exclusive patterns but rather probably represent different phases of the problem solving process.

In the following discussion of problem solving the work done with infants will be presented separately from that done with young children. Investigators have seemed to use either infants or preschool children in their studies so that in order to deal with the problem solving of three-year-olds it is necessary to introduce investigations that employ children beyond the age brackets which are our primary concern. Rather than leave out the work with the three-year-olds, the studies with preschool children will be included here.

Problem Solving in Infants and Toddlers

In a study using infants and toddlers from six to twenty-seven months of age McGraw (1942) observed a variety of behaviors which ranged from gross emotional reaction to trial-and-error and sudden "insightful" solutions. McGraw placed her subject on one side of a sheet of plate glass and put a desired object, such as a bell, on the other. In one situation the child had to reach *over* the glass to reach the goal, in another it had to reach *around* the glass and in the third situation it had to *climb* over the glass to attain the goal object. McGraw's careful observations of infants' performances in this situation not only point up the difficulties of studying problem solving at this age level but also in the wide variety of behaviors elicited by problem solving situations even in infancy and early childhood.

Using a different problem situation with children of approximately the same age, seven to twelve months, Richardson (1932) also found a wide variety of behaviors. Richardson employed a series of string problems of varying complexity. Each setting was constructed behind a screen so that the child could not view it until the task was presented. One of the strings was always attached to the goal object but the problem of determining which string was in fact attached was made easy or difficult depending upon the patterning of the strings.

In describing his results Richardson listed what he called five types of *perceptive attitudes* which he grouped according to whether or not insight was involved in the solution. Successful solutions without insight were described as (a) interest in the string rather than in the lure and (b) interest in the lure and apparent accidental contact with the string. Successful solutions with incomplete insight were described as (c) awareness of both lure and string without evident purposive utilization of the string and (d) experimentation. Success with insight was described as (e) definite utilization of the string as a means to bring the lure into reach. With increasing age more of the infants demonstrated "e" type solutions and were able to solve more complex string problems.

A rather simple problem, but interesting nonetheless, has been used by Gesell *et al.* (1950). It involves having the child put a pellet in a bottle and then retrieve it. The behavior trends noted by Gesell and his associates were as follows. At fifteen months the child dropped the pellet into the bottle and tried to extract it by shaking the bottle. If he did not succeed, he inserted his finger into the bottle so as to "hook it." Some children turned over the bottle to let the pellet fall out. Similar behavior was observed at eighteen months. At two years, however, the child dropped the pellet into the bottle and immediately tipped it over in order to retrieve the pellet. This is probably a situation where "learning sets" play an important role since pouring sand or water is a frequent play activity at this age level and the older children may have readily seen the relevance of this play activity to the problem situation.

What is important to note about these studies is how dependent the child's behavior is upon the nature of the problem situation as well as upon the age of the child. The one-year-old child who can solve Richardson's string problems straightaway might have trouble with getting the pellet out of the bottle. Whether or not the child shows insight or trial and error will thus depend upon the nature of the problem situation as much as it will upon the child's level of development. The child who fumbles and bumbles with one task may solve another quickly and with apparent insight. To be sure, the infant and toddler do not have all the mental tools available to the older child and adult, but with problems attuned to their capacities they can be quite efficient problem solvers.

Problem Solving in Preschool Children

The preschool child (two and a half to five) is energetic, verbal, and highly imaginative. Problem solving takes up not a little of his time. The adult is hardly aware of the many problems the child actually encounters. The writer, for example, was amazed to see the difficulty his thirty-four-month-old son had with a pair of children's scissors. He could not manage to hold the scissors with one hand and so had to use both. But then he had nothing to hold the paper with so that it never remained at right angles to the cut and merely slipped between the blades. After several attempts he became infuriated and nearly threw the scissors to the floor. He finally arrived at a solution by getting his father to hold the paper for him!

If there is any characteristic that seems at all unique to this age group with respect to problem solving it would seem to be that *although the preschooler can respond to and solve problems posed verbally, he is not able to verbalize his solutions.* This point, as well as others, is illustrated in a variety of studies some of which are described below.

In a study instigated by Piaget's (1951a) contention that children cannot make an exception to a rule, Hazlitt (1930) carried out several studies with three- to seven-year-old children. In one study she employed a Russian egg which is an egg-shaped shell made of wood and gaily painted to resemble a doll. The egg is hollow and opens at the midline. Within the shell is another egg which also opens up, and within it is still another egg, and so on. Some of these eggs contain as many as six or seven smaller eggs. Hazlitt took out all of the eggs and told the child to put all *except* the green one back together. The youngest child to respond successfully to this command was three to four years of age, well below the age Piaget had claimed children were unable to make an exception. In addition, when Hazlitt substituted the word *but* for the word *except* many more children made a successful response.

Using different materials Hazlitt asked children to put on a tray all those cards except the ones with a moon and a star. In this slightly more complex situation the youngest child to make a correct exception was 4.8 years. The difference between being able to respond to the instruction to make an exception and being able to verbalize the exception was shown in Hazlitt's third experiment. The material consisted of colored and ornamented cards mixed with twelve cards of the same size and five smaller black cards. Hazlitt told the child that he was to watch her and to notice what she called K__________. K__________ was called out when each of the large black cards was put down. After this experience the child was told to pick Ks out of the whole set of cards. Many children who succeeded on this task in the sense of being able to pick out the Ks were nonetheless unable to verbalize the solution, i.e., "All the black cards except the little ones are Ks." These children were then given the sentence, "All the black cards __________ the little ones are Ks" and were asked to fill the blank with one of the three words "although," "except" or "but not." Among the children who could not verbalize the rule not many were able to choose the word *except*. Similar results with respect to verbalization were obtained by Roberts (1932) who found that although children could find which of three doors aways led to the falling of an airplane, they could not verbalize the solution.

The ability of preschool children to understand verbal instructions without being able to verbalize problem solutions was also illustrated in the work of Harter (1930). Harter worked with three- to six-year-old children and adults. Three performance tests, an obstacle peg test, a canal box test, and a pulley test were the problems. In each of the tests the solution of the problem consisted in recognizing that for a solution to be attained, a series of moves had to be made in a prescribed sequence. Thus, in the pegboard test there was a grooved path with side alleys. At the end of one of the alleys and close to the center of the circular path was a red hole. With the path were three styluses, one of which was green, one of which was yellow, and one of which was red. The yellow and green styluses were placed in the groove between the red stylus and the red hole. A solution involved recognition of the fact that both the yellow and green styluses had to be sidetracked into the short alleys in order for the red stylus to be moved into the red hole.

Harter found that the average number of moves needed to attain a solution decreased from a mean of 36.0 at age three years to a mean of 17.6 at six years. This decrease in errors with age suggests that as the child grows older he relies increasingly upon mental manipulation of materials rather than upon overt trial and error behavior. It is not far-fetched to suppose that at least some of this mental manipulation is verbal in nature and that the ability to mentally manipulate things increases as the child is able to internalize and utilize inner language.

Several other examples point to the same conclusion. Hamilton (1911, 1916) presented subjects from twenty-six months to adulthood with an insoluble problem. The subjects were presented with four doors only one of which was unlocked on each trial. The task was to find which door was unlocked on succeeding trials. There was, however, no rational solution since Hamilton merely locked at random one of the three doors that had previously been unlocked. For the unwitting subject this procedure suggests that there is a pattern or rule governing the procedure and he must discover what it is. The subject thus had to set up hypotheses, or in modern parlance, strategies, in order to solve the problem. In such a situation the optimum strategy is the one that maximizes the possibility of success, namely, to choose any door but the one that was previously locked. Hamilton described five different strategies that were employed by his subjects:

A. Trying three inferentially possible doors and avoiding the one that was previously locked.
B. Trying all four doors, each in irregular order.
C. Trying all four doors in regular order, right to left or the reverse.
D. Trying a given door more than once with intervening attempts to open some other door.

E. Repeated attempts to open a given door without intervening attempts at other doors, or persistent avoidance of a given door while attempting all others.

As might be expected, "A" or maximizing solutions, tended to appear more frequently with increasing age with 76 to 86 per cent of the adult subjects giving the maximizing pattern. It is at least possible that the ability to formulate some verbal rule was of importance for a maximizing solution in this insoluble problem.

Before closing this section on the problem solving behavior of young children it might be well to present a study in which the actual behavior of the child in the situation is described in some detail. Such descriptions convey better than anything else the variety and complexity of the behaviors elicited by problem solving situations.

Alpert (1928) used problems that were very similar to those employed by Kohler with the exception that she substituted toy for fruit as the goal objects or lures. In her study Alpert employed stacking problems, construction problems in which the child had to put together sticks to reach a toy, and tool problems in which the child had to use some object such as a stick in order to reach the goal. Among her nineteen- to forty-nine-month-old subjects Alpert noted a wide variety of behaviors from trial-and-error to sudden insightful solutions. The latter type of solution, however, generally appeared only after the child had been exposed to the problem several times. Here is the record of a child of thirty-eight months whose behavior depicts such a transition with a two stick problem.

First exposure. S examined one of the sticks and tried to reach objective with it over the top of the pen; examined the other stick and used it in the same way, repeating "I can't" over and over; tried out the stick between the bars, over the top of the pen, finally striking it viciously against the floor; complained bitterly, and tried again to reach as before, stretching and straining; tried to climb out and whined, "I can't." E terminated exposure to avoid fatigue.

Second exposure. S reached for objective as above and in ten seconds said, "Look, I can't," but continued her efforts; fitted sticks up against bars of pen, banged them together, etc. S tried to reach objective with her hand through the spaces, to force her way out, to shake the pen, etc.; said, "Dolly does not want me to get him."

Third exposure. As above, complaining intermittently and finally giving up.

Fourth exposure. S stretched for objective over top of pen, striking out angrily with stick, complaining and asking E to move object closer. S said, "Let's try big stick on little one," picked up the other stick, examined ends

> carefully and succeeded in fitting them with a shout of "bang." S angled for objective, reached it exultantly and repeated stunt several times (Alpert, 1928, p. 11).

For Alpert the criteria of insight were changes of facial expression and posture, tempo and precision of work, and verbalization of the solution. The most important factor in problem solution, according to Alpert, was the nature of the problem situation. In addition, emotional, temperamental, and mental factors all seemed relevant to the successful attainment of a solution. To this catalogue it is only necessary to add the wealth of the child's previous experience in general and the experience with the particular problem in particular.

Problem solving in preschool children thus has many of the characteristics to be found in problem solving at all age levels. The one outstanding characteristic of this age seems to be the ability to follow instructions without at the same time being able to verbalize the problem solution effectively. This suggests that verbal mediation is not an important factor in problem solving at this age level.

Memory Processes

In the broadest sense, memory is involved in all psychological activity. To the extent that previous experience affects current behavior, to that extent is memory operative. Put differently, memory is the retention of what has been learned or experienced. Different forms of memory, however, need to be distinguished. One of the most elementary forms of memory and the form that is perhaps most prominent in infancy and early childhood is what might be called *sensorimotor memory.* This involves the retention of sensorimotor coordinations learned in the course of adapting to the immediate environment. Sensorimotor memory includes the retention of a variety of motor coordinations such as those involved in swimming and skating which may be learned in early childhood and not practiced for many years but whose traces nevertheless seem to remain and to facilitate relearning. Exposure to foreign languages in early childhood also seems to have lasting effects with respect to relearning and could be included in sensorimotor memory. Another type of memory that makes its appearance in the early years of life might be called *representative memory.* This type of memory involves the retention of certain means-end relationships. Representational memory is revealed, for example, when the child seeks objects that are no longer present to perception. Still a third form of memory that begins to appear as the child's language skills develop has often been called *rote memory.* This form of memory involves the repetition of heard language sequences which may but usually do not have an adaptive function. A fourth type of memory and one about which there is much disagreement as to whether it occurs in infancy might be called *historical memory* and involves the retention of experiences and images that fit within a spatial-temporal conceptual framework.

Sensorimotor Memory

It is generally observed that motor skills acquired in early childhood are retained to some degree throughout life. Youngsters who learn to swim, ski, sail, or ride horseback as youngsters are usually more proficient in these skills as adults than are persons who attempt to learn them once they are grown up. Although this is probably due in large part to motor coordinations which are established early, it is probably also attributable to the emotional attitudes established in childhood. The adult who is relearning a childhood skill probably has fewer fears and inhibitions than another adult who has had no previous experience with the skill in question. Although these foregoing generalizations are common observations there has been little systematic research in this area.

In the area of language learning, which in the very young child is as much sensorimotor as it is symbolic, the efficacy of early experience has been demonstrated by the classic studies of Burtt (1932, 1937, 1967). Burtt was concerned with the effects of reading Greek passages to his infant son who was then taught these passages and others of equivalent difficulty as a child and as an adolescent. Learning the passages to which he had been exposed as a toddler was significantly more easy than learning new passages of equivalent difficulty at eight or fourteen years of age, but the effects had dissipated by the age of eighteen. Since Burtt's son did not understand the Greek passages that were read to him as an infant, the memory could not have been mediated by content or meaning and must have been mediated by some form of sensorimotor process or trace left by the early exposure to the passages.

Representational Memory

One of the classic ways of studying representational memory in young children is by means of the delayed reaction experiment. This is a variant of a game that most parents spontaneously play with their children. In essence the game involves hiding an object which is currently attracting the interest of the child to determine whether he can make use of the cues provided in the situation in order to find the object. A variant of the delayed reaction experiment is found on many infant intelligence scales and is a second-year item on the Stanford-Binet (Terman & Merrill, 1960). In the Binet test the materials consist of three small pasteboard boxes and a small toy cat. The child is told, "I am going to hide the kitty and then see if you can find it again." The cat is then hidden under the right, left, and middle boxes on successive trials to insure that successful responses do not occur by chance. For one trial the probability of a chance success is ⅓ whereas for three trials the probability of success on all three on the basis of chance drops to $\frac{1}{27}$.

An early study of the delayed reaction was carried out by Hunter (1917) and is of interest primarily because Hunter postulated that prior to language acquisition the child employs "kinesthetic sensory ideas" as a means of finding the hidden object. An elaborate study by Miller (1934) gives some support to this kinesthetic ideas hypothesis in the sense that Miller's youngest subjects (11½ to 24 months) used position cues in bridging the delay interval whereas the older children tended to use color. Recent research by Soviet psychologists (Zaporozhets, 1965) and by Piaget and Inhelder (1956) seems also to suggest that kinesthetic imagery plays a considerable role in the mediation of problem solutions.

In addition to the question of what it is that mediates the delayed reaction, psychologists have been concerned with the length of the delay interval and its relation to age. Like so many similar psychological questions, the answer would seem to be that "it depends." First of all it seems to depend upon the nature of the discriminative response required. In a test employed by Charlotte Bühler (1930) a ball containing a chicken that pops out is given to the child. As the child plays with the ball, pressure upon it makes the chicken pop out. Later a second ball, similar to the first but without a chicken, is given to the child. If the child shows by its facial expression and associated behaviors that he expected something which did not occur, then the child is credited with a memory of the chicken. Using this procedure, the time interval between the presentation of the first and second balls can be varied to test for length of memory. The length of interval that produced a surprise reaction with this procedure ranged from one minute at ten or eleven months to seventeen minutes at twenty-one to twenty-four months.

Other studies concerned with the delay interval have used goal seeking responses as evidence of representative memory processes. In studies by Hunter (1917), Allen (1931), and Skalet (1931) the correct response was an actual uncovering of the hidden object. With this type of procedure, comparable to that of the Binet test described earlier, the length of the delay interval is somewhat shorter than that obtained by Bühler. One possible reason for this difference may lie in the fact that the Bühler procedure required only the recognition of an absent object whereas finding an object which is hidden, as in the standard delayed reaction procedure, requires a more active recall. The difference in underlying processes called upon in the two procedures could, in part at least, explain why children can delay longer on the Bühler apparatus. As children get older, longer delays are probably mediated by language elements as Skalet (1931) has suggested. Even some of her two-year-old subjects were able to find a goal object after a day-long interval. Such lengthy delays are unlikely to be mediated by transitory kinesthetic imagery, and a more likely possibility is verbal mediation.

It is perhaps well to point out here, however, that the utilization of language as a mediating mechanism is still quite elementary in the preschool child and is apparently used in a systematic way only after the age of five or six, by which time language has to some extent become internalized (Kendler & Kendler, 1959). Among young children verbal signs may only serve to prolong the kinesthetic imagery rather than replace it as it seems to do among older children. In complex problem solving situations (Harter, 1930; Kendler, Kendler, and Wells, 1960) young children seem unable to use verbal mediation in an effective way. Nevertheless, it is perhaps well to recognize that verbal mediation has levels and degrees and that far from being a phenomenon that occurs only after school age, some forms of it may appear as early as the second year.

Historical Memory

Few people seem to be able to remember experiences that occurred prior to the third, fourth, or even fifth years of their lives. Different sorts of explanations have been given for this routine observation. Psychoanalytic authors often write as if they assumed that early experiences are retained and play an important part in later development. Freud's (1953a) notions regarding fixation at the early levels of development and Rapaport's (1954) model of thinking both imply that experiences during infancy and early childhood affect later development and behavior and to that extent are retained or remembered. So long as such memories are regarded as sensorimotor in nature then there is some evidence, cited above, to the effect that such memories can be retained and have lasting effects.

On the other hand, if these memories are regarded as being comparable, say, to an adult's memory of his last visit to the dentist, or of a dinner at a restaurant, the developmental psychologist would raise serious objections. That early experiences are sometimes thought of in these terms is reflected in the fact that some hypnotists argue that they can "regress" adults back to a preverbal level and uncover memories of the infantile period. Such a view flies in the face of our knowledge about cognition during the infantile period. True historical memory requires a spatial-temporal conceptual context. One can only remember an event by associating it with some particular place or time. Questions about past experience invariably contain temporal or spatial cues. We ask

"What did you do last Saturday night?" or "How did you like Hawaii?" and without these spatial-temporal cues recall could not occur.

Now we know very well that the spatial-temporal concepts of infants and children are limited pretty much to the here and now or at most to the immediate past and the immediate future. It is not without reason that children's stories begin with the indefinite "Once upon a time" and usually occur in a spatial-temporal limbo. It is not until middle childhood that children begin to have a true appreciation of historical time (Ames, 1946) as is evidenced by their language, their reading interests, and by the age at which history is introduced in the school curriculum.

The developmental psychologist thus does not deny that sensorimotor memories such as fixations persist throughout life nor that they may determine later behavior. What he does insist upon is that such memories are not cognitive in the sense of being labeled and dated and that they exist outside of the spatial-temporal conceptual context of cognitive memory. Freud himself (1953b) came to recognize that memories of childhood brought to him by his adult patients were in fact untrue when these were checked by objective sources. Such memories turned out to be "screens" for the real events which had occurred. Even screen memories, however, seldom date back before the fourth year. In short, truly cognitive or historical memory is probably not present until the school years and is not fully developed until later childhood and adolescence.

Concept Formation in Infancy and Early Childhood

Within psychology, concepts are generally thought of in behavioral terms as "the acquisition or utilization or both of a common response to dissimilar stimuli" (Kendler, 1961, p. 447). A child, for example, who calls a variety of differently shaped and colored objects "apples" is said to have a concept of apple if his designations are correct by adult standards. It is in this behavioral sense that animals can be said to have concepts since they too make common responses to diverse stimuli. In fact, however, concepts always have two aspects. On the one hand they have an *extension,* the population of objects, properties, or relations, to which the concept applies. On the other hand, they also have an *intension,* which corresponds to the feature or element that all of the objects, properties, or relations have in common.

From a developmental point of view, it would appear that during infancy the extensive aspect of a concept is cognitive whereas the intension is affective. Infants can, for example, distinguish between different geometric forms even when these are varied in size and orientation (Zaporozhets, 1965). By the end of the first year, the child recognizes his parents and a variety of objects despite changes in their appearance and this reveals a large store of extensive concepts. The meaning of these concepts or their intension is, however, affective rather than cognitive. The mother means warmth and affection, and the intension of the many objects the child can distinguish is measured by their positive or negative feeling tone.

With the development of language the child's conceptual sphere expands tremendously as he discovers that everything has a name. During the second year the child also discovers another form of representation, namely, pictures. The response to pictures is of importance because it reflects a new level of conceptual differentiation and an expansion of conceptual thinking to include pictures as well as things. According to Karl Bühler (1930) there are three steps in the development of pictorial recognition. At first pictures are no more than pieces of paper to be grasped and torn. A little later the child begins to recognize the picture and to treat pictured objects as if they were real. During this stage the child may be as frightened by a picture of a dog as of a dog itself. At the third stage the child appears to clearly grasp the notion of the picture as a representation of the real object. Bühler claims that this occurs at about the same age as the use of language and represents still another manifestation of the representational function which emerges at about the end of the first year of life.

Although the child does form a wide range of extensional concepts during these years it is necessary to point out that these concepts are less differentiated than those of older children and adults, even though they may appear to be quite similar. Every word that stands for a concept is ambiguous in the sense that it can stand either for an example of the class or for the class as a whole. The word *man*, for example, can mean a particular person or men in general. Older children and adults use both words appropriately, depending upon the context. Young children, on the other hand, fail to distinguish between individual and class designations of words. Piaget (1951b), for example, tells of his daughter who, on a walk, saw a slug and said, "There is the slug we saw yesterday," demonstrating that she did not distinguish between the particular slug and other exemplars of the class. Mothers whose children call strange men "daddy" are victims of the same phenomenon.

As children grow older, the intention of concepts gradually expands to include functional as well as affective meanings. Objects come to mean "what they can be used for" in addition to "how they make one feel." While these functional meanings are already implicit in the child's behavior during the second and third years of the child's life, they are not verbalized until much later. Indeed, the verbalization of intensive meanings seems to follow a parallel development to that reflected by the child's behavior with the exception that it occurs several years after the extensive development which it parallels. When at the age of four and five, children begin to be able to give verbal definitions, these are often perceptual-affective in nature. An apple is "red" or "sweet" and only later is it "something to eat" or a bicycle something to ride. The verbalization of intensive meaning thus lags behind its comprehension as evidenced by behavior, but follows the same developmental sequence.

With these general remarks as a starting point it might be well to make them concrete by describing the development of a few concepts in more detail.

It appears that children have conceptualized a variety of simple geometric forms by the age of two, the age at which the three-hole form board (square, triangle, circle) is placed on the Stanford Binet (Terman & Merrill, 1960). Recognition of these same forms in a rotated position occurs at a slightly later age, indicated by its being placed at the two-year, six-month level on the Binet. The difference would seem to be entirely due to the triangle since neither the square nor the circle are altered by rotation. Simple number concepts are also present by about the age of two years and most children can utilize "one" and "two" without much difficulty (e.g., Beckmann, 1923; Giltay, 1936). These are primitive numerosity concepts, however, since animals can make similar discriminations (Dantzig, 1954) and no one would claim that animals have a concept of unit which is necessary for a true conception of number.

Of particular interest with respect to children's concepts are those which Kant called *categorical.* These concepts—space, time, and causality—are imperative in the sense that regardless of the nature of "true reality" we can only experience within these categories. These are not innate ideas but rather principles according to which our experience is organized without our being able to help it. From a developmental point of view it is of interest to inquire how these organizing concepts change with age. We owe to Piaget (1954) the first intensive study as to how these organizing concepts develop in infancy. Piaget's observations and interpretations are impressive and have engendered considerable research (e.g., Wolff, 1959; White *et al.*, 1964).

Categorical Concepts during the First Two Years of Life

Kant wrote that although one could not deny the reality of the external world, one could not at the same time deny that everything regarded as external was at least in part dependent upon the thinking and perceiving subject. That is to say, the world we know is always limited by our organs of knowing. There are, for example, many forms of stimulation, such as high-frequency sounds, that are not within the sphere of our experience but are very much within the sphere of certain animal species such as dogs. Many other such examples could be cited, and von Uexküll (1957) has given vivid illustrations of perceptual worlds other than our own. Our knowledge of the world is, according to Kant, not simply limited by the sense organs but also by the innate organizing tendencies of the mind. Just as we cannot escape being sensitive to some stimuli and not to others, so can we not escape organizing our world within a spatial, temporal, and causal framework. We have, in Kant's view, really no choice in the matter, and all our experience is organized within such a framework. Kant, however, wrote as if these categories were the same in children as in adults, i.e., *a priori.* What Jean Piaget (1954) has shown is that although the child does seem to organize his experience within the categories described by Kant, the categories themselves go through a gradual process of construction.

This is not to say that the infant is aware of space, time, and causality as we know them. Far from it! On the other hand, it is possible for an observer to see in the infant's actions a causal, temporal, or spatial framework of which the infant is probably not aware. The situation is a little analogous to certain computers that can be programmed to play chess or write novels. To say that the computer operates in an intelligent way is not the same thing as saying that it is conscious of its intelligence, but only that it behaves *as if* it were. The same is true for infants in whom we can observe intelligent behavior without, at the same time, having to assert that the infant is aware of its intelligence. With these cautionary remarks in mind we can proceed to the development of the categorical concepts in infancy.

The Concept of the Object. Although not a categorical concept in the Kantian sense, the concept of the object is one of the fundaments of all categorical thinking and of all later conceptual developments and so must be dealt with. Although to the adult objects such as a chair or a table seem to be "out there" and to be separate from their physical properties (hardness, texture, color) and action properties (to be sat upon, leaned against, stood on), this is not the case for the very young infant.

In his brilliant studies of his own three infants during their first years of life, Piaget (1954) has provided evidence of the enormous labors involved in constructing the concept of an object. For Piaget the infant does not have a true object concept until he can represent it as evidenced by his pursuit of that object in its absence. This point is not reached, however, until the middle of the second year of life and is prefaced by a series of behaviors that gradually lead up to this representation. During the first few months after birth objects are really not distinguished from the actions associated with them. An object is simply something to suck, to grasp, to push, or merely to look at. When the object disappears, say the bottle removed before the infant has satisfied his hunger, he may continue to suck as if the sucking would reconstitute the object. Likewise, when the mother leaves the room, the infant continues to look at the point where the mother disappeared as if watching would bring the image back.

Between the third and seventh and eighth months after birth the child comes increasingly to differentiate the object from his own actions and to recognize that it has movements or trajectories of its own. One evidence of this differentiation is that the older infant now looks at the place where an object will land, after it has been dropped, rather than its position before it has dropped. Still another evidence of this differentiation is the recognition at this age level of an object from seeing only a part of it. This does not mean, according to Piaget, that the child conceives of a whole object, part of which is hidden. On the contrary, Piaget believes that the infant regards an object emerging from behind a screen to be in the *process of formation.* Still another interesting behavior at this age level is the child's ability to neglect an object and then return to it after an interval of time. All of these observations suggest that the object is coming to be regarded as something which exists independently of the infant's perceptual and motor activity. This point is not fully reached, however, until several additional skills are mastered.

One of the next advances in attainment of the object concept, which Piaget says occurs between the age of eight and ten months, occurs when the infant begins to search for objects that have been hidden behind a screen. What marks this stage distinctively is that the child actually removes the screen himself. Piaget, for example, describes hiding a cigar case under a cushion after which his son Laurent (age nine months, seventeen days) immediately raised the cushion in order to find the object. (Prior to this age the child ceased to pursue an object hidden before his eyes.) The behavior of this stage was, however, governed by a very important restriction which Piaget describes as follows: "The child looks for and conceives of the object only in a special position, the first place in which it was hidden and found" (1954, p. 50). That is to say, if the object is first hidden under one screen and then under another, both displacements visible to the child, the child nonetheless looks under the first and not the second screen.

During the second year of life the child gradually advances to the stage where he can recognize the position of an object even after it has been hidden in three successive positions. This ability, to take account mentally of the successive displacements of a hidden object, is for Piaget the mark of a true object concept. Such behavior suggests that the child can deal with objects as independent of his own perception and as having independent positions and trajectories in space.

Space Concepts. Many of the observations that Piaget used to describe the development of the object concept can equally well be employed to illustrate the attainment of space concepts. In the development of space, however, Piaget makes it clear just how complex and intricate are the unconscious coordinations by which the infant gradually orients himself and things within a spatial framework. One can get a rough idea of some of the difficulties involved by imagining oneself in a strange city whose narrow, winding streets are unmarked and for which there is no map or guide available. To learn the plan of the city one would have to make one's own map in a sort of trial-and-error fashion. Constructing a kind of map space is only one of the many problems of space conceptualization faced by the infant.

According to Piaget (1954) the infant progresses during the first two years of life from an initial sense of *practical space* to a *subjective space* and eventually to an *objective space*. To understand the differences between these three forms of spatial concept, we must begin by noting that any spatial concept involves the coordination of positions in a systematic way. These positions may either be of things or of the child in relation to things. Practical, subjective, and objective space correspond to different levels of differentiation and integration of these two kinds of positions.

Up to about the age of three months the infant can be observed to move his glance from one to another of a series of objects. From the point of view of the observer this type of visual behavior already implies a notion of space. It is still a practical notion because the child deals with positions in terms of his activity, looking, and the positions of objects are still not separated from the movement needed to perceive those positions. There are thus as many practical spaces or positions as different activities. The child might thus be said to have a buccal (mouth) space, a tactile space, a visual space, kinesthetic space, and so on. Practical space is revealed whenever the child shows an awareness of the position or displacement of things with respect to a particular activity as in the following example:

> From 0;2 Laurent knows how to carry to his mouth an object grasped independently of sight and how to adjust it empirically. . . . At 0;3 he puts a clothespin in his mouth, adjusting its position so that he may suck it (Piaget, 1954, p. 107).[1]

[1] From *The Construction of Reality in the Child* by Jean Piaget, translated by Margaret Cook, Basic Books, Inc., Publishers, New York, 1954. This is published in Britain by Routledge & Kegan Paul Ltd.

Beginning about the fourth month of life and extending until about the tenth, the child evolves what Piaget calls a *subjective* space. What seems to occur is that the various practical spaces begin to be coordinated one with the other. In Piaget's view the central condition for the establishment of these coordinations is the development of prehension, the ability to deal with things manually. Once prehension emerges, the child begins to grasp for what he sees, tastes, or feels and in this way gradually begins to bring together the spatial information obtained from each of the different sensory avenues. As a result he is able "to relate certain of his own movements to those of the environment" (Piaget, 1954, p. 114). Here is an example of this type of coordination:

> At 0;6 he (Laurent) directs his eyes towards an object after having touched it. But he cannot see it because of various screens (Piaget, 1954, p. 115).

From the age of about nine or ten months and extending into the second year the young child gradually elaborates what Piaget calls an *objective space.* At this age the coordination of positions and displacements of subjective space come gradually to be divorced from the child's activity as a whole and to be regarded as independent of his own influence. At the same time the child also begins to see himself and his positions as but one object among many objects in the spatial field. The crucial evidence for the attainment of objective space is that the child's recognition of the movements and displacements of things in the environment is reversible. This means simply that the child recognizes that an object moved from one position to another can be moved back again. Although this would seem to be the simplest of ideas, it is in fact extraordinarily complex. When the child does demonstrate the recognition of this, it indicates a fundamental change in spatial organization. Instead of regarding objects as having fixed positions, he suddenly comes to see positions as more or less temporary *states* of objects. This transition from a static to a dynamic way of viewing reality also occurs in the verbal and logical plane, as we will see later.

Although Piaget derived his notions of practical, subjective, and objective space from his observations on his three infants, older children seem to manifest similar stages in the mastery of more complex spatial problems. In a study by Meyer (1940) preschool-age children's spatial concepts were tested by having them fit wooden forms together and by determining their understanding of the rotations of a pivoted bar. With these tasks Meyer claims to have found Piaget's stages repeated between the ages two and five. In brief, Meyer found that up to the age of two or two and a half, children showed only a *practical* space in the sense that they regarded objects only as something to satisfy their needs.

Between the ages of three and four Meyer found what resembled Piaget's *subjective* space in the sense that although the children were still centered on their own activities, they also manifested some interest in the objects themselves, independent of their own immediate needs. After the age of four Meyer reported that the children manifested *objective* space in the sense that they considered themselves as but one object among many and that they attempted to adjust their own behavior to the position of objects.

Time Concepts. As in the case of the object and space, Piaget finds a gradual development in infancy from a practical time to a subjective and finally to an objective time by the end of the second year of life. During the first few months of life the infant shows a practical grasp of time in the sense that it knows how to coordinate its movements in time and how to perform certain actions before others in a regular order. Piaget notes, for example, that even at two months the infant turns its head when it hears a sound and tries to perceive what it has heard. These sequences, although suggesting some temporal ordering, do not imply that the child has any awareness of time. This is to say, the actions could occur in a reflex fashion without any sense of causal connection between the two.

Beginning about the age of three months the infant begins to construct what Piaget calls a *practical* time series. By this he means that the child begins to perform a series of actions that are not determined entirely by the external stimulations acting upon him. At this stage the infant adds to the simple reflex series of the previous stage the results of his past experience so that his action no longer appears reflexive but rather purposive. Consider the following illustration:

> abs 169 At 0;3 (13) Laurent, already accustomed for several hours to shake a hanging rattle by pulling the chain attached to it . . . is attracted by the sound of the rattle (which I have just shaken) and looks simultaneously at the rattle and at the hanging chain. Then while staring at the rattle, (R), he drops from his right hand a sheet he was sucking, in order to reach with the same hand for the lower end of the hanging chain (C). As soon as he touches the chain, he grasps and pulls it, thus reconstructing the series C-R (Piaget, 1954, p. 330).

What this illustration shows is that the child knew he had to drop one thing in order to attain another just as he knew he had to pull the chain in order to shake the rattle. It is the use of intermediary means to a goal that differentiates the purposive from the pure reflexive act and that suggests a temporal as well as a causal series. Piaget also argues that it is at this stage that one sees the beginning of memory formations. The child now not only recognizes the mother but also localizes her in the recent past. If she comes into the room and sits down while the infant is playing, the infant may note this and return to its play. Moments later it may again look toward where the mother is seated giving evidence of a recent memory.

From about the eighth or ninth month and extending through the second year, the child gradually constructs what Piaget calls an *objective* time. Some of the observations relevant to the attainment of the object-time notion are significant here. First of all, the child begins to search for hidden objects, which means he has ordered a series of perceptions of memories within a temporal sequence. As he grows older the child begins to be able to take account of successive displacements of the hidden object which implies a still more elaborate temporal organization of perceptual memories. Finally, toward the middle of the second year a "true" objective sense of time appears to the extent that the child begins to symbolize or represent temporal sequences and durations by means of words.

Causality. The development of elementary causal conceptions in the infant follows the pattern we have become familiar with in discussing the other categories. Beginning as practical causality in which psychological efficacy and physical force are undifferentiated, the infant progresses to the more or less complete practical separation of physical and psychic causality by the end of the second year. During the first few months of life the child's reflex behaviors are the only indices of causal connections. As Piaget puts it.

> Whether the nursling at the age of one or two months succeeds in sucking his thumb after having attempted to put it into his mouth or whether his eyes follow a moving object, he must experience, though in different degrees, the same impression: namely, that without his knowing how a certain action leads to a certain result, in other words that a certain complex of efforts, tension, expectation, desire, etc., is charged with efficacy (1954, p. 229).

For Piaget then, causality during the early months of life consists in little more than the feeling of effort experienced in connection with the resistance of things.

From about the third to the seventh or eighth month a new level of causal behavior emerges which Piaget calls *magico-phenomenalistic causality*. This it will be recalled is the period in which active prehension emerges and, hence, the coordination of vision and other sense modalities with movement. At this stage the infant becomes witness to, and takes interest in, three types of action: movements of the body, movements which depend upon body movements, and movements that are independent of the child's actions. In Piaget's view, the infant begins to discover causal relations in the process of observing the three types of movement. Children at this level of development "find their hands" and begin to associate the feeling of efficacy with the visual perception of hand movements. Likewise, the feeling of effort when pushing or shaking a rattle comes to be associated with the sight of the movement of the rattle and with its sound. Piaget calls this stage "magico-phenomenalistic" because the child fails to distinguish the sphere of its efficacy from the sphere of movements wherein it lacks efficacy. The infant, for example, may shake its leg on seeing a doll at a distance as if this act would move the doll. Piaget also argues that even when the child of this stage pulls a string to make a toy rattle, he does not realize that there is an intermediary between his action and the effect. It is the fact that the infant behaves as if his gestures alone could produce physical results that accounts for Piaget's attaching the magico-phenomenalistic label to this period. (Traces of this kind of causality can still be seen in adults who try to influence the course of a bowling or billiard ball with "body English.")

Beginning at about the seventh or eighth months of life and extending to the latter half of the second year Piaget found what he regarded as the gradual attainment of an objective sense of causality. By objective with respect to causality Piaget means that the child comes to distinguish between physical and psychological causes. At the outset of this period the child only dissociates causality from his own actions without at the same time attributing causality to objects. Then, at about one year of age, "the child recognizes causes that are entirely external to his activity and for the first time he establishes among events perceived links of causality independent of the action itself" (Piaget, 1954, p. 279). Here is an example:

> At 1; (28) Jacqueline touches with her stick a push cat placed on the floor, but does not know how to pull it to her. The spatial contact between the stick and the cat seem to her sufficient to displace the object. . . . Finally at 1;3 (12) Jacqueline utilizes the stick correctly; objective and spatialized causality are therefore applied to the physical conditions of the problem (Piaget, 1954, p. 284).

Finally, toward the middle of the second year of life the objectification of causality is completed with the beginnings of representation. What this means is that at this point the child begins to take account of causes that are outside his immediate sphere of perception, something he did not do heretofore. He comes at this final point in the development of practical causality to be, in Piaget's words, "capable of reconstructing causes in the presence of their effects alone." Here is one example of representative causal behavior:

> At 1;4 (4) Laurent tries to open a garden gate but cannot push it forward because it is held back by a piece of furniture. He cannot account either visually or by any sound for the cause that prevents the gate from opening, but after having tried to force it he suddenly seems to understand; he goes around the wall, arrives at the other side of the gate, moves the armchair which holds it firm and opens it with a triumphant expression (Piaget, 1954, p. 296).

At this final point the child shows clearly that causality is no longer merely perceptual and that by means representation he can reconstruct causes and anticipate consequences not immediately present.

It might be well, at this point, to summarize in a general way the process of conceptualization that occurs during the first two years of life as disclosed by the work of Piaget. What Piaget argues is that the infant begins in an egocentric universe in which there is nothing other than that which directly concerns his own activities and needs. It is thus a world that, in terms of the object and space, does not transcend the images and spatial relations that are immediately given in perception. It is a world, moreover, in terms of time and causality, that does not go beyond the immediate present. With increasing age the child gradually becomes aware of a world that is independent of his perception and action and that manifests laws and sequences that are independent of his will. By the end of the period, with the aid of representation, he recognizes the existence of objects not present to perception and the relativity of spatial position while his notions of time and causality expand backward to the past and press forward to the future. The egocentric universe has thus been transformed into an objective world.

At the same time, however, it must be recognized that this achievement holds only on the plane of perception and action. For, at the age of about two, the child begins to construct a new universe on the level of verbalization and representation. And, in the process of constructing this verbal, representational universe he again begins egocentrically and only gradually extricates himself to discover the objective verbal and representational world. For Piaget, each new, higher level of conceptual functioning demands a new structuring of experience. This is as arduous as that undertaken at the previous level of mental functioning.

Summary

This chapter began with the distinction between two broad approaches to the study of cognition in infancy and early childhood. Within the mental measurement approach three separate issues were taken up. The first issue was that of the usefulness of predicting IQs on the basis of infant tests. It was concluded that there is some promise of effective prediction if projections are based on categories of ability rather than upon numerical scores *per se.* A second issue dealt with the hypothesis of infancy as a critical period in intellectual development. This hypothesis was held to be supported by the literature which suggests that infancy is a critical period with respect to healthy personality growth. Since the infant is particularly vulnerable to deviations from a normal expectable emotional environment and since, during infancy, cognitive and affective factors are closely intertwined it follows that if infancy is a critical period from the affective point of view, then it must as a matter of course also be a critical period from the point of view of cognition. The third issue taken up dealt with several of the many causative agents affecting infant intelligence including prematurity, malnutrition, and nursery school experience.

With respect to the second, developmental approach, three issues were again taken up. The first of these concerned problem solving in infancy and the issues of insight vs. trial and error as well as the role of verbal mediation. It was concluded that in infancy problem solving behavior was dependent pretty much upon the nature of the task and that both trial and error as well as insight could be observed and could well represent different phases of the problem solving process. It was suggested that verbal mediation in problem solving may operate at different levels and make its first appearance at about two years of life. A second developmental issue was that of memory. Different forms of memory were distinguished and work upon them summarized. It was concluded that all forms of memory, sensory, motor, rote, and representational were present in early childhood with the exception of historical memory which does not make its appearance until middle childhood. The final issue taken up was that of concept formation in infancy. After distinguishing between the intensive and extensive aspects of concepts, the emergence of several fundamental concepts was described. The section was closed with a detailed summary of Piaget's work on the development of the categorical concepts during the first years of life.

The aim of this review was not to summarize all the available literature on the topic of cognition in infancy, but rather to point to issues that seem worthy of further research and exploration. The scientific study of cognition is only in the present day coming into its own right as a field of investigation and dogmatism is unwarranted. As we saw with respect to the question of the value of infant tests in predicting later IQ levels, today's truth can be tomorrow's fallacy. It is in this spirit that the present discussion has been offered.

CONSOLIDATED BIBLIOGRAPHY
from Essays

Ahrens, R., "Beitröge zur Entwicklung des Physiognomie—und Mimikerkennens," *Z. Exp. Agnew. Psychol.* 2 (1954), pp. 412–454 and 599–633.

Ajuriaguerra, J. De., "Language et dominance cerebrale," *J. Francais d' Oto-Rino-Laringol.*, 6 (1957), pp. 489–499.

Allen, C. N., "Individual Differences in Delayed Reaction in Infants," *Arch. Psychol.*, 19 (1931), No. 127.

Allport, C. W., and T. F. Pettigrew, "Cultural Influence on the Perception of Movement: The Trapezoidal Illusion among Zulus," *Journal of Abnormal and Social Psychology*, 55 (1957), pp. 104–113.

Allport, F. H., *Theories of Perception and the Concept of Structure*. New York: Wiley, 1955.

Alm, I., "The Long Term Prognosis for Prematurely Born Children. A Follow-up Study of 999 Premature Boys Born in Wed-lock and of 1002 Controls," *Acta Paediat.*, Supplement 94 (1953).

Alpert, A., "The Solving of Problem Situations by Preschool Children," *Teach. Coll. Contrib. Educ.*, No. 323 (1928).

Alston, William, *Philosophy of a Language*. Englewood Cliffs, N.J.: Prentice-Hall, 1964, p. 68.

Alt, J., The Use of Vision in Early Reaching. Unpublished honors thesis, Department of Psychology, Harvard University, 1968.

Altman, P. L., and D. I. Dittmer, eds., *Growth Including Reproduction and Morphological Development*. Washington, D.C.: Federation of American Societies for Experimental Biology, 1962.

Ames, Elinor W., Stimulus Complexity and Age of Infants as Determinants of the Rate of Habituation of Visual Fixation. Paper read at Western Psychol. Ass., Long Beach, Calif., April 1966.

Ames, Elinor W., and Carole K. Silfen, Methodological Issues in the Study of Age Differences in Infants' Attention to Stimuli Varying in Movement, Complexity, and Novelty. Paper read at Soc. Res. Child Develpm., Minneapolis, April 1965.

Ames, L. B., "The Development of the Sense of Time in the Young Child," *J. Genet. Psychol.*, 68 (1946), pp. 97–125.

Anderson, J., "Empiricism," *Aust. J. Psychol. Philos.*, 5 (1927), pp. 241–254.

Anderson, John E., "The Limitations of Infant and Preschool Tests in the Measurement of Intelligence," *J. Psychol.*, 8 (1939), pp. 351–379.

Anderson, John E., "Freedom and Constraint or Potentiality and Environment," *Psychol. Bull.*, 41 (1944), pp. 1–29.

Anderson, John E., "Personality Organization in Children," *Amer. Psychologist*, 3 (1948), pp. 409–416.

Anderson, John E., "Methods of Child Psychology," in L. Carmichael, ed., *Manual of Child Psychology* (2d ed.) New York: Wiley, 1954, pp. 1–59.

Anderson, John E., Dynamics of Development: Systems in Process (in this volume) in Dale B. Harris, ed., *The Concept of Development.* Minneapolis: University of Minnesota Press, 1957.

André-Thomas, St. Anne-Dorgossies, *Etudes neurologiques.* Paris: 1952.

Anokhin, P. K., "The Role of the Orientation Reaction in Conditioning, the Orientation Reaction and Orienting-Investigating Activity." Moscow: Acad. Pedag. Sciences, R.S.F.S.R., 1958.

Arber, Agnes, *The Mind and the Eye.* New York: Cambridge University Press, 1954.

Archer, E. J., "On Verbalizations and Concepts," in A. W. Melton, ed., *Categories of Human Learning.* New York: Academic Press, 1964.

Archer, E. J., L. E. Bourne, and F. G. Brown, "Concept Identification as a Function of Irrelevant Information and Instructions," *J. Exp. Psychol.*, 49 (1955), pp. 153–164.

Armstrong, D. M., *Perception and the Physical World.* London: Routledge, 1961.

Aronson, E., and S. Rosenbloom, Spatial Coordination of Auditory and Visual Information in Early Infant Perception. Unpublished manuscript, Harvard University, 1969.

Aronson, E., and E. Tronick, Implications of Infant Research for Theory (in this volume), 1971.

Atkinson, R. C., and R. M. Shiffrin, "Human Memory: A Proposed System and Its Control Processes," in K. W. Spence and J. T. Spence, eds., *The Psychology of Learning and Motivation,* Vol. 2. New York: Academic Press, 1968.

Attneave, F., "Some Informational Aspects of Visual Perception," *Psych. Rev.*, 61 (1954), pp. 183–193.

Austerlitz, Robert, "Gilyak Nursery Words," *Word,* 12 (1956), pp. 260–279.

Ausubel, D. P., *Theory and Problems of Adolescent Development.* New York: Grune & Stratton, 1954.

Ausubel, D., "Historical Overview of Theoretical Trends" (in this volume), in D. Ausubel, *Theory and Problems of Child Development.* New York: Grune & Stratton, 1957, pp. 22–49.

Baldwin, A. L., *Behavior and Development in Childhood.* New York: Holt, Rinehart and Winston, Inc., 1955.

Baldwin, J. M., *Genetic Theory of Reality.* New York: Putnam, 1915.

Barham, E. G., W. B. Huckabay, R. Gowdy, and B. Burns, "Microvolt Electric Signals from Fishes and the Environment," *Science,* 164 (1969), pp. 965–968.

Bar-Hillel, Y., *Language and Information.* Reading, Mass.: Addison-Wesley, 1964.

Barker, R. G., and H. F. Wright, *Midwest and Its Children.* New York: Harper & Row, 1955.

Baron, M. R., and C. T. Vacek, "Generalization Gradients along Wave Length and Angularity Dimensions in Pigeons following Two Differential Training Procedures," *Psychon. Sci.,* 9 (1967), pp. 423–424.

Bartoshuk, A. K., "Response Decrement with Repeated Elicitation of Human Neonatal Cardiac Acceleration to Sound," *J. Comp. Physiol. Psychol.,* 55 (1962), pp. 9–13.

Bartoshuk, A. K., "Human Neonatal Cardiac Responses to Sound: A Power Function," *Psychon. Sci.,* 1 (1964), pp. 151–152.

Bartoshuk, A. K., and J. M. Tennant, "Human Neonatal EEG Correlates of Sleep—Wakefulness and Neural Maturation," *J. of Psychiat. Res.,* 2 (1964), p. 73.

Bayley, H., "Comparisons of Mental and Motor Test Scores for Ages 1–15 Months by Sex, Birth Order, Race, Geographical Location, and Education of Parents," *Child Develpm.,* 36 (1965), pp. 379–411.

Bayley, N., "Mental Growth in Young Children," *Yearb. Nat. Soc. Stud. Educ.,* 39 (1940), Part II.

Bayley, N., and E. Schaefer, "Correlations of Maternal and Child Behaviors with the Development of Mental Abilities: Data from the Berkeley Growth Study," *Monogr. Soc. Res. Child Develpm.,* 29 (1964), No. 97.

Beach, F. A., "The Snark Was a Boojum," *Amer. Psychologist,* 5 (1950), pp. 115–124.

Beach, F., and J. Jaynes, "Effects of Early Experience upon the Behavior of Animals," *Psychol. Bull.,* 51 (1954), pp. 240–263.

Bean, C. H., An Unusual Opportunity To Investigate the Psychology of Language," *Journal of Genetic Psychology,* 40 (1932), pp. 181–202.

Beauzée, N., *Grammaire générale, on exposition raisonée des elements necessaires du language.* Paris, 1767.

Beckmann, H., "Die Entwicklung der Zahlleistung bei 2–6 jährigen Kindern" ("The Development of Number Achievement in Two–Six Year Old Children"), *Z. angew. Psychol.,* 22 (1923), pp. 1–72.

Bell, R. Q., "Relations between Behavior Manifestations in the Human Neonate," *Child Develpm.,* 31 (1960), pp. 463–477.

Bell, R. Q., and Joan F. Darling, "The Prone Head Reaction in the Human Neonate: Relation with Sex and Tactile Sensitivity," *Child Develpm.,* 36 (1965), pp. 943–949.

Bellugi, U., The Acquisition of Negation. Unpublished doctoral dissertation, Harvard University, 1967.

Benjamin, R. M., and R. F. Thompson, "Differential Effects of Cortical Lesions in Infant and Adult Cats on Roughness Discrimination," *Exp. Neurol.,* 1 (1959), pp. 305–321.

Bergmann, G., "Psychoanalysis and Experimental Psychology: A Review from the Standpoint of Scientific Empiricism," *Mind,* 52 (1943), pp. 122–140.

Bergmann, G., and K. W. Spence, "The Logic of Psychophysical Measurement," *Psychol. Rev.,* 51 (1944), pp. 1–24.

Berkeley, G., *An Essay towards a New Theory of Vision*, 1709. London: Dent, 1957.

Berko, J., "The Child's Learning of English Morphology," *Word*, 14 (1958), pp. 150–177. Reprinted in S. Saporta, ed., *Psycholinguistics*. New York: Holt, Rinehart and Winston, Inc., 1961.

Berko, Jean, and R. Brown, "Psycholinguistic Research Methods," in P. Mussen, ed., *Handbook of Research Methods in Child Development*. New York: Wiley, 1960.

Berlyne, D. E., "Attention, Perception and Behavior Theory," *Psych. Rev.*, 68 (1951), pp. 137–146.

Berlyne, D. E., "Knowledge and Stimulus-Response Psychology," *Psychol. Rev.*, 61 (1954), pp. 245–254.

Berlyne, D. E., *Conflict, Arousal, and Curiosity*. New York: McGraw-Hill, 1960.

Berlyne, D., *Structure and Direction in Thinking*. New York: Wiley, 1965.

Berlyne, D., "The Delimitation of Cognitive Development," in Stevenson, ed., *Concept of Development*. Monograph for Society in Research and Child Development (107), 1966, pp. 71–81.

Bertalanffy, L., *Modern Theories of Development*. London: Oxford, 1933.

Bertalanffy, L., *Problems of Life*. New York: Wiley, 1952.

Best, J. B., "The Photosensitization of *Paramecium Aurelia* by Temperature Shock," *J. Exper. Zool.*, 126 (1954), pp. 87–99.

Bever, T. G., J. A. Fodor, and W. Weksel, "On the Acquisition of Syntax: A Critique of 'Contextual Generalization,'" *Psych. Rev.*, 72 (1965), pp. 467–482.

Beyrl, F., "Über die Grossenauffasung bei Kindern." *Z. Psychol.*, 100 (1926), pp. 344–371.

Bijou, S. W., and D. M. Baer, "Some Methodological Contributions from a Functional Analysis of Child Development," *Adv. Child Develpm. Behav.*, 1 (1963), pp. 197–231.

Bijou, S. W., and D. N. Baer, *Child Development*, Vol. 2, *Universal Stage of Infancy*. New York: Appleton, 1965, Ch. 9.

Bingham, W. E., and W. J. Griffiths, Jr., "The Effect of Different Environments during Infancy on Adult Behavior in the Rat." *J. Comp. Physiol. Psychol.*, 45 (1952), pp. 307–312.

Birch, H. B., "The Relation of Previous Experience to Insightful Problem-Solving," *J. Comp. Psychol.*, 38 (1954), pp. 367–383.

Birch, H. B., Ontogenetic Sources for Order in the Maternal Behavior of the Rat. Unpublished paper.

Birch, H. G., A. Thomas, E. Chess, and M. E. Hertzig, "Individuality in the Development of Children," *Develpm. Med. Child Neurol.*, 4 (1962), pp. 370–379.

Birns, Beverly, Marion Blank, W. H. Bridger, and Sibylle K. Escalona, "Behavioral Inhibition in Neonates Produced by Auditory Stimuli," *Child Develpm.*, 36 (1965), pp. 639–645.

Blauvelt, Helen, "Dynamics of the Mother-Newborn Relationship in Goats," in *Group Processes*, 1st Conference, 1954, pp. 221–258. New York: Macy, 1955.

Bloom, Lois Masket, Language Development: Form and Function in Emerging Grammers. Unpublished doctoral dissertation, Columbia University, 1968.

Bloomfield, L., *Language*. New York: Holt, Rinehart and Winston, Inc., 1933.

Blough, D. S., "Stimulus Generalization as Signal Detection in Pigeons," *Science*, 158 (1967), pp. 440–441.

Boas, F., *Handbook of American Indian Languages*. Washington, D.C.: Smithsonian Inst., Bureau of American Ethnology, Bulletin 40, 1911.

Bonner, John T., *Morphogenesis: An Essay on Development*. Princeton, N.J.: Princeton University Press, 1952.

Boring, E. G., *A History of Experimental Psychology*. New York: Appleton, 1929.

Boring, E. G., *The Physical Dimensions of Consciousness*. New York: Appleton, 1933.

Boring, E. G., "Visual Perception as an Invariance," *Psych. Rev.*, 59 (1952a), pp. 141–148.

Boring, E. G., "The Gibsonian Visual Field," *Psych. Rev.*, 59 (1952b), pp. 246–247.

Bourne, Lyle, "Human Conceptual Behavior" (in this volume). Boston: Allyn and Bacon, 1966.

Bourne, L. E., and F. Restle, "Mathematical Theory of Concept Identification," *Psych. Rev.*, 66 (1959), pp. 278–298.

Bousfield, W. A., "The Occurrence of Clustering in the Recall of Randomly Arranged Associates." *J. Gen. Psychol.*, 49 (1953), pp. 229–240.

Bousfield, W. A., and B. H. Cohen, "Clustering in Recall as a Function of the Number of Word-Categories in Stimulus-Word Lists," *J. Gen. Psychol.*, 54 (1956), pp. 95–106.

Bousfield, W. A., "The Problem of Meaning in Verbal Learning," in C. N. Cofer, ed., *Verbal Learning and Verbal Behavior*. New York: McGraw-Hill, 1961, pp. 81–91.

Bower, T. G. R., "Discrimination of Depth in Premotor Infants," *Psychon. Sci.*, 1 (1964), p. 368.

Bower, T. G. R., "The Determinants of Perceptual Unity in Infancy," *Psychon. Sci.*, 3 (1965a), pp. 323–324.

Bower, T. G. R., Perception in Infancy. Paper Read at Center for Cognitive Studies Colloquium, Harvard University, 1965b.

Bower, T. G. R., "Stimulus Variables Determining Space Perception in Infants," *Science*, 149 (1965c), pp. 88–89.

Bower, T. G. R., "Slant, Perception and Shape Constancy in Infants," *Science*, 151 (1966a), pp. 832–834.

Bower, T. G. R., "The Visual World of Infants," *Scientific American*, 215 (1966b), pp. 80–90.

Bower, T. G. R., "Phenomenal Identity and Form Perception in an Infant," *Perception and Psychophysics*, 2 (1967a), pp. 74–76.

Bower, T. G. R., "The Development of Object-Permanence: Some Studies of Existence Constancy," *Perception and Psychophysics*, 2 (1967b), pp. 411–418.

Bower, T. G. R., Recent Research with Infants. Center for Cognitive Studies Colloquium, Harvard University, 1969.

Boyd, W. C., *Genetics and the Races of Man.* Boston: Little, Brown, 1953.

Brackbill, Yvonne, "Extinction of the Smiling Response in Infants as a Function of Reinforcement Schedule," *Child Develpm.*, 29 (1958), pp. 115–124.

Brackbill, Yvonne, Gail Adams, D. H. Crowell, and Libbie Gray, "Arousal Level in Neonates and Preschool Children under Continuous Auditory Stimulation," *J. Exp. Child Psychol.* 4 (1966), pp. 178–188.

Braine, M. D. S., "Ontogeny of Certain Logical Operations," *Psych. mono.*, 5 (1959).

Braine, M. D. S., "On Learning the Grammatical Order of Words," *Psychol. Rev.*, 70 (1963a), pp. 323–348.

Braine, M. D. S., "The Ontogeny of English Phrase Structure: The First Phase," *Lang.*, 39 (1963b), pp. 1–13.

Braine, M. D. S., "On the Basis of Phrase Structure: A Reply to Bever, Fodor, and Weksel," *Psychol. Rev.*, 72 (1965), pp. 483–492.

Bramstedt, F., "Dressurversuche mit *paramecium caudatum* und *stylonchia mytilus*," *Z. Vergl. Physiol.*, 22 (1935), pp. 490–516.

Brante, G., "Studies on Lipids in the Nervous System; with Special Reference to Quantitative Chemical Determination and Topical Distribution," *Acta Physiol. Scand.*, 18 (1949), Suppl. 63.

Bregman, E. O., "An Attempt To Modify the Emotional Attitudes of Infants by the Condition Response Technique," *J. Genet. Psychol.*, 45 (1934), pp. 169–198.

Brennan, Wendy, Elinor W. Ames, and R. W. Moore, "Age Differences in Infants' Attention to Patterns of Different Complexities," *Science*, 151 (1966), pp. 345–356.

Brenner, M. W., "The Developmental Study of Apparent Movement," *Quart. J. Exp. Psychol.*, 9 (1954), pp. 169–174.

Brett, W. J., "Persistent Diurnal Rhythmicity in *Drosophila*," *Ann. Ent. Soc. Amer.*, 48 (1955), pp. 119–131.

Bridger, W. H., "Sensory Habituation and Discrimination in the Human Neonate," *Amer. J. Psychiat.*, 117 (1961), pp. 991–996.

Brodbeck, A. J., and O. C. Irwin, "The Speech Behavior of Infants without Families," *Child Development*, 17 (1946), pp. 145–156.

Bronshtein, A. I., and E. P. Petrova, "The Auditory Analyzer in Young Infants," in Y. Brackbill and G. C. Thompson, eds., *Behavior in Infancy and Early Childhood.* New York: Free Press, 1967, pp. 163–172.

Brooks, C., and M. E. Peck, "Effect of Various Cortical Lesions on Development of Placing and Hopping Reactions in Rats," *J. Neurophysiol.*, 3 (1940), pp. 66–73.

Brown, R. W., "Language and Categories," appendix in J. S. Bruner, Jacqueline J. Goodnow, and G. A. Austin, *A Study of Thinking.* New York: Wiley, 1956.

Brown, R. W., *Words and Things.* New York: Free Press, 1957.

Brown, R. W., *Personal Communication.* 1966.

Brown, R., and U. Bellugi, "Three Processes in the Child's Acquisition of Syntax," *Harvard Ed. Rev.*, 34 (1964a), pp. 133–151.

Brown, R., and U. Bellugi, "Three Processes in the Child's Acquisition of Syntax," *New Directions in the Study of Language,* E. Lenneberg, Ed., Cambridge, Mass.: M.I.T. Press, 1964b.

Brown, R. W., and Jean Berko, "Word Association and the Acquisition of Grammar," *Child Develpm.,* 31 (1960), pp. 1–14.

Brown, R., and C. Fraser, "The Acquisition of Language," *Monogr. Socy. Res. Child Devel.,* 29 (1964), pp. 43–79.

Brown, R. W., C. Fraser, and Ursula Bellugi, "Explorations in Grammar Evaluation," in Ursula Bellugi and R. W. Brown, eds., *The Acquisition of Language. Monogr. Soc. Res. Child Develpm.,* 29 (1964), pp. 79–92.

Brown, R. W., and E. H. Lenneberg, "A Study of Language and Cognition," *J. Abnorm. Soc. Psychol.,* 49 (1954), pp. 454–462.

Bruner, J. S., "Personality Dynamics and the Process of Perceiving," in R. R. Blake and G. U. Ramsay, eds., *Perception: An Approach to Personality.* New York: Ronald, 1951.

Bruner, J. S., *On Going beyond the Information Given.* Cambridge, Mass.: Harvard University Press, 1957a.

Bruner, J. S., "On Perceptual Readiness," *Psychol. Rev.,* 64 (1957b), pp. 123–152.

Bruner, J. S., *The Process of Education.* Cambridge, Mass.: Harvard University Press, 1960.

Bruner, J. S., "The Growth of Mind," *American Psychologist,* 20 (1965), pp. 1007–1017.

Bruner, J. S., "On Cognitive Growth II" (in this volume), in J. S. Bruner, *et al., Studies in Cognitive Growth: A Collaboration at the Center for Cognitive Studies.* New York: Wiley, 1966a.

Bruner, J. S., *Toward a Theory of Instruction.* Cambridge, Mass.: Harvard University Press, 1966b.

Bruner, J. S., "On the Conservation of Liquids," in J. S. Bruner, *et al., Studies in Cognitive Growth.* New York: Wiley, 1966c.

Bruner, J. S., R. D. Busiek, and A. L. Minturn, "Assimilation in the Immediate Reproduction of Visually Perceived Figures," *J. Exp. Psychol.,* 43 (1952), pp. 151–155.

Bruner, J. S., *et al., Studies in Cognitive Growth.* New York: Wiley, 1966.

Bruner, J. S., J. J. Goodnow, and G. A. Austin, *A Study of Thinking.* New York: Wiley, 1956.

Bruner, J. S., and L. Postman, "Perception, Cognition and Behavior." *J. Pers.,* 18 (1949), pp. 14–31.

Brunswik, E., "The Conceptual Focus of Some Psychological Systems," *J. Unified Sci.,* 8 (1939), pp. 36–49.

Brunswik, E., "Psychology in Terms of Objects," reprinted in M. Marx, ed., *Psychological Theory.* New York: Macmillan, 1951.

Brunswik, E., "The Conceptual Framework of Psychology," *Internat. Encyc. Unified Sci.,* 1, No. 10. Chicago: University of Chicago Press, 1952.

Brunswik, E., *Perception and the Representative Design of Psychological Experiments.* Berkeley, Calif.: University of California Press, 1956.

Bryden, M. P., "Tachistoscopic Recognition of Non-alphabetical Material," *Canadian Journal of Psychology,* 14 (1960), pp. 78–86.

Bryden, M. P., A. O. Dick, and D. J. K. Mewhort, "Tachistoscopic Recognition of Number Sequences," *Canadian Journal of Psychology*, 22 (1968), pp. 52–59.

Bühler, C., *The First Year of Life*. New York: John Day, 1930.

Bühler, C., *Kindheit und Jugend*. Leipzig: Nirzel, 1931, 3d ed.

Bühler, K., *The Mental Development of the Child*. London: Routledge, 1930.

Burtt, H. E., "An Experimental Study of Early Childhood Memory," *J. Genet. Psychol.*, 40 (1932), pp. 287–295.

Burtt, H. E., "A Further Study of Early Childhood Memory," *J. Genet. Psychol.*, 50 (1937), pp. 187–192.

Burtt, H. E., "An Experimental Study of Early Childhood Memory: Final Report," *J. Genet. Psychol.*, 58 (1941), pp. 435–439. Reprinted in Y. Brackbill and G. G. Thompson, eds., *Behavior in Infancy and Early Childhood: A Book of Readings*. New York: Free Press, 1967.

Bush, R. R., and F. Mosteller, *Stochastic Models for Learning*. New York: Wiley, 1955.

Butter, C. M., "Stimulus Generalization along One and Two Dimensions with Pigeons," *J. Exp. Psychol.*, 65 (1963), pp. 339–346.

Cantor, G. N., and Joan H. Cantor, "Discriminative Reaction Time Performance in Preschool Children as Related to Stimulus Familiarization," *J. Exp. Child Psychol.*, 2 (1965), pp. 1–9.

Cantor, G. N., and W. S. Meyers, Inter-observer Reliability in Judging Infants' Observing Responses. Paper read at Soc. Res. Child Develpm., Minneapolis, March 1965.

Capper, A., "The Fate and Development of the Immature and of the Premature Child," *Amer. J. Dis. Child.*, 35 (1928), pp. 262–288 and pp. 443–491.

Carmichael, L., "The Development of Behavior in Vertebrates Experimentally Removed From the Influence of External Stimulation," *Psychol. Rev.*, 33 (1926), pp. 51–58.

Carmichael, L., "A Further Study of the Development of Behavior in Vertebrates Experimentally Removed from the Influence of External Stimulation," *Psychol. Rev.*, 34 (1927), pp. 34–47.

Carmichael, L., "A Re-evaluation of the Concepts of Maturation and Learning as Applied to the Early Development of Behavior," *Psychol. Rev.*, 43 (1936), pp. 450–470.

Carmichael, L., "The Onset of Early Development of Behavior," in L. Carmichael, ed., *Manual of Child Psychology*. New York: Wiley, 1946, pp. 43–166.

Carmichael, L., "The Onset and Early Development of Behavior," in L. Carmichael, ed., *Manual of Child Psychology*. New York: Wiley, 1954, pp. 60–185.

Carmichael, L., H. P. Hogan, and A. A. Walter, "An Experimental Study of the Effect of Language on the Reproduction of Visually Perceived Form," *J. Exp. Psychol.*, 15 (1932), pp. 73–86.

Carmichael, L., and M. F. Smith, "Quantified Pressure Stimulation and the Specificity and Generality of Response in Fetal Life," *J. Genet. Psychol.*, 54 (1939), pp. 425–434.

Carpenter, G. C. and G. Stechler, "Selective Attention to Mother's Face From

Week 1 through Week 8," *Proceedings, 75th Annual Convention*. American Psychological Association, 1967, pp. 153–154.
Carr, W. J., and D. I. McGuigan, "The Stimulus Basis and Modification of Visual Cliff Performance in the Rat," *Anim. Behav.*, 13 (1965), pp. 25–29.
Carter, G. S., "The Theory of Evolution and the Evolution of Man," in A. L. Kroeber, ed., *Anthropology Today*. Chicago: University of Chicago Press, 1963, pp. 327–342.
Champreys, F. H., "Notes on an Infant," *Mind*, 6 (1881), pp. 104–107.
Charlesworth, W. R., "Persistance of Orienting and Attending Behavior in Infants as a Function of Stimulus-Locus Uncertainty." *Child Develpm.*, 37 (1966a), pp. 473–491.
Charlesworth, W. R., Development of the Object Concept: A Methodological Study. Paper read at Amer. Psychol. Assoc., September 1966b.
Charlesworth, W. R., Surprise and Expectancies in Cognitive Development. Unpublished Manuscript, 1967.
Charlesworth, W. R., "Cognition in Infancy" (in this volume), *The Merrill-Palmer Quarterly*, 14 (1968), pp. 25–46.
Charrow, V., D. Ingram, and A. Dil, A Study in the Analysis of the Child's 2- and 3-word Utterances without Contextual Information, 1969.
Chase, R. A., "Evolutionary Aspects of Language Development and Function," in F. Smith and G. A. Miller, eds., *The Genesis of Language*. Cambridge, Mass.: M.I.T. Press, 1966, pp. 253–268.
Chisholm, Roderick, *Perceiving: A Philosophical Study*. Ithaca, N.Y.: Cornell University Press, 1957.
Chomsky, N., *Syntactic Structures*. The Hague: Mouton, 1957.
Chomsky, N., "A Review of B. F. Skinner's *Verbal Behavior*," *Language*, 35 (1959), pp. 26–58.
Chomsky, N., *Aspects of the Theory of Syntax*. Cambridge, Mass.: M.I.T. Press, 1965.
Chomsky, N., *Current Issues in Linguistic Theory*. The Hague: Mouton, 1966.
Chomsky, N., "The Formal Nature of Language," in E. H. Lenneberg, *Biological Foundations of Language*. New York: Wiley, 1967, pp. 397–442.
Chomsky, N., "Language and the Mind," *Psych. Today*, 1, no. 9 (Feb. 1968a), pp. 48–68.
Chomsky, N., *Language and Mind*. New York: Harcourt, 1968b.
Chomsky, N., and M. Halle, *The Sound Pattern of English*. New York: Harper & Row, 1968.
Chow, K. L., and H. W. Nissen, "Interocular Transfer of Learning in Visually Naïve and Experienced Infant Chimpanzees," *J. Comp. Physiol.*, 48 (1955), pp. 229–232.
Christie, R., "Experimental Naïveté and Experiential Naïveté," *Psychol. Bull.*, 48 (1951), pp. 327–339.
Christie, R., "The Effect of Some Early Experiences in the Latent Learning of Rats," *J. Exper. Psychol.*, 43 (1952), pp. 381–388.
Clark, W. E., and P. B. Medawar, *Essays on Growth and Form*. London: Oxford, 1945.

Cofer, C. N., and M. H. Appley, *Motivation: Theory and Research.* New York: Wiley, 1964.

Coghill, G. E., *Anatomy and the Problem of Behavior.* New York: Macmillan, 1929.

Cohen, J., "The Synthetic-Analytic Character of Color Vision in the Pigeon," *Psychon. Sci.*, 9 (1967), pp. 429–430.

Cohen, Leslie B., "Observing Responses, Visual Preferences and Habituation to Visual Stimuli in Infants," *Diss. Abst.*, 27 (1966), p. 310.

Commission on Science Education. *Science—a Process Approach* (3d experimental edition) Parts 1–7. Washington: American Association for the Advancement of Science, 1965.

Conrad, R., "Acoustic Confusions in Immediate Memory," *British Journal of Psychology*, 55 (1964), pp. 75–84.

Cordemoy, Gide, *Discours physique de la parole.* Paris, 1666.

Cornman, James, "The Identity of Mind and Body," *Journal of Philosophy*, 59 (1962), p. 490.

Craig, W., "Appetites and Aversions as Constituents of Instinct," *Biol. Bull.*, 34 (1911), pp. 91–107.

Cromer, Richard F., The Development of Temporal Reference during the Acquisition of Language. Unpublished doctoral dissertation, Harvard University, 1968.

Cronbach, L. J., *Educational Psychology.* New York: Harcourt, 1962 (2d ed.).

Crutchfield, R., Instructing Children in Creative Thinking. Paper read at 72d Annual Convention, Amer. Psychol. Assoc., Los Angeles, California, 1964.

Cullen, J. K., *et al.*, "The Development of Auditory Feedback Monitoring: I. Delayed Auditory Feedback Studies on Infant Cry," *Journal of Speech and Hearing Research*, 11 (1968), pp. 85–93.

Cullen, J. K., N. L. Fargo, and P. Baker. "The Development of Auditory Feedback Monitoring: III. Delayed Auditory Feedback Studies of Infant Cry Using Several Delay Times," in *Annual Reports, 1968, Neurocommunications Laboratory.* Baltimore: Johns Hopkins University School of Medicine, 1968, pp. 77–93.

Dantzig, T., *Number, the Language of Science.* New York: Macmillan, 1954.

Darwin, C. I., "A Biographical Sketch of an Infant," *Mind*, 2 (1877), pp. 285–294.

David, P. R., and L. H. Snyder, "Genetic Variability and Human Behavior," in J. H. Rohrer and M. Sherif, eds., *Social Psychology at the Crossroads.* New York: Harper & Row, 1951, pp. 53–82.

Davidon, R. S., "Effects of Symbols, Shift, & Manipulation," *J. Exp. Psychol.*, 44 (1952), pp. 70–79.

Davis, K., "Final Note on a Case of Extreme Isolation," *Amer. J. Sociol.*, 52 (1947), pp. 432–437.

DeBeer, G. R., *Embryos and Ancestors.* Oxford: Clarendon Press, 1951.

Dember, W. N., *Psychology of Perception.* New York: Holt, Rinehart and Winston, Inc., 1960.

Denis-Prinzhorn, M., "Perception des distances et constance des grandeurs (étude génétique)," *Arch. Psychol., Genève*, 37 (1960), pp. 181–309.

Dennis, W., ed., "Developmental Theories," *Current Trends in Psychological Theory*. Pittsburgh, Pa.: University of Pittsburgh Press, 1951, pp. 1–20.

Dennis, W., and M. C. Dennis, "Development under Controlled Environmental Conditions," in W. Dennis, ed., *Readings in Child Psychology*. Englewood Cliffs, N.J.: Prentice-Hall, 1951.

Dennis, W., and P. Najarian, "Infant Development under Environmental Handicap," *Psychol. Monogr.*, 71 (1957).

De Rivera, J., "Some Conditions Governing the Use of the Cue-Producing Response as an Explanatory Device," *J. Exp. Psychol.*, 57 (1959), pp. 299–304.

de Saussure, F., Cours de Linguistique Générale. Paris: Payot, 1916.

Descartes, René, Meditations on First Philosophy, Vol. VI, in E. S. Haldane and G. R. T. Ross, eds., *The Philosophical Works of Descartes,* Vol. I. New York: Dover, 1955. Pp. 185–186, 190, 198–199, 345–346, *Meditation* V, p. 180, and *Meditation* III, p. 163.

Dick, A. O., "Relations between the Sensory Register and Short-term Storage in Tachistoscopic Recognition," *Journal of Experimental Psychology*. In press, 1970.

Dick, A. O., and D. J. K. Mewhort, "Order of Report and Processing in Tachistoscopic Recognition," *Perception and Psychophysics,* 2 (1967), pp. 573–576.

Dobzhansky, T., "Heredity, Environment, and Evolution," *Science,* III, No. 2877 (1950), pp. 161–166.

Dodwell, P. C., "Children's Understanding of Number and Related Concepts," *Canadian Journal of Psychology,* 15 (1961a), pp. 29–36.

Dodwell, P. C., "Coding and Learning in Shape Discrimination," *Psychol. Rev.*, 68 (1961b), pp. 373–382.

Donner, K. G., "The Visual Acuity of Some Passerine Birds," *Acta Zool. Fennica,* 66 (1951), pp. 1–40.

Doris, J., and L. Cooper, "Brightness Discrimination in Infancy," *J. Exp. Child Psychol.*, 3 (1966), pp. 31–99.

Doty, R. W., "Effects of Ablation of Visual Cortex in Neonatal and Adult Cats," *Abstracts, Comm. XIX Int. Physiol. Congr.*, 316 (1953).

Douglas, J. W. B., and J. M. Bloomfield, *Children under Five*. London: G. Allen, 1958.

Dick, A. O., The Complexity of the Perceptual Act: A Stage Analysis (in this volume), 1971.

Drake, Diana, *Annual Report*. Center for Cognitive Studies, Harvard University, 1964.

Drillien, C. M., "Growth and Development in a Group of Children of Very Low Birth Weight," *Arch. Dis. Childh.*, 33 (1958), pp. 10–18.

Drillien, C. M., *The Growth and Development of the Prematurely Born Infant*. Baltimore: Williams & Wilkins, 1964.

Ducasse, C. J., "Is Life after Death Possible?" in A. Flew, ed., *Body, Mind, and Death*. New York: Macmillan, 1964, pp. 223–224.

Ebert, E., and K. Simons, "The Brush Foundation Study of Child Growth and Development. I. Psychometric Tests," *Soc. Res. Child Develpm. Monogr.*, 8, no. 2 (1943).

Ehrenfeld, D. W., and A. L. Koch, "Visual Accommodation in the Green Turtle," *Science*, 155 (1967), pp. 827–828.

Ehrenfreund, D., "An Experimental Test of the Continuity Theory of Discrimination Learning with Pattern Vision," *J. Comp. Physiol. Psychol.*, 41 (1948), pp. 408–422.

Ehrenhardt, H., "Formenschen und Sehscharfebestimmungen bei Eidechsen," *Z. vergl. Physiol.*, 24 (1937), pp. 258–304.

Ehrlich, Annette, and W. H. Calvin, "Visual Discrimination in Galago and Owl Monkey," *Psychon. Sci.*, 9 (1967), pp. 509–510.

Eibl-Eibesfeldt, I., "Nahrungserwerb und Beuteschema der Erdkröte, (*Bufo bufo* 1.)," *Behav.*, 4 (1952), pp. 1–35.

Eibl-Eibsfeldt, I., and H. Hass, "Zum Project einer ethologisch orientierten Untersuchung menschlichen Verhaltens," *Mitteilunger aus der Max-Planck-Gesellschaft*, 6 (1966), pp. 383–396.

Eisenberg, R. B., Stimulus Significance as a Determinant of Newborn Responses to Sound. Paper presented at the Biennial Meeting of the Society for Research in Child Development, 1967.

Elkind, D., "Children's Discovery of the Conservation of Mass, Weight, and Volume: Piaget Replication Study II," *Journal of Genetic Psychology*, 98 (1961), pp. 219–227.

Elkind, O., "Cognition in Infancy and Early Childhood" (in this volume) in Yvonne Brackbill, ed., *Infancy and Early Childhood*. New York: Free Press, 1967.

Elkind, D., R. R. Koegler, and E. Go, "Studies in Perceptual Development: Part-Whole Perception," *Child Develpm.*, 35 (1964), pp. 81–90.

Emlen, J. T., "Determinants of Cliff Edge and Escape Responses in Herring Gull Chicks," *Behavior*, 22 (1963), pp. 1–15.

Engen, T., "Effect of Practice and Instruction on Olfactory Thresholds," *Percept. Mot. Skills*, 10 (1960), pp. 195–198.

Engen, T., and L. P. Lipsitt, "Decrement and Recovery of Responses to Olfactory Stimuli in the Human Neonate," *J. Comp. Physiol. Psychol.*, 59 (1965), pp. 312–316.

Engen, T., L. P. Lipsitt, and H. Kaye, "Olfactory Responses and Adaptation in the Human Neonate," *J. Comp. Physiol. Psychol.*, 56 (1963), pp. 73–77.

Epstein, W., "Attitudes of Judgment and the Size-Distance in Variance Hypothesis," *J. Exp. Psychol.*, 66 (1963), pp. 78–83.

Epstein, W., "Experimental Investigations of the Genesis of Visual Space Perception," *Psychol. Bull.*, 61 (1964), pp. 115–128.

Epstein, W., J. Park, and A. Casey, "The Current Status of the Size-Distance Hypotheses," *Psychol. Bull.*, 58 (1961), pp. 491–514.

Epstein, William, "Developmental Studies of Perception," (in this volume) in William Epstein, ed., *Varieties of Perceptual Learning*. New York: McGraw-Hill, 1967.

Eriksen, C. W., "Temporal Luminance Summation Effects in Backward and Forward Masking," *Perception and Psychophysics*, 1 (1966), pp. 87–92.

Erikson, E. H., *Childhood and Society*. New York: Norton, 1963, 2d ed.

Ervin, S. M., "Imitation and Structural Change in Children's Language," in

E. H. Lenneberg, ed., *New Directions in the Study of Language*. Cambridge, Mass.: M.I.T. Press, 1964.

Ervin, S., and W. Miller, 'Language Development," in H. W. Stevenson, ed., *Child Psychology, The Sixty-second Yearbook of the National Society for the Study of Education, Part I*. Chicago: University of Chicago Press, 1963.

Ervin-Tripp, Susan M., and D. I. Slobin, "Psycholinguistics," *Annu. Rev. Psychol.*, 17 (1966), pp. 435–474.

Escalona, Sibylle K., and H. H. Corman, The Validation of Piaget's Hypothesis concerning the Development of Sensorimotor Intelligence: Methodological Issues. Paper read at Soc. Res. Child Develpm., New York, March 1967.

Escalona, S. K., and A. Moriority, "Prediction of School Age Intelligence from Infant Tests," *Child Develpm.*, 32 (1961), pp. 597–605.

Estes, W. K., "The Statistical Approach to Learning Theory," in S. Koch, ed., *Psychology: A Study of a Science*, Vol. 2. New York: McGraw-Hill, 1959, pp. 380–491.

Eysenck, J. H., *Uses and Abuses of Psychology*. Baltimore: Penguin, 1953.

Fabricus, E., "Zur Ethologie junger Anatiden," *Acta Zool. Fennica*, 68 (1951), pp. 1–177.

Fantz, R. L., "A Method for Studying Early Visual Development," *Percept. Mot. Skills*, 6 (1956), pp. 3–15.

Fantz, R. L., "Pattern Vision in Young Infants," *Psychol. Rec.*, 8 (1958), pp. 43–47.

Fantz, R. L., "A Method for Studying Depth Perception in Infants under Six Months of Age," *Psychol. Rec.*, 11 (1961a), pp. 27–32.

Fantz, R. L., "The Origin of Form Perception," *Scientific American* (May 1961b), pp. 66–72.

Fantz, R. L., "Pattern Vision in Newborn Infants," *Science*, 140 (1963), pp. 296–297.

Fantz, R. L., "Visual Experience in Infants: Decreased Attention to Familiar Patterns Relative to Novel Ones," *Science*, 146 (1964), pp. 668–670.

Fantz, R. L., "Ontogeny of Perception," in A. M. Schrier, H. F. Harlow, and F. Stollnitz, eds., *Behavior of Nonhuman Primates*, Vol. 2. New York: Academic Press, Inc., 1965.

Fantz, R. L., "The Crucial Early Influence: Mother Love or Environmental Stimulation?" (abstract) *Amer. J. Orthopsychiat.*, 36, No. 2 (1966), pp. 330–331.

Fantz, R. L., and J. M. Ordy, "A Visual Acuity Test for Infants under Six Months of Age," *Psychol. Rec.*, 9 (1959), pp. 159–164.

Fantz, R. L., J. M. Ordy, and M. S. Udolf, "Maturation of Pattern Vision in Infants during the First Six Months," *J. Comp. Physiol. Psychol.*, 55 (1962), pp. 907–917.

Fargo, N. L., *et al.*, "The Development of Auditory Feedback Monitoring: IV. Delayed Auditory Feedback Studies on the Vocalizations of Children between 6 and 19 Months," in *Annual Report, 1968, Neurocommunications Laboratory*. Baltimore: Johns Hopkins University School of Medicine, pp. 95–118.

Farrell, B. A., "Experience," in V. C. Chappell, ed., *The Philosophy of the Mind.* Englewood Cliffs, N.J.: Prentice-Hall, 1962, pp. 23–27.

Feigenbaum, E., and H. A. Simon, "A Theory of the Serial Position Effect," *British Journal of Psychology,* 53 (1962), pp. 307–320.

Fields, P. E., "Studies in Concept Formation: I. The Development of the Concept of Triangularity by the White Rat," *Comp. Psychol. Monogr.,* 9, No. 42 (1932).

Fillmore, C. S., "The Case for Case, in E. Bach and R. T. Harms, eds., *Universals in Linguistic Theory.* New York: Holt, Rinehart and Winston, Inc., 1968, pp. 1–88.

Fischel, G., *Die Seele des Hundes.* Berlin: Paul Parey Verlag, 1950.

Fishman, R., and R. B. Tallarico, "Studies of Visual Depth Perception: I. Blinking as an Indicator Response in Prematurely Hatched Chicks," *Perc. Mot. Skills,* 12 (1961a), pp. 247–250.

Fishman, R., and R. B. Tallarico, "Studies of Visual Depth Perception: II. Avoidance Reaction as an Indicator Response in Chicks," *Perc. Mot. Skills,* 12 (1961b), pp. 251–257.

Fisichelli, R. M., *A Study of Prelinguistic Speech Development of Institutionalized Infants,* 1950. Quoted by McCarthy, 1954.

Fitzgerald, H., L. Lintz, Yvonne Brackbill, and Gail Adams, Conditioning of Pupillary Constriction and Dilation to Simple and Compound Stimuli. Paper presented at Colorado Psychol. Assoc., 1965.

Flavell, J. H., *The Developmental Psychology of Jean Piaget.* Princeton, N.J.: Van Nostrand, 1963.

Fodor, J. A., "How To Learn To Talk: Some Simple Ways," in F. Smith and G. A. Miller, eds., *The Genesis of Language.* Cambridge, Mass.: M.I.T. Press, 1966.

Folch-pi, J., "Composition of the Brain in Relation to Maturation," in H. Waelsch, ed., *Biochemistry of the Developing Nervous System: Proceedings of the First International Neurochemical Symposium.* New York: Academic Press, Inc., 1955.

Foley, J. P., "First Year Development of a Rhesus Monkey (Acca Mulay) Reared in Isolation," *J. Genet. Psychol.,* 44 (1934), pp. 390–413.

Fontana, V. J., D. Donovan, and R. J. Wong, "The 'Maltreatment Syndrom' in Children," *New England J. Medicine,* 269, No. 26 (1963), pp. 1389–1394.

Forgays, D. G., and Janet W. Forgays, "The Nature of the Effect of Free-Environment Experience in the Rat," *J. Comp. Physiol. Psychol.,* 45 (1952), pp. 322–328.

Forgus, R. H., "The Effect of Early Perceptual Learning on the Behavioral Organization of Adult Rats," *J. Comp. Physiol. Psychol.,* 47 (1954).

Forgus, R. H., *Perception: The Basic Process in Cognitive Development.* New York: McGraw-Hill, 1966.

Fortes, M., "Social and Psychological Aspects of Education in Taleland," Supplement to *Africa,* II, No. 4. Also Memorandum XVII of the *Int. Inst. African Languages & Cultures.* London: Oxford, 1938.

Fowler, W., "Cognitive Learning in Infancy and Early Childhood," *Psychol. Bull.,* 59, No. 2 (1962), pp. 116–152.

Fraser, C., U. Bellugi, and R. W. Brown, "Control of Grammar in Imitation, Comprehension, and Production," *J. Verb. Learn Verb. Behav.*, 2 (1963), pp. 121–135.

Frege, Gottlob, "On Sense and Reference," in P. Geach and M. Black, eds., *Translations from the Philosophical Writings of Gottlob Frege*. New York, 1952, pp. 56–57.

Freud, S., *The Ego and the Id*. London: Hogarth, 1935.

Freud, S., *The Problem of Anxiety*. New York: Norton, 1936.

Freud, S., "Character and Anal Eroticism," *Collected Papers,* Vol. II. London: Hogarth, 1953a.

Freud, S., "Screen Memories," *Collected Papers,* Vol. V. London: Hogarth, 1953b.

Frisch, K. von, *Bees: Their Vision, Chemical Senses, and Language*. Ithaca, N.Y.: Cornell University Press, 1950.

Frcebel, F., *The Education of Man*. New York: Appleton, 1896.

Fromme, A., "An Experimental Study of the Factors of Maturation and Practice in the Behavioral Development of the Embryo of the Frog, Rana Pipens," *Genet. Psychol. Monogr.*, 24 (1941), pp. 219–256.

Fry, D. B., "The Development of the Phonological System in the Normal and the Deaf Child," in F. Smith and G. A. Miller, *The Genesis of Language*. Cambridge, Mass.: M.I.T. Press, 1966, pp. 187–206.

Fuller, J. L., *Nature and Nurture: A Modern Synthesis*. New York: Doubleday, 1954.

Fuller, J., G. A. Easler, and E. M. Banks, "Formation of Conditioned Avoidance Responses in Young Puppies," *Amer. J. Physiol.*, 160 (1950), pp. 462–466.

Furth, H. G., "Concerning Piaget's View on Thinking and Symbol Formation," *Child Development,* 38 (1967), pp. 819–826.

Furth, H. G., *Piaget and Knowledge: Theoretical Foundations*. Englewood Cliffs, N.J.: Prentice-Hall, 1968.

Furth, H. G., "Piaget's Theory of Knowledge" (in this volume) *Psychol. Rev.*, 75 (1968).

Gagné, R. M., "Problem Solving," in H. W. Melton, ed., *Categories of Human Learning*. New York: Academic Press, 1964.

Gagné, R. M., *The Conditions of Learning*. New York: Holt, Rinehart and Winston, Inc., 1965.

Gagné, R. M., Contributions of Learning to Human Development" (in this volume), *Psychol. Rev.*, 75 (1968).

Gagné, R. M., and L. T. Brown, "Some Factors in the Programming of Conceptual Learning," *J. Exp. Psychol.*, 62 (1961), pp. 313, 321.

Gagné, R. M., *et al.*, "Factors in Acquiring Knowledge of a Mathematical Task," *Psychological Monographs,* 76, No. 526 (1962).

Galanter, E., "Contemporary Psychophysics," in T. M. Newcomb, ed., *New Directions in Psychology*. New York: Holt, Rinehart and Winston, Inc., 1962.

Garner, W. R., N. W. Hake, and C. W. Eriksen, "Operationalism and the Concept of Perception," *Psychol. Rev.*, 63 (1956), pp. 317–329.

Gesell, A., *Infancy and Human Growth*. New York: Macmillan, 1928.

Gesell, A., "Maturation and the Patterning of Behavior," in C. Murchinson, ed., *A Handbook of Child Psychology,* Worcester, Mass.: Clark University Press, 1933 (2d ed., rev.), pp. 209–235.

Gesell, A., *The Embryology of Behavior: The Beginnings of the Human Mind.* New York: Harper & Row, 1945.

Gesell, A., *Infant Development: The Embryology of Early Human Behavior.* New York: Harper & Row, 1952.

Gesell, A., and C. S. Amatruda, "Developmental Diagnosis: Normal and Abnormal Child Development," in *Clinical Methods and Pediatric Applications.* New York: Hoeber, 1947 (2d ed.).

Gesell, A., and L. B. Ames, "Tonic-Neck-Reflex and Symmetro-Tonic Behavior," *J. Pediatr.,* 36 (1950), pp. 165–176.

Gesell, A., and H. Thompson, "Learning and Growth in Identical Twin Infants," *Genetic Psychology Monographs,* 6 (1929), pp. 1–124.

Ghent, Lila, "Perception of Overlapping and Embedded Figures by Children of Different Ages," *Amer. J. Psychol.,* 69 (1956), pp. 575–587.

Ghent, Lila, "Recognition by Children of Realistic Figures Presented in Various Orientations," *Canad. J. Psychol.,* 14 (1960), pp. 249–256.

Ghent, Lila, "Form and Its Orientation: A Child's Eye View," *Amer. J. Psychol.,* 74 (1961), pp. 177–190.

Ghent, Lila, "Effect of Orientation on Recognition of Geometric Forms by Retarded Children," *Child Develpm.,* 35 (1964), pp. 1127–1136.

Ghent, Lila, and L. Bernstein, "Influence of the Orientation of Geometric Forms on Their Recognition by Children," *Percept. Mot. Skills,* 12 (1961), pp. 95–101.

Gibson, E., *Principles of Perceptual Learning and Development.* New York: Appleton, 1969.

Gibson, Eleanor J., "Perceptual Development," in H. W. Stevenson, ed., *Child Psychology. Yearb. Nat. Soc. Stud. Educ.,* 62 (1963a), Part 1.

Gibson, Eleanor J., "Perceptual Learning," *Annu. Rev. Psychol.,* 14 (1963b), pp. 29–56.

Gibson, E. J., *et al.,* "A Developmental Study of the Discrimination of Letter-like Forms," *J. Comp. Physiol. Psychol.,* 55 (1962), pp. 897–906.

Gibson, Eleanor J., and Vivian Odum, "Experimental Methods of Studying Perception in Children," in P. H. Mussen, ed., *Handbook of Research Methods in Child Development.* New York: Wiley, 1960, pp. 311–373.

Gibson, Eleanor J., and R. D. Wolk, "The 'Visual Cliff,'" *Scientific American,* 202 (1960), pp. 64–71.

Gibson, J. J., "Studying Perceptual Phenomena," in T. G. Andrews, ed., *Methods of Psychology.* New York: Wiley, 1948.

Gibson, J. J., *The Perception of the Visual World.* Boston: Houghton Mifflin, 1950.

Gibson, J. J., "The Visual Field and the Visual World: A Reply to Professor Boring," *Psychol. Rev.,* 59 (1952), pp. 149–151.

Gibson, J. J., "Visually Controlled Locomotion and Visual Orientation in Animals," *Brit. J. Psychol.,* 49 (1958), pp. 182–194.

Gibson, J. J., "Perception as a Function of Stimulation," in S. Koch, ed.,

Psychology: A Study of a Science, Vol. I. New York: McGraw-Hill, 1959, pp. 456–504.

Gibson, J. J., "The Survival Value of Sensory Perception," in *Biological Prototypes and Synthetic Systems,* Vol. I. New York: Plenum Press, 1962.

Gibson, J. J., *The Senses Considered as Perceptual Systems.* Boston: Houghton Mifflin, 1966, p. 272.

Gibson, J. J., and E. J. Gibson, "Perceptual Learning: Differentiation or Enrichment?" *Psychol. Rev.,* 62 (1955), pp. 32–41.

Gilinsky, Alberta S., "Perceived Size and Distance in Visual Space." *Psychol. Rev.,* 58 (1951), pp. 460–482.

Gilinsky, Alberta S., The Effect of Growth on the Perception of Visual Space. Paper read at East Psychol. Assoc., New York, April 1960.

Giltay, M., "Sur l'apparition et le développement de la notion du nombre chez l'enfant de deux à sept ans" ("Concerning the Appearance and Development of the Notion of Number in Children From 2–7 Years"), *J. Psychol. Norm. Pathol.,* 33 (1936), pp. 673–688.

Ginsberg, A., "A Reconstructive Analysis of the Concept 'Instinct,'" *J. Psychol.,* 33 (1952), pp. 235–277.

Glucksberg, S., *Symbolic Processes.* Dubuque, Iowa: William C. Brown Company, 1966.

Glucksberg, S., Thinking: A Phylogenetic Perspective (in this volume), 1971.

Gluecksohn-Waelsch, Salome, "Some Genetic Aspects of Development," in *Mammalian Fetus: Psysiological Aspects of Development.* Cold. Spr. Sympos. Quant. Biol., XIX (1954).

Goddard, H. H., *The Kallikak Family.* New York: Macmillan, 1912.

Goldfarb, W., "The Effects of Early Institutional Care on Adolescent Personality," *J. Exp. Educ.,* 12 (1943), pp. 106–129.

Goldfarb, W., "Effects of Psychological Deprivation in Infancy and Subsequent Stimulation," *Am. J. Psychiat.,* 102 (1945a), pp. 18–33.

Goldfarb, W., "Psychological Privation in Infancy and Subsequent Adjustment," *Amer. J. Orthopsychiat.,* 15 (1945b), pp. 247–255.

Goldstein, K., and M. Scheerer, "Abstract and Concrete Behavior," *Psychol. Monogr.,* 53, no. 2 (1941).

Gollin, Eugene, "Research Problems for Development," *Child Development,* 27 (1956), pp. 223–235.

Goodenough, F. L., "New Evidence of Environmental Influence on Intelligence," *Yearb. Nat. Soc. Stud. Educ.,* 39, no. 1 (1940), pp. 307–365.

Goodenough, F. L., *Mental Testing.* New York: Holt, Rinehart and Winston, Inc., 1949.

Gos, M., "Les reflexes conditionels chez l'embryon d'oiseau," *Bull. Soc. Sci.,* (Liège, 1935), 4me. Année (4–5), pp. 194–199, (6–7), pp. 246–250.

Goss, A. E., "Early Behaviorism and Verbal Mediating Responses," *Amer. Psychol.,* 16 (1961), pp. 285–298.

Goss, A. E., "Verbal Mediating Responses and Concept Formation," *Psychol. Rev.,* 68 (1961), pp. 248–274.

Goss, A. E., and Marie C. Moylan, "Conceptual Block-Sorting as a Function of Type and Degree of Mastery of Discriminative Verbal Responses," *J. Genet. Psychol.*, 93 (1958), pp. 191–198.

Gottlieb, G., and M. L. Simner, "Relationship Between Cardiac Rate and Non-nutritive Suckling in Human Infants," *J. Comp. Physiol. Psychol.*, 61 (1966), pp. 128–131.

Gouin Decarie, Theresa, *Intelligence and Affectivity in Early Childhood*. New York: International Universities, 1965.

Grabowski, U., "Experimentelle Untersuchungen uber das angebliche Lernvermögen mogen von paramaecium," *Z. Tierpsychol.*, 2 (1939), pp. 265–281.

Graham, C. H., "Behavior, Perception and the Psychophysical Methods," *Psychol. Rev.*, 57 (1950), pp. 108–120.

Grant, D. A., O. R. Jones, and B. Tallantis, "The Relative Difficulty of Number, Form, and Color Concepts of a Weigl-type Problem," *J. Exp. Psychol.*, 39 (1949), pp. 552–557.

Grastyan, E., "The Significance of the Early Manifestation of Conditioning in the Mechanism of Learning," in A. Fessard, *et al.*, eds., *Brain Mechanisms and Learning*. Oxford: Blackwell, 1961.

Green, E. J., "Concept Formation: A Problem in Human Operant Conditioning," *J. Exp. Psychol.*, 49 (1955), pp. 175–180.

Greenberg, J. H., ed., *Universals of Language*. Cambridge, Mass.: M.I.T. Press, 1963.

Greenspoon, J., "The Reinforcing Effect of Two Spoken Sounds on the Frequency of Two Responses," *Amer. J. Psychol.*, 68 (1955), pp. 409–416.

Griffen, D. R., "More about Bat 'Radar,'" *Scient. Amer.*, 199 (1958), pp. 40–44.

Grohmann, J., "Modifikation order Funktionsregung? Ein Beitrag zur Klärung der wechselseitigen Beziehungen qwischen Instinkthandlung und Erfahrung," *Z. Tierpsychol.*, 2 (1938), pp. 132–144.

Gruber, H. E., "Relation of Perceived Size to Perceived Distance," *Amer. J. Psychol.*, 67 (1958), pp. 411–426.

Gullickson, G. R., and D. H. Crowell, "Neonatal Habituation to Electrotactual Stimulation," *J. Exp. Child. Psychol.*, 1 (1964), pp. 388–396.

Haber, R. N., Perceptual Processes and General Cognitive Activity. Paper presented at the Conference on Learning Processes and Thought, University of Pittsburgh, 1966.

Haber, R. N., "Repetition as a Determinant of Perceptual Recognition Processes," in W. Wathen-Dunn, ed., *Symposium on Models for the Perception of Speech and Visual Form*. Cambridge, Mass.: M.I.T. Press, 1967.

Haber, R. N., "Information-Processing Analysis of Information Processing, an Introduction," in R. N. Haber, ed., *Information Processing-Approaches to Visual Perception*. New York: Holt, Rinehart and Winston, Inc., 1969.

Haith, M. M., "The Response of the Human Newborn to Visual Movement," *J. Exp. Child Psychol.*, 3 (1966), pp. 235–243.

Hall, G. S., Adolescence. New York: Appleton, 1904.

Hall, G. S., *Aspects of Child Life and Education*. New York: Appleton, 1921.

Hall, G. S., "The Genetics of Behavior," in S. S. Stevens, ed., *Handbook of Experimental Psychology*, pp. 304–329. New York: Wiley, 1951.

Hall, G. S., and P. H. Whiteman, "The Effects of Infantile Stimulation upon the Emotional Stability in the Mouse," *J. Comp. Physiol. Psychol.*, 44 (1951), pp. 61–66.

Hallowell, A. I., *Culture and Experience*. Philadelphia: University of Pennsylvania Press, 1955.

Hamilton, G. U. N., "A Study of Trial and Error Behavior in Mammals," *J. Animal Behav.* (1911), pp. 33–36.

Hamilton, G. U. N., "A Study of Perseverance Reactions in Primates and Rodents," *Behav. Monogr.*, 3, No. 2 (1916).

Hanson, N. R., *Patterns of Discovery: An Inquiry into the Conceptual Foundations of Science*. New York: Cambridge University Press, 1958.

Harcum, E. R., and S. M. Friedman, "Reversal Reading by Israeli Observers of Visual Patterns without Intrinsic Directionality," *Canadian Journal of Psychology*, 17 (1963), pp. 361–369.

Harker, Janet E., "The Diurnal Rhythm of Activity of Mayfly Nymphs," *J. Exp. Biol.*, 30 (1953), pp. 525–533.

Harlow, H., "The Formation of Learning Sets," *Psychol. Rev.*, 56 (1949), pp. 51–65.

Harlow, H. F., "The Evolution of Learning," in Anne Roe and G. G. Simpson, eds., *Behavior and Evolution*. New Haven, Conn.: Yale University Press, 1958.

Harlow, H. F., "Learning Set and Error Factor Theory," in S. Koch, ed., *Psychology, A Study of a Science*, Vol. I. New York: McGraw-Hill, 1959.

Harlow, H. F., K. Akert, and K. A. Schiltz, "The Effects of Bilateral Prefrontal Lesions on Learned Behavior of Neonatal Infant, and Preadolescent Monkeys," in T. M. Warren and K. Akert, eds., *The Frontal Granular Cortex and Behavior*. New York: McGraw-Hill, 1964.

Harlow, H. F., and M. K. Harlow, "Learning To Think," *Scientific American*, 181 (1949), pp. 36–39.

Harnly, M. H., "Flight Capacity in Relation to Phenotypic and Genotypic Variations in the Wings of Drosophila Melanogaster," *J. Exper. Zool.*, 88 (1941), pp. 263–274.

Harper, P. A., L. K. Fischer, and R. U. Rider, "Neurological and Intellectual Status of Prematures at Three to Five Years of Age," *J. Pediat.*, 55 (1959), pp. 679–690.

Harper, R. S., and E. G. Boring, "Cues," *Amer. J. Psychol.*, 41 (1948), pp. 119–123.

Harris, Z. S., *Methods in Structural Linguistics*. Chicago: University of Chicago Press, 1951.

Harter, G. L., "Overt Trial and Error in Problem-Solving of Preschool Children," *J. Genet. Psychol.*, 38 (1930), pp. 361–372.

Hartline, H. K., "Visual Receptors and Retinal Interaction," *Science*, 164 (1969), pp. 270–277.

Harway, N. I., "Judgement of Distance in Children and Adults," *J. Exp. Psychol.*, 65 (1963), pp. 385–390.

Hasler, A. D., and W. J. Wisby, "Discrimination of Stream Odors by Fishes and Its Relation to Parent Stream Behavior," *Amer. Nat.*, 85 (1951), pp. 223–238.

Hayakawa, S. I., *Language in Thought and Action.* New York: Harcourt, 1941.
Hayes, W. N., and E. I. Saiff, "Visual Alarm Reactions in Turtles," *Animal Beh.*, 15 (1967), pp. 102–106.
Haynes, H., B. L. White, and Richard Held, "Visual Accommodation in Human Infants," *Science*, 148 (1963), pp. 528–530.
Hazlitt, U., "Children's Thinking," *Brit. J. Psychol.*, 20 (1930), pp. 354–360.
Hebb, D. O., *The Organization of Behavior.* New York: Wiley, 1949.
Hebb, D. O., "Heredity and Environment in Mammalian Behavior," *Brit. J. Anim. Behav.*, 1 (1953), pp. 43–47.
Hebb, D. O., "Drives and the Central Nervous System," *Psychol. Rev.*, 62 (1955), pp. 243–254.
Hebb, D. O., "Concerning Imagery," *Psychol. Rev.*, 75 (1968), pp. 466–477.
Hécaen, H., and J. De Ajuriaguerra, *Les gauchers, prévalence manuelle et dominance cérébralle.* Paris: Presses Universitaires de France, 1963.
Heinstein, M. I., "Behavioral Correlates of Breast-Bottle Regimes under Varying Parent-Infant Relationships," *Soc. Res. Child Develpm. Monogr.*, 28, No. 4 (1963).
Held, R., "Motor-Sensory Feedback and the Geometry of Visual Space," *Science*, 141 (1963), pp. 722–723.
Held, R., "Plasticity in Sensory-Motor Systems," *Scientific Amer.*, 213, No. 5 (1965), pp. 84–94.
Held, R., and S. J. Freedman, "Plasticity in Human Sensorimotor Control," *Science*, 142 (1963), pp. 455–462.
Held, R., and A. Hein, "Movement-Produced Stimulation in the Development of Visually-Guided Behavior," *J. Comp. Physiol. Psychol.*, 56 (1963), pp. 872–876.
Helson, J., "Perception," in H. Helson, ed., *Theoretical Foundations of Psychology.* New York: Van Nostrand, 1951.
Hempelmann, F., *Tierpsychologie vom standpunkte des biologen.* Leipzig: Akad. Verlagsges., 1926.
Herman, D. T., R. H. Lawless, and R. W. Marshall, "Variables in the Effect of Language on the Reproduction of Visually Perceived Forms," *Percept. Mot. Skills*, 7, Monogr. Suppl. 2 (1957), pp. 171–186.
Hernández-Péon, R. H. Scherrer, and M. Jouvet, "Modification of Electrical Activity in Cochlear Nucleus during 'Attention' in Unanesthetized Cats," *Science*, 123 (1956), pp. 331–332.
Herrick, C. J., *Neurological Foundations of Animal Behavior.* New York: Holt, Rinehart and Winston, Inc., 1924.
Herrnkind, W., "Queuing Behavior of Spiny Lobsters," *Science*, 164 (1969), pp. 1425–1427.
Hershenson, M., "Visual Discrimination in the Human Newborn," *J. Comp. Physiol. Psychol.*, 58 (1964), pp. 270–276.
Hershenson, M., Form Perception in the Human Newborn. Paper read at the Second Annual Symposium, Center for Visual Science, University of Rochester, June 1965.
Hershenson, M., H. Munsinger, and W. Kessen, "Preference for Shapes of Intermediate Variability in the Newborn Human," *Science*, 147 (1965), pp. 630–631.

Herskovits, M., *Man and His Works*. New York: Knopf, 1948.

Hess, E. H., "Development of the Chick's Responses to Light and Shade Cues of Depth," *J. Comp. Physiol. Psychol.*, 43 (1950), pp. 112–122.

Hess, E. H., "Space Perception in the Chick," *Scient. Amer.*, 195 (1956), pp. 71–80.

Hess, E. H., "Ethology: An Approach towards the Complete Analysis of Behavior," in R. W. Brown, *et al.*, *New Directions in Psychology*. New York: Holt, Rinehart and Winston, Inc., 1962.

Hess, W., "Reactions to Light in the Earthworm, *Lumbricus Terrestris, L.*, *Morphol. Physiol.*, 39 (1924), pp. 515–542.

Hilgard, E. R., *Theories of Learning*. New York: Appleton, 1956.

Hilgard, E. R., and D. G. Marquis, *Conditioning and Learning*. New York: Appleton, 1940.

Hill, A. A., *Introduction to Linguistic Structures: From Sound to Sentence in English*. New York: Harcourt, 1958.

Hill, W. F., *Learning*. San Francisco: Chandler, 1963.

Hirsch, J., R. H. Lindley, and E. C. Tolman, "An Experimental Test of an Alleged Sign-Stimulus, *J. Comp. Physiol. Psychol.*, 48 (1955), pp. 278–280.

Hochberg, J. E., "Nativism and Empiricism in Perception," in L. Postman, ed., *Psychology in the Making*. New York: Knopf, 1962.

Hockett, C., *A Course in Modern Linguistics*. New York: Macmillan, 1958.

Hockett, C., "The Origin of Speech," *Sci. Amer.*, 203 (1960), pp. 88–96.

Hockett, C., "The Problem of Universals in Language," in J. H. Greenberg, ed., *Universals of Language*. Cambridge, Mass.: M.I.T. Press, 1963, pp. 1–22.

Holland, J. G., "The Influence of Previous Experience and Residual Effects of Deprivation on Hoarding in the Rat," *J. Comp. Physiol. Psychol.*, 47 (1954), pp. 244–247.

Holt, E. B., *Animal Drive and the Learning Process*. London: Williams & Norgate, 1931.

Holway, A. H., and E. G. Boring, "Determinants of Apparent Visual Size with Distance Variant," *Amer. J. Psychol.*, 54 (1941), pp. 21–37.

Honigmann, H., "The Visual Perception of Movements by Toads," *Proc. Roy. Zool. Soc.*, Series B, 132 (1945), pp. 291–307.

Honzik, M. P., J. W. MacFarlane, and L. Allen, "The Stability of Mental Test Performance between Two and Eighteen Years," *J. Exp. Educ.*, 17 (1948), pp. 309–324.

Hooker, D., Evidence of Prenatal Function of the Central Nervous System in Man. James Arthur Lecture, New York, 1958. Quoted by Peiper, 1961.

Horowitz, Francis D., "Theories of Arousal and Retardation Potential," *Ment. Ret.*, 3 (1965), pp. 20–23.

Horowitz, Francis D., "Infant Learning and Development" (in this volume), *Merrill-Palmer Quarterly*, 14 (1968), pp. 101–120.

Hospers, John, "Knowledge: Concepts" (in this volume) in *An Introduction to Philosophical Analysis*, Englewood Cliffs, N.J., Prentice-Hall, Inc., 1967.

Houston, S. H., "Inquiry into the Structure of Mentation Processes," *Psych. Reports*, 21 (1967), pp. 649–659.

Houston, S. H., "A Diachronic Examination of Linguistic Universals," *ASHA Journal,* 10 (1968), pp. 247–249.

Houston, S. H., Studies of Language (in this volume), 1971.

Hovland, C. I., "The Generalization of Conditioned Responses. I. The Sensory Generalization of Conditioned Responses with Varying Frequencies of Tone," *J. Gen. Psychol.,* 17 (1937), pp. 125–148.

Hovland, C. I., "A 'Communication Analysis' of Concept Learning," *Psychol. Rev.,* 59 (1952), pp. 461–472.

Hovland, C. I., "Computer Simulation of Thinking," *Amer. Psychologist,* 15 (1960), pp. 687–693.

Howe, M. J. A., "Intra-list Differences in Short-term Memory," *Quarterly Journal of Experimental Psychology,* 17 (1965), pp. 338–342.

Hubel, D. H., "Integrative Processes in Central Visual Pathways of the Cat," *Journal of the Optical Society of America,* 53 (1963), pp. 58–66.

Hubel, D. H., and T. N. Wiesel, "Receptive Fields of Single Neurones in the Cat's Striate Cortex," *Journal of Physiology,* 148 (1959), pp. 574–591.

Hubel, D. H., and T. N. Wiesel, "Receptive Fields, Binocular Interaction and Functional Architecture in the Cat's Visual Cortex," *Journal of Physiology,* 160 (1962), pp. 106–154.

Hubel, D. H., and T. N. Wiesel, "Receptive Fields and Functional Architecture of Monkey Striate Cortex," *Journal of Physiology,* 195 (1968), pp. 215–243.

Hull, Clark L., *A Behavior System.* New Haven, Conn.: Yale University Press, 1952.

Hull, C. L., "Quantitative Aspects of the Evolution of Concepts," *Psychol. Monogr.,* 28 (1920).

Hull, C. L., "Knowledge and Purpose as Habit Mechanisms," *Psychol. Rev.,* 37 (1930), pp. 511–525.

Hull, C. L., *Principles of Behavior.* New York: Appleton, 1943.

Hume, David, *An Enquiry concerning Human Understanding,* 1751, Section 2, Paragraphs 4 and 5. Many editions.

Hume, David, "Of Abstract Ideas," in Hume, *A Treatise of Human Nature,* L. A. Selby-Bigge, ed., Book I, Oxford, 1888, pp. 1, 17–25, 253, 261.

Hunt, J. McV., "The Effects of Infant Feeding-Frustration upon Adult Hoarding in the Albino Rat," *J. Abnor. and Soc. Psychol.,* 36 (1941), pp. 338–360.

Hunter, W. S., "The Delayed Reaction in Animals and Children," *Behavior Monographs,* 2 (1913).

Hunter, W. S., "The Delayed Reaction in a Child," *Psychol. Review,* 24 (1917), pp. 74–87.

Hunter, W. S., "Summary Comments on the Heredity Environment Symposium," *Psychol. Rev.,* 54 (1947), pp. 348–352.

Hymovitch, B., "The Effects of Experimental Variations on Problem Solving in the Rat," *J. Comp. Physiol. Psychol.,* 45 (1952), pp. 312–320.

Illingworth, R. S., "The Predictive Value of Developmental Tests in the First Year, with Special Reference to the Diagnosis of Mental Subnormality," *J. Child Psychol. Psychiat.,* 2 (1961), pp. 210–215.

Ingram, D., Transitivity in Child Language. Unpublished manuscript, 1969.

Inhelder, Börbel, and J. Piaget, *The Growth of Logical Thinking from Childhood to Adolescence*. New York: Basic Books, 1958.

Inhelder, Börbel, and J. Piaget, *The Early Growth of Logic in the Child*. New York: Harper & Row, 1964.

Irwin, O. C., "Infant Speech," *J. Speech and Hearing Disorders*, 13 (1948), pp. 224–225 and 320–326.

Irwin, O. C., "Phonetical Description of Speech Development in Childhood," in L. Kaiser, ed., *Manual of Phonetics*. Amsterdam: North Holland Publishing Co., 1957, pp. 403–425.

Jacobsen, C. F., M. M. Jacobsen, and J. G. Yoshioka, "Development of an Infant Chimpanzee during Her First Year," *Comp. Psychol. Monogr.*, 9, No. 41 (1932), pp. 1–94.

Jakobson, R., "Two Aspects of Language and Two Types of Aphasic Disturbances, Part II," in R. Jakobson and M. Halle, *Fundamentals of Language*. The Hague: Mouton, 1956.

Jakobson, R., *Kindersprache, Aphasie und Allgemeine Lautgesetze* (Child Language, Aphasia, and Phonological Universals), 1941. Translated by A. R. Keiler. The Hague: Mouton, 1968.

Jakobson, R., and M. Halle, *Fundamentals of Language*. The Hague: Mouton, 1956.

James, W., *Principles of Psychology*. London: Macmillan, 1890.

James, W. T., and D. J. Cannon, "Conditioned Avoiding Responses in Puppies," *Amer. J. Physiol.*, 168 (1952), pp. 251–253.

Janos, O., "Development of Higher Nervous Activity in Premature Infants," *Pavlov J. High Nerv. Act.*, 9 (1959), pp. 760–767.

Jenkins, J. J., W. D. Mink, and W. A. Russell, "Associative Clustering as a Function of Verbal Associative Strength," *Psychol. Rep.*, 4 (1958), pp. 127–136.

Jenkins, J. J., and D. S. Palermo, "Mediation Processes and the Acquisition of Linguistic Structure," in U. Bellugi and R. Brown, eds., *The Acquisition of Language. Monogr. Socy. Res. Child Dev.*, 29, 1 (1964), pp. 141–149.

Jerison, H. J., "Brain to Body Ratios and the Evolution of Intelligence," *Science*, 121 (1955), pp. 447–449.

Jones, H., "The Environment and Mental Development," in L. Carmichael, ed., *Manual of Child Psychology*. New York: Wiley, 1954.

Jones, H. E., "The Retention of Conditioned Emotional Reactions in Infancy," *J. Genet. Psychol.*, 37 (1930), pp. 485–498.

Jones, H. E., "The Conditioning of Overt Emotional Responses," *J. Educ. Psychol.*, 22 (1931), pp. 127–130.

Kagan, J., Personal Communication, 1966.

Kagan, J., and M. Freeman, "Relation of Childhood Intelligence, Maternal Behaviors and Social Class to Behavior during Adolescence," *Child Develpm.*, 34 (1963), pp. 899–911.

Kagan, J., and Barbara Henker, Assimilation and Habituation. Paper read at Amer. Psychol. Assoc., Chicago, September 1965.

Kahn, M. W., "Infantile Experience and Mature Aggressive Behavior of Mice: Some Maternal Influences," *J. Genet. Psychol.*, 84 (1954), pp. 65–75.

Kalmus, H., "The Discrimination by the Nose of the Dog of Individual Human Odours and in Particular of the Odours of Twins," in S. C. Ratner and M. R. Denny, *Comparative Psychology: Research in Animal Behavior.* Homewood, Ill.: The Dorsey Press, 1964, pp. 116–128.

Kantrow, R. W., "An Investigation of Conditioned Feeding Responses and Concomitant Adaptive Behavior in Young Infants," *Univer. Iowa Stud. Child Welfare,* 13 (1937), p. 64.

Kaplan, E. L., "The Role of Intonation in the Acquisition of Language." Unpublished Ph.D. Thesis, Cornell University, 1969.

Kaplan, Eleanor, and George, The Prelinguistic Child (in this volume), 1971.

Kappers, C. U. A., G. C. Huber, and E. C. Crosby, *The Comparative Anatomy of the Nervous System of Vertebrates, Including Man.* New York: Macmillan, 1946.

Kasatkin, N. I., "The Development of Visual and Acoustic Conditioned Reflexes and Their Differentiation in Infants," *Sovelsk. Pediat.* (1935), pp. 127–237.

Kasatkin, N. I., N. S. Mirzoints, and A. Khokhitva, "Ob orientirovochnykh useovnykh refleksakh u detie pervogo goda Zhiznie" (Oriented, Conditioned Reflexes in Infants during the First Year of Life), *Zh. vyssh. ner Deiat.,* 3 (1953), pp. 192–202. Reprinted in *The Central Nervous System and Behavior,* translations from the Russian medical literature collected for participants of the J. J. Macy conference on the central nervous system and behavior, Princeton, February 1960, pp. 343–358.

Kasatkin, N. I., and A. M. Levikova, "On the Development of Early Conditioned Reflexes and Differentiations of Auditory Stimuli in Infants," *J. exp. Psychol.,* 18 (1935), pp. 1–9.

Katz, J. J., *Philosophy of Language.* New York: Harper & Row, 1966.

Katz, J. J., and J. A. Fodor, "The Structure of a Semantic Theory," *Language,* 39 (1963), pp. 170–210.

Kaye, H., "Skin Conductance in the Human Neonate," *Child Develpm.,* 35 (1964), pp. 1297–1305.

Kaye, H., "The Conditioned Bobkin Reflex in Human Newborns," *Psychon. Sci.,* 2 (1965), pp. 287–288.

Kaye, H., "The Effects of Feeding and Tonal Stimulation on Nonnutritive Sucking in the Human Newborn," *J. Exp. Child Psychol.,* 3 (1966), pp. 131–145.

Keen, Rachel E., "Effects of Auditory Stimuli on Sucking Behavior in the Human Neonate," *J. Exp. Child Psychol.,* 1 (1964), pp. 348–354.

Keen, Rachel E., H. H. Chase, and F. K. Graham, "Twenty-four Hour Retention by Neonates of an Habituated Heart Rate Response," *Psychon. Sci.,* 2 (1965), pp. 265–266.

Kendler, H. H., "'What Is Learned'—A Theoretical Blind Alley," *Psychol. Rev.,* 59 (1952), pp. 269–277.

Kendler, H. H., "Learning," *Ann. Rev. Psychol.,* 10 (1959a), pp. 43–88.

Kendler, H. H., "Teaching Machines and Psychological Theory," in E. Galanter, ed., *Automatic Teaching: The State of the Art.* New York: Wiley, 1959b, pp. 177–186.

Kendler, H. H., "Problems in Problem-Solving Research," in *Current Trends in Psychological Theory.* Pittsburgh, Pa.: University of Pittsburgh Press, 1961, pp. 180–207.

Kendler, Howard, "The Concept of the Concept" (in this volume) from Melton, *Categories of Human Learning.* New York: Academic Press, 1964.

Kendler, H. H., S. Glucksburg, and R. Keston, "Perception and Mediation in Concept Learning," *J. Exp. Psychol.,* 61 (1961), pp. 186–191.

Kendler, H. H., and T. S. Kendler, "Inferential Behavior in Young Children," *J. Exp. Psychol.,* 51 (1956), pp. 311–314.

Kendler, H. H., and T. S. Kendler, "Effect of Verbalization on Reversal Shifts in Children," *Science,* 134 (1961), pp. 1619–1620.

Kendler, H. H., and T. S. Kendler, "Vertical and Horizontal Processes in Problem Solving," *Psychol. Rev.,* 69 (1962), pp. 1–16.

Kendler, H. H., and T. S. Kendler, "Mediation and Conceptual Behavior," in K. W. Spence and J. T. Spence, eds., *The Psychology of Learning and Motivation,* Vol. 2. New York: Academic Press, Inc., 1968.

Kendler, T. S., "Concept Formation," in P. R. Farnsworth, O. McNemar, and Q. McNemar, eds., *Annual Review of Psychology,* 12 (1961), pp. 447–472.

Kendler, T. S., and H. H. Kendler, "Reversal and Non-reversal Shifts in Kindergarten Children," *J. Exp. Psychol.,* 58 (1959), pp. 56–60.

Kendler, T. S., H. H. Kendler, and D. Wells, "Reversal and Non-reversal Shifts in Nursery School Children," *J. Comp. Physiol. Psychol.* 53 (1960), pp. 83–88.

Kenny, Anthony, *Descartes, A Study of His Philosophy.* New York, 1968, pp. 148–149.

Kessen, W., "Research Design in the Study of Developmental Problems," in P. H. Mussen, ed., *Handbook of Research Methods in Child Development.* New York: Wiley, 1960, pp. 35–70.

Kessen, W., "Research in the Psychological Development of Infants: An Overview," *Merrill-Palmer Quart.,* 9 (1963), pp. 83–94.

Kessen, W., *The Child.* New York: Wiley, 1965.

Kessen, W., "Questions for a Theory of Cognitive Development" (in this volume), in H. W. Stevenson, ed., *Concepts of Development. Monogr. Soc. Res. Child Develpm.,* 31, No. 3 (1966), pp. 55–70.

Kessen, W., "Sucking and Looking: Two Organized Congenital Patterns of Behavior in the Human Newborn," in H. W. Stevenson, E. H. Hess and H. L. Rheingold, eds., *Early Behavior: Comparative and Developmental Approaches.* New York: Wiley, 1967.

Kessen, W., Louise S. Hendry, and Ann-Marie Leutzendorff, "Measurement of Movement in the Human Newborn: A New Technique," *Child Develpm.,* 32 (1961), pp. 95–105.

Kessen, W. E., and Ann-Marie Leutzendorff, "The Effect of Non-nutritive Sucking on Movement in the Human Newborn," *J. Comp. Physiol. Psychol.,* 56 (1963), pp. 69–72.

Kessen, W., E. Jane Williams, and Joanna P. Williams, "Selection and Test of Response Measures in the Study of the Human Newborn," *Child Develpm.,* 32 (1961), pp. 7–24.

Kinder, E. F., "A Study of the Nest Building Activity of the Albino Rat., *J. Exper. Zool.*, 47 (1927), pp. 117–161.

Kirkpatrick, E. A., *Fundamentals of Child Study*. London: Macmillan, 1910.

Kistyokovskaya, M. J., "Stimuli Evoking Positive Emotions in Infants," *Voprosy, Psikhologü*, 2 (1965), pp. 129–140.

Knobloch, H., and B. Pasamanick, "Further Observations on the Development of Negro Children," *J. Genet. Psychol.*, 83 (1953), pp. 137–157.

Knobloch, H., and B. Pasamanick, "Predicting Intellectual Potential in Infancy," *Amer. J. Dis. Childh.*, 107, No. 1 (1963), pp. 43–51.

Knobloch, H., *et al.*, "Neuropsychiatric Sequelae of Prematurity, a Longitudinal Study," *J. Amer. Med. Ass.*, 161 (1956), pp. 581–585.

Koch S., C. L. Hull, in W. K. Estes, *et al., Modern Learning Theory*. New York: Appleton, 1954, pp. 1–176.

Koehler, O., "Wolfskinder," Affen im Hous und vergleichenda Verhaltensforschung, *Folia Phoniatrica*, 4 (1952), pp. 29–53.

Koffka, K., *Principles of Gestalt Psychology*, New York: Harcourt, 1935.

Koffka, K., *The Growth of the Mind*. Tatawa, N.J.: Littlefield, Adams, & Company, 1959.

Kohler, W., Mentality of Apes. London: Routledge, 1925.

Köhler, W., and H. Wallach, "Figural After-Effects: On Investigation of Visual Processes," *Proc. Amer. Philos. Soc.*, 88 (1944), pp. 269–357.

Kolers, P. A., "Intensity and Contour Effects in Visual Masking," *Vision Research*, 2 (1962), pp. 277–294.

Kol'tsova, M. M., "The Role of Temporary Connexions of Associative Type in the Development of Systematized Activity," *Pavlov. J. High. New. Act.*, 11 (1961), pp. 56–60.

Korner, Anneliese F., and Rose Grobstien, "Visual Alertness as Related to Soothing in Neonates: Implications for Maternal Stimulation and Early Deprivation," *Child Develpm.*, 37 (1966), pp. 867–876.

Krames, L., and W. J. Carr, "The Effect of Previous Experience Upon the Response to Depth in Domestic Chickens," *Psychon. Sci.*, 10 (1968), pp. 249–250.

Krech, D., "Dynamic Systems as Open Neurological Systems," *Psychol. Rev.*, 57 (1950), pp. 345–361.

Krech, D., "Dynamic Systems, Psychological Fields and Hypothetical Constructs," *Psychol. Rev.*, 57 (1950a), pp. 283–290.

Krech, D., and G. S. Klein, eds., *Theoretical Models and Personality Theory*. Durham, N.C.: Duke University Press, 1952.

Krechevsky, I., "An Experimental Investigation of the Principle of Proximity in the Visual Perception of the Rat," *J. Exp. Psychol.*, 22 (1938), pp. 497–453.

Kroeber, A. L., "The Speech of a Zuñi Child," *Amer. Anthrop.*, 18 (1916), pp. 529–534.

Kroeber, A. L., and C. Kluckhohn, "Culture: A Critical Review of Concepts and Definitions," *Papers of the Peabody Museum of American Archaeology and Ethnology*, Vol. XLVII. Cambridge, Mass.: The Peabody Museum, 1952.

Kugelmass, I. N., L. E. Poull, and E. L. Samuel, "Nutritional Improvement of Child Mentality," *Amer. J. Med. Sci.*, 208 (1944), pp. 631–633.

Kuhlman, Clementira, Visual Imagery in Children. Unpublished doctoral dissertation, Harvard University, 1960.

Kummer, B., "Untersuchungen über die Entwicklung der Schädelform des Menschen und einiger Anthropoiden," *Abhandlungen zur exakten Biologie*. Borntraeger, Berlin: L. V. Bertalanffy, 1953 (fasc. 3d ed.).

Kuo, Z. Y., "A Psychology without Heredity," *Psychol. Rev.*, 31 (1924), pp. 427–448.

Kuo, Z. Y., "The Genesis of the Cat's Response to the Rat," *J. Comp. Psychol.*, 11 (1930), pp. 1–30.

Kuo, Z. Y., "Ontogeny of Embryonic Behavior in Apes. II. The Mechanical Factors in the Various Stages Leading to Hatching," *J. Exper. Zool.*, 13 (1932a), pp. 395–430.

Kuo, Z. Y., "Ontogeny of Embryonic Behavior in Apes. III. The Structure and Environmental Factors in Embryonic Behavior," *J. Comp. Psychol.*, 13 (1932b), pp. 245–272.

Kuo, Z. Y., "Ontogeny of Embryonic Behavior in Apes. IV. The Influence of Embryonic Movements upon Behavior after Hatching," *J. Comp. Psychol.*, 14 (1932c), pp. 109–112.

Kuo, Z. Y., "Ontogeny of Embryonic Behavior in Apes. II. The Mechanical Factors in the Various Stages Leading to Hatching," *J. Exper. Zool.*, 61 (1943), pp. 395–430.

Lachman, R., "The Model in Theory Construction," *Psychol. Rev.*, 67 (1960), pp. 113–130.

Lado, R., *Language Teaching: A Scientific Approach*. New York: McGraw-Hill, 1964.

Lambercier, M., "La constance des grandeurs en comparisons sériales," *Arch. Psychol., Genève*, 31 (1946), pp. 79–282.

Lancelot, C., and A. Arnauld, *Grammaire generale et raisonee*. Paris, 1660.

Landau, W. M., R. Goldstein, and L. K. Kleffner, "Congenital Aphasia: A Clinicopathological Study," *Neurology*, 10 (1960), pp. 915–921.

Landauer, T. K., "Rate of Implicit Speech," *Perceptual and Motor Skills*, 15 (1962), pp. 646–647.

Landreth, C., *The Psychology of Early Childhood*. New York: Knopf, 1962.

Lang, A., "Personal Communication," 1967.

Lashley, K. S., "The Mechanism of Vision. I. A Method for Rapid Analysis of Pattern-Vision in the Rat," *J. Genet. Psychol.*, 37 (1930), pp. 353–460.

Lashley, K. S., "The Mechanism of Vision: XV. Preliminary Study of the Rat's Capacity for Detail Vision," *J. Gen. Psychol.*, 18 (1938), pp. 123–193.

Lashley, K. S., "The Problem of Serial Order in Behavior," in L. A. Jeffress, ed., *Cerebral Mechanisms in Behavior: The Hixon Symposium*. New York: Wiley, 1951.

Latif, I., The Physiological Basis of Linguistic Development and of the Ontogeny of Meaning. Part I. *Psychol. Rev.*, 41 (1934), pp. 55–85.

Lee, I. J., *Language Habits in Human Affairs*. New York: Harper & Row, 1941.

Lehrman, D. S., "A Critique of Lorenz's 'Objectivistic' Theory of Animal Behavior," *Quart. Rev. Biol.*, 28 (1953), pp. 337–363.

Lehrman, D. S., "The Physiological Basis of Parental Feeding Behavior in the Ring Dove, *Streptopelia Risoria*," *Behav.*, 7 (1955), pp. 241–286.

Lehrman, D. S., "Induction of Broodiness by Participation in Courting and Nest-Building in the Ring Dove (*Streptopelia Risoria*)," *J. Comp. Physiol.*, 51 (1958a), pp. 32–36.

Lehrman, D. S., "Effect of Female Sex Hormones on Incubation Behavior in the Ring Dove (*Streptopelia Risoria*)," *J. Comp. Physiol.*, 51 (1958b), pp. 142–145.

Lenneberg, E. H., "Understanding Language without Ability To Speak: A Case Report. *J. Abnorm. Soc. Psychol.*, 65 (1962), pp. 419–425.

Lenneberg, E. H., "A Biological Perspective of Language," in E. H. Lenneberg, ed., *New Directions in the Study of Language*. Cambridge, Mass.: M.I.T. Press, 1964a.

Lenneberg, E. H., "Speech as a Motor Skill with Special Reference to Nonaphasic Disorders," *Monogr. Soc. Res. Child Devlpm.*, 1964b.

Lenneberg, E. H., "Language Disorders in Childhood," *Harvard Educational Review* (1964c), pp. 152–177.

Lenneberg, E. H., "Speech Development: Its Anatomical and Physiological Concomitants," in *Speech, Language, and Communication. Brain and Behavior*, from UCLA Forum in Medical Sciences, 3, No. 4, edited by Edward C. Carterette. Berkeley: University of California Press, 1966a.

Lenneberg, E. H., "The Natural History of Language" (in this volume), from Frank Smith and George A. Miller, eds., *The Genesis of Language*. Cambridge, Mass.: M.I.T. Press, 1966b.

Lenneberg, E. H., *Biological Foundations of Language*. New York: Wiley, 1967.

Lenneberg, E. H., I. A. Nichols, and E. F, Rosenberger, "Primitive Stages of Language Development in Mongolism," in *Proceedings of the Assoc. for Res. in Nervous and Mental Disease*. 42d Annual Meeting, New York, 1964.

Lenneberg, E. H., Freda G. Rebelsky, and I. A. Nichols, "The Vocalizations of Infants Born to Deaf and Hearing Parents," *Human Develpm.*, 8 (1965), pp. 23–27.

Leopold, W. F., "Speech Development of a Bilingual Child: A Linguist's Record," in Vol. III. *Grammar and General Problems in the First Two Years*. Evanston, Ill.: Northwestern University Press, 1949.

Leventhal, Alice S., and L. P. Lipsitt, "Adaptation, Pitch Discrimination, and Sound Localization in the Neonate," *Child Develpm.*, 35 (1964), pp. 759–767.

Levin, G. R., and H. Kaye, "Non-nutritive Sucking by Human Neonates," *Child Develpm.*, 35 (1964), pp. 749–758.

Lewis, M., Exploratory Studies in the Development of a Face Schema. Paper read at Amer. Psychol. Assoc., Chicago, September 1965.

Lewis, M. M. *Infant Speech*. London: Routledge, 1951.

Lewis, M. M., *Language, Thought, and Personality*. New York: Basic Books, 1963.

Lewis, M., D. Fodel, B. Bartels, and H. Campbell, Infant Attention: The Effect of Familiar and Novel Stimuli as a Function of Age. Paper read at Eastern Psychol. Assoc., New York, April 1966.

Lewis, M., Susan Goldberg, and Marilyn Rausch, Novelty and Familiarity as Determinants of Infant Attention within the First Year. Unpublished manuscript, 1967.

Lewis, M., J. Kagan, and J. Kalafat, "Patterns of Fixations in the Young Infant," *Child Develpm.*, 37 (1966a), pp. 331–341.

Lewis, M., J. Kagan, H. Campbell, and J. Kalafat, "The Cardiac Response as a Correlate of Attention in Infants," *Child Develpm.*, 37 (1966b), pp. 63–71.

Lewis, M., W. J. Myers, J. Kagan, and Ruth Grossberg, Attention to Visual Patterns in Infants. Paper presented at Symposium on Studies of Attention in Infants: Methodological Problems and Preliminary Results. Amer. Psychol. Ass., Philadelphia, August 1963.

Lieberman, P. "On the Acoustic Basis of the Perception of Intonation by Linguists," *Word*, 21, (1965), pp. 40–54.

Lieberman, P. *Intonation, Perception, and Language*. Cambrdge, Mass.: M.I.T. Press, 1967.

Lillie, R. S., *General Biology and Philosophy of Organism*. Chicago: University of Chicago Press, 1945.

Lindsley, D. B., "Brain Potentials in Children and Adults," *Science*, 84 (1936), p. 354.

Ling, B. C., "Form Discrimination as a Learning Cue in Infants," *Comp. Psychol. Monogr.*, 17, No. 2 (1941).

Lipsitt, L. P., "Learning in the First Year of Life," in L. P. Lipsitt and C. C. Spiker, eds., *Advances in Child Development and Behavior*, Vol. I. New York: Academic Press, Inc., 1963, pp. 147–195.

Lipsitt, L. P., "Learning Process of Human Newborns," *Merrill-Palmer Quart.*, 12 (1966), pp. 45–71.

Lipsitt, L. P., and C. DeLucia, "An Apparatus for the Measurement of Specific Response and General Acting in the Human Neonate," *Amer. J. Psychol.*, 73 (1960), pp. 630–632.

Lipsitt, L. P., T. Engen, and H. Kaye, "Developmental Changes in the Olfactory Threshold of the Neonate," *Child Develpm.*, 34 (1963), pp. 371–376.

Lipsitt, L. P., and H. Kay, "Conditioned Sucking in the Human Newborn," *Psychon. Sci.*, 1 (1964), pp. 29–30.

Lipsitt, L. P., H. Kaye, and T. N. Bosack, "Enhancement of Neonatal Sucking through Reinforcement," *J. Exp. Child Psychol.*, 4 (1966), pp. 163–168.

Lipsitt, L. P., Linda J. Pederson, and C. A. DeLucia, "Conjugate Reinforcement of Operant Responding in Infants," *Psychon. Sci.*, 4 (1966), pp. 67–68.

Lipton, E. L., A. Steinschneider, and J. B. Richmond, "Autonomic Function in the Neonate: VII. Maturational Changes in Cardiac Control," *Child Develpm.*, 37 (1966), pp. 1–16.

Liss, P., "Does Backward Masking by Visual Noise Stop Stimulus Processing?" *Perception and Psychophysics*, 4 (1968), pp. 328–330.

Long, L., and L. Welch, "The Development of the Ability To Discriminate and Match Numbers," *J. Genet. Psychol.*, 59 (1941), pp. 377–387.

Lorenz, K., "Der kumpan in der umwelt des vogels," *j. fur Ornithologie*, 83 (1935), pp. 137–213.

Lorenz, K., "The Comparative Method of Studying Innate Behavior Patterns," in *Physiological Mechanisms in Animal Behavior*. New York: Academic Press, Inc., 1950, pp. 221–268.

Lorenz, K., "Comparative Study of Behavior," in C. H. Schiller, ed., *Instinctive Behavior*. New York: International Universities, 1957.

Lorenz, K. Z., "The Evolution of Behavior," *Sci. Amer.*, 199, No. 6 (1958), pp. 67–78.

Lorenz, K., *Evolution and Modification of Behavior*. Chicago: University of Chicago Press, 1965.

Lovell, K., *The Growth of Basic Mathematical and Scientific Concepts in Children*. New York: Philosophical Library, 1961.

Lubchenco, L. O., *et al.*, "Sequelae of Premature Birth," *Amer. J. Dis. Child.*, 106 (1963), pp. 101–115. Reprinted in Y. Brackbill and G. G. Thompson, eds., *Behavior in Infancy and Early Childhood: A Book of Readings*. New York: Free Press, 1967.

Luchins, A. S., and R. H. Forgus, "The Effect of Differential Post Weaning Environment on the Rigidity of an Animal's Behavior," *J. Genet. Psychol.*, 86 (1955), pp. 51–58.

Luchsinger, R., and C. E. Arnold, *Lehrbuch der Stimm-und Sprachheilkunde*, Wein: Springer, 1959, 2d ed.

Luria, A. R., "The Role of Language in the Formation of Temporary Connections," in B. Simon, ed., *Psychology in the Soviet Union*. Stanford, Calif.: Stanford University Press, 1957, pp. 115–129.

Luria, A. R., *The Role of Speech in the Regulation of Normal and Abnormal Behavior*. New York: Pergamon, 1961.

Lynn, R., *Attention, Arousal, and Orientation Reaction*. New York: Pergamon, 1966.

Lyons, J., *Introduction to Theoretical Linguistics*. New York: Cambridge University Press, 1968.

MacCorquodale, K., and P. E. Meehl, "Preliminary Suggestions as to the Formulation of Expectancy Theory," *Psychol. Rev.*, 60 (1953), pp. 55–63.

MacCorquodale, K., and P. E. Meehl, "Edward Chace Tolman," in W. K. Estes *et al.*, eds., *Modern Learning Theory*. New York: Appleton, 1955.

MacRae, J. M., "Retests of Children Given Mental Tests as Infants," *J. Genet. Psychol.*, 87 (1955), pp. 111–119.

McCall, R. B., and J. Kagan, Attention in the Infant: Effects of Complexity, Contour, Perimeter, and Familiarity. Unpublished manuscript, 1967.

McCarthy, Dorothea, "Language Development in Children," in L. Carmichael, ed., *Manual of Child Psychology*. New York: Wiley, 1954 (2d edition), pp. 492–630.

McDougall, W., *An Introduction to Social Psychology*. London: Methuen, 1908. Univ. Paperbacks ed., 1960, p. 25.

McGeoch, John, and Arthur Irion, *The Psychology of Human Learning.* New York: Longmans, 1956, pp. 228–230.

McGinnies, J. M., "Eye-Movements and Optic Mystagmus in Early Infancy," *Genetic Psychology Monographs,* 8 (1930), pp. 320–427.

McGrade, Betty Jo, and W. Kessen, "Activity in the Human Newborn as Related to Delivery Difficulty," *Child Develpm.,* 36 (1965), pp. 73–79.

McGranahan, D. V., "Some Remarks on the Human Implications of Change in Underdeveloped Areas," *Soc. Problems,* 1 (1963), pp. 13–16.

McGraw, M. B., "Appraising Test Responses of Infants and Young Children," *J. Psychol.,* 14 (1942), pp. 89–100.

McGraw, M. B., *The Neuromuscular Maturation of the Human Infant.* New York: Hafner, 1963.

McGraw, Myrtle, "Maturation of Behavior," in L. Carmichael, ed., *Manual of Child Psychology.* New York: Wiley, 1946, pp. 332–369.

McNeill, D., Development of the Semantic System. Unpublished paper, Center for Cognitive Studies, Harvard University, 1965, pp. 1–3.

McNeill, D., Developmental Psycholinguistics. Mimeographed by Center for Cognitive Studies, Harvard University, 1966a.

McNeill, D., "Developmental Psycholinguistics," in F. Smith, and G. A. Miller, eds., *The Genesis of Language.* Cambridge, Mass.: M.I.T. Press, 1966b.

McNeill, D., "On Theories of Language Acquisition," in T. R. Dixon, and D. L. Horton, eds., *Verbal Behavior and General Behavior Theory.* Englewood Cliffs, N.J.: Prentice-Hall, 1968, pp. 406–420.

McNeill, D., "The Development of Language." Chapter to appear in P. A. Mussen, ed., *Carmichael's Manual of Child Psychology* (forthcoming, 1970).

McNeill, David, and Nobuko B. McNeill, A Question in Semantic Development: What Does a Child Mean When He Says "No"? Paper presented to the Society for Research in Child Development, 1967.

McNiven, M. A. Responses of the Chicken, Duck and Pheasant to a Hawk and Goose Silhouette: A Controlled Replication of Tinbergen's Study." Unpublished thesis, University of Pennsylvania, 1954.

Maccoby, Eleanor E., and Helen L. Bee, "Some Speculations Conceiving the Lag between Perceiving and Performing," *Child Develpm.,* 36 (1965), pp. 367–377.

Maier, N. R. F., "Reasoning in Humans: I. On Direction," *J. Comp. Psychol.,* 10 (1930), pp. 115–143.

Maier, N. R. F., "Age and Intelligence in Rats," *J. Comp. Psychol.,* 13 (1932), pp. 1–6.

Maier, N. R., "Mechanisms in Conditioning," *Psychol. Rev.,* 49 (1942), pp. 117–134.

Maier, N. R. F., and J. B. Klee, "Studies of Abnormal Behavior in the Rat. XII. The Pattern of Punishment and Its Relation to Abnormal Fixations," *J. Exper. Psychol.,* 32 (1943), pp. 377–398.

Maier, N. R. F., and T. C. Schneirla, *Principles of Animal Psychology.* New York: Dover, 1935.

Major, D. R., *First Steps in Mental Growth: A Series of Studies in Infancy.* New York: Macmillan, 1906.

Maltzman, I., "On the Training of Originality," *Psychol. Rev.*, 67 (1960), pp. 229–242.

Mandler, G., "Organization and Memory," in K. W. Spence and J. T. Spence, eds., *The Psychology of Learning and Motivation.* Vol. I. New York: Academic Press, Inc., 1967.

Mandriota, F. J., J. Siegel, and R. Gallon, "Electrical Interactions among Mormyrid Fish," *Personal Communication.* 1969.

Mandriota, F. J., R. L. Thompson, and M. U. L. Bennett, "Avoidance Conditioning of the Rate of Electric Organ Discharge in Mormyrid Fish," *Animal Beh.*, 16 (1968), pp. 448–455.

Marler, P., and W. J. Hamilton, *Mechanisms of Animal Behavior.* New York: Wiley, 1966.

Marquis, D. P., "Can Conditioned Responses Be Established in the Newborn Infant?" *J. Genet. Psychol.*, 39 (1931), pp. 479–492.

Martin, E., and K. H. Roberts, "Gramatical Factors in Sentence Retention," *J. Vbl. Lng. Vbl. Behav.*, 5 (1966), pp. 211–218.

Marx, M. H., W. W. Murphy, and A. J. Brownstein, "Recognition of Complex Visual Stimuli as a Function of Training with Abstracted Patterns," *J. Exp. Psychol.*, 62 (1961), pp. 456–460.

Medawar, P., "Onwards from Spencer: Evolution and Evolutionism, *Encounter*, 21 (1963), pp. 35–43.

Meehl, P. E., and K. MacCorquodale, "Some Methodological Comments concerning Expectancy Theory," *Psychol. Rev.*, 58 (1951), pp. 230–233.

Meili, R., and E. Tobler, "Stroboscopic Movement in Children," *Arch. Psychol.*, Genève, 23 (1931), pp. 131–156.

Melcher, R. T., "Development within the First Two Years of Infants Prematurely Born," *Child Develpm.*, 8 (1937), pp. 1–14.

Mentzer, T. L., "Early versus Late Introduction of Stimula in Psychophysical Discrimination by Pigeons, *Psychon. Sci.*, 10 (1968), pp. 9–10.

Menyuk, P., "Syntactic Structure in the Language of Children," *Child Development*, 34 (1963), pp. 407–422.

Menyuk, P., *Sentences Children Use.* Cambridge, Mass.: M.I.T. Press, 1969.

Meumann, E., *Die Entstehung der ersten wortbedentungen beim Kinde.* Leipzig: Engelmann, 1908 (2d edition).

Mewhort, D. J. K., P. M. Merikle, and M. P. Bryden, "On the Transfer from Iconic to Short-term Memory," *Journal of Experimental Psychology.* 1969, in press.

Meyer, L. E., "Comprehension of Spatial Relations in Pre-school Children," *J. Genet. Psychol.*, 57 (1940), pp. 119–151.

Michels, K. M., and Anne W. Schumacher, "Color Vision in Tree Squirrels," *Psychon. Sci.*, 10 (1968), pp. 7–8.

Michotte, A., *La perception de la causalité.* Louvain, 1946.

Mill, J., *Analysis of the Phenomena of the Human Mind.* London: Longmans, 1878 (2d ed.)

Miller, G. A., "The Magical Number Seven, Plus or Minus Two: Some Limits on our Capacity for Processing Information," *Psychol. Rev.*, 63 (1956), pp. 81–97.

Miller, G. A., "Some Preliminaries to Psycholinguistics," *Amer. Psychologist*, 20 (1965), pp. 15–20.

Miller, G. A., "Some Psychological Studies of Grammar," in L. A. Jacobovits and M. S. Miron, eds., *Readings in the Psychology of Language*. Englewood Cliffs, N.J.: Prentice-Hall, 1967, pp. 201–218.

Miller, G. A., and N. Chomsky, "Finitary Models of Language Users," in D. Luce, R. Bush, and E. Galanter, *Handbook of Mathematical Psychology*. Vol. II. New York: Wiley, 1963.

Miller, G. A., E. Galanter, and K. H. Pribram, *Plans and the Structure of Behavior*. New York: Holt, Rinehart and Winston, Inc., 1960.

Miller, N. E., "The Perception of Children: A Genetic Study Employing the Critical Choice Delayed Reaction," *J. Genet. Psychol.*, 44 (1934), pp. 321–339.

Miller, W., and S. Ervin, "The Development of Grammar in Child Language," in U. Bellugi and R. Brown, eds., *The Acquisition of Language. Monogr. Socy. Res. Child Develpm.*, 29, No. 1 (1964), pp. 9–34.

Milner, B., Neuropsychological Evidence for Differing Memory Processes. XVIII. *International Congress of Psychology*, Moscow, 1966.

Moffitt, A. R., Speech Perception by Infants. Unpublished thesis, University of Minnesota, 1968.

Mohr, G. J., and P. Bartelme, "Mental and Physical Development of Children Prematurely Born," *Amer. J. Dis. Childhood.*, 40 (1930), pp. 1000–1015.

Montessori, M., *The Montessori Method*. New York: Schocken, 1964.

Moon, L. E., and H. F. Harlow, "Analysis of Oddity Learning by Rhesus Monkeys," *J. Comp. Physiol. Psychol.*, 48 (1955), pp. 188–194.

Moore, K., "The Effect of Controlled Temperature Changes on the Behavior of the White Rat," *J. Exper. Psychol.*, 34 (1944), pp. 70–79.

Moore, K. C., "The Mental Development of a Child," *Psych. Rev.*, Monograph supplement, No. 3, 1896.

Moray, N., and Anne Jordan, "Vision in Chicks with Distorted Visual Fields," *Psychon. Sci.*, 9 (1969), p. 303.

Morley, M. E., *The Development and Disorders of Speech in Childhood*. Edinburgh: Livingstone, 1967.

Morrow, J. E., and B. L. Smithson, "Learning Sets in an Invertebrate," *Science*, 164 (1969), pp. 850–857.

Mouton, David, Principal Elements of Thinking (in this volume), 1971.

Mowrer, D. H., "Hearing and Speaking: An Analysis of Language Learning," *Journal of Speech and Hearing Disorders*, 23 (1958), pp. 143–151.

Mowrer, O. H., *Learning Theory and the Symbolic Processes*. New York: Wiley, 1960.

Mowrer, O. H., and A. D. Ullman, "Time as a Determinant in Integrative Learning," *Psychol. Rev.*, 52 (1945), pp. 61–90.

Müller, J., *Elements of Physiology*. Vol. II (trans. by W. Baly). London: Taylor & Walker, 1842.

Mumford, L., "Speculations in Prehistory," *The American Scholar*, 36 (Winter 1966–1967), pp. 43–53.

Munn, N. L., *The Evolution and Growth of Human Behavior*. Boston: Houghton Mifflin, 1955.

Murai, J., "Speech Development of Infants—Analysis of Speech Sounds by Sona-graph," *Psychologia*, 3 (1960), pp. 27–35.

Murai, J. "The Sounds of Infants: Their Phonemicization and Symbolization," *Studia Phonologica*, 3 (1963), pp. 18–34.

Murdock, B. B., and W. vom Saal, "Transportation in Short-term Memory," *Journal of Experimental Psychology*, 74 (1967), pp. 137–143.

Murdock, G. P., The Common Denominator of Cultures in R. Linton, ed., *The Science of Man in the World Crisis*. New York: Columbia University Press, 1945, pp. 123–142.

Murray, H. A., "Towards a Classification of Interaction," in T. Parsons and E. Shils, eds., *Towards a General Theory of Action*. Cambridge, Mass.: Harvard University Press, 1951.

Nakazima, S., "A Comparative Study of the Speech Developments of Japanese and American English in Childhood (1)—A Comparison of the Developments of Voices at the Pre-linguistic Period," *Studia Phonologica*, 2 (1962), pp. 27–39.

Nakazima, S., "A Comparative Study of the Speech Developments of Japanese and American English in Childhood (2)—The Acquisition of Speech," *Studia Phonologica*, 4 (1966), pp. 38–55.

Natsoulas, T., "What Are Perceptual Reports About?" *Psychological Bulletin*, 67 (1967), pp. 249–272.

Needham, J., *The Sceptical Biologist*. London: Chatto, 1929.

Needham, J., *Order and Life*. New Haven, Conn.: Yale University Press, 1936.

Neisser, U., "Cultural and Cognitive Discontinuity," in the Anthropological Society of Washington, *Anthropology and Human Behavior*. Washington, D.C.: Gaus, 1962, pp. 54–71.

Neisser, U., *Cognitive Psychology*. New York: Appleton, 1967.

Newell, A., and H. A. Simon, in *Current Trends in Psychological Theory*. Pittsburgh, Pa.: University of Pittsburgh Press, 1961a, pp. 153–179.

Newell, A., and H. A. Simon, "Computer Simulation of Human Thinking," *Science*, 134 (1961b), pp. 2011–2017.

Nissen, H. W., "Phylogenetic Comparison," in S. S. Stevens, ed., *Handbook of Experimental Psychology*. New York: Wiley, 1951, pp. 347–386.

Nissen, H. W., "The Nature of the Drive as Innate Determinant of Behavioral Organization," in *Nebraska Symposium on Motivation*. Lincoln, Nebraska: University of Nebraska Press, 1954.

Nixon, M. C., Perceived Shape and Slant. Unpublished thesis, University of Sydney, 1958.

O'Neil, W. M., "Basic Issues in Perceptual Theory" (in this volume), *Psychol. Rev.*, 65 (1958), pp. 348–361.

Orton, Grace, "The Role of Ontogeny in Systematics and Evolution," *Evol.*, 9 (1955), pp. 75–83.

Oseas, L., and B. J. Underwood, "Studies of Distributed Practice: V. Learning and Retention of Concepts," *J. Exp. Psychol.*, 43 (1952), pp. 143–148.

Osgood, C. E., *Method and Theory in Experimental Psychology*. New York: Oxford University Press, 1953.

Osgood, C. E., "Comments on Professor Bousfield's Paper," in C. N. Cofer, ed., *Verbal Learning and Verbal Behavior*. New York: McGraw-Hill, 1961, pp. 91–106.

Osgood, C. E., G. J. Suci, and P. H. Tannenbaum, *The Measurement of Meaning*. Urbana, Ill.: University of Illinois Press, 1957.

Paivio, A., "Mental Imagery in Associative Learning and Memory," *Psychol. Rev.*, 76 (1969), pp. 241–263.

Papoušek, Hanuš, "A Method of Studying Conditioned Food Reflexes in Young Children up to the Age of 6 Months," *Pavlov J. High. Nerv. Act.*, 9 (1959), pp. 136–140.

Papoušek, Hanuš, "On the Development of the So-called Voluntary Moments in the Earliest Stages of the Child's Development," *Cesk. Pediat.*, 17 (1962), pp. 588–591.

Papoušek, Hanuš, Experimental Studies of Appetitional Behavior in Human Newborns and Infants. Unpublished paper, 1966a.

Papoušek, Hanuš, Learning in infants. Unpublished paper, 1966b.

Papoušek, Hanuš, "Conditioning during Early Postnatal Development," in Yvonne Brackbill and G. Thompson, eds., *Behavior in Infancy and Early Childhood*. New York: Free Press, 1967, pp. 259–274.

Patrick, J. R., and R. M. Laughlin, "Is the Wall-Seeking Tendency in the White Rat an Instinct," *J. Genet. Psychol.*, 44 (1934), pp. 378–389.

Pavlov, I. P., *Lectures on Conditioned Reflexes, Vol. II: Conditioned Reflexes and Psychiatry* (W. H. Gantt, trans.). New York: International, 1941.

Pavlov, I. P., *Conditioned Reflexes* (G. V. Anrep, trans.). London: Oxford University Press, 1927 and 1953.

Pedersen, H., *The Discovery of Language*. Bloomington, Ind.: Indiana University Press, 1931.

Peel, E. A., "Experimental Examination of Some of Piaget's Schemata concerning Children's Perception and Thinking, and a Discussion of Their Educational Significance," *British Journal of Educational Psychology*, 29 (1959), pp. 89–103.

Peiper, A., "Die Schreit- und Steig-bewegungen der Neugeborenen," *Arch. Kinderhk.*, 147 (1953), pp. 135 ff.

Peiper, A., *Die Eigevart der kindlichen Hirntätigkeit*. Leipiz: G. Thieme, 1961 (3d ed.).

Peiper, N., *Cerebral Function in Infancy and Childhood*. New York: Consultations Bureau, 1963.

Penfield, W., "Some Observations on the Functional Organization of the Human Brain," *Proc. Amer. Phil. Soc.*, 98 (1954), pp. 293–297.

Penfield, W., "The Permanent Record of the Stream of Consciousness," *Acta Psychologica*, 11 (1956), pp. 47–59.

Perry, R. B., *Present Philosophical Tendencies*. New York: Longmans, 1912.

Perry, William, *Forms of Ethical Development.* Cambridge, Mass.: Harvard University Press, 1969.

Pestalozzi, J. H., *Leonard and Gertrude.* Boston: Heath, 1895.

Peterson, L. R., and M. Peterson, "Short-term Retention of Individual Verbal Items," *Journal of Experimental Psychology,* 58 (1959), pp. 193–198.

Piaget, J., *The Child's Conception of Physical Causality.* New York: Harcourt, 1930.

Piaget, J., *La psychologie de l'intelligence.* Paris: Presses Univer. de France, 1946. [*The Psychology of Intelligence.* New York: Routledge, 1950. Trans. from 2d French ed.]

Piaget, J., *Judgement and Reasoning in the Child.* London: Routledge, 1951a.

Piaget, J., *Play, Dreams, and Imitation in Childhood.* New York: Norton, 1951b.

Piaget, J., *The Origins of Intelligence in Children.* New York: Norton, 1952.

Piaget, J., *Logic and Psychology.* Manchester, England: Manchester University Press, 1953.

Piaget, J., *The Construction of Reality in the Child.* New York: Basic Books, 1954.

Piaget, J., *The Language and Thought of the Child.* New York: Meridian, 1955.

Piaget, J., "Essai d'une nouvelle interprétation probabiliste des effets de centration, de la loi de Weber et de celle des centrations relatives." *Arch. Psychol.,* 35 (Genéve, 1955a), pp. 1–24.

Piaget, J., "Les stades du développement intellectual de l'enfant et de l'adolescent." In P. Osterrieth *et al., Le probleme des stades en psychologie de l'enfant.* Paris: Presses Univer. de France, 1956, pp. 33–49.

Piaget, J. Logique et équilibre dans les comportements du sujet." In L. Apostel *et al., Logique et équilibre* (Études d'epistémolgie génétique, 2), Paris: Presses Univer. de France, 1957, pp. 27–117.

Piaget, J., *Les mécanismes perceptifs.* Paris: Presses Universitaires de France, 1961.

Piaget, J., *Six Psychological Studies.* New York: Random House, Inc., 1967.

Piaget, J., and B. Inhelder, *The Child's Conception of Space.* New York: Humanities Press, 1956.

Piaget, J., and Báibel Inhelder, *Le developpement des quantitiés physiquéaes chez l'enfant.* Neuchâtel, Switzerland: Delachaux & Niestlé, 1962.

Piaget, J., and B. Inhelder, *The Early Growth of Logic in the Child.* New York: Harper & Row, 1964.

Piaget, J., and B. Inhelder, *La psychologie de l'enfant.* Collection "Que sais-je" No. 369. Paris: Presses Universitaires de France, 1966a.

Piaget, J., and B. Inhelder, *L'image mental chez l'enfant.* Paris: Presses Universitaires de France, 1966b.

Piaget, J. et M. Lambercier, "Transpositions perceptives et transitivité opértoire dans les comparaisons en profondeur." *Arch. Psychol.,* 31 (Genéve, 1946), pp. 325–368.

Piaget, J. et A. Morf, "Les isomorphismes partiels entre les structures logiques et les structures perceptives," in J. S. Bruner, *et al., Logique et perception* (Études d'epistémologie génétique, 6). Paris: Presses Univer. de France, 1958a, pp. 49–116.

Piaget, J. et A. Morf, "Les préinférences perceptives et leurs relations avec les schémes sensorimoteurs et operatoires," in J. S. Bruner *et al.*, *Logique et perception* (Études d'epistémologie génétique, 6). Paris: Presses Univer. de France, 1958b, pp. 117–155.

Piaget, J. et S. Taponier, "L'estimation des longeurs de deux droites horizontales et parallèles extrémités décalies." *Arch. Psychol.*, 35 (Genéve, 1956), pp. 369–400.

Pick, J., "The Evolution of Homeostasis," *Proc. Amer. Phil. Soc.*, 98 (1954), pp. 298–303.

Pike, K., *Language in Relation to a Unified Theory of the Structure of Human Behavior.* The Hague: Mouton, 1967.

Place, U. T., "Is Consciousness a Brain Process?" in V. C. Chappell, ed., *The Philosophy of Mind.* Englewood Cliffs, N.J.: Prentice-Hall, 1962, p. 105.

Plato, *Meno.* Many editions.

Polikanina, R. I., "The Relationship between Autonomic and Semantic Components of a Defensive Conditioned Reflex in Premature Children," *Pavlov, J. High. Nerv. Act.*, 11 (1961), pp. 72–82.

Polikanina, R. I., and L. E. Probatova, "Development of the Conditioned Motor Alimentary Reflex to Light in Prematurely Born Children," *Zh. vyssh. nerv. Diet.*, 7 (1957), pp. 673–682.

Pollack, R. H., Some Implications of Ontogenetic Changes in Perception," in D. Elkind and J. H. Flavell, eds., *Studies in Cognitive Development.* New York: Oxford University Press, 1969.

Posner, M. I., *et al.*, "Retention of Visual and Name Codes of Single Letters," *Journal of Experimental Psychology,* 79 (1969).

Posner, M. I., and R. F. Mitchell, "Chronometric Analysis of Classification," *Psychol. Rev.*, 74 (1967), pp. 392–409.

Posner, M. I., and E. Rossman, "Effect of Size and Location of Informational Transforms upon Short-term Retention," *Journal of Experimental Psychology,* 70 (1965), pp. 496–505.

Postman, L., "Toward a General Theory of Cognition," in J. H. Rohrer and M. Sherief, eds., *Social Psychology at the Crossroads.* New York: Harper & Row, 1951.

Postman, L., "Perception, Motivation and Behavior," *J. Pers.*, 22 (1953), pp. 17–31.

Poull, L. E., "The Effect of Improvement in Nutrition on the Mental Capacity of Young Children," *Child Develpm.*, 9 (1938), pp. 123–126.

Prechtl, H., "The Directed Head Turning Response and Allied Movements of the Human Baby," *Behavior,* 13 (1958), pp. 212–242.

Prentice, W. C. H., Visual recognition of verbally labeled figures. *Amer. J. Psychol.* 67 (1954), pp. 315–320.

Preyer, W., *The Mind of the Child.* New York: Appleton, 1888.

Preyer, W., *The Mind of the Child. Part II. The Development of the Intellect* (translated by H. W. Brown). New York: Appleton, 1890.

Preyer, W., *Mental Development in the Child.* New York: Appleton, 1893.

Price, H. H., *Thinking and Experience,* Cambridge, Mass., 1953, pp. 234–237.

Rambusch, N. M., *Learning To Learn.* Baltimore: Helicon Press, Inc., 1962.

Rapaport, D., "On the Psychoanalytic Theory of Thinking," in R. P. Knight and

C. R. Friedman, eds., *Psychoanalytic psychiatry and psychology*. New York: International Universities, 1954.

Rashevsky, N., *Mathematical Biophysics*. Chicago: University of Chicago Press, 1948.

Raskin, D. C., H. Kotses, and J. Bever, "Autonomic Indicators of Orienting and Defensive Reflexes," *Journal of Experimental Psychology*, 80 (1969), pp. 423–433.

Rasmussen, Elizabeth A., and E. J. Archer, "Concept Identification as a Function of Language Pretraining and Task Complexity," *J. Exp. Pyschol.*, 61 (1961), pp. 437–441.

Rasmussen, K., *The Netsilik Eskimos: Social Life and Spiritual Culture*. Report of the 5th Thule Expedition: 1921–1924, Vol. VIII. Nordick Verlag-Copenhagen, Denmark: Gyldendalske, Boghandel, 1931.

Redfield, R., ed., *Levels of Integration in Biological and Social Systems*. Biological Symposia, Vol. VIII. Tempe, Ariz.: Jaques Cattell Press, 1942.

Reitman, W. R. *Cognition and Thought: An Information Processing Approach*. New York: Wiley, 1965.

Rendle-Short, J., "The Puff Test. An Attempt To Assess the Intelligence of Young Children by Use of a Conditioned Reflex," *Arch. Dis. Childh.*, 36 (1961), pp. 50–57.

Restle, F., "Toward a Quantitative Description of Learning Set Data." *Psychol. Rev.*, 65 (1958), pp. 77–91.

Restle, F. A., "A Theory of Discrimination Learning," *Psychol. Rev.*, 62 (1955), pp. 11–19.

Rheingold, H. L., "The Modification of Social Responsiveness of Institutional Babies," *Monogr. Soc. Res. Child Develpm.*, 21 (1956).

Rheingold, H. L., "The Effect of Environmental Stimulation upon Social and Exploratory Behavior in the Human Infant," in B. M. Foss, ed., *Determinants of Infant Behavior*, Vol. I. New York: Wiley, 1961.

Rheingold, H. L., J. L. Gewirtz, and H. W. Ross, "Social Conditioning of Vocalizations in the Infant. *Journal of Comparative Physiological Psychology*, 52 (1959), pp. 68–73.

Rheingold, H. L., W. C. Stanley, and J. A. Cooley, "Method for Studying Exploratory Behavior in Infants," *Science*, 136, No. 3521 (1962), pp. 1054–1055.

Ribble, M. A., *The Rights of Infants*. New York: Columbia University Press, 1943.

Richardson, H. M., "The Growth of Adaptive Behavior in Infants: An Experimental Study of Seven Age Levels," *Genet. Psychol. Monogr.*, 12 (1932), pp. 195–359.

Riesen, A. H., "Arrested Vision," *Scient. Amer.*, 183 (1950), pp. 16–19.

Riesen, A. H., and L. Aarons, "Visual Movement and Intensity Discrimination in Cats after Early Deprivation of Pattern Vision," *J. Comp. Physiol. Psychol.*, 52 (1959), pp. 142–149.

Riess, B., "The Isolation of Factors of Learning and Native Behavior in Field and Laboratory Studies," *Ann. N.Y. Acad. Sci.*, 51 (1950), pp. 1093–1102.

Rivers, W. M., *The Psychologist and the Foreign-Language Teacher*. Chicago: University of Chicago Press, 1964.

Rivers, W. M., *Teaching Foreign-Language Skills*. Chicago: University of Chicago Press, 1968.

Roberts, K. E., "The Ability of Preschool Children To Solve Problems in Which a Simple Principle of Relationship Is Kept Constant," *J. Genet. Psychol.*, 40 (1932), pp. 118–135.

Robins, R. H., *Ancient and Mediaeval Grammatical Theory in Europe*. London: Bell, 1951.

Robinson, E. S., *Association Theory Today*. New York: Appleton, 1932.

Roeper, A., and I. Sigel, "Finding the Clue to Children's Thought Processes," *Young Children*, 21 (1966), pp. 335–349.

Roffward, H. P., J. N. Muzio, and W. C. Dement, "Ontogenetic Development of the Human Sleep Dream Cycle," *Science*, 152 (1966), pp. 604–619.

Rosenblatt, J., *et al.*, Analytical Studies on Maternal Behavior and Litter Relations in the Domestic Cat. II. From Birth to Weaning. Unpublished paper.

Rousseau, J. J., *Emile*. New York: Appleton, 1895.

Routh, D. K., "Conditioning of Vocal Response Differentiation in Infants," *Journal of Developmental Psychology*, 1 (1969), pp. 219–226.

Routtenberg, A., and S. E. Glickman, "Visual Cliff Behavior in Undomesticated Rodents, Land and Aquatic Turtles, and Cats (*Panthera*)," *J. Comp. Physiol. Psychol.*, 58 (1964), pp. 143–146.

Rovee, Carolyn K., and G. R. Levin, "Oral 'pacification' and Arousal in the Human Newborn," *J. Exp. Child Psychol.*, 3 (1966), pp. 1–17.

Rush, Grace P., "Visual Grouping in Relation to Age," *Arch. Psychol.*, 31, No. 217 (1937).

Russell, Bertrand, *The Problems of Philosophy*. London, 1912, p. 93.

Ryle, Gilbert, *The Concept of Mind*. London: Hutchinson House, 1949.

Saayman, G., Elinor Wordwell Ames, and Adrienne Moffett, "Response to Novelty as an Indicator of Visual Discrimination in the Human Infant," *J. Exp. Child Psychol.*, 1 (1964), pp. 189–198.

Sachs, Jacqueline, Developmental Studies of Language (in this volume), 1971.

Sackett, G. P., "Monkees Reared in Isolation with Pictures as Visual Input: Evidence for an Innate Releasing Mechanism," *Science*, 154 (1966), pp. 1468–1473.

Salapatek, P., and W. Kessen, "Visual Scanning of Triangles by the Human Newborn," *J. Exp. Child Psychol.*, 3 (1966), pp. 155–167.

Salzen, E. A., "Visual Stimuli Eliciting the Smiling Response in the Human Infant," *J. Genet. Psychol.*, 102 (1963), pp. 51–54.

Sameroff, A. J., "An Experimental Study of the Response Components of Sucking in the Human Newborn," *Dissert. Abstr.*, 26 (1965), p. 2341.

Sapir, E., *Language*. New York: Harcourt, 1921.

Sapir, E., *Selected Writings in Language, Culture, and Personality*. Berkeley: University of California Press, 1949.

Saugstad, P., "Problem Solving as Dependent on Availability of Functions," *Brit. J. Psychol.*, 46 (1955), pp. 191–198.

Schachtel, E. G., "On Memory and Childhood Amnesia," *Psychiatry*, 10 (1947), pp. 1–26.

Schaffer, H. R., and P. E. Emerson, "The Development of Social Attachments in Infancy," *Soc. Res. Child Develpm., Monogr.*, 29, No. 3 (1964).

Scharlock, D. P., T. S. Tucker, and N. L. Strominger, "Auditory Discrimination by the Cat after Neonatal Ablation of Temporal Cortex," *Science*, 20 (1963), pp. 1197–1198.

Scheerer, M., "Cognitive Theory," in G. Lindzey, ed., *Handbook of Social Psychology*. Reading, Mass.: Addison-Wesley, 1954.

Scheffler, I., *Conditions of Knowledge*, Chicago: Scott, Foresman, 1965.

Schiff, W., The Perception of Impending Collision. Thesis, Cornell University, 1964.

Schiff, W., "Perception of Impending Collision: A Study of Visually Directed Avoidant Behavior," *Psychol. Monogr.*, 79, No. 604 (1965), pp. 1–26.

Schiff, William, Comparative Study of Sensory and Perceptual Processes (in this volume), 1971.

Schiffman, H. R., "Physical Support with and without Optical Support: Reaction to Apparent Depth by Chicks and Rats," *Science*, 159 (1968), pp. 892–894.

Schiffman, H. R., *et al.*, "Comparison of Tactual and Visual Cues on the Visual Cliff," *Psychon. Sci.*, 9 (1967), pp. 437–438.

Schiller, C. H., ed., *Instinctive Behavior: The Development of a Modern Concept*. New York: International Universities, 1957.

Schiller, P. H., "Monoptic and Dichoptic Visual Masking by Patterns and Flashes," *Journal of Experimental Psychology*, 69 (1965), pp. 193–199.

Schneider, K. M., "Aus der Jungendent-wicklung einer kunstlich aufgezogenen Schimpansin. III. Vom Verhalten," *Z. Tierpsychol.*, 7 (1950), pp. 485–558.

Schneirla, T. C., "A Theoretical Consideration of the Basis for Approach-Withdrawal Adjustments in Behavior," *Psychol. Bull.*, 37 (1939), pp. 501–502.

Schneirla, T. C., "Ant Learning as a Problem of Comparative Psychology," in P. Harriman, ed., *Twentieth Century Psychology*. New York: Philosophical Library, 1945, pp. 276–305. See also "Contemporary American Animal Psychology," in the same volume, pp. 306–316.

Schneirla, T. C., "Problems in the Biopsychology of Social Organization," *J. Abn. and Soc. Psychol.*, 41 (1946), pp. 385–402.

Schneirla, T. C., "Levels in the Psychological Capacities of Animals," in R. W. Sellars *et al.*, *Philosophy for the Future*. New York: Macmillan, 1949.

Schneirla, T. C., "A Consideration of Some Conceptual Trends in Comparative Psychology," *Psychol. Bull.*, 49 (1952), pp. 559–597.

Schneirla, T. C., "Interrelationships of the Innate and the Acquired in Instinctive Behavior," in *Coloque Int. sur l'Instinct Animale*. Paris: Masson et Cie, 1956, pp. 387–432.

Schneirla, T. C., "The Concept of Development in Comparative Psychology," (in this volume), in Dale B. Harris, ed., *The Concept of Development*. Minneapolis: University of Minnesota Press, 1957.

Schneirla, T. C., "An Evolutionary and Developmental Theory of Biphasic Processes underlying Approach and Withdrawal," in Maier, N. R. F., and T. C. Schneirla, 1964, pp. 511–554.

Schultz, A. H., "Growth and Development of the Chimpanzee," *Carnegie Inst. Wash. Pub. 518, Contrib. to Embryol.*, 28 (1940), pp. 1–63.

Schultz, A. H., "Postembryonic Age Changes," in H. Hofer, A. H. Schultz, and D. Starck, eds., *Primatalogia: Handbook of Primatology.* Basel: Karger, 1956.

Scott, J. P., "The Process of Primary Socialization in Canine and Human Infants," *Soc. Res. Child Develpm. Monogr.*, 28, no. 1 (1963).

Seitz, A., "Untersuchungen über Verhaltensweisen bei Caniden," *Z. Tierpsychol.*, 7 (1950), pp. 1–46.

Seitz, P. F. D., "The Effects of Infantile Experience upon Adult Behavior in Animal Subjects: Effects of Litter Size during Infancy upon Adult Behavior in Rats," *Amer. J. Psychiat.*, 110 (1954), pp. 916–927.

Semb, G., and L. Lipsett, "Effects of Acoustic Stimulation on Cessation and Initiation of Non-nutritive Sucking in Neonates," *Journal of Experimental Child Psychology*, 6 (1969), pp. 585–597.

Shepard, R. N., C. I. Hovland, and H. M. Jenkins, "Learning and Memorization of Classifications," *Psychol. Monogr.*, 75, Whole No. 517 (1961).

Sherif, M., "Introduction," in J. H. Rohrer and M. Sherif, eds., *Social Psychology at the Crossroads.* New York: Harper & Row, 1951.

Sherrington, C., *The Integrative Action of the Nervous System.* New Haven, Conn.: Yale University Press, 1906.

Shiffrin, R. M., and R. C. Atkinson, "Storage and Retrieval Processes in Long-term Memory," *Psych. Rev.*, 76 (1969), pp. 179–193.

Shinn, M. W., "Notes on the Development of a Child," *University of California Publications in Education,* 1 (1893), pp. 10–24.

Shirley, M., "Development of Immature Babies during Their First Two Years," *Child Develpm.*, 9 (1938), pp. 347–360.

Shirley, Mary M., *Postural and Locomotor Development,* Vol. I of *The First Two Years: A Study of Twenty-five Babies.* Minneapolis: University of Minnesota Press, 1931.

Shvachkin, N., "Razvitiye fonematichaskogo vospriyatiya rechi v rannam vozraste," *Izvestiya Akad. Pedag: Nank, RSFSR,* 13 (1948), pp. 102–132.

Siegel, A. I., "Deprivation of Visual Form Definition in the Ring Dove: Discriminatory Learning," *J. Comp. Physiol. Psychol.*, 46 (1953), pp. 115–119. See also "Perceptual Motor Transfer," in the same volume, pp. 250–252.

Simmons, Mae Williamson, "Operant Discrimination Learning in Human Infants," *Child Develpm.*, 35 (1964), pp. 737–748.

Simon, A. J., and L. G. Bass, "Toward a Validation of Infant Testing," *Amer. J. Orthopsychiat.*, 26 (1956), pp. 340–350.

Simon, H. A., "An Information Processing Theory of Intellectual Development," in W. Kessen and Clementina Kuhlman, eds., *Thought in the Young Child. Monogr. Soc. Res. Child Develpm.*, 27, (2), Serial No. 83 (1962), pp. 150–155.

Singh, T. A. L., and R. M. Zingg, *Wolf Children and Feral Man*. New York: Harper & Row, 1942.

Siqueland, E. R., "Operant Conditioning of Head-Turning in Four-Month Infants," *Psychon. Sci.*, 1 (1964), pp. 223–224.

Siqueland, E. R., and L. P. Lipsitt, "Conditioned Head-Turning in Human Newborns," *J. Exp. Child Psychol.*, 3 (1966), pp. 356–376.

Skalet, M., "The Significance of Delayed Reaction in Young Children," *Comp. Psychol. Monogr.*, 7, No. 34 (1931).

Skinner, B. F., *Science and Human Behavior*. New York: Macmillan, 1953.

Skinner, B. F., *Verbal Behavior*. New York: Appleton, 1957.

Slobin, D. I., Grammatical Transformations in Childhood and Adulthood. Unpublished dissertation, Harvard University, 1963.

Slobin, D. I., "The Acquisition of Russian as a Native Language," in F. Smith, and G. A. Miller, eds., *The Genesis of Language*. Cambridge, Mass., M.I.T. Press, 1966, pp. 129–148.

Slobin, D. I., "Imitation and Grammatical Development in Children," in N. S. Endler, L. R. Boulter, and H. Osser, eds., *Contemporary Issues in Developmental Psychology*. New York: Holt, Rinehart and Winston, Inc., 1968.

Slobin, D. I., "Early Grammatical Development in Several Languages, With Special Attention to Soviet Research." Unpublished manuscript.

Smedslund, J., Learning and Equilibration: A Study of the Acquisition of Concrete Logical Structures. Prepublication, Draft, Oslo, 1959.

Smedslund, J., "The Acquisition of Conservation of Substance and Weight in Children. I. Introduction," *Scandinavian Journal of Psychology*, 2 (1961a), pp. 11–20.

Smedslund, J., "The Acquisition of Conservation of Substance and Weight in Children. II. External Reinforcement of Conservation of Weight and of the Operations of Addition and Subtraction," *Scandinavian Journal of Psychology*, 2 (1961b), pp. 71–84.

Smedslund, J., "The Acquisition of Conservation of Substance and Weight in Children. III. Extinction of Conservation of Weight Acquired 'Normally' and by Means of Empirical Controls on a Balance Scale," *Scandinavian Journal of Psychology*, 2 (1961c), pp. 85–87.

Smedslund, J., "The Acquisition of Conservation of Substance and Weight in Children. IV. An Attempt at Extinction of the Visual Components of the Weight Concept," *Scandinavian Journal of Psychology*, 2 (1961d), pp. 153–155.

Smedslund, J., "The Acquisition of Conservation of Substance and Weight in Children. V. Practice in Conflict Situations without External Reinforcement," *Scandinavian Journal of Psychology*, 2 (1961e), pp. 156–160.

Smith, M. E., "An Investigation of the Development of the Sentence and the Extent of Vocabulary in Young Children," *University of Iowa Stud. Child Welfare*, 3, No. 5 (1926).

Snyder, L. H., *The Principles of Heredity*. Boston: Heath, 1950.

Sokolov, E. H., "The Modeling Properties of the Nervous System," in M. Cole and I. Maltzman, eds., *A Handbook of Contemporary Soviet Psychology*. New York: Basic Books, 1969.

Sokolov, E. N., "Neuronal Models and the Orienting Reflex," in M. A. Brazier, ed., *The Central Nervous System and Behavior.* New York: John Macy, 1960.

Soltis, Jonas F., "The Language of Visual Perception" (in this volume), in B. P. Komisar and C. J. B. MacMillan, eds., *Psychological Concepts in Education.* Chicago, Rand McNally, 1967.

Sonstroem, A. M., "On the Conservation of Solids," in J. S. Bruner, *et al.*, *Studies in Cognitive Growth.* New York: Wiley, 1966.

Spears, W. C., "Assessment of Visual Preference and Discrimination in the Four-Month-Old Infant," *J. Comp. Physiol. Psychol.*, 57 (1964), p. 381.

Spears, W. C., and R. H. Hohle, "Sensory and Perceptual Processes in Infants," in Y. Brackbill, ed., *Infancy and Early Childhood.* New York: Free Press, 1967, pp. 51–121.

Spencer, H., *Principles of Psychology,* Vol. I. New York: Appleton, 1899.

Sperling, G., "The Information Available in Brief Visual Presentations," *Psychological Monographs,* 74, No. 11 (1960).

Spiro, M. E., "Culture and Personality: The Natural History of a False Dichotomy," *Psychiatry,* 14 (1951), pp. 19–56.

Spitz, R. A., *Hospitalism: An Inquiry into the Genesis of Psychiatric Conditions in Early Childhood on the Psychoanalytic Study of the Child.* New York: International Universities, 1945.

Spitz, R. A., "Ontogenisis: The Proleptic Function of Emotion," in P. H. Knapp, ed., *Expressions of the Emotions in Man.* New York: International Universities, 1963.

Staats, A. W., and C. K. Staats, *Complex Human Behavior.* New York: Holt, Rinehart and Winston, Inc., 1964.

Standing, L. G., B. D. Sales, and R. N. Haber, "Repetition *vs.* Luminance as a Determinant of Recognition," *Canadian Journal of Psychology,* 22 (1968), pp. 442–448.

Staples, Ruth, "The Response of Infants to Color," *J. Exp. Psychol.*, 15 (1932), pp. 119–141.

Stavrianos, B. K., "The Relation of Shape Perception to Explicit Judgements of Inclination," New York, *Arch. Psychol.*, No. 296 (1945).

Stechler, G., Susan Bradford, and H. Levy, "Attention in the Newborn: Effect on Motility and Skin Potential," *Science,* 151 (1966), pp. 1246–1248.

Steinschneider, A., E. Lipton, and J. B. Richmond, "Autonomic Function in the Neonate: IV. Discriminability, Consistency, and Slope Measures of an Individual's Cardiac Responsitivity," *J. Genetic Psychol.*, 105 (1964), pp. 295–310.

Steinschneider, A., E. Lipton, and J. B. Richmond, "Auditory Sensitivity in the Infant: Effect of Intensity on Cardiac and Motor Responsitivity," *Child Develpm.*, 37 (1966), pp. 233–252.

Stern, C., and W. Stern, "Die Kindersprache: Eine psychologische und sprach theoretische Untersuchung," *Monogr. seel. entwick, Kindes,* Vol. I. Leipzig: Barth, 1907.

Stern, E. R., and W. E. Jeffrey, Operant Conditioning of Non-nutritive Sucking in the Neonate. Paper presented at meeting of Soc. Res. Child Develpm., 1965.

Stern, W., "Psychology of Early Childhood up to the Sixth Year of Age (translated by A. Borwell). New York: Holt, Rinehart and Winston, Inc., 1924.

Steward, J. H., "Evolution and Process," in A. L. Kroeber, ed., *Anthropology Today*. Chicago: University of Chicago Press, 1953, pp. 313–326.

Stirnimann, F., *Schweiz. med. Wschr.* 1938, p. 1374 ff.

Stock, M. B., and P. M. Smyth, "Does Undernutrition during Infancy Inhibit Brain Growth and Subsequent Intellectual Development?" *Arch. Dis. Childh.*, 38 (1963), pp. 546–552.

Stone, G. R., "The Effect of Negative Incentive on Serial Learning: The Spread of Variability under Electric Shock," *J. Exper. Psychol.*, 36 (1946), pp. 137–142.

Stott, L. H., and Rachel S. Ball, "Infant and Pre-school Mental Tests: Review and Evaluation." *Monogr. Soc. Res. Child Develpm.*, 30, No. 3 (1965).

Stout, G. F., *Manual of Psychology*. London: Univer. Tutorial Press, 1929, 4th ed.

Tallarico, R. B., and W. W. Farrell, "Studies of Visual Depth Perception: An Effect of Early Experience on Chicks on the Visual Cliff. *J. Comp. Physiol. Psychol.*, 57 (1964), pp. 94–96.

Tauber, Edward S., and Sandra Koffler, "Optomotor Response in Human Infants to Apparent Motion: Evidence of Innateness," *Science*, 152 (1966), pp. 382–383.

Taylor, Richard, *Action and Purpose*. Englewood Cliffs, N.J.: Prentice-Hall, 1966, pp. 113, 163–164.

Teas, D. C., and M. E. Bitterman, "Perceptual Organization in the Rat," *Psychol. Rev.*, 59 (1952), pp. 130–140.

Terman, L. M., and M. A. Merrill, *Stanford-Binet Intelligence Scale. Manual for the Third Revision Form L—M*. Boston: Houghton Mifflin, 1960.

Teuber, H. L., "Effects of Brain Wounds Implicating Right or Left Hemisphere in Man: Hemisphere Differences and Hemisphere Interaction on Vision, Audition, and Somesthesis," in V. B. Mountcastle, ed., *Interhemispheric Relations and Cerebral Dominance*. Baltimore: The Johns Hopkins Press, 1962.

Thiessen, D. D., *et al.*, "Visual Behavior of the Mongolian Gerbil (*Meriones Unguicalatis*)," *Psychon. Sci.*, 11 (1968), pp. 23–24.

Thomas, E., and Fr. Schaller, "Das Spiel der optisch isolierten, jungen Kasper-Hauser-Katze," *Naturwiss*, 41 (1954), pp. 557–558.

Thompson, D'Arcy W., *On Growth and Form*. New York: Cambridge University Press, 1952, 2d ed.

Thompson, J. A., *The Minds of Animals: An Introduction to the Study of Animal Behavior*. London: George Newnes Ltd., 1927, p. 97.

Thompson, W. R., "The Effects of Restricting Early Experience on the Problem Solving Capacity of Dogs," *Canad. J. Psychol.*, 8 (1954), pp. 17–31.

Thorndike, E. L., *Man and His Works*. Cambridge, Mass.: Harvard University Press, 1943.

Thorndike, E. L., *et al.*, *The Measurement of Intelligence*. New York: Teachers College, 1926.

Tinbergen, N., *The Study of Instinct*. New York: Oxford University Press, 1931.

Tinbergen, N. *A Study of Instinct*. London: Oxford University Press, 1951.

Tinbergen, N., and D. J. Kuenen, "Feeding Behavior in Young Thrushes; Releasing and Directing Stimulus Situations in "Turdus m. Mercula L. und T. E. ericetorum Turton," *Z. Tierpsychol.*, 3 (1939) in C. H. Schiller, ed., 1957, pp. 209–238.

Titchener, E. B., *A Text-book of Psychology*. New York: Macmillan, 1909.

Titchener, E. B., *Systematic Psychology: Prolegomena*. New York: Macmillan, 1929.

Tolman, E. C. *Purposive Behavior in Animals and Men*. New York: Appleton, 1932.

Tolman, E. C., and E. Brunswik, "The Organism and the Causal Texture of the Environment," *Psychol. Rev.*, 42 (1933), pp. 43–77.

Trevarthen, C., and M. Richards, Communicative Behavior in Infants. Center for Cognitive Studies Colloquim, Harvard University, 1968.

Urmson, J. O., and G. J. Warnock, eds., *Philosophical Papers by the Late J. L. Austin*. London: Oxford University Press, 1961, p. 187.

Usol'tsev, A. N., and N. T. Terekhova, "Functional Peculiarities of the Skin Temperature Analyzer in Children during the First Six Months of Life," *Pavlov. J. High. Nerv. Act.*, 8 (1958), pp. 174–184.

Uzgiris, Ina C., and J. McV. Hunt, An Instrument for Assessing Infant Psychological Development. Unpublished manuscript, February 1966.

Valentine, C. W., "The Colour Perception and Colour Preferences of an Infant during Its Fourth and Eighth Months," *Brit. J. Psychol.*, 6 (1913–1914), pp. 363–386.

Van der Kloot, G., and C. M. Williams, "Cocoon Construction by the Cerropia Silkworm: The Role of the Internal Environment," *Behav.*, 5 (1953), pp. 157–174.

Velten, H. U., "The Growth of Phonemic and Lexical Patterns in Infant Language," *Language*, 19 (1943), pp. 281–292.

Vernon, Margaret D., *Backwardness in Reading*. New York: Cambridge University Press, 1957.

von Uexküll, J., "A Stroll through the World of Animals and Men" (1934), in C. H. Schiller, ed., *Instinctive Behavior*. New York: International Universities, 1957.

Vurpillot, E., "Étude génétique sur la formation d'un concept; role données perceptives," *Psychol. Franc.*, 5 (1960), pp. 135–152.

Vurpillot, Elaine, "Piaget's Law of Relative Concentration," *Acta. Psychol.*, Amsterdam, 16 (1959), pp. 403–430.

Vygotsky, L. S., *Thought and Language*. New York: Wiley, 1962.

Waddington, C. H., *Organizers and Genes*. New York: Cambridge University Press, 1940.

Wahler, R. G., "Infant Social Development: Some Experimental Analyses of an Infant-Mother Interaction during the First Year of Life," *Journal of Experimental Child Psychology*, 1 (1969), pp. 101–113.

Walk, R. D., and Sue H. Dodge, "Visual Depth Perception of a 10-Month-Old Monocular Human Infant," *Science*, 137 (1962), pp. 529–530.

Walk, R. D., and E. J. Gibson, "A Comparative and Analytical Study of Visual Depth Perception," *Psychol. Monogr.*, 75, No. 519 (1961).

Wapner, S., and H. Werner, *Perceptual development.* Worcester, Mass.: Clark University Press, 1957.

Warden, C. J., T. N. Jenkins, and L. Warner, *Plants and Invertebrates,* Vol. II of *Comparative Psychology.* New York: Ronald Press, 1940.

Warkentin, J., and K. U. Smith, "The Development of Visual Acuity in the Cat," *J. Genet. Psychol.*, 50 (1937), pp. 371–399.

Warnock, G. J., "Seeing," *Proceedings of the Aristotelian Society.* New Series, LV.

Washburn, S. L., and F. C. Howell, "Human Evolution and Culture," in S. Tox, *The Evolution of Man,* Vol. II. Chicago: University of Chicago Press, 1960.

Watson, J. B., *Psychology from the Standpoint of a Behaviorist.* Philadelphia: Lippincott, 1924.

Watson, J. B., *Psychological Care of Infant and Child.* New York: Norton, 1928. Reprinted in part in Y. Brackbill and G. G. Thompson, eds., *Behavior in Infancy and Early Childhood: A Book of Readings.* New York: Free Press, 1967.

Watson, J. B., and R. A. Rayner, "Conditioned Emotional Reactions," *J. Exp. Psychol.*, 3 (1920), pp. 1–4.

Watson, J. S., "The Development and Generalization of 'Contingency Awareness' in Early Infancy: Some Hypotheses," *Merrill-Palmer Quart.*, 12 (1966a), pp. 123–135.

Watson, J. S., "Perception of Object Orientation in Infants," *Merrill-Palmer Quart.*, 12 (1966b), pp. 73–94.

Weir, R. H., "Some Questions on the Child's Learning of Phonology," in F. Smith and G. A. Miller, eds., *The Genesis of Language.* Cambridge, Mass.: M.I.T. Press, 1966, pp. 153–172.

Weir, Ruth H., *Language in the Crib.* The Hague: Mouton, 1962.

Weisberg, P., "Social and Non-social Conditioning of Infant Vocalizations, *Child Development,* 34 (1963), pp. 377–388.

Weisburg, P., and E. Fink, "Fixed Ratio and Extinction Performance of Infants in the Second Year of Life," *J. Exp. Anal. of Beh.*, 9 (1966), pp. 105–109.

Weiss, P., "Some Introductory Remarks on the Cellular Basis of Differentiation," *J. Embryol. Exper. Morphol.*, 1 (1954), pp. 181–211.

Weisstein, Naomi, "Backward Masking and Models of Perceptual Processing," *Journal of Experimental Psychology,* 72 (1966), pp. 232–240.

Weitzman, E. E., and L. Graziani, "Auditory Evoked Responses from Newborn Infants during Sleep," *Pediatrics,* 35 (1965), p. 458.

Weller, G. M., and R. Q. Bell, "Basal Skin Conductance and Neonatal State," *Child Develpm.*, 36 (1965), pp. 647–658.

Wendt, G. R., "Auditory Acuity of Monkeys," *Comp. Psychol. Monogr.*, 10 (1934), pp. 1–51.

Wenger, M. A., "An Investigation of Conditioned Response in Human Infants," *Univ. Iowa Stud. Child Welf.*, 12 (1936), pp. 9–90.

Werner, E. E., and N. Bayley, "The Reliability of Bayley's Revised Scale of

Mental and Motor Development during the First Year of Life," *Child Develpm.*, 37 (1966), pp. 39–50.

Werner, H., *Comparative Psychology of Mental Development.* New York: Harper & Row, 1940, rev. ed., 1957.

Werner, H., and Edith Kaplan, "The Acquisition of Word Meanings: A Developmental Study, *Monogr. Soc. Res. Child Develpm.*, 15, No. 51 (1950).

Werner, H., and S. Wapner, "Toward a General Theory of Perception," *Psychol. Rev.*, 59 (1952), pp. 324–338.

Wertheimer, M., "Untersuchungen zur Lehre von de Gestalt," II. *Psych. Forsch.*, 4 (1923), pp. 301–350.

Wertheimer, M., *Productive Thinking.* New York: Harper & Row, 1959.

Wertheimer, M., "Psychomotor Coordination of Auditory and Visual Space at Birth," *Science*, 134 (1967), p. 1692.

Wheeler, W. M., *The Social Insects.* New York: Harcourt, 1928.

White, B. L., "An Experimental Approach to the Effects of Experience on Early Human Behavior," in J. P. Hill, ed., *Minnesota Symposia on Child Psychology*, Vol. I. Minneapolis: University of Minnesota Press, 1967.

White, B. L., P. Castle, and R. Held, "Observations on the Development of Visually Directed Reaching," *Child Develpm.*, 35 (1964), pp. 333–348.

White, B. L., and K. R. Clark, "An Apparatus for Eliciting and Recording the Eyeblink," in C. H. Ammons, ed., *Psychological Reports: Perceptual and Motor Skills.* In press.

White, B. L., and R. M. Held, "Plasticity of Sensori-motor Development in the Human Infant," in Judy R. Rosenblith and W. Allinsmith, eds., *The Causes of Behavior: Readings in Child Development and Educational Psychology.* Boston: Allyn & Bacon, 1966, 2d ed., pp. 60–70.

White, L. A., *The Science of Culture.* New York: Farrar, Straus, 1949.

Whorf, B. L., *Language, Thought & Reality.* Cambridge, Mass.: M.I.T. Press, 1956.

Williams, Joanna P., and W. Kessen, "Effect of Hand-Mouth Contacting on Neonatal Movement," *Child Develpm.*, 32 (1961), pp. 243–248.

Willier, B. H., "Phases in Embryonic Development," *J. Cell. Comp. Physiol.*, 43 (1954), pp. 307–317.

Winitz, H., *Articulatory Acquisition and Behavior.* New York: Appleton, 1968.

Winnick, W. A., and R. L. Dornbush, "Pre- and Post-exposure Processes in Tachistoscopic Identification," *Perceptual and Motor Skills*, 20 (1965), pp. 107–113.

Witkin, H. A., *et al.*, *Psychological Differentiation: Studies of Development.* New York: Wiley, 1962.

Wittgenstein, Ludwig, *Philosophical Investigations.* New York: Macmillan, 1953. P. 161.

Wodinsky, J., *et al.*, The Establishment of Stable Individual Nursing Adjustments by Neonate Kittens. Unpublished paper.

Wohlwill, J. F., "Developmental Studies of Perception," *Psychol. Bull.*, 57 (1960), pp. 249–288.

Wohlwill, J. F., "**From Perception to Inference: A Dimension of Cognitive Development**" (in this volume), in W. Kessen and Clementina Kuhlman, eds., "Thought in the Young Child: Report of a Conference on Intellective Development with Particular Attention to the Work of Jean Piaget," *Monogr. Soc. Res. Child Develpm.*, 27, No. 2 (1962a), pp. 87–112.

Wohlwill, J. F., "The Perspective Illusion: Perceived Size and Distance in Fields Varying in Suggested Depth, in Children and Adults," *J. Exp. Psychol.*, 64 (1962), pp. 300–310.

Wohlwill, J. F., "The Development of 'Overconstancy' in Space Perception," in L. P. Lipsitt and C. C. Spiker, eds., *Advances in Child Development and Behavior*." New York: Academic Press, 1963.

Wohlwill, J. F., "Changes in Distance Judgments as a Function of Corrected and Noncorrected Practice," *Percept. Mot. Skills*, 2 (1964), pp. 87–112.

Wohlwill, J. F., "Texture of the Stimulus Field and Age as Variables in the Perception of Relative Distance in Photographic Slides," *J. Exp. Child Psychol.*, 2 (1965), pp. 163–177.

Wohlwill, J. F., and R. C. Lowe, "An Experimental Analysis of the Development of the Conservation of Number," *Child Development*, 33 (1962), pp. 153–167.

Wolff, P. H., "Observations on Newborn Infants," *Psychosom. Med.*, 21 (1959), pp. 110–118.

Wolff, P. H., "Observations on the Early Development of Smiling," in B. M. Foss, ed., *Determinants of Infant Behavior*, Vol. II. New York: Wiley, 1963.

Wolff, P. H., "The Development of Attention in Young Infants," *Ann. N.Y. Acad. Science*, 118 (1965), pp. 783–866.

Wolff, P. H., "The Causes, Controls and Organization of Behavior in the Newborn," *Psychol. Issues*. New York: International Universities, 1966a.

Wolff, P. H., "The Natural History of Crying and Other Vocalizations in Early Infancy," in B. M. Foss, ed., *Determinants of Infant Behavior*, Vol. IV. London: Metheun, 1966b.

Wolk, R. D., and Eleanor J. Gibson, "A Comparative and Analytical Study of Visual Depth Perception," *Psychol. Monogr.*, 75, No. 519 (1961), pp. 1–44.

Woodger, J. H., *Biology and Language*. New York: Cambridge University Press, 1952.

Woodger, J. H., "What We Mean by 'Inborn,'" *Brit. J. Phil. Sci.*, 3 (1953), pp. 319–326.

Woodworth, R. S., "A Revision of Imageless Thought," *Psychol. Rev.*, 22 (1915), pp. 1–27.

Woodworth, R. S., and H. Schlosberg, *Experimental Psychology*. New York: Holt, Rinehart and Winston, Inc., 1954.

Yale University, Clinic of Child Development, *The First Five Years of Life: A Guide to the Study of the Preschool Child*. New York: Harper & Row, 1950.

Yerkes, R. M., "Space Perception of Tortoises," *J. Comp. Neurol. Psychol.*, 14 (1904), pp. 17–26.

Zaporozhets, A. V., "The Development of Perception in the Preschool Child," in P. H. Mussen, ed., "European Research in Cognitive Development," *Monogr. Soc. Res. Child Develpm.*, 30, No. 2 (1965), pp. 82–101.

Zeigler, H. P., and H. Leibowitz, "Apparent Visual Size as a Function of Distance for Children and Adults," *Amer. J. Psychol.*, 70 (1957), pp. 106–109.

Zhurova, L. E., "The Development of Analysis of Words into Their Sounds by Preschool Children," *Soviet Psychology and Psychiatry,* 2 (1963), pp. 17–27.

Zippelius, Hanna-Maria, and F. Goethe, "Ethologische Beobachtungen an Haselmausen (*Mascardinus a. avellanarius* L.)," *Z. Tierpsychol.*, 8 (1951), pp. 348–367.

Index